Lecture Notes in Computer Science 15526

The series Lecture Notes in Computer Science (LNCS), including its subseries Lecture Notes in Artificial Intelligence (LNAI) and Lecture Notes in Bioinformatics (LNBI), has established itself as a medium for the publication of new developments in computer science and information technology research, teaching, and education.

LNCS enjoys close cooperation with the computer science R & D community, the series counts many renowned academics among its volume editors and paper authors, and collaborates with prestigious societies. Its mission is to serve this international community by providing an invaluable service, mainly focused on the publication of conference and workshop proceedings and postproceedings. LNCS commenced publication in 1973.

Anirban Dasgupta · Rage Uday Kiran ·
Radwa El Shawi · Satish Srirama ·
Mainak Adhikari
Editors

Big Data and Artificial Intelligence

12th International Conference, BDA 2024
Hyderabad, India, December 17–20, 2024
Proceedings

Springer

Editors
Anirban Dasgupta
Indian Institute of Technology
Gandhinagar, Gujarat, India

Radwa El Shawi
University of Tartu
Tartu, Estonia

Mainak Adhikari
Indian Institute of Science Education and Research
Thiruvananthapuram, Kerala, India

Rage Uday Kiran
University of Aizu
Aizu-Wakamatsu, Fukushima, Japan

Satish Srirama
University of Hyderabad
Hyderabad, Telangana, India

ISSN 0302-9743 ISSN 1611-3349 (electronic)
Lecture Notes in Computer Science
ISBN 978-3-031-81820-2 ISBN 978-3-031-81821-9 (eBook)
https://doi.org/10.1007/978-3-031-81821-9

This Springer imprint is published by the registered company Springer Nature Switzerland AG
The registered company address is: Gewerbestrasse 11, 6330 Cham, Switzerland

Preface

Human data generation and capturing capabilities are increasing rapidly with each passing year. As a consequence, the scope of Big Data Analytics is also expanding to cover all domains including healthcare, agriculture, social development, climate change, manufacturing, etc. Data feeds multiple downstream tasks that attempt to uncover the hidden information in data characterized by the 6Vs of Big Data. Big Data and AI are complementary technologies and their synergy is helping researchers solve many longstanding complex socio-economic and scientific problems.

The 12th International Conference on Big Data and Artificial Intelligence (BDA 2024) was held in physical mode from December 17–20, 2024 at the University of Hyderabad (UoH). These proceedings include 28 peer-reviewed research papers. The program included 3 talks by distinguished keynote speakers, 3 invited industry talks, 2 workshops, and 5 tutorial sessions related to Big Data and Artificial Intelligence. This edition of BDA covered a wide range of topics related to big data and artificial intelligence, such as data science, deep learning, image classification, graph analytics, big data analytics, healthcare analytics, wild-life interactions, traffic management, eLearning, and environment monitoring.

The conference received 106 valid submissions. The review process of research papers followed a single-blind two-tiered review system following the tradition of BDA. We employed Microsoft's Conference Management Toolkit (CMT) for the paper submission and review processes. The Program Committee (PC) consisted of competent data analytics and AI researchers from academia and industry coming from several countries. Each submission was reviewed by at least two reviewers from the Program Committee (PC) and discussed by the PC chairs before making a decision. Based on the above review process, the Program Committee accepted 16 full papers (about 15% acceptance ratio) and 12 short papers (about 26% acceptance ratio, including both full and short papers).

The three keynote talks were delivered by Rajkumar Buyya (University of Melbourne, Australia), Giuliano Casale (Imperial College London, UK), and Santanu Chaudhury (Indian Institute of Technology Delhi, India). The invited industry talks were delivered by Asoke Talukder (ARAI Solutions, India), Rajeev Gupta (Microsoft, India) and Sailesh Patra (Cognida.ai, India).

The workshop co-chairs selected two workshops to be held in conjunction with BDA 2024; 1. GenAI-Driven Big Data Innovations in the Edge-Cloud Continuum, by Chinmaya Dehury (University of Tartu, Estonia). 2. Secure, Reliable, and Ethical Generative AI Implementation, by Tathagata Chakrabarty (TCS, Lucknow, India).

The tutorial chairs selected five tutorials, which were delivered in a parallel session to the main conference. 1. Self-Supervised Learning: A Next Step in AI to Learn Representations from Unlabeled Data Generating Implicit Labels, by Mayuri Mehta (Sarvajanik College of Engineering and Technology, India). 2. Deploying a Scalable Cloud Infrastructure for Big Data Management Challenges and Research Issues, by Vengadeswaran and Shajulin Benedict (Indian Institute of Information Technology, Kottayam, India). 3. Mastering ETL and Data Pipelines with Big Data from Basics to Advanced Techniques, by G. Victor Daniel (Anurag University, India). 4. Social and Environmental Implications of the AI Boom, by Vishnu S. Pendyala (San José State University, USA). 5. Data Fabric: Enabling Technologies and Applications, by Kamalakar Karlapalem (IIIT Hyderabad, India), P. Radha Krishna (National Institute of Technology Warangal, India) and Satyanarayana R. Valluri (Databrics Inc., USA).

We want to thank all the PC members for their hard work in providing us with thoughtful and comprehensive reviews and recommendations. Many thanks to the authors who submitted their papers to the conference. The Steering Committee and the Organizing Committee deserve praise for their support. Several volunteers contributed to the success of the conference. We thank all keynote speakers, invited speakers, workshop speakers, and panelists for their insights on the various themes of the conference. We also thank our sponsors, IoE Directorate, University of Hyderabad, and Cognida.ai, India (Platinum sponsor). Lastly, we also acknowledge the University of Hyderabad, India, for hosting the conference. We hope that the readers of the proceedings find the content interesting, contemporary, rewarding, and beneficial to their research.

December 2024

Anirban Dasgupta
Rage Uday Kiran
Radwa El Shawi
Satish Srirama
Mainak Adhikari

Organization

Patrons

Basuthkar Jagadeeshwar Rao	University of Hyderabad, India
P. J. Narayanan	IIIT Hyderabad, India

Honorary Chairs

Atul Negi	University of Hyderabad, India
Vadlamani Ravi	IDRBT, India

General Chair

Satish Narayana Srirama	University of Hyderabad, India

Steering Committee Liaison

P. Krishna Reddy	IIIT Hyderabad, India

Steering Committee Members

P. Krishna Reddy	IIIT Hyderabad, India
Srinath Srinivasa	IIIT Bangalore, India
Mukesh Mohania	IIIT Delhi, India
Sharma Chakravarthy	University of Texas at Arlington, USA
Philippe Fournier-Viger	Shenzhen University, China
Sanjay Kumar Madria	Missouri University of Science and Technology, USA
Masaru Kitsuregawa	University of Tokyo, Japan
Raj K. Bhatnagar	University of Cincinnati, USA
Vasudha Bhatnagar	University of Delhi, India
Ladjel Bellatreche	ISAE-ENSMA, France
H. V. Jagadish	University of Michigan, USA
Ramesh Kumar Agrawal	Jawaharlal Nehru University, India

Divyakant Agrawal	University of California at Santa Barbara, USA
Arun Agarwal	University of Hyderabad, India
Jaideep Srivastava	University of Minnesota, USA
Sanjay Chaudhary	Ahmedabad University, India
Jerry Chun-Wei Lin	Western Norway University of Applied Sciences, Norway
S. K. Gupta	IIIT Delhi, India
Krithi Ramamritham	IIT Bombay, India

Program Committee Chairs

Anirban Dasgupta	IIT Gandhinagar, India
Rage Uday Kiran	University of Aizu, Japan
Radwa El Shawi	University of Tartu, Estonia

Program Committee Members

Aadarsh Vadlamudi	University of Hyderabad, India
Acquah Hackman	Changhua Christian Hospital, Taiwan
Ali Ghazal	University of Tartu, Estonia
Aneesh S. Chivukula	BITS Pilani, India
Anil Vuppala	IIIT Hyderabad, India
Anitya Kumar Gupta	Illinois Institute of Technology, USA
Anjeneya Swami Kare	University of Hyderabad, India
Ashly Joseph	Cisco Systems, USA
Bijay Prasad Jaysawal	National Cheng Kung University, Taiwan
Chandra Mohan Dasari	IIIT Sri City, India
Deepika Saxena	University of Aizu, Japan
P. Deepthi Jordhana	PVKK Institute of Technology, India
Suresh Kumar	Manav Rachna International Institute of Research and Studies, India
Durga Bhavani S.	University of Hyderabad, India
Erick Martin Fiestas Sorogastua	University of Tartu, Estonia
Jan Timko	University of Tartu, Estonia
José María Luna	University of Córdoba, Spain
Karthik Rajashekaran	Avalara, USA
P. Krishna Reddy	IIIT Hyderabad, India
Mohamed Maher	University of Tartu, Estonia
Narendra Babu Unnam	IIIT Hyderabad, India
Naveen Kumar	University of Delhi, India

Neeraj Goel	IIT Ropar, India
Nidhya Rangarajan	Madanapalle Institute of Technology & Science, India
Praveen Borra	Florida Atlantic University, USA
Praveen Rao	University of Missouri-Columbia, USA
Siba K. Udgata	University of Hyderabad, India
Ramakrishna Garine	University of North Texas, USA
Ravi Kiran Kodali	Cognizant Technology Solutions, USA
Ravi Kiran Magham	Lyft, USA
Rukma Rekha Nalukurthy	University of Hyderabad, India
Sai Prasad P. S. V. S.	University of Hyderabad, India
Satish N. Srirama	University of Hyderabad, India
Simon Pierre Dembele	University of Tartu, Estonia
Sobha Rani T.	University of Hyderabad, India
Sreekanth Bonthala	Madanapalle Institute of Technology & Science, India
Subba Rao Y. V.	University of Hyderabad, India
Swarupa K. Rani	University of Hyderabad, India
Vadlamudi Ch. Venkaiah	University of Hyderabad, India
Vaishali Mishra	Expedia, USA
Vasudha Bhatnagar	University of Delhi, India
Veena Pamalla	SBRIT College, India
Yutaka Watanobe	University of Aizu, Japan

Publication Chair

Mainak Adhikari	IISER Thiruvananthapuram, India

Industry Track Chair

Tathagata Chakrabarty	TCS Lucknow, India

Sponsorship Chair

Bhupathi Shanti	TCS Hyderabad, India

Workshop Chairs

Chinmaya Kumar Dehury	University of Tartu, Estonia
Saurabh Shukla	IIIT Lucknow, India
Prasan Kumar Sahoo	Chang Gung University, Taiwan

Tutorial Chairs

Jimson Mathew	IIT Patna, India
Abhishek Hazra	IIIT Sri City, India

Publicity Chairs

Praveen Kumar Donta	TU Wien, Austria
G. Geethakumari	BITS Pilani, India
Tilottama Goswami	Vasavi College of Engineering, India

Panel Chairs

Chakravarthy Bhagvati	University of Hyderabad, India
J. V. R. Sagar	ANURAG Lab, DRDO, India

Diversity and Inclusion Chair

P. S. V. S. Sai Prasad	University of Hyderabad, India

Student Chair

Aditya Kumar	University of Hyderabad, India

Web Chair

Rahul Siyanwal	University of Hyderabad, India

Local Organizing Committee

Satish Narayana Srirama	University of Hyderabad, India
Atul Negi	University of Hyderabad, India
Durga Bhavani S.	University of Hyderabad, India
Salman Abdul Moiz	University of Hyderabad, India
Rukma Rekha N.	University of Hyderabad, India
Saifulla Md. Abdul	University of Hyderabad, India
Nagamani M.	University of Hyderabad, India
Srinivasa Rao Battula	University of Hyderabad, India
Sobha Rani T.	University of Hyderabad, India
Subba Rao Y. V.	University of Hyderabad, India
Rajendra Prasad Lal	University of Hyderabad, India
Avatharam Ganivada	University of Hyderabad, India
Sai Prasad P. S. V. S.	University of Hyderabad, India
Naveen Nekuri	University of Hyderabad, India
Anjeneya Swami Kare	University of Hyderabad, India
Venkaiah V. C.	University of Hyderabad, India
Digambar Pawar	University of Hyderabad, India
Siba K. Udgata	University of Hyderabad, India
Alok Singh	University of Hyderabad, India
Chakravarthy Bhagvati	University of Hyderabad, India
P. Krishna Reddy	IIIT Hyderabad, India
Rahul Mishra	IIIT Hyderabad, India
Arun Kumar Das	University of Hyderabad, India

Contents

Applications-1

Data Science

Health-Care Analytics

Image Classification

Efficient Detection of Acute Lymphoblastic Leukemia in Microscopic Blood Cell Images

Bhupathi Vishva Pavani, Gaddam Thanmai, and Rimjhim Padam Singh(✉)

Department of Computer Science and Engineering, Amrita School of Computing, Amrita Vishwa Vidyapeetham, Bengaluru, India
ps_rimjhim@blr.amrita.edu

Abstract. Acute Lymphoblastic Leukemia (ALL), a type of blood cell cancer, rapidly progresses until it affects the white blood cells, compromising the human body's ability to fight infections. Untreated Acute Lymphoblastic Leukemia (ALL) can quickly infiltrate vital organs, leading to life-threatening situations and its early treatment is essential to prevent the destructive spread and improve the survival chances. Recent deep-learning techniques help in timely diagnosis of the fatal disease with Convolutional Neural Networks (CNN) detecting Acute Lymphoblastic Leukemia (ALL) in patients by automatically analyzing microscopic blood cell images. Hence, this paper proposes a novel CNN-based approach employing the fine-tuned EfficientNet-B0 model enhanced using an optimally placed self-attention layer with HardSwish activation function. Experimentational analysis has also been done by employing self-attention layers at different architectural positions and with recent activation functions to develop an efficient and optimal model architecture. Several other CNN models such as MobileNet, EfficientNetB0, DenseNet, NASNet, InceptionV3, and Xception have also been trained to classifying blood cell images into normal and ALL classes for comparison purposes. The proposed model outperformed all other techniques with a huge margin of 5% and had the highest accuracy of 86.85%, precision of 87%, recall of 87%, and F1-Score of 87%.

Keywords: ALL · CNN · EfficientNetB0 · Self-Attention Layers · Activation Functions

1 Introduction

Human anatomy is composed of billions of cells that undergo mitosis, the process of cells splitting for DNA replication, during the course of their existence. Cancer, a significant life-threatening health problem in the modern world, develops with new cells growing abnormally or when the old ones do not die. The World Health Organization asserts that cancer affects around one in every five persons in their lifetime, with one in every nine men and one in every twelve women dying from

A. Dasgupta et al. (Eds.): BDA 2024, LNCS 15526, pp. 3–17, 2025.
https://doi.org/10.1007/978-3-031-81821-9_1

it. There are more than 100 different types of cancer. Leukaemia, a type of cancer malignancy, impacts white blood cells in humans rapidly and leads to dysfunction of the immune system.

Leukemia affects two kinds of blood cells: myeloid and lymphoid cells. Myeloid cells are red blood cells and platelets, whereas lymphocytes are a type of white blood cells. Acute leukaemia occurs when cancer grows rapidly and visibly in the body, whereas chronic leukaemia occurs when cancer grows tardily. The bone marrow is the site of genesis for acute lymphoblastic leukaemia (ALL), often referred to as acute lymphocytic leukaemia, which results in an abundance of immature lymphocytes inhibiting the development of red blood cells and platelets vital for defending the immune system. Lymphoma, another kind of cancer, develops in lymphocytes spreading across lymph nodes and other organs. If malignant lymphoblasts make up 20% of the bone marrow, it is regarded as leukaemia; otherwise, it is known as lymphoma.

Children are more likely to get affected by ALL and unfortunately to this day, the primary cause of ALL remains unclear. Medical diagnostics such as bone marrow tests, complete blood count tests(CBC), and a few others are preferred for the detection of ALL and their result analysis needs expertise from healthcare professionals. However, there are only a few hematologists and medical oncologists in India.

Considering the lack of doctors and the severity of the disease, there is a need for a system that automates the detection of ALL in blood cell images. Deep learning has recently made substantial advances in medical diagnosis, allowing for more efficient image processing. Convolutional Neural Networks (CNN), one of the deep learning models, provide early diagnosis and treatment planning. The use of deep learning with personalised medicine brought up new opportunities in healthcare. Although the previous works have few limitations that are to be considered. Earlier models lack the use of advanced techniques which focus on the relevant features, such as attention mechanism. Integration of a self-attention layer enhances the model's interpretability.

This paper provides an automated deep learning-based method for detecting cancer in blood cell microscopy images.

The main contributions of this paper are as follows:

- A novel fine-tuned EfficientNet-B0 convolutional neural network with optimally placed self-attention layer using HardSwish activation function for the task of classifying microscopic blood cell images.
- Detailed performance analysis of the fine-tuned EfficientNet-B0 model with multiple self-attention layers and different state-of-art activation functions.
- Detailed comparative analysis of state-of-art fine-tuned CNN models against the proposed model.

2 Literature Survey

Gupta et al. [1] includes existing research on computer-aided cancer diagnosis systems, with a focus on the use of deep learning techniques. An analysis of past work on deep learning applications in medical imaging, datasets used for cancer research, and approaches for diagnosing B-Lineage ALL is done.

Bhuiyan et al. [2] discusses image enhancing pre-processing approaches, which are then used to ML models, RF, SVM, LR and DT in which SVM performed the best. Using microscopic blood image analysis, Shafique et al. [3] suggest a computerised method for identifying ALL that includes extraction of features, segmentation, and SVM classification. This method surpasses manual techniques on the ALL-IDB-1 dataset, offering a quicker and more accurate means of diagnosis. Genovese et al. [4] address a gap in focus quality analysis of ALL images, proposing a machine learning-based adaptive unsharpening strategy to enhance image quality before classification. Their approach improves classification accuracy, validated through CNNs on a public ALL image dataset. These works employed a machine learning approach, that requires more human involvement. Deep learning is more effective at image recognition than machine learning.

Gupta et al. [5] proposes a novel method, GCTI-SN, which uses the contours of the underlying colour vector space to adjust the variations in illumination, stain chemical, and amount in medical imagery. The research examines the diagnostic effectiveness of GCTI-SN in detection of breast cancer using a CNN-based classifier, emphasising its clinical significance. Mathur et al. [6] presents a novel approach called the MMA-MTL, which combines Pointwise Attention Convolution Layers with Local Spatial Attention blocks to gather both global and local features at the same time. Rastogi et al. [7] presents a two-step approach called "LeuFeatx," which uses a deep learning model which adapted from the VGG16 model to reliably classify leukocytes for leukemia diagnosis. The research gap in these works is scalability and the exploration of computational efficiency. It would be more beneficial if different staining protocols and other feature extraction methods are explored.

Raza et al. [8] empirically evaluates and compares several saliency methods, such as activation-based, gradient-based, region-based, and perturbation-based approaches, employing CNNs on the CNMC dataset. The study's goal is to analyze the visualisation of pathological features and their impact on classification outcomes, hence improving interpretability in cancer cell classification tasks.

Putri et al. [9] compares DL and wide DL architectures, namely DenseNet201 and Wide-ResNet50-2, for classifying segmented blood cell images, using ResNet50 as a baseline. Transfer learning is employed to reduce the number of training steps. The results show that DenseNet201 outperforms other architectures, while histogram equalization has a negative effect on performance. In their thorough analysis of current developments in ALL detection. Das et al. [10] divide methods into three categories: deep learning based methods like CNNs, RNNs, and Autoencoder, conventional machine learning, and strategies based on processing of images and signals. Saeed et al. [11] employs pre-trained ResNet50, Inception-V3, and VGG-16 architectures with data augmentation. Genovese et al. [12] introduces DL4ALL, a groundbreaking multi-task learning deep learning model for Acute Lymphoblastic Leukemia (ALL) detection. To increase classification accuracy, DL4ALL trains simultaneously on both target and source databases using a cross-dataset approach. Although these works used transfer learning, they can focus more on segmentation and employ more effective classification layers.

Kumar et al. [13] uses the SN-AM dataset to offer an accurate method for diagnosing multiple myeloma (MM) and ALL. By using a Dense CNN, the model outperforms traditional machine learning techniques, achieving a high accuracy-97.2%. Khandekar et al. [14] presents a machine-learning approach that uses artificial intelligence to identify blast cells in images of microscopic blood samples in order to diagnose acute lymphoblastic leukemia (ALL). With object detection and classification using the YOLOv4 algorithm, the method obtains MAP scores. Liu et al. [15] addresses the issue of low programming skills, which restricts the use of deep learning techniques in microscopic picture classification by clinical users. The created software, called AIMIC (artificial intelligence-based microscope image classifier), combines deep learning techniques with data preparation to provide a code-free way. Jain et al. [16] introduces CollateNet, a convolutional neural network containing collation blocks which solves the prompt diagnosis of ALL. CollateNet is successful in differentiating ALL from normal cells in single-cell blood smear pictures. Its ability to address vanishing gradient and overfitting problems makes it a viable method for early-phase ALL detection. Oguluri et al. [17] aims to improve the classification of ALL using computer vision and deep learning, notably MobileNet and EfficientNetB3. This study explores deep learning-based approaches like input-based methods, GANs and methods based on loss, on the C-NMC and ALLIDB-2 datasets. Depto et al. [18] concludes that when it comes to reducing imbalanced class problems in leukemia diagnosis, loss-based techniques perform better than GANs and input-based techniques. The performance of the models employed in these papers can be improved by hybridization of conventional models.

Mondal et al. [19] makes use of deep CNNs to propose a weighted ensemble model on the C-NMC-2019 ALL dataset. The proposed model, in contrast to individual candidate models, shows focused learnt regions. Dayata et al. [20] aims to improve ALL detection by means of a stacking ensemble of five CNN models: DenseNet121, ConvNeXtTiny, InceptionV3, EfficientNetV2B3, and MobileNetV2. When compared to individual CNN models, the ensemble supports clinical judgements for ALL detection more effectively, potentially helping pathologists lower error rates and increase patient survival rates. Several other methods have been proposed for medical image classification and these papers can explore class re-balancing and noise sensitivity and show how deep learning can improve medical image representation [21,22].

To overcome the research gaps of the above papers, the proposed architecture employs EfficientNet-B0 with custom dense layers and HardSwish activation, incorporating optimally placed self-attention layers to enhance feature relevance. Key contributions include a comprehensive performance analysis of self-attention layers at different places, different activation functions and comparison with state-of-the-art models.

3 Dataset Description

This paper primarily focuses on the CNMC-2019 dataset provided by the National Cancer Institute under the Cancer Imaging Programme. The CNMC-2019 dataset includes 15,135 images from 118 individuals with two class labels: leukemic B-lymphoblast cells (cancer cells) and normal B-lymphoid precursors (normal cells). The dataset is carefully labelled by an expert oncologist. These images had staining noise and luminosity defects, which were corrected using stain colour normalisation.

The data is organised into three folds, with no repeated members in each fold. For example, fold1 has all of the information from subjects 1, 2, and 3, whereas fold2 contains all of the information from subjects 4, 5, and 6. As the labels for test data were not provided, out of 15,135 images, only 12,528 images were considered. The three folds of train and validation data were combined and were split into new train, test and validation data in the ratio of 3:1:1. The train dataset now consists of 7516 images, validation dataset consists of 2507 images and test dataset consists of 2505 images. Few images from the dataset are shown in Fig. 1.

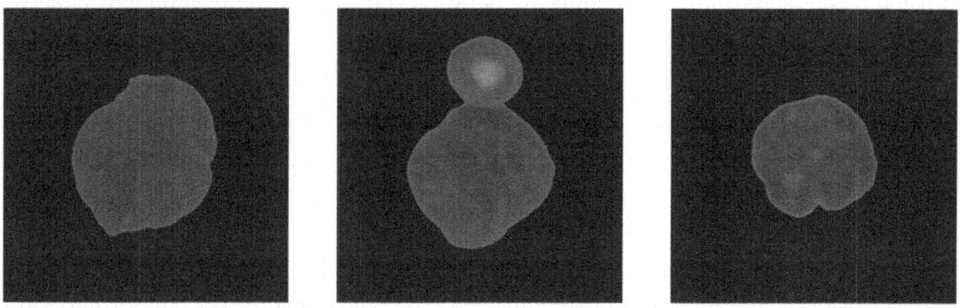

Fig. 1. Sample images from CNMC 2019 dataset

4 Proposed Methodology

The over-all system architecture of the proposed system is presented in Fig. 2 that describes the steps followed in this work. Initially the raw CNMC-2019 dataset is taken and the images of pre-processed to have a standard size of $224 \times 224 \times 3$ pixels to be considered as the input to the model. The images are then distributed randomly across training, validation and testing dataset in the ratio of 60:20:20. Batches 32 images are constructed and are used for training the model. In order to select a good model, initially, the work focuses on implementing and evaluating several state-of-art convolution neural networks namely, MobileNet, EfficientNetB0, DenseNet, NASNet, Inceptionv3 and Xception model for detecting acute lymphoblastic leukemia. Later, for the best baseline model i.e. EfficientNetB0, the model is fine-tuned and leveraged using attention layers at different places between the dense layers to obtain an optimal model without blindly increasing on parameters. The proposed work also analyzes different state-of-art activation functions in the proposed model to find the nest model fit and model learning for detecting acute lymphoblastic leukemia.

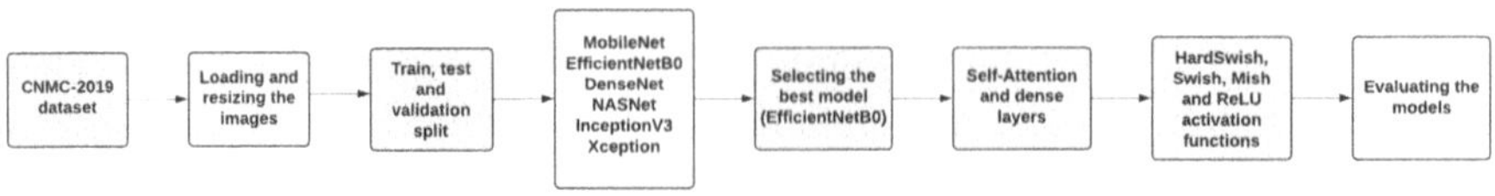

Fig. 2. Architecture Diagram

As mentioned earlier, the proposed model relies on the EfficientNet-B0 model architecture and leverages its learning capacity with the integration of optimal attention mechanism and various recent activation functions.

4.1 EfficientNetB0 Model

This light-weight convolutional neural network is known for its unique feature - compound scaling, which uniformly scales depth, width and resolution based on a fixed set of scaling coefficients to balance the accuracy and efficiency of the model. A compound coefficient Φ is used to control the scaling. Given a baseline network, compound scaling scales the depth 'd', width 'w', and resolution 'r' of the network according to a set of fixed constants:

$$d = \alpha^{\phi}$$
$$w = \beta^{\phi}$$
$$r = \gamma^{\phi}$$

where α, β, and γ are constants. The compound scaling approach ensures that the network scales efficiently. EfficientNetB0 is the baseline variant of the EfficientNet family, with a relatively small number of parameters compared to larger variants like B1, B2, etc. However, it still achieves competitive performance on various image classification benchmarks, making it suitable for applications with limited computational resources. By fine-tuning EfficientNetB0, a pre-trained ImageNet model, for the particular purpose of categorising microscopic blood cell images to identify Acute Lymphoblastic Leukemia (ALL), the model is built on a solid foundation of feature extraction abilities developed on a diverse dataset. EfficientNetB0 model has several advanced features. It is based on the MobileNetV2 backbone, which is known for its efficiency and effectiveness. It employs depthwise separable convolutions, which reduces the number of parameters and computations. In EfficientNet, the network architecture is made up of efficient building blocks, such squeeze-and-excitation(SE) and inverted residual blocks with linear bottleneck. Using fewer parameters, EfficientNet can achieve state-of-the-art performance due to these blocks that optimize the trade-off between model complexity and accuracy. SE blocks are built into the architecture to adjust and enhance feature responses, making the network better at recognizing patterns.

4.2 Fine-Tuning EfficientNet-B0 Model

In the proposed approach, the architecture of the base CNN model, EfficientNetB0 has been modified. The top classification layer is removed and three intermediate dense layers are added to the architecture. First dense layer has 1024 neurons and the rest two dense layers have 512 and 128 neurons respectively. The layers initially also employ max pooling layers for down-sampling and Rectified Linear Unit (ReLU) activation layer. These layers are added to enhance feature extraction as they aid in capturing fine features and hierarchical patterns in blood cell images, which are critical for identifying healthy and leukemic cells (Fig. 3).

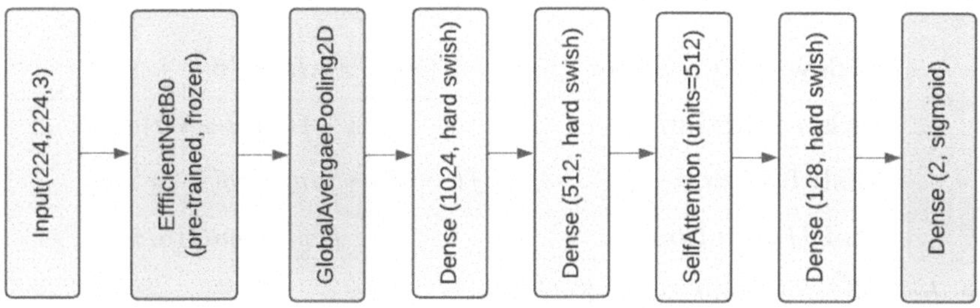

Fig. 3. Fine-tuned EfficientNet-B0 model with self-attention layers

4.3 Self-attention Mechanism

This fine-tuning of the model is further leveraged by the addition of self-attention layer between dense layers. This further elevates the performance of the model by capturing long-range dependencies and global context. This layer allows the model to focus on the most relevant parts of the feature map, ensuring that critical features related to ALL detection are emphasized. The layer comprises learnable weights initialized for transformation and attention score computation and dynamically adapts to input features, enabling the model to focus on salient regions of the feature map. Through a forward pass, the layer computes attention scores via a Softmax operation, facilitating the selection of relevant information while preserving spatial and depth dimensions. This mechanism not only improves the model's interpretability but also enhances its ability to capture long-range dependencies and subtle features within images. As a result, the incorporation of the self-attention layer augments the model's overall effectiveness. The layer helps reduce the model's sensitivity to irrelevant or noisy features. It allows for efficient utilization of computational resources, as it selectively allocates attention to the most relevant features. Dropout layers are placed after the dense layers to prevent overfitting. These layers ensure that the model does not extremely depend on specific features and learns to recognise patterns in the data and classify new data accurately.

4.4 Activation Functions

The proposed model is trained with four different activation functions namely, ReLU, Swish, Hard Swish, Mish activation function. A comparative analysis is done for the performance metrics of these four activation functions. ReLU (rectified Linear Unit) is popular due to its simplicity and efficiency but can lead to inactive neurons. Swish, being smooth and non-monotonic, typically provides better performance, though it is more computationally demanding. HardSwish, a more efficient approximation of Swish, is ideal for environments with limited computational resources. Mish, also smooth and non-monotonic, may deliver superior performance in certain tasks, albeit with higher computational costs. Equations (1), (2), (3), and (4) represents Hard Swish, Swish, Mish and ReLU activation function equations respectively.

Hardswish function $$f(x) = x * ReLU6(x + 3)/6 \quad (1)$$

Swish function $$f(x) = x * \sigma(x) \quad (2)$$

Mish function $$f(x) = x * tanh(log(1 + e^x)) \quad (3)$$

ReLU function $$f(x) = max(0, x) \quad (4)$$

Here, $ReLU6 = min(max(0, x), 6)$.

The proposed solution incorporates EfficientNet-B0 as the backbone, leveraging pre-trained ImageNet weights for feature extraction due to its balance between model complexity and performance. While the use of EfficientNet-B0 is not novel, the solution introduces new aspects, including the Hard Swish activation function, which improves gradient flow and non-linearity, enhancing convergence. Additionally, a self-attention layer is added between dense layers to focus on the most relevant features and refine spatial relationships within the feature space. These modifications address both performance limitations and interpretability challenges in ALL detection, distinguishing this approach from previous works.

5 Experimental Setup and Results

Firstly, six CNN models, MobileNet, EfficientNetB0, DenseNet, NASNet, InceptionV3 and Xception, are implemented for classifying the blood cell images into ALL and HEM. All these models are transfer learned and fine-tuned using the same architecture obtained by removing the top layer of each pre-trained model followed by adding the same set of intermediate dense layers as in the proposed model. Same set of parameters and settings have been employed for training all these models, as presented in Table 1.

Table 1. Standard Parameter Settings

Parameter	Value
Learning rate	0.001
Epochs	50
Loss	categorical_crossentropy
Optimizer	Adamax
Activation function	Softmax

All the models have been trained using 50 epochs and have used early stopping to get a suitable model fit with a learning rate of 0.001. For comprehensive comparison of the imbalanced binary classification, standard performance metrics namely, Precision, Recall, F-score, Accuracy and Specificity have been used [23].

Table 2. Performance metrics for fine-tuned EfficientNet-B0 architecture with different self-attention layers (SA layers) between different dense layers

Model	Accuracy	Precision	Recall	F1 Score	Specificity
SA layer between DL-1 and DL-2	82.85	83	83	83	69
SA layer between DL-2 and DL-3	85.89	86	86	86	72
SA layer between DL-1 & DL-2 and DL-2 & DL-3	84.05	85	84	83	56

Different unique architectures are formed for experimenting self-attention layers at different places in the fine-tuned base model's architecture i.e. fine-tuned EfficientNet-B0 model. The first architecture has a self-attention layer placed between the first two dense layers (DL-1 and DL-2) while the second architecture has the self-attention layer between the last two dense layers (DL-2 and DL-3). Two self-attention layers are placed between each set of dense layers forming the third architecture (between DL-1 & DL-2 and DL-2 & DL-3). It must be noted that these experimentation with placement of Self-attention layers at different positions in the architecture is done to get the optimal placement of the self-attention layer yielding the highest accuracy with minimum number of increase in parameters. All these different combinations of layers in model architectures are trained on the dataset.

As shown in Table 2, the best performing model is the fine-tuned EfficientNet-B0 model with the second self-attention layer i.e. when a single self-attention layers is placed between the second and the third dense layer. This architecture obtained an accuracy of 85.89% and the highest F1 score of 86%. It can be seen that the this architecture also successfully obtained a good balance between the Precision, Recall and Specificity values. Also, these model were able to achieve

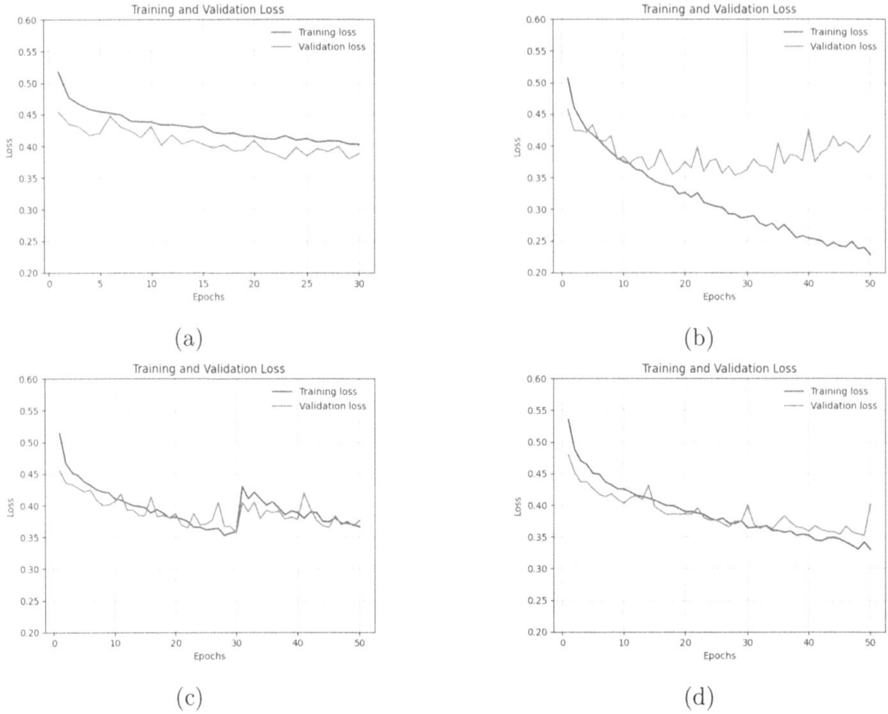

Fig. 4. Loss plots for different variations of fine-tuned EfficientNet-B0 model (a) without self attention layer (b) with self attention layer between DL-1 and DL-2 (c) self attention between DL-2 and DL-3 and (c) self-attention between DL-1 & DL-2 and DL-2 & DL-3

the improved evaluation metrics when compared against the other baseline techniques (will be discussed in subsequent subsection). This indicates that placing self-attention layer between the dense layers of EfficientNetB0-model has improved the performance.

Figure 4 presented the learning trends of the fine-tuned EfficientNet-B0 model when trained without self-attention layers and with self attention layer at different locations in the model architecture. It can be seen that fine-tuned EfficientNet-B0 model converged for both training and validation but at a higher loss value. While loss plots obtained for self-attention placed between initial two dense layers (Fig. 4b) and the double attention layers placed between DL-1 & DL-2 and DL-2 & DL-3 (Fig. 4c) kept fluctuating even after reaching the 50th epoch. The plot obtained for a single attention layer between DL-1 and DL-2 clearly indicates model over-fitting with a huge gap in training and validation loss. While for the best model obtained till now that is the fine-tuned EfficientNet with self attention layers between 2nd and 3rd dense layers achieved the best model convergence for both training and validation plots at a loss value of 0.37 approximately.

Table 3. Performance Metrics for EfficientNetB0-different activation functions

Model	Accuracy	Precision	Recall	F1 Score	Specificity
EfficientNetB0-ReLU	85.89	86	86	86	56
EfficientNetB0-Swish	85.89	86	86	85	25
EfficientNetB0-Hard Swish	86.85	87	87	87	77
EfficientNetB0-Mish	85.65	85	86	85	72

Next, this model is further experimented with new activation functions, Swish, Hard Swish and Mish along with ReLU. The results of the use of these activation functions are shown in Table 3. Hard Swish function with self-attention mechanism for EfficientNetB0 model gave the best results when compared to the remaining three activation functions. The Hard Swish function improves the performance metrics by providing a smoother gradient flow compared to traditional ReLU. This function enhances network's ability to learn complex patterns without suffering from issues like vanishing gradients and exploding gradients, thus facilitating better convergence and improved performance. Swish and Mish activation functions also gave results as good as the ReLU function.

Table 4 provides the analysis of training, validation and testing loss and accuracy values for the fine-tuned EfficientNet-B0 with self attention between DL-1 and DL-2 and different activation functions. It can be seen clearly that the proposed model achieved the minimum validation and test loss of approximately 0.349 and a competitive test accuracy of 86.85%, thereby, supporting the suitability of the proposed model with Hardswish activation function. The precision, recall, F1 score and specificity of the proposed model are 87%, 87%, 87% and 76.5% respectively. Figure 5 provides the loss plots for the proposed model with different activation functions. It can be seen clearly that for different activation functions, the model is able to converge in 50 epochs in almost similar way but the minimum validation accuracy is obtained by the corresponding model with HardSwish activation function. While for other activation functions, the models converge but have a higher validation loss.

Table 4. Accuracy and Loss for fine-tuned EfficientNetB0 model with different activation functions

Model	Tr_Acc	Val_Acc	Te_Acc	Tr_Loss	Val_Loss	Te_Loss
EfficientNetB0-ReLU	92.22	85.57	85.89	0.1926	0.415	0.4364
EfficientNetB0-Swish	90.7	85.53	85.89	0.2232	0.3697	0.3841
EfficientNetB0-Hard Swish	91.5	85.61	86.85	0.2199	0.3493	0.349
EfficientNetB0-Mish	92.18	86.41	85.65	0.1985	0.3623	0.3725

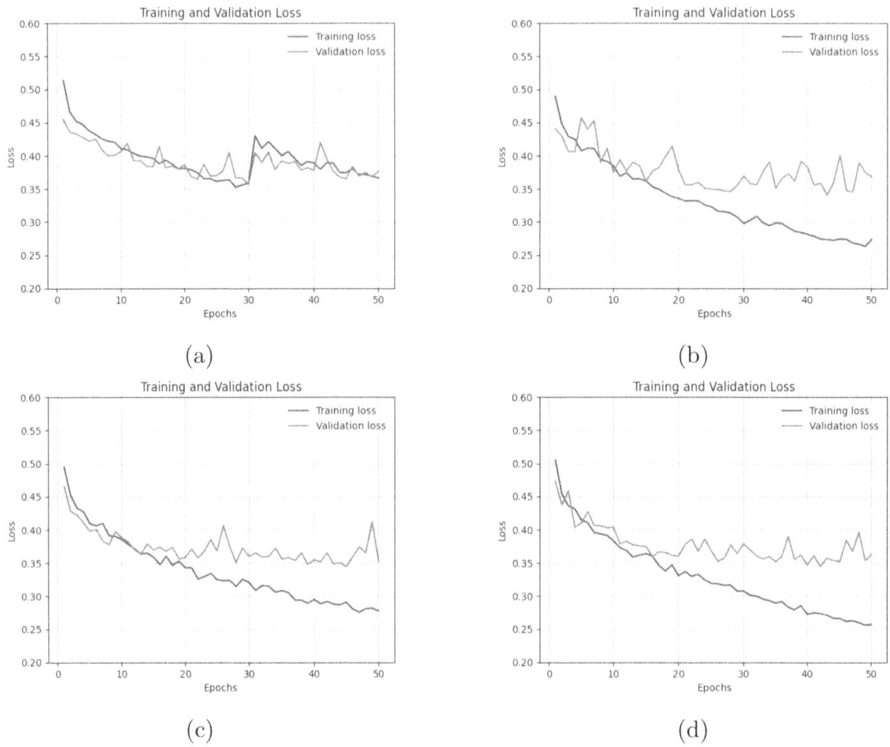

Fig. 5. Loss plots for fine-tuned EfficientNetB0 model with self attention layer between DL-1 and DL-2 layers and different activation functions: (a) ReLU activation, (b) Swish activation, (c) HardSwish activation and (d) Mish activation

Table 5. Performance Comparisons of the proposed model against other state-of-art fine-tuned CNN Models

Model	Class	Precision	Recall	F-Score	Accuracy	Loss
EfficientNet	All	84	92	88	82.85	0.3951
	Hem	79	64	71		
	Over-all	83	83	82		
DenseNet	All	80	95	87	80.56	0.4552
	Hem	82	51	63		
	Over-all	81	81	79		
NASNet	All	81	91	86	79.25	0.4752
	Hem	75	54	63		
	Over-all	79	79	78		
Xception	All	81	88	84	77.92	0.5068
	Hem	69	58	63		
	Over-all	77	78	77		
MobileNet	All	83	91	87	81.33	0.449
	Hem	77	61	68		
	Over-all	81	81	81		
InceptionV3	All	78	96	86	78.96	0.4849
	Hem	83	44	58		
	Over-all	80	79	77		
Proposed	All	89	92	90	**86.85**	**0.349**
	Hem	82	76	79		
	Over-all	**87**	**87**	**87**		

Table 5 presents the class-wise performance metrics of all the base CNNs along with the proposed model. It can be observed that the proposed model outperforms the traditional CNNs in terms of accuracy, precision, recall and F1 score for CNMC-2019 dataset for both class-wise results and the over-all combined results. The fine-tuned EfficientNet-B0 model with self attention between 2nd and 3rd dense layers and employing HardSwish as activation function performs better than all the other state-of-art models with an accuracy of 86.85%, precision, recall and F1-Score of 87%. It must be noted that it has a very high Recall of 92% for All class, which is very crucial for the underlying Leukemia detection problem. The test loss produced by the proposed model is also very low when compared against other models at 0.349. The increase 5% in F-score by the proposed model against all other models clearly indicates the superiority of the proposed model against all other models for detecting Acute Lymphoblastic Leukemia. IT must be noted that all other models performed average on the dataset images with their F-score values restricted to merely 81%.

6 Conclusion

This paper proposed a CNN model for detecting Acute Lymphoblastic Leukemia from microscopic blood cell images using the fine-tuned EfficientNet-B0 architecture. The work explored fine-tuned EfficientNet-B0 model with self-attention mechanism and different activation functions- Swish, Mish and Hard Swish. The combination of the Hard Swish activation function along with second self-attention layer configuration for fine-tuned EfficientNet-B0 model can classify microscopic blood cell image accurately various CNN architectures were experimented. Other the six base CNN models were also fine-tuned and implemented using the similar setup for comparison purposes but all failed to outperform the proposed methodology. Future work involves fine-tuning model hyperparameters, exploring advanced data augmentation techniques to enhance the dataset diversity and implementing ensemble methods for improved accuracy.

References

1. Gupta, R., Gehlot, S., Gupta, A.: C-NMC: B-lineage acute lymphoblastic leukemia: a blood cancer dataset. Med. Eng. Phys. **103**, 103793 (2022)
2. Bhuiyan, M.N.Q., Rahut, S.K., Tanvir, R.A., Ripon, S.: Automatic acute lymphoblastic leukaemia detection and comparative analysis from images. In: 2019 6th International Conference on Control, Decision and Information Technologies (CoDIT), Paris, France, pp. 1144–1149 (2019)
3. Shafique, S., Tehsin, S., Anas, S., Masud, F.: Computer-assisted acute lymphoblastic leukaemia detection and diagnosis. In: 2019 2nd International Conference on Communication, Computing and Digital systems (C-CODE), Islamabad, Pakistan, pp. 184–189 (2019)
4. Genovese, A., Hosseini, M.S., Piuri, V., Plataniotis, K.N., Scotti, F.: Acute lymphoblastic leukaemia detection based on adaptive unsharpening and deep learning. In: ICASSP 2021 2021 IEEE International Conference on Acoustics, Speech and Signal Processing (ICASSP), Toronto, ON, Canada, pp. 1205–1209 (2021)
5. Gupta, A., Duggal, R., Gehlot, S., et al.: GCTI-SN: geometry-inspired chemical and tissue invariant stain normalisation of microscopic medical images. Med. Image Anal. **65**, 101788 (2020)
6. Mathur, P., Piplani, M., Sawhney, R., Jindal, A., Shah, R.R.: Mixup multi-attention multi-tasking model for early-stage leukemia identification. In: ICASSP 2020 - 2020 IEEE International Conference on Acoustics, Speech and Signal Processing (ICASSP), Barcelona, Spain, pp. 1045–1049 (2020)
7. Rastogi, P., Khanna, K., Singh, V.: LeuFeatx: deep learning-based feature extractor for the diagnosis of acute leukemia from microscopic images of peripheral blood smear. Comput. Biol. Med. **142**, 105236 (2022)
8. Raza, M.U., et al.: Visual explanations: activation-based acute lymphoblastic leukemia cell classification. In: 2023 IEEE International Conference on Development and Learning (ICDL), Macau, China, pp. 61–66 (2023)
9. Putri, D.R., Jamal, A., Septiandri, A.A.: Acute lymphoblastic leukemia classification in nucleus microscopic images using convolutional neural networks and transfer learning. In: 2021 2nd International Conference on Artificial Intelligence and Data Sciences (AiDAS), IPOH, Malaysia, pp. 1–6 (2021)

10. Das, P.K., Diya, V.A., Meher, S., Panda, R., Abraham, A.: A systematic review on recent advancements in deep and machine learning based detection and classification of acute lymphoblastic leukemia. IEEE Access **10**, 81741–81763 (2022)
11. Saeed, A., Rehman, F., Ashfaq, A., Sharif, H., Zeeshan, R., Zahid, A.: Acute lymphoblastic leukemia prediction from microscopic blood smear images using advanced machine learning. In: 2022 14th International Conference on Mathematics, Actuarial Science, Computer Science and Statistics (MACS), Karachi, Pakistan, pp. 1–6 (2022)
12. Genovese, A., Piuri, V., Plataniotis, K.N., Scotti, F.: DL4ALL: multi-task cross-dataset transfer learning for acute lymphoblastic leukemia detection. IEEE Access **11**, 65222–65237 (2023)
13. Kumar, D., et al.: Automatic detection of white blood cancer from bone marrow microscopic images using convolutional neural networks. IEEE Access **8**, 142521–142531 (2020)
14. Khandekar, R., Shastry, P., Jaishankar, S., Faust, O., Sampathila, N.: Automated blast cell detection for acute lymphoblastic leukemia diagnosis. Biomed. Signal Process. Control **68**, 102690 (2021). ISSN 1746-8094
15. Liu, R., Dai, W., Wu, T., Wang, M., Wan, S., Liu, J.: AIMIC: deep learning for microscopic image classification. Comput. Methods Programs Biomed. **226**, 107162 (2022)
16. Jain, S., Vishnawat, P., Shukla, P.K., Khatri, N.: Detection of acute lymphoblastic leukemia using CollateNet. In: Chaudhary, N. (ed.) Proceedings - International Conference on Technological Advancements in Computational Sciences, ICTACS 2023, pp. 1095–1100. Institute of Electrical and Electronics Engineers (2023)
17. Oguluri, V.B., Raju, M.S.N., Vaka, J., Murukutla, V.: Classification of acute lymphoblastic leukemia using MobileNet and EfficientNetB3. In: 2023 IEEE International Conference on Blockchain and Distributed Systems Security (ICBDS), New Raipur, India, pp. 1–7 (2023)
18. Depto, D.S., Rizvee, M.M., Rahman, A., Zunair, H., Rahman, M.S., Mahdy, M.R.C.: Quantifying imbalanced classification methods for leukemia detection. Comput. Biol. Med. **152**, Art. no. 106372 (2023)
19. Mondal, C., et al.: Ensemble of convolutional neural networks to diagnose acute lymphoblastic leukemia from microscopic images. Inform. Med. Unlocked **27**, 100794 (2021). ISSN 2352-9148
20. Dayata, W.M.A., Yap, S.Y.C., Bandalan, C.D.: Automated blast cell detection for acute lymphoblastic leukemia using a stacking ensemble of convolutional neural networks. In: 2023 IEEE International Conference on Communication, Networks and Satellite (COMNETSAT), Malang, Indonesia, pp. 95–102 (2023)
21. Ganguly, T., Singh, R.P., Kumar, P.: Self-attention based ResNet model for cervical cancer detection. In: 2023 Second International Conference on Informatics (ICI), pp. 1–6. IEEE (2023)
22. Boyina, K., Reddy, G.M., Akshita, G., Singh, R.P., Kumar, P., Duraisamy, P.: Detection and analysis of tuberculosis disease from chest X-ray (CXR) images using machine learning and deep learning techniques. In: 2024 3rd International Conference for Innovation in Technology (INOCON), pp. 1–6. IEEE (2024)
23. Singh, R.P., Sharma, P.: Instance-vote-based motion detection using spatially extended hybrid feature space. Vis. Comput. **37**(6), 1527–1543 (2021)

Neutrosophic Clustering Method with Local Spatial Information for Dermoscopic Image Segmentation

Avni Mishra(✉), R. K. Agrawal, and Pinki Kumari

School of Computer and Systems Sciences, Jawaharlal Nehru University, New Delhi, India
avni.mtech24@gmail.com

Abstract. The segmentation of dermoscopic images is a crucial phase in computer-aided diagnostic systems. Dermoscopic images are often affected by artifacts and noise. Also, many images have unclear boundaries between lesion and background pixels. Hence, under such conditions, the segmentation of dermoscopic images becomes challenging. To overcome these challenges, we introduce a novel image segmentation method in the neutrosophic set framework with local spatial information (NCM_S) for dermoscopic images. The neutrosophic set-based approach used in the proposed NCM_Sgives a more nuanced representation of the image pixels, effectively addressing the vagueness present due to unclear boundaries. Incorporating local spatial information into the objective function of traditional neutrosophic c-means makes the proposed NCM_Smethod less sensitive to artifacts and noise. We demonstrate the effectiveness of the proposed NCM_Smethod on two publicly available datasets. We compare the performance of $NCM_S$with six existing non-fuzzy-based clustering methods for skin-lesion segmentation and eight fuzzy-based clustering methods in terms of Dice score and Jaccard index. The results of the experiments indicate that the proposed NCM_Smethod is superior to the six skin-lesion segmentation and eight fuzzy-based clustering methods.

Keywords: Image segmentation · Skin cancer · Fuzzy c-means clustering · Neutrosophic c-means clustering · Dermoscopic images

1 Introduction

Over the last several decades, there has been a significant rise in the incidence of skin cancer. It occurs when abnormal cells in the skin grow out of control [30]. Melanoma and non-melanoma are the two types of skin cancer. Malignant melanoma is the deadliest type of skin cancer and arises from melanocytes, pigment-producing skin cells. Squamous cell carcinoma (SCC) and basal cell carcinoma (BCC) are two types of non-melanoma skin cancers that usually happen because of sun exposure [29]. In 2022, there were around 1.5 million new cases

A. Dasgupta et al. (Eds.): BDA 2024, LNCS 15526, pp. 18–33, 2025.
https://doi.org/10.1007/978-3-031-81821-9_2

of skin cancer reported, making it the most common disease diagnosed globally, according to the International Agency for Research on Cancer. In 2022, there were around 60,000 melanoma deaths, and it is expected that 330,000 additional cases of melanoma will be identified [23].

A non-invasive imaging diagnostic technique called dermoscopy is used to spot skin cancer. It helps dermatologists detect melanoma in its early stages that would not be visible to the naked eye [5]. Early diagnosis of melanoma can significantly improve survival rates. However, early and accurate diagnosis in healthcare has become considerably more challenging with the increasing number of patients. The primary cause of this challenge is the dependence on the knowledge and skills of medical professionals. Furthermore, dermoscopic images contain inherent artifacts such as unclear boundaries between lesion and background pixels, hair, skin lines, air bubbles, etc., further complicating the diagnostic process. These challenges in the case of dermoscopic images raise the need for computer-aided diagnostic (CAD) systems [25] in the diagnosis of skin cancer.

Image segmentation [16] is one of the essential phases of CAD systems. Several dermoscopic image segmentation methods have been suggested in the literature, such as thresholding-based [4,20,21,35,42], region-based [2,8,11,17], edge-based [43], deformable model-based [46], and clustering-based [6,9,18,19,22,24, 34,47] methods.

Xu et al. [42] proposed a thresholding-based lesion segmentation technique. This technique determines the threshold value by calculating the mean intensity of high-gradient pixels. A hybrid thresholding-based segmentation method was presented by Garnavi et al. [20] for dermoscopic images. This approach detects the lesion border by estimating the optimal color channel through histogram thresholding and applying morphological operations. Gomez et al. [21] employed an independent histogram thresholding algorithm, that estimates linear combinations of spectral bands to improve the structure of skin lesions. Sforza et al. [35] proposed an adaptive thresholding technique for skin lesion segmentation, which adjusts the varied statistical characteristics of the data to determine the threshold value adaptively. This technique dynamically modifies the threshold interval based on the histogram's skewness, which allows for more precise detection of gray areas specific to melanoma. Al-Azawi et al. [4] suggested an adaptive automatic thresholding approach that combines the benefits of both adaptive thresholding and k-means clustering. Thresholding-based segmentation methods are simple and computationally efficient. However, they do not perform well on low-contrast images and overlapping regions. Additionally, they are sensitive to noise.

To segment skin lesions, Celebi et al. [11] used region-growing and region-merging methods. This unsupervised approach detects borders in skin lesion images using an enhanced version of the JSEG algorithm [17]. Agarwal et al. [2] presented an automatic lesion segmentation using the region-growing method. This method includes pre-processing, segmentation of lesion, and post-processing steps. In the pre-processing step, images are converted into XYZ-color space, and the inverted Z-channel images are found. This step also applied median filtering for noise reduction and histogram equalization to improve contrast. The lesion segmentation task is done using the region-growing method. Finally, in the post-processing step, the final contour selection is based on the solidity and extent of binary images generated with multiple threshold values. Bama et al. [8] proposed another region-growing lesion segmentation technique where the image is initially converted into superpixels, with the darkest superpixel serving as the seed point for region growing. Region-based methods are appreciated for their simplicity but often result in over-segmentation and are highly noise-sensitive. Furthermore, a significant challenge is the requirement for seed initialization.

Yasmin et al. [43] employed a combination of thresholding and the Laplacian of Gaussian (LOG) edge detection technique for skin lesion segmentation. It detects edges with the LOG method and traces the lesion's border by identifying border pixels in the binary image, which is obtained using the thresholding method. Zhou et al. [46] used a revised snake model based on region-based techniques to identify the lesion borders. Complex structures in the image can be captured using edge- and contour-based approaches. However, the appropriate segmentation using these approaches requires parameter tuning, which is a cumbersome task. Also, these approaches are not robust to noise.

Clustering-based segmentation is the most popular unsupervised technique due to small sample space constraints and strong universality [41]. Two primary clustering categories suggested in the literature are k-means [24] and fuzzy c-means (FCM) [9]. A k-means clustering method is a well-known example of a hard clustering method, which is suitable for well-separable groups of similar samples and spherically distributed samples. Garg et al. [19] employed a combination of k-means clustering and firefly algorithm to segment the skin lesion. This technique uses pre-processing for image enhancement and morphological operations to reduce artifacts. However, the method is noise-sensitive and cannot effectively handle the vagueness in the lesion borders.

To overcome the problems of k-means clustering, the research work [9] suggested fuzzy c-means (FCM) clustering, which utilizes the fuzzy set (FS) theory [44]. Schmid [34] suggested a modified FCM method for segmenting skin lesions in $L*u*v$ color space and uses a 2-dimensional histogram to find the number of clusters. Mean-shift FCM is proposed by Zhou et al. [47]. It involves a mean-field component to improve the precision of cluster centroids in the skin lesion segmentation process. Another approach suggested by Devanathan et al. [18] uses the combination of two clustering methods to segment the skin lesions: superpixel-based segmentation (SPS) and the superpixel-based fast fuzzy c-means method (SPFFCM).

Traditional FCM relies on an ordinary fuzzy set (a type-1 fuzzy set) and mitigates image vagueness through partial membership functions. In this, membership of a sample point to a cluster is inversely proportional to its distance from the centroid of the cluster. However, the membership of the samples depends on the choice of the distance measure used. Also, the computation of membership is affected by the noise. Thus, clustering methods based on a type-1 fuzzy set cannot effectively address the uncertainty in decision-making and noise in an image.

To address these limitations, Zadeh [45] introduced the extension of the type-1 fuzzy set named type-2 fuzzy set that overcomes the issues of the type-1 fuzzy set by grading the fuzzy membership that provides a more flexible approach to model uncertainty. The intuitionistic fuzzy set (IFS) [7] is the extension of the concept of the fuzzy set. IFS includes details beyond membership degrees, such as hesitation degrees and non-membership degrees, which helps to mitigate the ambiguity and uncertainty inherent in real data. Picture fuzzy set (PFS) [14] extends the concept of the IFS by incorporating a refusal degree, which reflects the degree to which a data point neither belongs to a cluster nor strongly opposes it.

F. Smarandache [36–38] introduced neutrosophic set (NS), that generalizes the concepts of fuzzy, intuitionistic fuzzy, and picture fuzzy sets by introducing three independent components: truth, indeterminacy, and falsity. This generalization allows for an even more nuanced handling of uncertainty. Guo and Sengur [22] suggested a neutrosophic c-means (NCM) clustering approach that utilizes an NS to obtain more accurate segmentation. NCM considers three membership values for each data point: (i) True membership (T), representing belongingness to determinant clusters; (ii) Indeterminate membership (I), utilized to indicate uncertainty or ambiguity in cluster assignment; and (iii) False membership (F), utilized for detecting outliers provides a robust framework for clustering in environments with complex uncertainty and imprecise information.

Utilizing the advantages of NS, the research [1] applied NCM to dermoscopic images, which demonstrated the superior performance of NCM compared to FCM. Ashour et al.(2018) [6] used the NCM method to segment the lesion in the dermoscopic image, where the image is first transformed into neutroscopic domain in terms of true membership, indeterminate and false subsets. The histogram-based clustering method estimates the number of clusters required for initial input in NCM. Furthermore, intensity and morphological features are used to detect the skin lesion. However, the existing segmentation methods in the neutrosophic set framework do not consider spatial information in the objective function for the segmentation of dermoscopic images. It has been demonstrated in the research works [3] that the performance of the segmentation method improves with the incorporation of spatial information into the objective function. Motivated by this, we propose a novel and robust segmentation method within the neutrosophic set framework that includes local spatial information, named *NCM_S*, for the segmentation of dermoscopic images.

We compare the proposed *NCM_S*method with the six existing non-fuzzy-based clustering methods for skin-lesion segmentation [4,17,19,27,35], to evaluate its performance. Also, we compare its performance with FCM and its variants such as spatial constrained fuzzy c-means (FCM_S) [3], spatial constrained fuzzy c-means using mean/median filter (FCM_S1/S2) [12], enhanced fuzzy c-means (EnFCM) [39], fast generalized fuzzy c-means (FGFCM) [10], fuzzy local information c-means (FLICM) [26], and neutrosophic c-means (NCM) in terms of two evaluation criteria (Dice score and Jaccard index). To evaluate and compare the proposed NCM_Smethod, we used ISIC 2018 [13], and PH2 [28], well-known publicly available dermoscopic datasets.

The contributions of this proposed NCM_Smethod are as follows:

(i) We introduce a novel robust segmentation method in a neutrosophic set framework to handle inherent artifacts, noise, and unclear boundaries present in dermoscopic images,
(ii) We analyze the proposed NCM_Smethod on ISIC 2018 and PH2, two publicly accessible datasets, and compare its performance with six existing non-fuzzy-based clustering methods for skin-lesion segmentation and eight existing fuzzy-based clustering methods in terms of Dice score and Jaccard index.

This paper is structured as follows: Sect. 2 outlines the preliminaries and notations used in our work. Section 3 introduces the proposed NCM_Smethod. Section 4 describes the datasets and discusses the results. The paper concludes with Sect. 5.

2 Preliminaries and Notations

This section describes the notations utilized in this research, along with relevant algorithms.

2.1 Fuzzy C-Means Algorithm (FCM)

FCM uses a fuzzy set to cluster sample points, assigning each sample point to multiple clusters with degrees of membership in the [0, 1] interval.

Given a set of samples $X = \{x_l; l = 1, 2, \ldots, N\}$ where each x_l represents a sample in a d-dimensional space, which belongs to C number of clusters. Each cluster is characterized by a d-dimensional space centroid $\mathbf{v}_k$ $\{\mathbf{v}_k; k = 1, 2, ..., C\}$.

The objective function with constraints conditions is defined as [9]:

$$\min J_m(U, V : X) = \sum_{l=1}^{N} \sum_{k=1}^{C} \mu_{kl}^{m} ||\mathbf{x}_l - \mathbf{v}_k||^2 \tag{1}$$

$$\text{subject to:} \quad \sum_{k=1}^{C} \mu_{kl} = 1, \quad 1 \leq l \leq N$$

where m is a fuzzification parameter, a real number > 1, μ_{kl} is the membership degree of $\mathbf{x}_l$ in cluster k, and the term $||.||$ calculates the euclidean distance between the sample $\mathbf{x}_l$ and centroid $\mathbf{v}_k$.

The cluster centroid $\mathbf{v}_k$ and membership μ_{kl} are determined using the Lagrange method of undermined multiplier and are given as [9]:

$$\mathbf{v_k} = \frac{\sum_{l=1}^{N} \mu_{kl}^m \mathbf{x_l}}{\sum_{l=1}^{N} \mu_{kl}^m} \tag{2}$$

$$\mu_{kl} = \frac{1}{\sum_{r=1}^{C} \left(\frac{\|\mathbf{x}_l - \mathbf{v}_k\|^2}{\|\mathbf{x}_l - \mathbf{v}_r\|^2} \right)^{\frac{1}{m-1}}} \tag{3}$$

However, the performance of FCM declines in the presence of noise in the image. To overcome this issue, a variant called FCM_S [3] incorporates local spatial information of the pixels, making it less noise-sensitive. The objective function of FCM_S is as follows [3]:

$$\min J_m(U, V : X) = \sum_{l=1}^{N} \sum_{k=1}^{C} \mu_{kl}^m \|\mathbf{x}_l - \mathbf{v}_k\|^2 + \frac{\alpha}{N_R} \sum_{l=1}^{N} \sum_{k=1}^{C} \mu_{kl}^m \sum_{r \in N_l} \|\mathbf{x}_r - \mathbf{v}_k\|^2 \tag{4}$$

For each sample point $\mathbf{x}_l$, define N_l as the neighboring pixels around it, N_R represents the count of sample pixels in this neighborhood, and α is a spatial regularization parameter.

The cluster centers and membership are computed as [3]:

$$\mathbf{v}_k = \frac{\sum_{l=1}^{N} \mu_{kl}^m \left(\mathbf{x}_l + \frac{\alpha}{N_R} \sum_{r \in N_l} \mathbf{x}_r \right)}{(1+\alpha) \sum_{l=1}^{N} \mu_{kl}^m}, 1 \leq k \leq C \tag{5}$$

$$\mu_{kl} = \left(\frac{\left(\|\mathbf{x}_l - \mathbf{v}_k\|^2 + \frac{\alpha}{N_R} \sum_{r \in N_l} \|\mathbf{x}_r - \mathbf{v}_k\|^2 \right)}{\sum_{p=1}^{C} \left(\|\mathbf{x}_l - \mathbf{v}_p\|^2 + \frac{\alpha}{N_R} \sum_{r \in N_l} \|\mathbf{x}_r - \mathbf{v}_p\|^2 \right)} \right)^{\frac{-1}{(m-1)}} \tag{6}$$

2.2 Neutrosophic C-Means Clustering Algorithm (NCM)

Because of the incomplete representation of uncertain information in the type-1 fuzzy set, clustering methods based on the type-1 set face difficulties in handling artifacts, noise, and unclear boundaries in an image. Guo and Sengur [22] proposed a novel clustering algorithm named NCM, based on NS to address the limitations of a type-1 fuzzy set. The objective function of NCM is given by [22]:

$$\begin{aligned} \min N_m(T, I, F, V) = & \sum_{l=1}^{N} \sum_{k=1}^{C} (\omega_1 T_{lk})^m \|\mathbf{x}_l - \mathbf{v}_k\|^2 + \sum_{l=1}^{N} (\omega_2 I_l)^m \|\mathbf{x}_l - \mathbf{v}_l^{\max}\|^2 \\ & + \sum_{l=1}^{N} \delta^2 (\omega_3 F_l)^m \end{aligned} \tag{7}$$

$$\mathbf{v}_l^{\max} = \frac{\mathbf{v}_{p_l} + \mathbf{v}_{q_l}}{2} \tag{8}$$

$$p_l = \arg\max_{k=1,2,3,\ldots,C}(T_{lk}) \tag{9}$$

$$q_l = \arg\max_{k \neq p_l \cap k \in [1,C]}(T_{lk}) \tag{10}$$

The symbols T_{lk}, I_l, and F_l represent the values of membership associated with the determinate clusters, boundary pixels, & noisy data sets, respectively. ω_1, ω_2, and ω_3 are the weighting parameters. p_l and q_l are the cluster numbers with the highest and second-highest values of T_{lk}. Once p_l and q_l are identified, $\mathbf{v}_l^{\max}$ is computed, and this value is fixed for each data point l and remains unchanged thereafter, while δ is a noise distance. Also, $0 < T_{lk}, I_l, F_l < 1$. The three memberships are constrained by the following condition:

$$\sum_{k=1}^{C} T_{lk} + I_l + F_l = 1 \tag{11}$$

The objective function (7) is minimized using the Lagrange multiplier method, the following equations are obtained $\mathbf{v}_k$, T_{lk}, I_l, and F_l as follows [22]:

$$\mathbf{v}_k = \frac{\sum_{l=1}^{N}(\omega_1 T_{lk})^m \mathbf{x}_l}{\sum_{l=1}^{N}(\omega_1 T_{lk})^m} \tag{12}$$

$$T_{lk} = \frac{K}{\omega_1}(\mathbf{x}_l - \mathbf{v}_k)^{-2/(m-1)} \tag{13}$$

$$I_l = \frac{K}{\omega_2}(\mathbf{x}_l - \mathbf{v}_l^{\max})^{-2/(m-1)} \tag{14}$$

$$F_l = \frac{K}{\omega_3}\delta^{-2/(m-1)} \tag{15}$$

where,

$$K = \left[\frac{1}{\omega_1}\sum_{k=1}^{C}(\mathbf{x}_l - \mathbf{v}_k)^{-\frac{2}{m-1}} + \frac{1}{\omega_2}(\mathbf{x}_l - \mathbf{v}_l^{\max})^{-\frac{2}{m-1}} + \frac{1}{\omega_3}\delta^{-\frac{2}{m-1}}\right]^{-1} \tag{16}$$

3 Proposed Neutrosophic Clustering Method with Local Spatial Information (*NCM_S*)

Motivated by FCM_S and NCM, we introduce a novel Neutrosophic Clustering Method with Local Spatial Information (*NCM_S*) to segment the dermoscopic images, specifically to handle inherent artifacts, noise, and unclear boundaries present in dermoscopic images.

3.1 Spatial Neighborhood Information

NCM objective function is modified by introducing local spatial information, which is given by:

$$\min N(T,I,F,C) = \sum_{l=1}^{N}\sum_{k=1}^{C}(\omega_1 T_{lk})^m\left(\|\mathbf{x}_l-\mathbf{v}_k\|^2+\frac{\alpha}{N_R}\sum_{r\in\mathbb{N}_l}\|\mathbf{x}_r-\mathbf{v}_k\|^2\right)$$
$$+\sum_{l=1}^{N}(\omega_2 I_l)^m\|\mathbf{x}_l-\mathbf{v}_l^{\max}\|^2+\sum_{l=1}^{N}\delta^2(\omega_3 F_l)^m \tag{17}$$

$$\mathbf{v}_l^{\max}=\frac{\mathbf{v}_{p_l}+\mathbf{v}_{q_l}}{2} \tag{18}$$

$$p_l=\arg\max_{k=1,2,3,\ldots,C}(T_{lk}) \tag{19}$$

$$q_l=\arg\max_{k\neq p_l\cap k\in[1,C]}(T_{lk}) \tag{20}$$

$$\delta^2=\frac{\sum_{l=1}^{N}\sum_{k=1}^{C}\|\mathbf{x}_l-\mathbf{v}_k\|^2}{C} \tag{21}$$

The parameter α represents spatial regularization. For each sample point $\mathbf{x}_l$, N_l indicates its neighboring pixels, and N_R is the number of pixels in this neighborhood. The value of δ is determined using a scheme based on the average interpoint distance [15], which captures the structural relationships among data points. Also, $0 < T_{lk}, I_l, F_l < 1$. The three memberships are constrained by the following condition:

$$\sum_{k=1}^{C}T_{lk}+I_l+F_l=1 \tag{22}$$

The Lagrange problem of the objective function (17) is expressed as:

$$L=\sum_{l=1}^{N}\sum_{k=1}^{C}(\omega_1 T_{lk})^m\left(\|\mathbf{x}_l-\mathbf{v}_k\|^2+\frac{\alpha}{N_R}\sum_{r\in\mathbb{N}_l}\|\mathbf{x}_r-\mathbf{v}_k\|^2\right)$$
$$+\sum_{l=1}^{N}(\omega_2 I_l)^m\|\mathbf{x}_l-\mathbf{v}_l^{\max}\|^2+\sum_{l=1}^{N}\delta^2(\omega_3 F_l)^m$$
$$-\sum_{l=1}^{N}\lambda_l\left(\sum_{k=1}^{C}T_{lk}+I_l+F_l-1\right) \tag{23}$$

To minimize the Lagrange function (23), the partial derivatives of L must be zero. Hence,

$$\frac{\partial L}{\partial v_k}=0,\ \frac{\partial L}{\partial T_{lk}}=0,\ \frac{\partial L}{\partial I_l}=0,\ \frac{\partial L}{\partial F_l}=0 \tag{24}$$

Solving (24), we obtain:

$$\mathbf{v}_k=\frac{\sum_{l=1}^{N}\omega_1^m T_{lk}^m(\mathbf{x}_l+\frac{\alpha}{N_R}\sum_{r\in\mathbb{N}_l}x_r)}{(1+\alpha)\sum_{l=1}^{N}\omega_1^m T_{lk}^m} \tag{25}$$

$$T_{lk} = \frac{K}{\omega_1}[(\mathbf{x}_l - \mathbf{v}_k)^2 + \frac{\alpha}{N_R}\sum_{r\in\mathbb{N}_l}(\mathbf{x}_r - \mathbf{v}_k)^2]^{-\frac{1}{m-1}} \tag{26}$$

$$I_l = \frac{K}{\omega_2}(\mathbf{x}_l - \mathbf{v}_l^{\max})^{-2/(m-1)} \tag{27}$$

$$F_l = \frac{K}{\omega_3}\delta^{-2/(m-1)} \tag{28}$$

where,

$$K = \left[\frac{1}{\omega_1}\sum_{k=1}^{C}\left((\mathbf{x}_l - \mathbf{v}_k)^2 + \frac{\alpha}{N_R}\sum_{r\in\mathbb{N}_l}(\mathbf{x}_r - \mathbf{v}_k)^2\right)^{-\frac{1}{m-1}} + \frac{1}{\omega_2}(\mathbf{x}_l - \mathbf{v}_l^{\max})^{-\frac{2}{m-1}} + \frac{1}{\omega_3}\delta^{-\frac{2}{m-1}}\right]^{-1} \tag{29}$$

Algorithm 1. The proposed *NCM_S*algorithm

1. Initialize the values of T^0, I^0, F^0, and the parameters C, ϵ, m, δ, ω_1, ω_2, ω_3.
2. Obtain the center vectors $\mathbf{v}_{(j)k}$ at the jth step using Eq. (25).
3. **while** $|T^{j+1} - T^j| > \epsilon$ **do**
   3.1. Compute $\mathbf{v}_l^{\max}$ based on the indexes of the highest and 2nd highest values of T using a comparison process.
   3.2. Update T^j to T^{j+1} according to Eq. (26), I^j to I^{j+1} according to Eq. (27), and F^j to F^{j+1} according to Eq. (28).
   3.3. Recalculate $\mathbf{v}_{(j)k}$.
4. **end**
5. Label each data point $\mathbf{x}(l)$ to the class with the highest $TM = [T, I, F]$: $\mathbf{x}(l)$ is allocated to the jth class if $j = \arg\max_{k=1,2,\ldots,C+2}(TM_{kl})$.

4 Experimental Results and Discussion

We analyze the quantitative effectiveness of the proposed *NCM_S*method by comparing it with six existing non-fuzzy-based clustering methods for skin-lesion segmentation and eight existing fuzzy-based clustering methods, using the Dice score and Jaccard index on two publicly available datasets. The results for six existing non-fuzzy-based clustering methods for skin-lesion segmentation, as shown in Table 1, are taken from [31–33].

4.1 Datasets

The ISIC 2018 collection [13], derived from the HAM10000 dataset [40], was unveiled at the 2018 ISIC Challenge with three primary objectives: lesion segmentation (Task 1), identifying lesion attributes (Task 2), and disease classification (Task 3). Each task used three sets of images: training, validation, and testing. Similar to the research work [1], 200 images are randomly selected from 2594 dermoscopic images comprising Tasks 1 and 2 of the training sets, with corresponding ground truth images. The PH2 dataset [28] was collected at the Pedro Hispano Hospital's Dermatology Centre in Portugal. This includes 200 dermoscopic images featuring 40 melanoma skin cancers, 80 atypical nevi, and 80 common nevi. Each 8-bit RGB color image with 768×560 pixels resolution, with ground truth images available for all.

4.2 Evaluation Criteria

Quantitative measures like the Dice Score (***DS***) and Jaccard Index (***JI***) are used to assess the performance of the proposed NCM_Smethod in comparison with existing methods. These are defined as follows:

$$DS = \frac{2 \cdot TP}{(2 \cdot TP + FP + FN)} \tag{30}$$

$$JI = \frac{DS}{(2 - DS)} \tag{31}$$

where TP = True Positive, FP = False Positive, TN = True Negative, and FN = False Negative.

4.3 Results on PH2 and ISIC 2018 Datasets

Table 1 depicts the comparison of the proposed NCM_Smethod with six existing non-fuzzy-based clustering methods for skin-lesion segmentation and eight existing fuzzy-based clustering methods in terms of average DS & average JI on PH2 and ISIC 2018 datasets. We can draw the following observations from Table 1:

- The performance of the proposed NCM_Smethod is better than all the six existing non-fuzzy-based clustering methods for skin-lesion segmentation and the eight existing fuzzy-based clustering methods.
- The proposed NCM_Smethod is more efficient than FCM_S due to the neutrosophic set, which represents vagueness and uncertainty in a better way in comparison to the type-1 fuzzy set.
- The proposed NCM_Smethod performs better than NCM due to the incorporation of a local spatial information term, which reduces the effect of the artifacts and noise.

Table 1. Comparison of segmentation performance of the proposed *NCM_S* method with six existing non-fuzzy-based clustering methods for skin-lesion segmentation and eight existing fuzzy-based clustering methods for *DS* and *JI* on both the PH2 & ISIC 2018 datasets.

Methods	PH2		ISIC 2018	
	DS	*JI*	*DS*	*JI*
Non-Fuzzy-Based Clustering Methods				
JSEG	0.7554	0.6069	-	-
Level-set	0.6456	0.4766	-	-
SRG	0.6123	0.4412	-	-
Skew-corrected thresholding	0.7000	0.6800	0.7000	0.7300
Adaptive automatic thresholding	0.6800	0.6300	0.6500	0.6900
Optimized firefly algorithm	0.6900	0.6800	0.7100	0.7000
Fuzzy-Based Clustering Methods				
FCM	0.7721	0.6522	0.8689	0.7745
FCM_S	0.7822	0.6688	0.8722	0.7797
FCM_S1	0.7766	0.6581	0.8694	0.7752
FCM_S2	0.7770	0.6587	0.8695	0.7755
EnFCM	0.7471	0.6339	0.8478	0.7432
FGFCM	0.7471	0.6341	0.8480	0.7434
FLICM	0.7650	0.6629	0.8749	0.7840
NCM	0.8009	0.6936	0.9152	0.8471
NCM_S	**0.8348**	**0.7376**	**0.9194**	**0.8549**

*Bold values indicate the best results

- For the PH2 dataset, the maximum performance in terms of *DS* and *JI* are 0.8348 and 0.7376 respectively, which is achieved with *NCM_S*. The minimum performance in terms of *DS* and *JI* are 0.6123 and 0.4412 respectively, which is achieved with SRG.
- For the ISIC 2018 dataset, the maximum performance in terms of *DS* and *JI* are 0.9194 and 0.8549 respectively, which is achieved with *NCM_S*. The minimum performance in terms of *DS* and *JI* are 0.6500 and 0.6900 respectively, which is achieved with adaptive automatic thresholding.

Figure 1 shows the segmented image of the proposed *NCM_S*method in comparison to eight existing fuzzy-based clustering methods. It can be observed that the segmented image using FCM is more deviant from the ground truth image for both datasets. In contrast, the variants of FCM such as FCM_S/S1/S2, EnFCM, FGFCM, and FLICM, perform better than FCM to mitigate the inherent noise in the images. The performance of these methods improves due to the incorporation of spatial information in the objective function of FCM to handle inherent noise present in the image. However, type-1 fuzzy set-based methods still struggle to handle the vagueness and uncertainty present in images. It can

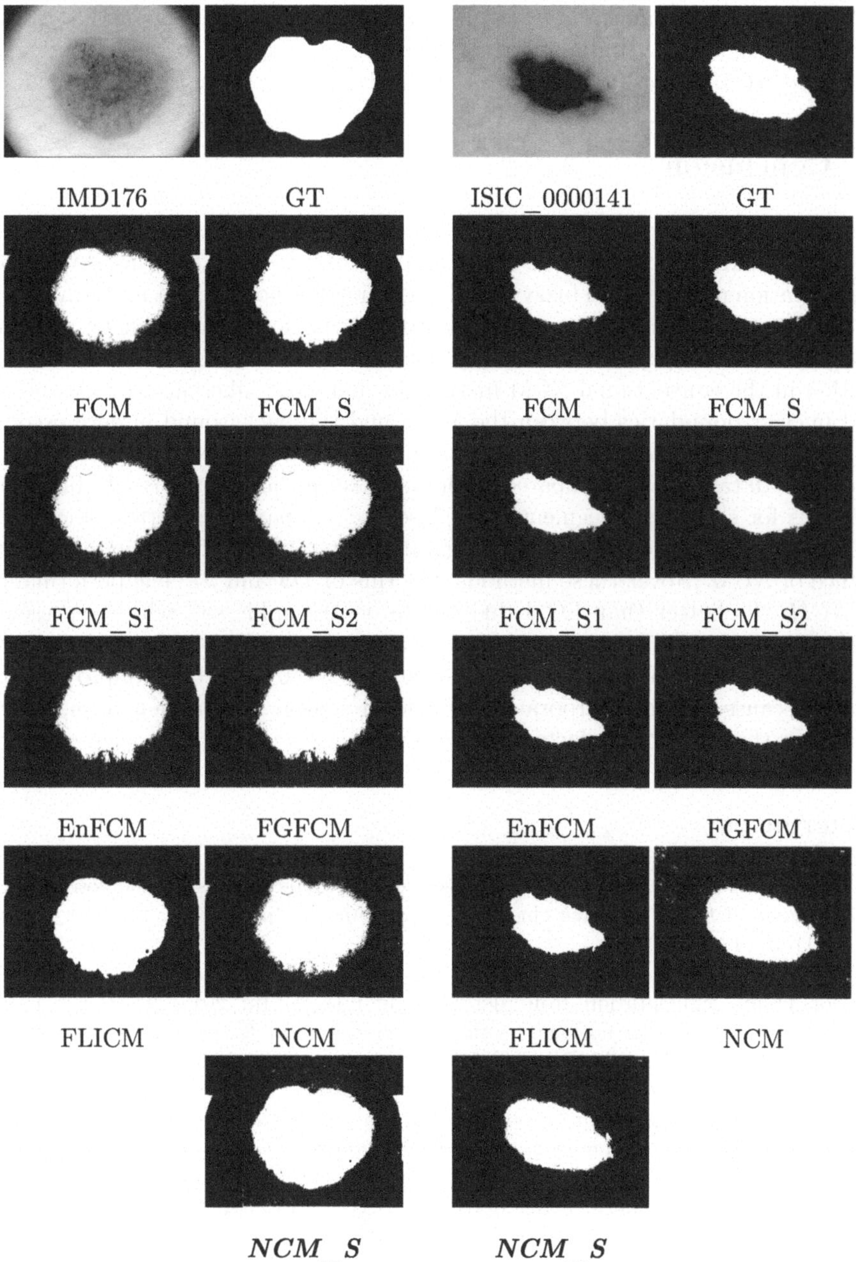

Fig. 1. Qualitative comparison of the proposed *NCM_S* method with the eight existing fuzzy-related methods on the IMD176 image of the PH2 dataset and the ISIC_0000141 image of the ISIC 2018 dataset.

be observed from Fig. 1 that the segmented image obtained using the proposed NCM_Smethod is much closer to the ground truth image. This is due to the incorporation of local spatial information and neutrosophic set framework in the proposed NCM_Smethod, which can handle both inherent noise and vagueness.

5 Conclusion

Dermoscopic images are often characterized by inherent artifacts, noise, and unclear boundaries between the lesion and the background. Existing skin lesion segmentation methods and fuzzy-based clustering methods do not perform better in the presence of the abovementioned problems of dermoscopic images. In this research, we have proposed the NCM_Smethod incorporating local spatial information in the conventional NCM method for handling inherent artifacts, noise, and unclear boundaries between the lesion and the background in dermoscopic images. The performance of the proposed $NCM_S$is assessed on the ISIC 2018 and PH2 datasets and compared with six existing non-fuzzy-based clustering methods for skin-lesion segmentation. Also, we compared its performance with FCM and variants of FCM. Experimental results indicate the superior performance of NCM_Sover these methods in terms of *DS* and *JI*. The performance of $NCM_S$is better than FCM due to the neutrosophic set, which represents vagueness and uncertainty in a better way in comparison to the type-1 fuzzy set. Further, the performance of the suggested NCM_Smethod improves over NCM because of the incorporation of a local spatial information term, which mitigates the effect of artifacts and noise.

References

1. Abdelhafeez, A., Mohamed, H.K.: Skin cancer detection using neutrosophic C-means and fuzzy C-means clustering algorithms. J. Intell. Syst. Internet Things **8**(1), 33–42 (2023)
2. Agarwal, A., Issac, A., Dutta, M.K.: A region growing based imaging method for lesion segmentation from dermoscopic images. In: 2017 4th IEEE Uttar Pradesh Section International Conference on Electrical, Computer and Electronics (UPCON), pp. 632–637. IEEE (2017)
3. Ahmed, M.N., Yamany, S.M., Mohamed, N., Farag, A.A., Moriarty, T.: A modified fuzzy C-means algorithm for bias field estimation and segmentation of MRI data. IEEE Trans. Med. Imaging **21**(3), 193–199 (2002)
4. Al-azawi, R.J., Abdulhameed, A.A., Ahmed, H.M.: A robustness segmentation approach for skin cancer image detection based on an adaptive automatic thresholding technique. Am. J. Intell. Syst. **7**(4), 107–112 (2017)
5. Argenziano, G., Soyer, H.P.: Dermoscopy of pigmented skin lesions-a valuable tool for early. Lancet Oncol. **2**(7), 443–449 (2001)
6. Ashour, A.S., Guo, Y., Kucukkulahli, E., Erdogmus, P., Polat, K.: A hybrid dermoscopy images segmentation approach based on neutrosophic clustering and histogram estimation. Appl. Soft Comput. **69**, 426–434 (2018)
7. Atanassov, K.T., Atanassov, K.T.: Intuitionistic Fuzzy Sets. Springer (1999)

8. Bama, S., Velumani, R., Prakash, N., Hemalakshmi, G., Mohanarathinam, A.: Automatic segmentation of melanoma using superpixel region growing technique. Mater. Today Proc. **45**, 1726–1732 (2021)
9. Bezdek, J.C., Ehrlich, R., Full, W.: FCM: the fuzzy C-means clustering algorithm. Comput. Geosci. **10**(2–3), 191–203 (1984)
10. Cai, W., Chen, S., Zhang, D.: Fast and robust fuzzy C-means clustering algorithms incorporating local information for image segmentation. Pattern Recogn. **40**(3), 825–838 (2007)
11. Celebi, M.E., Aslandogan, Y.A., Bergstresser, P.R.: Unsupervised border detection of skin lesion images. In: International Conference on Information Technology: Coding and Computing (ITCC 2005)-Volume II, vol. 2, pp. 123–128. IEEE (2005)
12. Chen, S., Zhang, D.: Robust image segmentation using FCM with spatial constraints based on new kernel-induced distance measure. IEEE Trans. Syst. Man Cybern. Part B (Cybern.) **34**(4), 1907–1916 (2004). https://doi.org/10.1109/TSMCB.2004.831165
13. Codella, N., et al.: Skin lesion analysis toward melanoma detection 2018: a challenge hosted by the international skin imaging collaboration (ISIC). arXiv preprint arXiv:1902.03368 (2019)
14. Cuong, B.C., Kreinovich, V.: Picture fuzzy sets. J. Comput. Sci. Cybern. **30**(4), 409–420 (2014)
15. Dave, R.N.: Characterization and detection of noise in clustering. Pattern Recogn. Lett. **12**(11), 657–664 (1991)
16. Dawant, B.M., Zijdenbos, A.P.: Image segmentation. Handb. Med. Imaging **2**, 71–127 (2000)
17. Deng, Y., Manjunath, B.S.: Unsupervised segmentation of color-texture regions in images and video. IEEE Trans. Pattern Anal. Mach. Intell. **23**(8), 800–810 (2001)
18. Devanathan, K., et al.: Lesion segmentation in dermoscopic images using superpixel based fast fuzzy C-means clustering. In: 2020 IEEE Congreso Bienal de Argentina (ARGENCON), pp. 1–6. IEEE (2020)
19. Garg, S., Jindal, B.: Skin lesion segmentation using K-mean and optimized fire fly algorithm. Multimedia Tools Appl. **80**, 7397–7410 (2021)
20. Garnavi, R., Aldeen, M., Celebi, M.E., Varigos, G., Finch, S.: Border detection in dermoscopy images using hybrid thresholding on optimized color channels. Comput. Med. Imaging Graph. **35**(2), 105–115 (2011)
21. Gomez, D.D., Butakoff, C., Ersboll, B.K., Stoecker, W.: Independent histogram pursuit for segmentation of skin lesions. IEEE Trans. Biomed. Eng. **55**(1), 157–161 (2007)
22. Guo, Y., Sengur, A.: NCM: neutrosophic C-means clustering algorithm. Pattern Recogn. **48**(8), 2710–2724 (2015)
23. IARC: Skin cancer (2022). https://www.iarc.who.int/cancer-type/skin-cancer/#:~:text=Skin%20cancers%20are%20the%20most,people%20died%20from%20the%20disease. Accessed 12 July 2024
24. Jain, A.K.: Data clustering: 50 years beyond K-means. Pattern Recogn. Lett. **31**(8), 651–666 (2010)
25. Khattar, S., Kaur, R.: Computer assisted diagnosis of skin cancer: a survey and future recommendations. Comput. Electr. Eng. **104**, 108431 (2022)
26. Krinidis, S., Chatzis, V.: A robust fuzzy local information C-means clustering algorithm. IEEE Trans. Image Process. **19**(5), 1328–1337 (2010)
27. Ma, Z., Tavares, J.M.R.: Segmentation of skin lesions using level set method. In: Computational Modeling of Objects Presented in Images. Fundamentals, Methods,

and Applications: 4th International Conference, CompIMAGE 2014, Pittsburgh, PA, USA, 3–5 September 2014, pp. 228–233. Springer (2014)
28. Mendonça, T., Ferreira, P.M., Marques, J.S., Marcal, A.R., Rozeira, J.: Ph 2-a dermoscopic image database for research and benchmarking. In: 2013 35th Annual International Conference of the IEEE Engineering in Medicine and Biology Society (EMBC), pp. 5437–5440. IEEE (2013)
29. World Health Organization: Radiation: Ultraviolet (UV) radiation and skin cancer (2017). https://shorturl.at/VBkq1. Accessed 12 July 2024
30. Pacheco, A.G., Krohling, R.A.: The impact of patient clinical information on automated skin cancer detection. Comput. Biol. Med. **116**, 103545 (2020)
31. Pennisi, A., Bloisi, D.D., Nardi, D., Giampetruzzi, A.R., Mondino, C., Facchiano, A.: Skin lesion image segmentation using delaunay triangulation for melanoma detection. Comput. Med. Imaging Graph. **52**, 89–103 (2016)
32. Rehman, M., et al.: Machine learning based skin lesion segmentation method with novel borders and hair removal techniques. PLoS ONE **17**(11), e0275781 (2022)
33. Roy, A., Pal, A., Garain, U.: JCLMM: a finite mixture model for clustering of circular-linear data and its application to psoriatic plaque segmentation. Pattern Recogn. **66**, 160–173 (2017)
34. Schmid, P.: Segmentation of digitized dermatoscopic images by two-dimensional color clustering. IEEE Trans. Med. Imaging **18**(2), 164–171 (1999)
35. Sforza, G., et al.: Using adaptive thresholding and skewness correction to detect gray areas in melanoma in situ images. IEEE Trans. Instrum. Meas. **61**(7), 1839–1847 (2012)
36. Smarandache, F.: Neutrosophy, neutrosophic logic, neutrosophic set, neutrosophic probability and statistics. ProQuest, Ann Arbor (1998)
37. Smarandache, F.: Neutrosophic set-a generalization of the intuitionistic fuzzy set. J. Defense Resour. Manag. (JoDRM) **1**(1), 107–116 (2010)
38. Smarandache, F.: Neutrosophic set is a generalization of intuitionistic fuzzy set, inconsistent intuitionistic fuzzy set (picture fuzzy set, ternary fuzzy set), pythagorean fuzzy set, spherical fuzzy set, and q-rung orthopair fuzzy set, while neutrosophication is a generalization of regret theory, grey system theory, and three-ways decision (revisited). J. New Theory **29**, 1–31 (2019)
39. Szilagyi, L., Benyo, Z., Szilágyi, S.M., Adam, H.: Mr brain image segmentation using an enhanced fuzzy c-means algorithm. In: Proceedings of the 25th Annual International Conference of the IEEE Engineering in Medicine and Biology Society (IEEE Cat. No. 03CH37439), vol. 1, pp. 724–726. IEEE (2003)
40. Tschandl, P., Rosendahl, C., Kittler, H.: The HAM10000 dataset, a large collection of multi-source dermatoscopic images of common pigmented skin lesions. Sci. Data **5**(1), 1–9 (2018)
41. Wei, D., Wang, Z., Si, L., Tan, C., Lu, X.: An image segmentation method based on a modified local-information weighted intuitionistic fuzzy C-means clustering and gold-panning algorithm. Eng. Appl. Artif. Intell. **101**, 104209 (2021)
42. Xu, L., et al.: Segmentation of skin cancer images. Image Vis. Comput. **17**(1), 65–74 (1999)
43. Yasmin, J.J., Sathik, M.M., Beevi, S.Z.: Robust segmentation algorithm using log edge detector for effective border detection of noisy skin lesions. In: 2011 International Conference on Computer, Communication and Electrical Technology (ICCCET), pp. 60–65. IEEE (2011)
44. Zadeh, L.A.: Fuzzy sets. Inf. Control **8**(3), 338–353 (1965)
45. Zadeh, L.A.: The concept of a linguistic variable and its application to approximate reasoning–I. Inf. Sci. **8**(3), 199–249 (1975)

46. Zhou, H., et al.: Skin lesion segmentation using an improved snake model. In: 2010 Annual International Conference of the IEEE Engineering in Medicine and Biology, pp. 1974–1977. IEEE (2010)
47. Zhou, H., Schaefer, G., Sadka, A., Celebi, M.E.: Anisotropic mean shift based fuzzy C-means segmentation of skin lesions. In: Proceedings of the 5th International Conference on Soft Computing as Transdisciplinary Science and Technology, pp. 438–443 (2008)

DLMLP with Elastic-Net Regularization for Hyperspectral Image Classification

Madhuri Gadwal(✉) and Atul Negi

School of Computer and Information Sciences, University of Hyderabad, Hyderabad 500046, Telangana, India
{17mcpc05,atul.negi}@uohyd.ac.in
https://www.scis.uohyd.ac.in/

Abstract. Huge spectral-spatial information available in 3d-hyperspectral images is exploited to learn optimal dictionaries for classification. Unlike deep neural network-based methods, sparse representation for hyperspectral image processing requires low computational resources. In this work, specific constraints on the coefficients and the dictionary result in two convex objective functions that guarantee unique optimal solutions for the dictionary and the coefficient matrix. When coded with a single shared discriminative dictionary, pixels from the same region share the same sparsity profile. The proposed method uses the sparse representation of HSI data with adaptive elastic-net regularization over the coefficient matrix for robust spatial sparse feature extraction. These spatial features are inputs to a multilayer perceptron of a few hidden layers for classification through supervised learning. The experimental evaluation of the model on five HSI datasets shows that the model exhibits competitive performance with lower complexity compared to the state-of-the-art methods.

Keywords: HSI classification · Discriminative Dictionary learning · Regularization · Spatial reduction · Perceptron

1 Introduction

Big data requires energy-hungry Artificial Intelligence (AI) models to extract useful information. So, the thrust has been on using energy-efficient models to process. Real-life applications resulting in large dimensional big data include Hyperspectral Image (HSI) data. HSI has been used to monitor the changes due to urbanization in agriculture, aqua cover, forest cover, etc., which further impact sustainability. Hyperspectral images are acquired as tensors containing spectral bands collected through sensors or satellites on different dates and are preprocessed with image normalization, converting digital numbers to reflectance values. Classes are identified, and labels are given for each image pixel [3] in the tensor. Spectral bands are highly correlated, and reducing spectral features using principal components might result in the loss of crucial information for classification. For spatial feature reduction, spatial groups are acquired through, for

A. Dasgupta et al. (Eds.): BDA 2024, LNCS 15526, pp. 34–51, 2025.
https://doi.org/10.1007/978-3-031-81821-9_3

example, superpixels. In the existing literature, low-rank and sparsity conditions are applied as regularization techniques for spectral-spatial reduction in HSI classification.

1.1 Challenges in HSI Classification

Classification of HSI data is challenging due to different spectral values for identical regions and similar spectral values for different regions. Each pixel corresponds to a class label, and each class (spatial region) has a different number of pixels, i.e., class imbalance as in Indian Pines data given in Table 1.

Table 1. Class imbalance in HSI: Indian Pines

Class	No. of samples	Class	No. of samples
C1	46	C9	20
C2	1428	C10	972
C3	830	C11	2455
C4	237	C12	593
C5	483	C13	205
C6	730	C14	1265
C7	28	C15	386
C8	478	C16	93

Also, sufficient spectral bands are required to extract relevant features for classification. For example, the Indian Pines image has 21025 pixels, of which only less than half are labeled (10249). The vast area captured in HSI results in low spatial resolution of the sensors, resulting in mixed pixels. Mixed pixels could also be due to a composite of different materials in a single pixel. In either case, the resulting hyperspectral images are nonlinear [7]. Same region pixels do not have a fixed value but have values within a range, and these values sometimes overlap with the neighboring pixels of different regions [5]. Thus, a classification method that can capture the nonlinear dependencies among the features to achieve better performance on HSI datasets is the motivation. The spectral bands, when sufficient, are generally highly correlated due to redundant information. This redundancy in data is used to design a dictionary emphasizing feature selection and variable shrinkage. Dictionary Learning (DL) model is based on Least Squares Minimization (LSM) to maintain data fidelity for representation and classification applications. Regularization is applied to the corresponding weights to improve DL performance. Generally, l_1 and l_2 penalties are commonly used in HSI applications. l_1-regularization for variable shrinkage and feature selection corresponds to assuming the Laplacian prior over the coefficients (or weights) but selects highly correlated dictionary atoms. The Gaussian prior over the coefficients corresponds to l_2-regularization of coefficients and groups the correlated dictionary atoms. Thus, spatial structure information (e.g., edges) from neighboring pixels is derived using l_2-regularization. These two penalties are combined in elastic-net regularization proposed by Zou and Hastie [14]. In this work, we leverage the best properties of elastic-net regularization for learning a discriminative dictionary. The strengths of each of these penalties are controlled using two regularization parameters μ_1, μ_2.

HSI classification models using deep convolutional neural networks are heavily parameterized and, hence, require vast data for training. On the other hand, a Multi-Layer Perceptron (MLP) with few hidden layers can approximate the functional relationship among the features. Thus, multilayer perceptrons provide the required tradeoff between model complexity and performance. Further, model complexity could be reduced by using sparse input features, which also results in better generalization [10]. Here, we train an MLP with three hidden layers with the derived sparse coefficients of training data as input features. Sparse coefficients as inputs result in sparse weight connections in the neural network, improving the generalization performance of DLMLP. The classifier matrix learned as a byproduct in the DL process is used as the weight matrix to initialize the MLP. This helps the classifier to converge faster.

2 Sparse Representation for HSI Classification: Related Work

The sparse representation model has been used for hyperspectral image classification as it represents high dimensional data with very few dictionary columns (similar to the basis of a vector space). Sparse representation for HSI classification has been first applied in [1]. The authors have used Joint Sparse Representation (JSR) to extract spectral and spatial information for classification. The authors in [2] elucidated the problems faced in classifying HSI images using sparse representation and dictionary learning. Due to high correlations among samples from different classes, the Orthogonal Matching Pursuit (OMP) algorithm for sparse coding gives decreasing residuals with increasing sparsity thresholds. Hence, the authors recommended using the same sparsity thresholds for all the classes. However, the same sparsity level in HSI is useful when the label is determined based on the minimal residual error as in [16].

Bin et al. proposed a multi-objective SRC for HSI classification [12] optimizing the error residual and the coefficient vector in parallel. However, parallel computation of the error residual and the coefficient vectors does not reflect the changes of one in another, slowing down the convergence. Multifeature-based sparse representation gives equal weight to all the features. However, HSI images with class imbalance do not have equal pixels for all classes. Yong Ma et al. proposed a multi-feature label consistent $k-$svd method to construct a dictionary from multi-features with optimal feature combination strategies in [9]. Such an optimal combination of features can be attained as sparse coefficients when coded with a fixed discriminative dictionary. Also, regularization in DL-based HSI classification avoids model overfitting. Dictionary learning with elastic net regularization is proposed in [13] to address local smoothness and sparsity of features. Imposing l_0-norm penalty on the coefficient matrix, JSM-DKSVD [15] uses discriminative K-SVD on spatial pixel neighborhoods with a labeled pixel as the center. Each pixel in the spatial neighborhood of the central pixel gets the same label as the central pixel. However, this method increases the data size to be processed, and the borderline pixels are also included in some pixel neighborhoods. Ziyu et al. have proposed a semi-supervised joint sparse model with matrix l_0-norm regularization on the coefficient matrix and using discriminative K-SVD for HSI classification. However, joint sparsity and local smoothness assumptions do not hold when the data has nonlinear correlations. Yong et al. extracted multiple features from each HSI image with a similar objective and applied a label consistent with K-SVD for HSI classification. Extraction of multiple features from spectral bands is computationally demanding. LCKSVD and Discriminative K-SVD derive a classifier matrix for classification, but the classifier matrix is not robust to the training sequence.

In the proposed Discriminative Dictionary Learning and MLP (DLMLP) based HSI classification, the smoothness and sparsity properties of features extracted with elastic net regularization are robust to the training sequence. With sparse features and weight connections, DLMLP requires relatively low computational resources compared to the models using GPUs for HSI processing [4]. The steps involved in classifying HSI data using DLMLP are summarized as follows:

1. A novel combination of Discriminative K-SVD with elastic-net regularization and classification using a simple three-layered MLP is proposed to address the nonlinear HSI data classification problem.
2. The proposed DLMLP method uses labeled pixels to learn the dictionary and extracts relevant sparse features from HSI datasets.
3. Using label information in the objective function gives a better subspace segmentation.
4. In step 1, a discriminative classifier matrix is learned that preserves the spatial structure of the data, and the dictionary is learned utilizing the structural correlations among different bands (spectral similarity). The elastic net regularization helps extract meaningful features by balancing sparsity (l_1) and local smoothness (l_2). The dictionary learned in Step 1 is used to derive discriminative sparse features. The learned classifier matrix is the initial weight matrix connecting the inputs to the outputs in the neural network.
5. In step 2, a multilayer perceptron is used to classify highly nonlinear HSI data using sparse coefficients as inputs and the corresponding labels as targets with the weight connections derived from step 1.
6. The sparsity of inputs leads to sparse weight connections from input to hidden layers. It thus lowers the model's complexity, improving the model's generalization on datasets.

2.1 Notation

The notation for the proposed Dictionary Learning and MLP (DLMLP) is in Table 2.

Table 2. Notation for DLMLP

M	Dictionary size
τ	Sparsity threshold
d	Number of features (bands)
N	Number of samples
$\|\vec{a}\|_p = \sum_i (\|a_i\|^p)^{\frac{1}{p}}$	vector l_p−norm
$\|A\|_F = \sqrt{\sum_i \sum_j a_{ij}^2}$	Matrix Frobenious Norm
$\|A_{q\times r}\|_1 = \max_{1\leq j\leq r} \sum_i^q \|a_{ij}\|$	Matrix 1-Norm induced by vector l_1−norm

Section 3 explains the proposed DLMLP method, Sect. 4 gives the results of applying DLMLP on standard HSI datasets, Sect. 5 explains how each module of the DLMLP contributes to the classification performance, finally Sect. 6 concludes the article with future directions.

3 Proposed Method: DLMLP with Elastic-Net Regularization

Consider an HSI data matrix $H^{d\times L}$ constructed spanning L labeled pixels from d spectral bands. Let $D \in \mathbb{R}^{d\times M}$ be the dictionary of M atoms constructed from labeled pixels. D could also be a random projection but takes longer to converge.

The training set $Y = [y_1, y_2, \ldots, y_N]$ is constructed by randomly selecting N labeled pixels from the data matrix H. Each $y_i \in \mathbb{R}^d$ denotes the ith sample.

3.1 Step1: Discriminative Dictionary Learning with Elastic-Net Regularization

Sparse coefficients of test samples extracted w.r.t. a discriminative dictionary form equivalence classes (subspace segmentation), i.e., same class samples have similar sparsity profiles, and those of different classes have different sparsity profiles [6]. A classifier matrix $W \in \mathbb{R}^{C\times M}$ is learned along with the dictionary, and the corresponding sparse coefficients are learned to approximate the test data and the classifier matrix. This is similar to adding a structural prior constraint on the coefficients in [11]. The objective function of sparse representation finds the optimal $D \in \mathbb{R}^{d\times M}$, and the corresponding sparse coefficient matrix $X \in \mathbb{R}^{M\times N}$ to approximate Y i.e.,

$$\min_{D,X} \|X\|_0 \text{ such that } Y = DX.$$

The problem is jointly non-convex and NP-Hard due to the pseudonorm l_0. The convex relaxation form of the objective replaces l_0-norm with the convex but non-smooth l_1-norm i.e.,

$$\min_{D,X} \|X\|_1, \text{ such that } Y = DX.$$

Block Coordinate Descent method [8] is used to alternatingly fix one variable and update another resulting in *sparse coding problem*

$$\min_{X} \|X\|_1 \text{ such that } Y = DX$$

and *dictionary learning problem*

$$\min_{D} \|Y - DX\|_F^2, \text{ such that } \|d_j\|_2^2 = 1 \forall j = 1, 2, \ldots, M$$

Given a dataset $Y^{d\times N}$, *find the optimal* dictionary $D_{d\times M}$ and the corresponding coefficient matrix $X^{M\times N}$, such that $Y = DX$ where X is sparse and columns of D are normalized to have unit norm. The sparsity profile of a sample is the set of dictionary atoms used to represent the sample with minimal residual error.

Problem Formulation. A shared dictionary is obtained using Discriminative KSVD by adding label information to the objective function. $T \in \mathbb{R}^{C\times N}$ is the label matrix of training samples Y. W is the classifier matrix to be determined. α, β are the regularization parameters. Elastic-net regularization [14] is imposed on the coefficient matrix X with two parameters $0 < \mu_1, \mu_2 < 1$ to get

$$< D, W, X >= \underset{D,W,X}{\arg\min} \|Y - DX\|_F^2 + \alpha\|T - WX\|_F^2 + \beta\|W\|_F^2 + \mu_1\|X\|_1 + \mu_2\|X\|_F^2 \tag{3.1}$$

Elastic-net regularization removes the constraint on the number of selected atoms from the dictionary and thus forms groups. This helps in selecting the dictionary atoms corresponding to the image under question. The parameters μ_1, μ_2 drive the sparsity of features.

The matrix $\begin{pmatrix} D \\ \sqrt{\alpha}W \end{pmatrix}$ is always normalized column-wise, so the regularization penalty $\|W\|_F^2$ and β are dropped to get

$$< D, W, X >= \underset{D,W,X}{\arg\min} \left\| \begin{pmatrix} Y \\ \sqrt{\alpha}T \end{pmatrix} - \begin{pmatrix} D \\ \sqrt{\alpha}W \end{pmatrix} X \right\|_F^2 + \mu_1\|X\|_1 + \mu_2\|X\|_F^2. \tag{3.2}$$

That is

$$< D^1, X >= \underset{D^1,X}{\arg\min} \left\| Y^1 - D^1 X \right\|_F^2 + \mu_1\|X\|_1 + \mu_2\|X\|_F^2,$$

where $Y^1 = \begin{pmatrix} Y \\ \sqrt{\alpha}T \end{pmatrix}$ is the augmented data matrix, and $D^1 = \begin{pmatrix} D \\ \sqrt{\alpha}W \end{pmatrix}$ is the augmented dictionary. Using the Block Coordinate Descent technique, this jointly non-convex problem is split into two convex problems given by

$$g(X) = \arg\min_{X} \left\| Y^1 - D^1 X \right\|_F^2 + \mu_1 \|X\|_1 + \mu_2(\|X\|_F^2), \tag{3.3}$$

The mathematical argument given in [13] proves that with fixed D^1 and LSM with elastic-net regularization in (3.3), X is achieved as a unique optimum.

$$g(D^1) = \arg\min_{D^1} \left\| Y^1 - D^1 X \right\|_F^2 \; s.t. \|d_{.j}\|_2^2 \leq 1, \forall j = 1, 2, \ldots, M. \tag{3.4}$$

Similarly, fixing X in (3.4), the unit norm constraint on the dictionary atoms and LSM gives a unique optima D^1 [13]. The first d rows of matrix D^1 are extracted to get the shared dictionary D, and the sparse coefficients of the test set are generated with D. The last C rows of D^1 give the classifier matrix W.

From [13], the solution of (3.3) is

$$x_{ji} = \frac{1}{1+\mu_2}\{\max(P_{ji}, \mu_1/2) + \min(P_{ji}, -\mu_1/2)\}, \forall j = 1, 2, \ldots, M, i = 1, 2, \ldots, N, \tag{3.5}$$

where $P_{ji} = [{D^1}^T Y^1]_{ji} - \sum_{l=1, l\neq j}^{M} [{D^1}^T D^1]_{jl}) x_{li}$.

Similarly, ignoring the terms independent of D^1 in (3.4), from [13], the optimal solution of (3.4) is

$${D^1}^T_{.\mathbf{j}} = \frac{Y^1 X_{j.}^T - \tilde{D}_{.\mathbf{r}} X X_{j.}^T}{\|Y^1 X_{j.}^T - \tilde{D}_{.\mathbf{r}} X X_{j.}^T\|_2}, \tag{3.6}$$

where $\tilde{D}_{.\mathbf{r}} = {D^1}_{.\mathbf{r}}$ when $r \neq j$ and $\mathbf{0}$, otherwise. The first d rows of D^1 give the shared discriminative dictionary, and the last C rows of D^1 give the classifier matrix. D and W are normalized as in [17]. The classifier matrix W forces the sparse coefficient matrix X to correspond to the label matrix T. The updated discriminative coefficient matrix X is coded using the shared discriminative dictionary D. Thus, the sparse coefficients become discriminative features for HSI classification.

3.2 Step2: Classification of HSI Using a Lightweight MLP

Multilayer perceptron (MLP) with few hidden layers offers a better solution balancing model complexity and generalization. The proposed DLMLP method uses the discriminative sparse coefficients X as inputs and the corresponding labels T to train a neural network with three hidden layers. A neural network with back-propagation requires a good initial weight matrix for faster convergence.

DLMLP uses the classifier matrix W as the initial weight matrix between X and T for faster network convergence. Due to sparse inputs, the number of weights to be updated in each iteration is relatively low compared to deep CNN-based methods that use raw features to train.

Classification of Test Samples. Given a query $y_q \in \mathbb{R}^d$, the sparse coefficient vector using D is

$$x_q = (D^T D)^{-1} D^T y_q.$$

Similarly, given a set of test samples $Y_{test} \in \mathbb{R}^{d \times n}$,

$$X_{test} = (D^T D)^{-1} D^T Y_{test}.$$

$X_{test} \in \mathbb{R}^{M \times n}$ is classified using the trained MLP. Algorithm 1 describes the steps in classifying HSI data using DLMLP.

Algorithm 1. DLMLP with Elastic-net regularization

procedure DLMLP(Input: Normalized augmented data matrix Y^1, D_0, $X^{(0)} = (D_0^T D_0)^{-1} D_0^T Y$, μ_1, μ_2, labels T, error tolerance tol, Y_{test})
  Output: $D, W, labels(Y_{test})$
  $Y^1 \leftarrow \begin{pmatrix} Y \\ \sqrt{\alpha} T \end{pmatrix}$, $D^1 \leftarrow \begin{pmatrix} D \\ \sqrt{\alpha} W \end{pmatrix}$.
  while $\|Y^1 - D^1 X\|_F^2 \geq tol$ **do**
    X from (3.5)
    D^1 from (3.6)
  end while
  $D \leftarrow D^1[0:d,:]$ ▷ the first d rows
  $W \leftarrow D^1[d:d+C,:]$ ▷ the last C rows
  $D = \{\frac{d_1}{\|d_1\|}, \frac{d_2}{\|d_2\|}, \dots, \frac{d_M}{\|d_M\|}\}$
  $X \leftarrow (D^T D)^{-1} D^T Y$
  Train a MLP(X, W, T) with three hidden layers using X as inputs, corresponding labels T as targets, W the weight matrix.
  $X_{test} = (D^T D)^{-1} D^T Y_{test}$
  return (labels(Y_{test}) $\leftarrow MLP(X, W, T)$).
end procedure

Dictionary learning with elastic-net regularization imposes weighted l_1, l_2 penalties on each pixel's coefficients of dictionary atoms. Each spatial pixel is a vector whose elements are spectral band values corresponding to that pixel location. While l_1 regularization helps feature selection and variable shrinkage, l_2 regularization groups the spatially correlated dictionary atoms and shrinks the coefficients accordingly. Spatial structures like edges are recovered using l2-regularization. The sparse coefficients are the weights given to each dictionary atom to represent each pixel in the feature space. DLMLP extracts the discriminative sparse coefficients as input features to train an MLP with the corresponding one-hot encoded labels as the output layer.

4 Experiments and Results

The performance of DLMLP is compared with SPCLSR-DID [11] where the authors have used only $1\% to 5\%$ of the labeled samples for training. Overcomplete dictionary learning needs minimal representation from each class to allow redundancy and improved selection of atoms. Different from [11], in DLMLP, the number of training samples used for dictionary learning is $10\% to 30\%$ of the available labeled pixels. The remaining labeled pixels are used for testing. Thus, the better performance of our method is attributed to the number of training samples used for learning the dictionary. The accuracy values and prediction times presented here are averages of 3−fold cross-validation performed on labeled samples. The classification performance measures of HSI datasets are overall accuracy, average accuracy, and kappa values.

$$OverallAccuracy(OA) = \frac{\text{No. of correctly classified pixels}}{\text{No. of test pixels}}$$

$$AverageAccuracy(AA) = \frac{1}{C}\sum_{i=1}^{C} \frac{\text{No. of correctly classified pixels of the ith class}}{\text{No. of test pixels of the ith class}}$$

$$Kappa(\kappa) = \frac{OA - (\sum_{i=1}^{C}(pb_i \times pb_i^t))}{1 - (\sum_{i=1}^{C}(pb_i \times pb_i^t))} \times 100,$$

where pb_i is the probability of pixels being predicted as the ith class and pb_i^t is the probability of pixels from the ith class.

Figure 1(f) demonstrates that DLMLP predicted labels closely match with the ground truth labels of the Salinas data.

The results of DLMLP on Salinas, as shown in Table 3, show the superior performance of DLMLP both class-wise and overall. In this case, the prediction time of DLMLP is less than that of MF-LCKSVD.

Figure 2(f) gives the subtraction of ground truth from DLMLP predicted labels.

Table 4 shows accuracies of DLMLP on Pavia center data close to those of SPCLSR-DID. However, the prediction time is relatively low using DLMLP, indicating lower computational overhead.

Figure 3(f) demonstrates that DLMLP-predicted labels closely match the ground truth labels.

Table 5 shows the comparable performance of DLMLP to the Multi-feature based LCKSVD. DLMLP uses sparse coefficients as features for classification, whereas MF-LCKSVD computes a combination of features, which makes it computationally expensive.

Figure 4(f) shows that the difference between DLMLP predicted and ground truth labels is scarce.

Figure 5(e) gives the DLMLP predicted labels on the Botswana image. Table 6 and Table 7 show that the DLMLP classwise and overall accuracies are better than the other three methods.

Table 3. Overall accuracy, average accuracy, and Kappa values on Salinas dataset

Class	JSMDKSVD [15]	MF-LCKSVD [9]	SPCLSR-DID [11]	DLMLP
C1	**100 ± 0.0**	**100 ± 0.0**	**100 ± 0.0**	**100 ± 0.0**
C2	**100 ± 0.0**	**100 ± 0.0**	**100 ± 0.0**	**100 ± 0.0**
C3	98.43 ± 0.30	98.32 ± 0.50	97.59 ± 2.54	**98.97 ± 0.22**
C4	98.43 ± 0.25	**99.43 ± 0.30**	98.52 ± 0.48	**99.1 ± 0.17**
C5	98.55 ± 1.03	99.43 ± 0.30	99.54 ± 0.26	**100 ± 0.0**
C6	98.43 ± 0.30	**100 ± 0.0**	99.79 ± 0.06	**100 ± 0.0**
C7	**100 ± 0.0**	**100 ± 0.0**	99.99 ± 0.02	**100 ± 0.0**
C8	98.9 ± 0.54	99.24 ± 0.29	**99.62 ± 0.23**	99 ± 0.19
C9	**100 ± 0.0**	98.8 ± 1.08	**100 ± 0.0**	**100 ± 0.0**
C10	97.84 ± 0.65	**99.42 ± 0.20**	98.01 ± 0.98	99.31 ± 0.04
C11	99.76 ± 0.21	**100 ± 0.0**	99.8 ± 0.04	**100 ± 0.0**
C12	97.59 ± 1.05	**100 ± 0.0**	99.98 ± 0.05	100 ± 0.0
C13	99.1 ± 0.16	98.96 ± 2.1	99.67 ± 0.01	**99.9 ± 0.01**
C14	98.69 ± 0.40	**100 ± 0.0**	97.43 ± 1.73	99.58 ± 0.18
C15	98.64 ± 0.01	99.46 ± 0.05	98.69 ± 0.78	**99.71 ± 0.2**
C16	98.97 ± 0.13	99.35 ± 0.08	**99.8 ± 0.28**	99.3 ± 0.32
OA	97.93 ± 0.65	99.3 ± 0.8	99.4 ± 0.20	**99.6 ± 0.13**
AA	97.45 ± 0.04	99.01 ± 0.06	99.3 ± 0.23	**99.5 ± 0.2**
Kappa	0.984	0.992	0.993	**0.995**
time(s)	279	127	248	102.3

Table 4. Overall accuracy, average accuracy, and Kappa values on Pavia Center dataset

Class	JSMDKSVD [15]	MF-LCKSVD [9]	SPCLSR-DID [11]	DLMLP
C1	98.12 ± 0.75	98.53	**100 ± 0.0**	**100 ± 0.0**
C2	97.6 ± 0.30	97.56	99.7 ± 0.19	**100 ± 0.0**
C3	98.6 ± 0.78	99.44	**100 ± 0.0**	99 ± 0.15
C4	98.74 ± 0.6	98.93	99.32 ± 0.46	**99.46 ± 0.09**
C5	99.02 ± 0.21	99.32	**100 ± 0.0**	**100 ± 0.0**
C6	98.07 ± 0.56	98.9	**100 ± 0.0**	**100 ± 0.0**
C7	98.67 ± 0.53	98.89	99.8 ± 0.12	**100 ± 0.0**
C8	99.56 ± 0.8	99.12 ± 0.08	99.87 ± 0.40	**100 ± 0.0**
C9	**100 ± 0.0**	99.54	**100 ± 0.0**	99.1 ± 0.20
OA	98.53 ± 0.12	98.97	99.05 ± 0.15	**99.92 ± 0.01**
AA	98.8 ± 043	98.85 ± 0.12	98.8 ± 0.09	**99.71 ± 0.02**
Kappa	0.989	0.98	0.991	**0.998**
Prediction time	97.2	43.3	65.3	40.6

Table 5. Overall accuracy, average accuracy, and Kappa values on Indian Pines dataset using 30% as training samples and 70% as test samples

Class	JSMDKSVD [15]	MF-LCKSVD [9]	SPCLSR-DID [11]	DLMLP
C1	97.96 ± 0.29	**100 ± 0.0**	**100 ± 0.0**	**100 ± 0.0**
C2	**98.31 ± 0.98**	97.02	95.98 ± 1.23	96.2 ± 0.6
C3	98.55	**98.64 ± 0.17**	95.57 ± 2.0	97.4 ± 1.9
C4	**98.58 ± 1.23**	97.88	96.18 ± 2.33	97.1 ± 1.1
C5	**99.78 ± 0.22**	99.22	97.21 ± 1.78	97.9 ± 0.5
C6	99.71	**100 ± 0.0**	**100 ± 0.0**	**100 ± 0.0**
C7	86.96	96	**100 ± 0.0**	97 ± 2.3
C8	**100 ± 0.0**	**100 ± 0.0**	**100 ± 0.0**	**100 ± 0.0**
C9	50	**100 ± 0.0**	93.68 ± 6.16	96.9 ± 0.12
C10	96.36	97.1	95.33 ± 1.78	**97.9 ± 0.6**
C11	99.06	99.6	99.13 ± 0.55	**99.6 ± 0.38**
C12	**99.10 ± 1.2**	98.53	98.19 ± 1.35	98.7 ± 0.76
C13	98.96	**100 ± 0.0**	**100 ± 0.0**	**100 ± 0.0**
C14	99.23	99.8	**100 ± 0.0**	**100 ± 0.0**
C15	99.71	98.38	98.25 ± 1.34	100 ± 0
C16	97.67	**100 ± 0.0**	99.09 ± 0.52	99 ± 0.9
OA	98.68	**98.8 ± 0.1**	98.04 ± 0.36	98.6 ± 0.23
AA	95.0	**98.85 ± 0.12**	98.04 ± 0.92	98.61 ± 0.21
Kappa	0.985	**0.99**	97.74 ± 0.64	98.41 ± 0.56
time(s)	37	36.2	35	31

Table 6. Overall accuracy, average accuracy and Kappa values on Pavia University dataset using 30% training pixels

Class	JSMDKSVD [15]	MF-LCKSVD [9]	SPCLSR-DID [11]	DLMLP
C1	93.19	96.21	98.92 ± 0.8	**100 ± 0.0**
C2	98.44	99.1	99.92 ± 0.07	**100 ± 0.0**
C3	77.33	97.99	98.52 ± 2.18	**99 ± 0.20**
C4	89.55	97.85	97.92 ± 0.48	**100 ± 0.0**
C5	99.62	99.62	99.92 ± 0.08	**100 ± 0.0**
C6	74.89	84.9	99.52 ± 0.48	**100 ± 0.0**
C7	87.84	90.12	99.92 ± 0.08	**99 ± 0.21**
C8	76.84	92.45	99.05 ± 1.08	**100 ± 0.0**
C9	98.51	97.9	98.42 ± 0.28	**99.8 ± 0.14**
OA	91.04	94.92	99.43 ± 0.22	**99.76 ± 0.16**
AA	88.47	94.02 ± 0.21	99.15 ± 0.0	**99.53 ± 0.21**
Kappa	0.88	0.956	0.962	**0.996**
time(s)	192	159	132	113

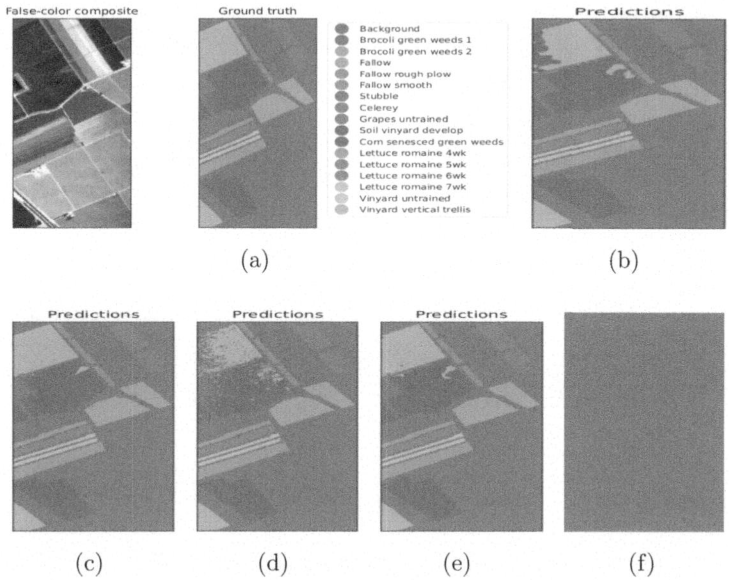

Fig. 1. Results of DLMLP on Salinas: (a) Ground Truth (b) JSMDKSVD (c) MF-LCKSVD (single feature, 5%) (d) SPCLSR-DID (e) DLMLP (f) Difference between DLMLP and GT

Table 7. Overall accuracy, average accuracy, and Kappa values on Botswana dataset using 30% as training samples and 70% as test samples

Class	JSMDKSVD [15]	MF-LCKSVD [9]	SPCLSR-DID [11]	DLMLP
C1	97.96 ± 1.1	$\mathbf{100 \pm 0.0}$	$\mathbf{100 \pm 0.0}$	99.47 ± 0.20
C2	96.31 ± 0.13	97.82 ± 0.76	96.19 ± 0.1	$\mathbf{99.21 \pm 0.40}$
C3	98.55 ± 0.51	$\mathbf{100 \pm 0.0}$	$97.1 \pm 0.98 \pm 0.0$	$\mathbf{100 \pm 0.0}$
C4	98.58 ± 0.6	$\mathbf{100 \pm 0.0}$	96.32 ± 0.71	$\mathbf{100 \pm 0.0}$
C5	89.78 ± 0.36	89.87 ± 0.32	$\mathbf{98.12 \pm 0.91}$	96.1 ± 0.32
C6	99.1 ± 0.26	97.1 ± 0.9	99.08 ± 0.13	$\mathbf{100 \pm 0.0}$
C7	96.6 ± 0.34	$\mathbf{100 \pm 0.0}$	97.67 ± 0.55	98.97 ± 0.10
C8	96.9 ± 0.96	$\mathbf{100 \pm 0.0}$	99.16 ± 0.56	99.75 ± 0.08
C9	94 ± 0.61	97.57 ± 0.23	98.19 ± 0.23	$\mathbf{99.49 \pm 0.51}$
C10	97.36 ± 0.19	100 ± 0.7	$\mathbf{100 \pm 0.0}$	$\mathbf{100 \pm 0.0}$
C11	99.06 ± 0.5	$99.2 \pm 0.6 \pm 0.12$	99.17 ± 0.21	$\mathbf{99.6 \pm 0.8}$
C12	99.10 ± 0.25	99.8 ± 0.29	$\mathbf{100 \pm 0.0}$	99.7 ± 0.3
C13	97.96 ± 0.23	99.69 ± 0.27	$\mathbf{100 \pm 0.0}$	$\mathbf{100 \pm 0.0}$
C14	$\mathbf{97.83 \pm 0.31}$	97.7 ± 0.65	95.95 ± 0.35	94.76 ± 0.08
OA	97.18	98.49	99.21 ± 0.27	$\mathbf{99.38 \pm 0.14}$
AA	95.20	98.41	98.38 ± 0.5	$\mathbf{98.85 \pm 0.51}$
Kappa	0.95	0.98	0.989	**0.993**
time(s)	121	55.21	132	89.7

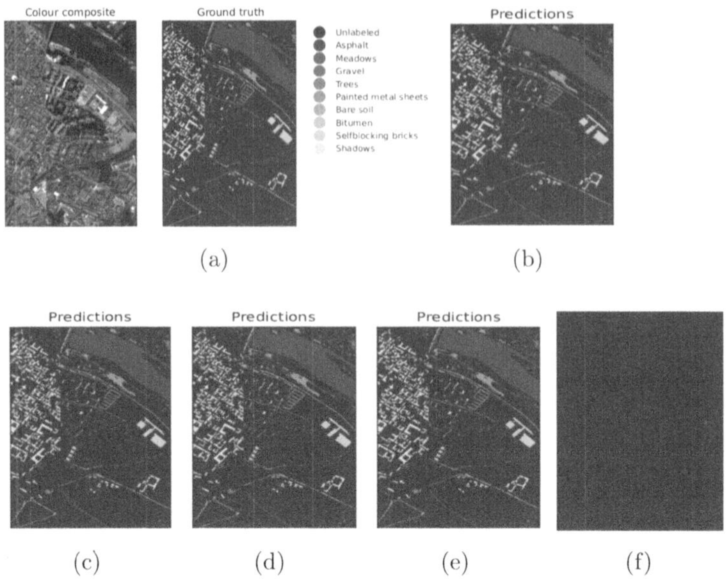

Fig. 2. Results of DLMLP on Pavia Center: (a) Ground Truth (b) JSMDKSVD (c) MF-LCKSVD (single feature, 5%) (d) SPCLSR-DID (e) DLMLP (f) Difference between DLMLP and GT

All the experiments have been conducted using Python 3.8.10 on Intel®Core^{TM} i7-10510U CPU @ 1.80 GHz × 8 with 31 GiB memory.

5 Discussion

The generalization performance of the proposed DLMLP is tested by varying the test ratios in the experiments. From Fig. 6, it is evident that the model performs well when the test ratio is around 0.3 for all the datasets. When the dictionary is trained using at least 10% labeled pixels, the performance of DLMLP is better compared to JSM-DKSVD, MF-LCKSVD, and SPCLSR-DID on the remaining four datasets.

The proposed DLMLP is computationally less demanding than the other methods. The complexity of the proposed DLMLP is given module-wise in Table 8. Class imbalance is a challenge faced in HSI classification. Reducing the size of the dictionary has helped achieve relatively better performance on the Indian Pines dataset. Increasing the test and decreasing the training ratios in an imbalanced dataset drops the performance levels, as demonstrated in Fig. 6(b). However, with a higher training ratio, DLMLP performs relatively better. The performance of Indian Pines reduces when the test ratio increases. This is due to the low representation of minority classes in the training due to class imbalance in the Indian Pines dataset. The average accuracies on the Salinas dataset are

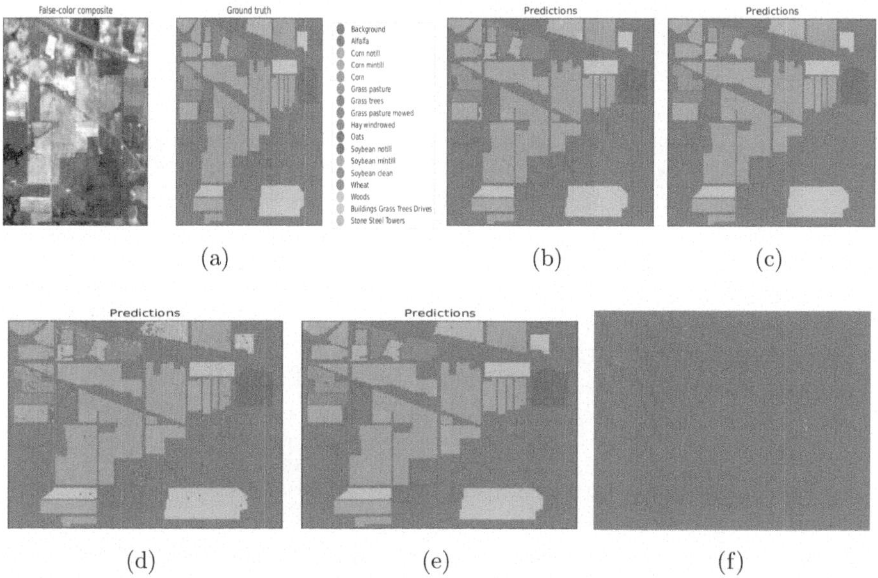

Fig. 3. Results of DLMLP on Indian Pines: (a) Ground Truth (b) JSMDKSVD (c) MF-LCKSVD (single feature, 5%) (d) SPCLSR-DID (e) DLMLP (f) Difference between DLMLP and GT

Table 8. Computational complexity of DLMLP explained, h_1, h_2, h_3 correspond to the number of neurons in each hidden layer.

Discriminative k-SVD with elastic-net (D^1)	$O((d+C)MN)$
Computation of X	$O(d^2M)$
No. of weights in 3-MLP	$\tau * h_1 + h_1 * h_2 + h_2 * h_3 + h_3 * C$

higher than the overall accuracies as shown in Fig. 6(a) as there is a balanced representation of each class pixel in the dataset.

5.1 Parameter Sensitivity Analysis

The regularization parameter α is empirically set to 0.2 for discriminative dictionary learning. The elastic net is controlled by μ_1 and μ_2. The parameter μ_1 controls the tradeoff between the data reconstruction and the coefficient matrix's sparsity. As μ_1 increases, sparse discriminative coefficients are acquired. However, the reconstruction error increases, leading to sparse features that may not represent the data correctly and are not applicable for classification. In this work, μ_1 is set to 0.15 as better performance is achieved.

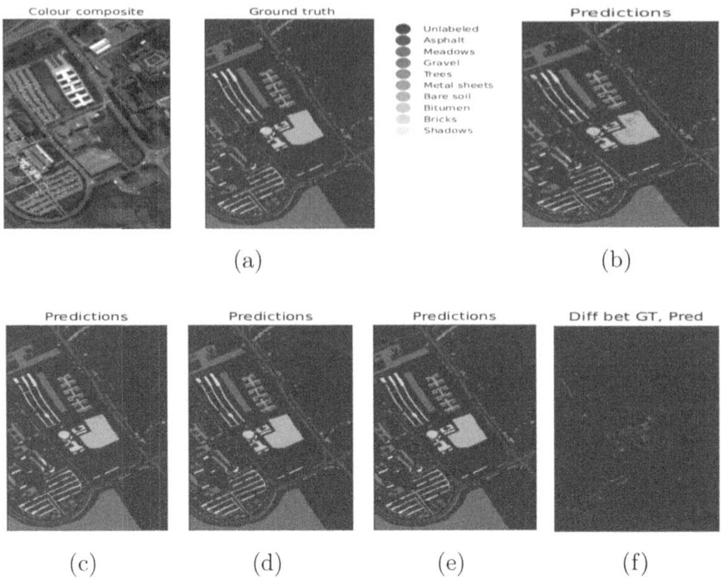

Fig. 4. Results of DLMLP on PaviaU: (a) Ground Truth (b) JSMDKSVD (c) MF-LCKSVD (single feature, 5%) (d) SPCLSR-DID (e) DLMLP (f) Difference between DLMLP and GT

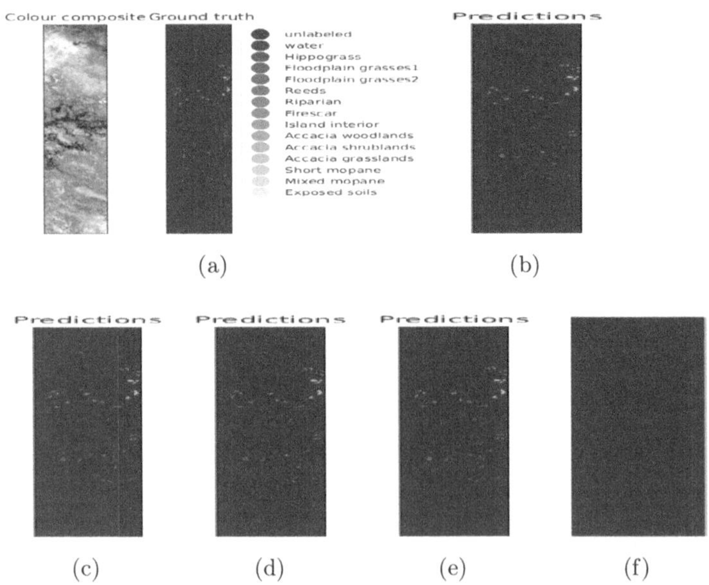

Fig. 5. Results of DLMLP on Botswana HSI: (a) Ground Truth (b) JSMDKSVD (c) MF-LCKSVD (single feature, 5%) (d) SPCLSR-DID (e) DLMLP (f) Difference between DLMLP and GT

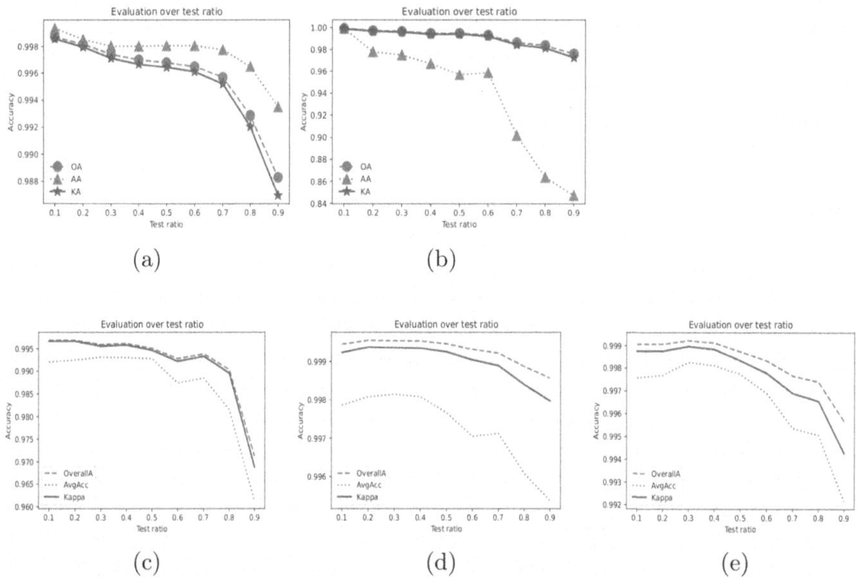

Fig. 6. The decrease in overall and average accuracies of DLMLP over increasing test ratios: (a) Salinas, (b) Indian Pines, (c) Botswana, (d) Pavia Center, (e) PaviaU. At least 30% labeled pixels are required to get the maximum performance

The parameter μ_2 helps in grouping the atoms based on their mutual correlations retrieving spatial information like edges. As $\mu 2$ increases, a wider choice of atoms from the dictionary results in discriminative sparse coefficients. Thus, elastic-net regularization, with proper tuning of hyperparameters, provides a tradeoff between sparsity and spatially correlated features relevant for classification. Hence, the method's performance is analyzed with μ_2 in $\{0.01, 0.015, 0.02, 0.025, 0.03, 0.035, 0.04\}$ while $\mu_1 = 0.15$. For the Pavia Center data, $\mu_2 = 0.02$ gives maximum accuracy value; for the Pavia University and Botswana datasets, $\mu_2 = 0.015$ gives better classification. With $\mu_2 = 0.025$, a better classification of Salinas and Indian Pines is observed.

6 Conclusions

This article proposes a novel discriminative dictionary learning method for HSI data classification. Spatial dimensionality reduction of the feature space in the form of discriminative sparse coefficients is achieved using the discriminative-KSVD with elastic-net regularization. Unlike other spectral dimensionality reduction-based dictionary learning methods, DLMLP preserves the spectral information, thus enriching the dictionary atoms with minor components important for classification. The sparse coefficients used to train a simple three-layered perceptron result in a simpler architecture of the neural network classifier. With relatively less complexity, DLMLP demonstrates superior classification results.

However, when border pixels from different classes share the same features, spatial dimensionality reduction is quite challenging when the labels are missing. DLMLP with augmented dictionary atoms to accommodate the border pixels and address the missing label data problem could be considered a future direction.

References

1. Chen, Y., Nasrabadi, N.M., Tran, T.D.: Hyperspectral image classification using dictionary-based sparse representation. IEEE Trans. Geosci. Remote Sens. **49**(10), 3973–3985 (2011). https://doi.org/10.1109/TGRS.2011.2129595
2. Cui, M., Prasad, S.: Class-dependent sparse representation classifier for robust hyperspectral image classification. IEEE Trans. Geosci. Remote Sens. **53**(5), 2683–2695 (2015). https://doi.org/10.1109/TGRS.2014.2363582
3. Gumma, M.K., Mohammad, I., Nedumaran, S., Whitbread, A., Lagerkvist, C.J.: Urban sprawl and adverse impacts on agricultural land: a case study on Hyderabad, India. Remote Sens. **9**(11) (2017). https://doi.org/10.3390/rs9111136. https://www.mdpi.com/2072-4292/9/11/1136
4. Guo, Y., Qu, F., Yu, Z., Yu, Q.: Deep LSTM with guided filter for hyperspectral image classification. Comput. Inform. **39**(5), 973–993 (2021). https://doi.org/10.31577/cai-2020-5-973. https://www.cai.sk/ojs/index.php/cai/article/view/2020-5-973
5. Han, T., Goodenough, D.G.: Investigation of nonlinearity in hyperspectral imagery using surrogate data methods. IEEE Trans. Geosci. Remote Sens. **46**(10), 2840–2847 (2008). https://doi.org/10.1109/TGRS.2008.2002952
6. Jiang, Z., Lin, Z., Davis, L.S.: Label consistent K-SVD: learning a discriminative dictionary for recognition. IEEE Trans. Pattern Anal. Mach. Intell. **35**(11), 2651–2664 (2013). https://doi.org/10.1109/TPAMI.2013.88
7. Keshava, N., Mustard, J.: Spectral unmixing. IEEE Signal Process. Mag. **19**(1), 44–57 (2002). https://doi.org/10.1109/79.974727
8. Liu, B.D., Wang, Y.X., Shen, B., Li, X., Zhang, Y.J., Wang, Y.J.: Blockwise coordinate descent schemes for efficient and effective dictionary learning. Neurocomputing **178**, 25–35 (2016). https://doi.org/10.1016/j.neucom.2015.06.096. http://www.sciencedirect.com/science/article/pii/S0925231215016148. Smart Computing for Large Scale Visual Data Sensing and Processing
9. Ma, Y., Zhang, Y., Mei, X., Dai, X., Ma, J.: Multifeature-based discriminative label consistent k-SVD for hyperspectral image classification. IEEE J. Sel. Top. Appl. Earth Obs. Remote Sens. **12**(12), 4995–5008 (2019). https://doi.org/10.1109/JSTARS.2019.2949621
10. Madhuri, G., Kashyap, M.N.L., Negi, A.: Telugu OCR using dictionary learning and multi-layer perceptrons. In: 2019 International Conference on Computing, Power and Communication Technologies (GUCON), pp. 904–909 (2019)
11. Nie, X., Xue, Z., Lin, C., Zhang, L., Su, H.: Structure-prior-constrained low-rank and sparse representation with discriminative incremental dictionary for hyperspectral image classification. IEEE Trans. Geosci. Remote Sens. **62**, 1–19 (2024). https://doi.org/10.1109/TGRS.2024.3353370
12. Pan, B., Shi, Z., Xu, X.: Multiobjective-based sparse representation classifier for hyperspectral imagery using limited samples. IEEE Trans. Geosci. Remote Sens. **57**(1), 239–249 (2019). https://doi.org/10.1109/TGRS.2018.2853268

13. Shen, B., Liu, B.D., Wang, Q.: Elastic net regularized dictionary learning for image classification. Multimedia Tools Appl. **75**, 8861–8874 (2016). https://doi.org/10.1007/s11042-014-2257-y
14. Soomro, B.N., Xiao, L., Huang, L., Soomro, S.H., Molaei, M.: Bilayer elastic net regression model for supervised spectral-spatial hyperspectral image classification. IEEE J. Sel. Top. Appl. Earth Obs. Remote Sens. **9**(9), 4102–4116 (2016). https://doi.org/10.1109/JSTARS.2016.2559524
15. Wang, Z., Liu, J., Xue, J.H.: Joint sparse model-based discriminative k-SVD for hyperspectral image classification. Signal Process. **133**, 144–155 (2017). https://doi.org/10.1016/j.sigpro.2016.10.022. https://www.sciencedirect.com/science/article/pii/S0165168416302882
16. Wright, J., Yang, A.Y., Ganesh, A., Sastry, S.S., Ma, Y.: Robust face recognition via sparse representation. IEEE Trans. Pattern Anal. Mach. Intell. **31**(2), 210–227 (2009). https://doi.org/10.1109/TPAMI.2008.79
17. Zhang, Q., Li, B.: Discriminative k-SVD for dictionary learning in face recognition. In: 2010 IEEE Computer Society Conference on Computer Vision and Pattern Recognition, pp. 2691–2698 (2010). https://doi.org/10.1109/CVPR.2010.5539989

Graph Analytics

D*is*E*v*D: A Benchmark Dataset for Disruptive Event Identification from Online Social Network

Aditi Seetha[1(✉)], Jitendra Parmar[2], Satyendra Singh Chouhan[1], Emmanuel S. Pilli[1], and Abhinav Jain[3]

[1] MNIT Jaipur, Jaipur 302017, India
{2021rcp9548,sschouhan.cse,espilli.cse}@mnit.ac.in
[2] VIT Bhopal University, Kothrikalan, Sehore, India
jitendraparmar@vitbhopal.ac.in
[3] IIIT Kota, Jaipur, India
2020kucp1018@iiitkota.ac.in

Abstract. This paper introduces a new Disruptive Event Dataset, D*is*E*v*D, for developing machine learning models to predict multiple disruptive events that can help to monitor and mitigate such events with appropriate actions. Earlier datasets did not include all the event categories that might lead to disruption, like crime, disaster, cyber-attacks, rumours, and crime-based events. Some of these datasets suffer from data scarcity, while others suffer low coverage. There is no dataset consisting of disruptive events tweets of the Indian subcontinent. Also, no disruptive event dataset contains geotagged tweets for identifying the location of the occurred event. Our dataset D*is*E*v*D contains all the major disruptive events that happened in the bounding box of India from 1st April 2022 to 15th Jan 2023. It best describes disruptive events with the help of geotagged tweets. The experimental results, using different clustering algorithms, show the effectiveness and usefulness of the dataset. Further, we also present the methodology to extract the disruptive event information from the events cluster.

Keywords: Disruptive Events · Twitter Social Network · Event Classification

1 Introduction

A *disruptive event* is an event that prevents the routine process from completing its objectives. Numerous unofficial sources direct these events, which could result in civil disorder by upsetting law and order [1,2]. These events are unpredictable; they could last a few days or even a few weeks [2]. The success rate of such events relies on the size of the gathering. Social media data is used for event detection, which records a variety of current and future events. The event detection presents three crucial questions: *what* (the event's name), *where* (the event's location), and *when* (time of the event).

A. Dasgupta et al. (Eds.): BDA 2024, LNCS 15526, pp. 55–74, 2025.
https://doi.org/10.1007/978-3-031-81821-9_4

The social media platform helps people to discuss varying topics around the world. Social media has established itself as a significant informational resource for such events. According to statistics for 2022, almost 486 million people use the social media platform Twitter around the world [3]. Disruptive events could include political protests, cultural and religious riots, crimes, attacks, natural disasters, etc. For example, Twitter has broken out the protest against the Agneepath scheme in India introduced by the government. The false information spread over social media platforms has the potential to turn a peaceful protest into violence. Therefore, it becomes crucial to identify any signs of disruptive events and take preventive actions against them [4].

Researchers have published various datasets for detecting events using social media [5–8]. These events belong to social events like sports, politics, traffic, real-time event, and aggressive events like riots, protests, and disaster events. Most earlier researchers proposed datasets that identified public social events like sports, politics, etc. Mcminn et al. [5] published a dataset called Event2012, consisting of Tweets for detecting social events. In another research, Wang et al. [6] published a MAVEN dataset consisting of Wikipedia documents for detecting general domain events. Later, Alnazzawi et al. [7] published a corpus HateMotiv for identifying hate crimes, including verbal and physical assault. In another work, Giorgi et al. [8] published a Twitter corpus of the Black Lives Matter movement and its counter-protest.

These datasets did not help predict and identify disruptive events. Only a few datasets were available for identifying events like riots, protests, and crises. However, these datasets did not include all the event categories that might lead to disruption, like crime, disaster, cyber-attacks, rumors, and crime-based events. Some researchers acquired data for limited months or days, leading to the problem of data scarcity, while some suffered from low coverage. There is no dataset consisting of disruptive events Tweets of the Indian subcontinent.

In light of the above works, this paper proposed a benchmark dataset for disruptive event identification from the online social network Twitter. It contains geo-tagged Tweets of the Indian subcontinent. The machine learning models based on this dataset can be helpful for government agencies to monitor events and mitigate them before they turn violent. This dataset can be considered the benchmark for developing state-of-the-art machine learning-based disruptive event detection models.

Contributions. The major contribution of the paper is as follows:

- **Dataset** D*is*E*v*D is the first disruptive event dataset containing the Tweets of the Indian subcontinent. This dataset contains geo-tagged Tweets, which help in determining the location of the event about to happen.
- **Methodology** We proposed a methodology *DiEvD*, for disruptive event detection from the Tweets collected from the social media Twitter.
- **Validation** We performed experiments to validate the proposed D*is*E*v*D dataset using the proposed methodology *DiEvD*.

2 Literature Survey

This section provides a summary of the latest research on datasets related to disruptive events, encompassing their discoveries and limitations.

McMinn et al. [5] proposed a dataset, Event2012, for detecting social events on the Twitter platform. The dataset comprises 500 news events sampled over a span of four weeks, incorporating tweet-level relevance judgments for thousands of tweets related to these events. Wang et al. [6] proposed a dataset, MAVEN, for event detection consisting of 4,480 event mention Wikipedia documents, 118,732 event mention instances, and 168 event types of the general domain. However, both datasets are primarily used for general event detection and cannot be used directly for disruptive event detection.

In [7], Alnazzawi et al. proposed, HateMotiv corpus, for identifying hate crimes like cyber bullying, trolling, harrasement from Twitter social media. This dataset is publicly available and contains 5000 manually annotated Tweets between January 2010 and December 2019.

Alsaedi et al. [2], proposed a corpus of 40 millions Tweeter messages, for detecting the disruptive events. This dataset was collected from 1st Oct, 2015 till 30th November, 2015 using Twitter streaming API. This dataset contains both Tweets related to disruptive event and non-disruptive events. Additionally, researcher proposed a dataset of England riots 2011 that encompasses total of 1.6 million Tweets. These Tweets were extracted using some popular hashtags used during the England riots. Wang et al. [9] collected real time datasets from both Twitter and Instagram social media on the 'protest' events consisting of 295,643 Tweets and 5,688 photos from Instagram.

Huang et al. [10] proposed a dataset of 4.2 million Tweets over a period of 12 months from four U.S. Midwestern college towns and proposed a framework to detect social events. Modha et al. [11] created a corpus of 15,001 aggression annotated Facebook posts and comments in both English and Hindi language and classified the comments in three categories: 'Overtly Aggressive (OAG)', 'Covertly Aggressive (CAG)' and 'Non-aggressive (NAG)'. Morabia et al. [12] used a dataset of 11M Tweets collected during 11th October 2017 to 17th October 2021 from Twitter social media platform.

Shah et al. [13] collected a corpus of 11.7M English language Tweets of 100 cities from the Twitter social media and retrieved 1% of the sample Tweets which is freely available for research.

Liu et al. [14] proposed a multilingual text dataset consisting of 6.4B Tweets and Weibos, 7.8M news articles and 15.6M forum messages. Alsmadi et al. [15] collected 1.5M Tweets related to Mizzou protest from Twitter using popular protest hashtags.

Chen et al. [16] proposed a public dataset containing 570M Tweets of the Ukraine and Russia war from February 22, 2022, to January 8, 2023, extracted using Twitter API. Suwaileh et al. [17] proposed a dataset, IDRISI-RE, containing Tweets of 19 disaster events like floods, earthquakes, etc., from 22 English-speaking countries consisting of 20.5k Tweets. Alduailaj et al. [18] collected a dataset containing 30K Arabic comments related to cyberbullying on February 5, 2021, from Twitter and YouTube APIs.

However, all the datasets did not help predict and identify disruptive events. Only a few datasets were available for identifying events like riots, protests, and crises. However, these datasets did not include all the event categories that might lead to disruption, like crime, disaster, cyber-attacks, rumors, and crime-based events. Some researchers acquired data for limited months or days, leading to the problem of data scarcity, while some suffered from low coverage. There is no dataset consisting of disruptive events Tweets of the Indian subcontinent.

3 Methodology

This section comprises two sub-sections: Disruptive Event Dataset (D*is*E*v*D) Curation and Disruptive Event Identification (D*is*E*v*D) Framework. The first subsection discusses the D*is*E*v*D collection pipeline, data description and comparison with existing benchmark datasets, while the second subsection discusses the methodology of identifying the disruptive events from the collected D*is*E*v*D dataset.

3.1 D*is*E*v*D Curation

In this sub-section, we establish a Tweet collection Pipeline to gather D*is*E*v*D from Twitter social media platform. A portion of the collected dataset has been released and available now.

D*is*E*v*D Collection Pipeline. We collected our dataset, D*is*E*v*D, by following a pipeline shown in Fig. 1. The pipeline begins by capturing input search parameters. Three specific input search parameters have been identified: English-language Tweets, Tweets whose location service is enabled, and Tweets belonging to the bounding box fixed to India. We have used Python's Twitter API (Application Programming Interface) 'Tweepy' for tweet extraction. Consequently, the Tweepy API was invoked based on the above input search parameters by keeping a time-window of 1 day (equivalent to 24 h). Thus, our pipeline script is executed continuously for 24 h to facilitate the extraction of Tweets.

In addition to the Tweets, several other parameters were extracted, including the time at which Tweets were created, text content, number of retweets, follower count, the geographical location from which the tweet was posted, status count, and user name. All the extracted Tweets were then saved in a CVS file. The CSV file serves as an input to the proposed Tweet classifier, a pre-trained DNN (Deep Neural Network) trained on the disruptive event and non-disruptive event Tweets. The complete details of the Tweet classifier is discussed in Sub-sect. 3.2 (Fig. 2).

The Tweet classifier classifies the Tweets in a CSV file into disruptive event and non-disruptive event Tweets. Subsequently, the disruptive event Tweets are stored in the D*is*E*v*D dataset. This pipeline operates to collect Tweets for a

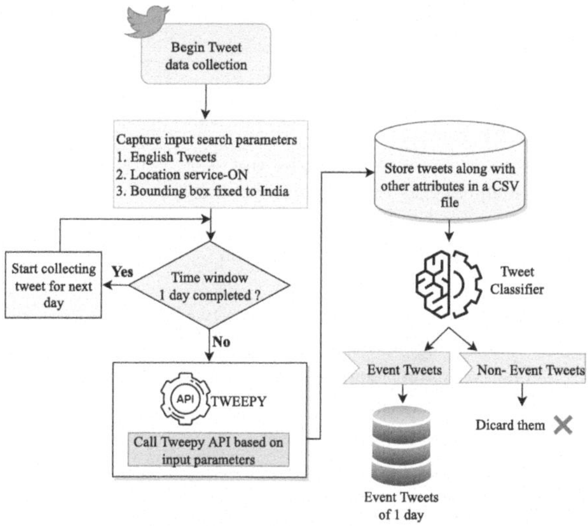

Fig. 1. D*is*E*v*D Collection Pipeline

Fig. 2. Disruptive event and Non-disruptive event Keywords

single time-window (24-h). Upon finishing the Tweet collection for the allotted time window, it starts collecting Tweets for the subsequent time-window.

Data Description. Our proposed dataset, D*is*E*v*D, comprises a total of 12.5 million records from 1st April 2022 to 15th January 2023. A total of ten folders corresponding to ten months are present in the dataset, where each folder contains 30 or 31 CSV files depending on the number of days in a month, which contains around 35–45K records. Each CSV file contains seven attributes. The complete detail of the attributes along with its format is shown in Table 1.

The format of data is textual, numeric, or alphanumeric. This data can be used for analyses based on prediction, classification, clustering, and association algorithms. Some instances from the dataset are shown in Table 2. Tweets along with their date and time are shown in Table 2, and the rest all other attributes can be accessible from the dataset present at the GitHub repository (link is omitted for anonymity).

Table 1. Attribute Table

S.No	Attribute	Detail	Format
1	created-at	The time at which a tweet is posted by the user	YYYY-MM-DD HH-MM-SS
2	tweet	A short message posted by the user which consists of text, images, emojis, videos, and links	Textual
3	follower count	The total number of people who follows a tweeter's account	Numeric
4	retweet count	The number of time a tweeter's tweet is retweeted	Numeric
5	location	The estimated location from where the tweet was posted	Textual
6	username	The twitter handle of a user	Alphanumeric
7	statues count	The number of Tweets possessed by a user including retweets	Numeric

Table 2. Instances of Disruptive Event Tweets

Created_at	Tweet
2022-06-16 0:36:17	Protest is not a crime! Let's start protests #PetrolDieselPriceHike
2022-06-16 0:43:26	#AgnipathRecruitmentScheme spells yet another disaster for the future of #India and our youth
2022-06-16 4:50:30	Pls join all hindus peacefull protest of VHP&bajrangdal
2022-06-16 7:16:27	Desi People coming to Twitter to protest about #FuelPrice who were gone at bed early night
2022-06-16 11:38:14	Agneepath started from #Bihar and #UP. Now in #Haryana and #Uttrakhand. https://t.co/hisfG06Hbd
2022-06-16 11:45:44	Don't play with youths emotions #AgnipathRecruitmentScheme

Example: Timeline of Agnipath Protest. This dataset helps detect upcoming disruptive events. For example, our dataset contains enough Tweets to detect the protests against the government's Agneepath scheme. The government announced the scheme on the $14th$ of June. There were no Tweets related to this scheme on Twitter before the announcement date. People started posting Tweets regarding the Agneepath scheme on the morning of the $14th$ of June. The timeline also shows the number of Tweets available in our dataset from $14th$ of June to $17th$.

For example, 642 Tweets of $14th$ of June are present in our dataset. They kept on posting till the press conference ended. The Tweets posted till the press

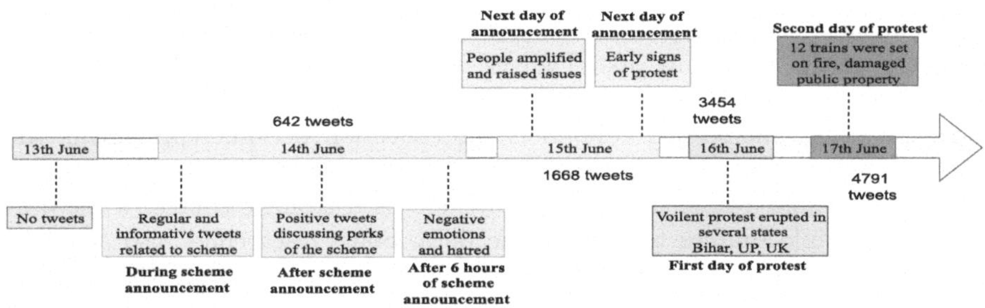

Fig. 3. Timeline of Agnipath Protest

conference were regular and informative. Later on, people started sharing their positive emotions related to this scheme. They praised the perks offered in the scheme; however, some started sharing negative emotions and hatred with the hashtags #boycottagnipath, #agneepathprotest, #No_Rollback_For_Agneepath, after 6 h of the announcement. The next day people criticized and raised severe concerns about the scheme. They placed their demands and asked the scheme to be put on hold until the government accepted their demands. However, on the 16th of June, people expressed aggression in the form of protests held in multiple districts of Bihar, Uttar Pradesh, and Uttrakhand. By the 17th of June, some trains were set on fire, and some were canceled and diverted. The complete timeline is shown in Fig. 3.

Comparison with Existing Disruptive Event Detection Datasets. In this section, we have compared various existing disruptive event detection datasets. We have chosen datasets from the last decade. The comparison has been made on several parameters, as shown in Table 3.

Table 3. Comparison of Existing Disruptive Event Detection Datasets

References	Corpus Size	Collection-Period	Sources	Geo-Tagged	English Language	Disruptive Event	Publicly Available
Ahmad et al. 2016 [19]	0.49M photos of 14 events	7th-20th Sept 2015	Flicker	✗	✗	NA	✗
Alsaedi et al. 2017 [2]	40M Tweets	Oct-Nov 2015	Twitter	Not all tweets are geo-tagged	✓	Multiple	✗
Wang et al. 2017 [9]	5699 instagram post and 2.9M Tweets	1st-29th Feb 2016	Instagram & Twitter	✗	✓	NA	✗
Huang et al. 2018 [10]	4.2 M Tweets	2014-2015	Twitter	✓	✓	NA	✗
Modha et al. 2018 [11]	0.015M	-	Facebook Post, Twitter	yes	✓	Multiple	✗
Morabia et al. 2019 [12]	11M Tweets	11th-17th Oct 2017	Twitter	✗	✓	NA	✓
George et al.2019 [20]	0.2M Tweets & 100M photo	Jan-Dec 2017	Twitter, Flicker	✓	✓	NA	✗
Shah et al.2019 [13]	11.7 M Tweets	13th July- 31st Dec 2018	Twitter	✓	✓	NA	✗
Liu et al. 2020 [14]	7.8 M news, 15.6M forum messages	Feb 2016- 2020	Twitter, News	✗	✓	NA	✗
Wang et al. 2020 [6]	4480 Wikipedia documents of 168 types	-	wikipedia	✗	✓	NA	✓
Alsmadi et al. 2021 [15]	1.5M Tweets	Nov 2015 - Dec 2016	Twiiter	✗	✓	Single	✗
Ristea et al. 2021 [21]	9.5M Tweets	Oct 2012 -April 2014	Twitter	✓	✓	Single	✗
Alnazzawi et al. 2022 [7]	5K manually annotated Tweets	Jan 2010-Dec 2019	Twitter	✗	✓	Single	✓
Suwaileh et al. 2023 [17]	20.5K Tweets	-	Twitter	✓	✓	Multiple	✓
Chen et al. 2023 [16]	570M Tweets	February 2022 - Jan 2023	Twitter	✗	✓	Single	✓
Alduailaj et al. 2023 [16]	30K Tweets	-	Twitter	✗	✗	Single	✓
DisEvD	12.5M+ Tweets	April 2022-Jan 2023	Twitter	✓	✓	Multiple	✓

To make a large corpus, the researchers have collected data from various social media platforms, like Twitter, Weibo, Instagram, and Flicker. The first column is corpus size, representing the total number of instances in the dataset/corpus. The second column is the collection period, representing the time in which the data is acquired. Some researchers acquired data for a month, while some acquired it for a year. Researchers used a single source, while others used multiple sources to acquire the data. Some researchers also acquired the location where the Tweets, Facebook, and Instagram posts were posted. Most of the researchers acquired instances of unilingual language that is the English language only, while others acquired instances of multilingual languages like English, Hindi, French, Japanese, and Russian. We have shown the datasets that contain English-language Tweets. The next column in the Table 3 shows the datasets that contain the tweet of the Indian subcontinent. As we can see, only our dataset D*is*E*v*D contains Tweets from the Indian region. The last column shows whether the dataset is a disruptive event dataset or not. It also indicates that the disruptive event dataset contains single or multiple disruptive events. Three datasets contain Tweets of single disruptive events, while three datasets include tweets of multiple disruptive events.

Our dataset D*is*E*v*D aims to identify the multiple disruptive events occurring in the Indian subcontinent, which makes it different from the other disruptive event datasets. It contains all the major disruptive events that happened in the bounding box of India from 1st April 2022 to 15th Jan 2023. It best describes disruptive events with the help of geo-tagged Tweets.

3.2 Disruptive Event Detection (*DiEvD*) Framework

This proposed framework detects the disruptive events from Tweets present in the D*is*E*v*D dataset in a time-window manner. The proposed framework comprises of three modules: Preprocessing, Identification of Disruptive Events, Disruptive Event Information Extraction. The detailed description of each module, shown in Fig. 4, is discussed below.

DET Prepocessing Module. The module collects a CSV file of Tweets acquired over 24-hour-time-window of our proposed *DisEvD* dataset. The preprocessing of the Tweets is performed by removing the stopwords, emoticons, punctuations, non-ASCII characters, and hyperlinks. These preprocessed Tweets are then fed to the *Tweet Embedding Block* for generating the embeddings corresponding to each Tweet using the *sentence transformer*. The sentence transformer uses *Bidirectional Encoder Re*presentation from Transformers (BERT), to create contextualized word embeddings, during the first step. In second step, BERT applies the pooling operation on the generated word embeddings to get the sentence embedding.

Formally, given s preprocessed Tweets, m_1, m_2, $\ldots$, m_s, are input to the BERT model to calculate the token level representation.

$$[H_{i,0}; \ldots; H_{i,j}; \ldots; H_{i,L}] = Sentence\_encoder(m_i) \tag{1}$$

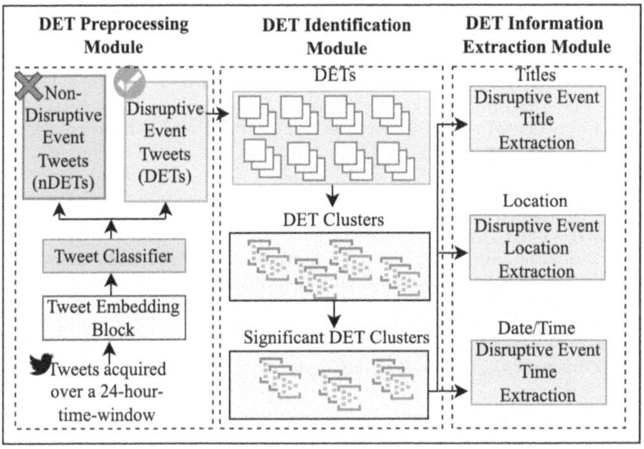

Fig. 4. *DiEvD* Framework

where,

$$H_{i,k} \in \mathbb{R}^{len(m_i) \times d} \tag{2}$$

where $0 \leq j \leq L$, L represents the number of hidden layers, d represents the size of hidden representation and len(m_i) represents the length of tokenized sentence (m_i). Thereafter, the mean pooling P is applied to $H_{i,k}$ to generate fixed length sentence embedding $h_{i,L} \in \mathbb{R}^d$ from all the layers i.e., $h_{i,L} = P(H_{i,j})$.

$$R^L = \{h_{i,j} \, |0 \leqslant j \leq L|\} \tag{3}$$

The output R^L is given input to the Tweet classifier.

The Tweet classifier consists of a *DNN (Dense Neural Network)*, which contains two hidden layers having 64 and 32 neurons each followed by a classification layer. It is trained on the disruptive events and non disruptive event dataset (Sect. 4.2). The preprocessed Tweets are classified as Disruptive Event Tweets (DETs) and non Disruptive Event Tweets (nDETS) using the Tweet Classifier. The DETs are subjected to the second module, Disruptive Event Identification, for further processing.

DET Identification Module. This module performs clustering to create clusters of the DETs. The Louvain and Pycombo community detection techniques are used to generate communities (DET clusters). The DET clusters having number of Tweets greater than threshold are considered as significant clusters as shown in Fig. 4. The community detection techniques are chosen because it creates the optimized communities for the social media networks. Additionally, It detects sub-communities within a community and it does not require the cluster count before applying the clustering technique like K-means algorithm. The psuedo code of the second module is shown in Algorithm 1. It takes embedding of DETs produced by the BERT sentence transformer together with a list of

Algorithm 1: DET Preprocessing and Identification Module

```
Input: Twitter message stream for time window T=0,...,t  Output: Set of DET clusters
Tweet Classifier ← DNN(DETandNon − DETDataset) for T = 0 to t do
    for m_i in TweetStream do
        m'_i ← Preprocess(m_i) if TweetClassifier(m'_i)== 1) then
            D_T ← D_T ∪ m'_i
        end
    end
    Emb ← BERT_Transformer(D_T);
    for thres in thres'list do
        match_index_list, similarity_list ← Neighborhood-search(emb, thres);
        Emb' ← Neighborhood_blending(emb,match_index_list, similarity_list);
    end
    G ← Graph_Creation(match_index);
    Membership ← Louvain_Community_Algorithm(G);
    C'_T ← |(Membership)| C_T ← Refine_Clustering(Emb');
    for c in C_T do
        if |c| < threshold then
            C_T = C_T - c;
        end
    end
end
End
```

threshold values for the[1] *neighborhood blending* (Algorithm 1). To identify other DET embeddings that are similar to DET embedding, we use neighborhood search (Algorithm 1). The neighborhood search takes *embeddings* and *threshold values* as input and results *match index list* and *similarity list.* The cosine similarity metric is used in the neighborhood search to compare every pair of embeddings.

Neighbor embeddings are those whose cosine similarity is equal to a threshold value. All of the embeddings' similarity scores are updated in the *similarity list*, along with the embeddings of their neighbors and the *matching index list* includes the embedding index of each neighboring embedding. The *Neighborhood Blending approach* (Algorithm 1), is used to update and modify the embeddings in the subsequent step. The *Neighborhood Blending* takes *embeddings*, *match index list*, and *similarity list* as input and outputs the updated embedding. The updated embedding is computed by multiplying the original embedding by the weighted sum of its neighborhood embedding, where the weights are similarity scores. E.g., A embedding e_1 matches with the four of its neighbors embeddings like e_2, e_3, e_4, e_5. After applying neighborhood blending, the resulted embedding is shown in Eq. 3.2.

$$e_1 = N(e_1+e_2*cos(e_1, e_2)+e_3*cos(e_1, e_3)+e_4*cos(e_1, e_4)+e_5*cos(e_1, e_5)) \quad (4)$$

where N() represents a normalization function. To improve the DET embedding, the *neighborhood search* and *neighborhood blending strategies* for the two distinct threshold values have been used (Algorithm 1). The following step constructs the graph by inputting the *match index list.* Furthermore, the graph created by the *create graph function* is subjected to Louvain community detection. Different

[1] https://pypi.org/project/NeighborBlend/.

communities are assigned to each graph node via community detection. The modularity gain is computed for each node by deleting the current node and adding it in the neighbor's community. If there is a positive modularity gain, the node will be added to the community; if not, it will stay in the current community. This step continues till there is no changes in the graph and maximum modularity is achieved.

Following the creation of the communities, we choose one community as an instance, perform all the aformentioned steps using Louvain community detection, and determine whether the number of communities following *refining clustering* is smaller than the number of communities prior to it, indicating that some of the communities have been blended. If not, there will be the same number of communities as there were before to the *refine clustering* (Algorithm 1). The refine clustering results the final communities. Next, we determine the significant communities by determining if the community's membership exceeds the threshold. Only then we keep them as the significant communities.

DET Information Extraction Module. This module presents a methodology for extracting a disruptive event's information (title, location, and time). First, the Tweets collected over a time window of one day are given to a community detection algorithm for creating communities (disruptive event clusters). After that, the disruptive event communities are given input to Algorithm 2, which generates the title, location, and time of the disruptive event. We applied the stringify function to extract a sentence ($clusterString$) from a disruptive event cluster. The sentence ($clusterString$) is given input to the keyword_extraction function to generate keywords. These keywords are generated using the spacy library applied to the cluster of Tweets. The $keywords$ is a list of pairs where each item is a tag along with the label.

The $keywords$ are then given input to the DisruptiveEventTitles function, and return a multi-set of titles for the disruptive events. It uses the YAKE library to filter out the keywords with lower confidence. We then randomly select nine keywords from this new filtered list of keywords along with the word with the highest confidence percentage and then push them in another pre-trained model, keytotext[2]. The model, keytotext converts a bunch of keywords into a coherent sentence. Hence, we get a set of probable titles for our event. We repeat this process ten times to ensure titles concerning the ten top keywords.

The second function, DisruptiveEventLocation extracts the location information of the disruptive event. It takes sentences ($clusterString$) as input. The BERT model [22] has been fine-tuned to perform entity recognition notions at a sentence level. The sentences are then passed through the BERT model, which results in a list of precise location mentions in the sentence. These raw text locations are given to the spacy library, which categorizes raw locations into GPE and Non-GPE labels. After this, the locations are geocoded using the APIs, OpenStreetMap (OSM), and Google Places offered by Google. These return the location's longitude and latitude for that place.

[2] https://pypi.org/project/keytotext/.

Algorithm 2: Algorithm for Disruptive Events Information Extraction

```
Input: Disruptive Event Cluster
Output: Disruptive Event information (Event Name, Location, Time);
begin
clusterString = stringify(cluster);
keywords = keyword_extraction(clusterString);
Function DisruptiveEventTitles(keywords):
    key = ConfidenceChecker(keywords);
    for item in key do
        list = item,Random(keys,9);
        title ← keytotext(list);
    end
    return Multiset of titles for the disruptive event;
End Function
Function DisruptiveEventLocation(keywords):
    for item in keywords do
        if item → label == "LOC" then
            object = item
        end
    end
    return Multiset of locations for the disruptive event;
End Function
Function DisruptiveEventTime(keywords,clusterString):
    timeTuple = timeofhumans (clusterString);
    for item in keywords do
        if item → label = "DATE" then
            timeKeywordList = item
        end
    end
return timeTuple ∪ timeKeywordList;
End Function
returnOutput = "Disruptive Event Titles", "Disruptive Event Location", "Disruptive
Event Time" ;
End
```

The third function, DisruptiveEventTime, extracts the time and date of the disruptive event. The *keywords* and sentences (*clusterString*) are given input to this function. We require an API that can convert the tweet describing a particular day into its ISO date format. For example, The event will take place next Monday. To extract time and date information from such Tweets the *TimeofHuman* API is used and stores the information in a list called *timeTuple*. There are some Tweets that contain date and time information within the Tweets. For example, We will gather in Delhi at 6.00 pm on $12th$ OCT. So, to extract information from such Tweets, the *keywords* are generated using the spacy library and the items whose label equals DATE are added to the *timekeywordList*; otherwise, items are discarded. At last, the two lists *timekeywordList* and *timeTuple* are merged to get the time and date information of the disruptive event cluster.

4 Performance Evaluation

This section discusses the experimental details and performance metrics used in the paper. We created a number of research queries (RQs) in the experimental analysis and tried to find out the solutions pertaining to these questions.

4.1 Implementation Setup

Python 3.0 is used for the implementation. The threshold value of 0.6 is kept for the sigmoid function used in the tweet classifier. The communities were refined in two iterations by keeping the threshold value of 0.4 and 0.5 in the first and second iterations. The minimum number of Tweets present in the community is set to 200. The YAKE[3] library is used to filter out the keywords with lower confidence. The APIs, OpenStreetMap(OSM)[4], and Google Places[5] have been used to get the location The TimeofHuman[6] API is used for extracting time and date information. The random search strategy is applied to select all the parameters. All the source code and dataset are available at GitHub repository[7].

4.2 Dataset and Performance Metrics Used

We used two datasets, DiEvD, and DET and Non-DET dataset in this work. The former dataset is used to evaluate our proposed disruptive event detection framework whereas the latter is used to train the Tweet classifier. The details of DisEvD dataset is discussed in Sect. 2.1 and DET and Non-DET dataset is as follows:

- **DET and Non-DET Dataset**[8] This dataset is collected painstakingly by our research team from the Twitter social media using the set of disruptive event ('riot', 'calamities', 'protest', 'disaster', 'disruption', etc.) and non-disruptive event ('holiday', 'football', 'apparel', 'wardrobe', etc.). The Tweepy library[9] (open-source) has been used to extract the Tweets. The dataset contains 94,885 DETs and 168,676 non-DETs, resulting a total of 263,561 Tweets. We created a training set of 8000 samples by extracting 4000 Tweets each from the DET and non-DET categories using systematic sampling. The Tweets were painstakingly annotated by four human annotators who maintained Cohen's kappa coefficient at 0.80. The Tweet classifier was trained using this dataset.

4.3 Research Queries (RQs)

We created the following three research queries and tried to find out the solution of the same for experimental analysis (Table 4).
RQ1: How well does the *DisEvD* dataset helps in detecting the disruptive events effectively?
RQ2: How well does the DET information extraction module work?
RQ3: How accurate and effective is the proposed disruptive event detection framework in detecting disruptive events?

[3] https://pypi.org/project/yake/.
[4] https://pypi.org/project/openstreetmap/.
[5] https://pypi.org/project/python-google-places/.
[6] https://pypi.org/project/timefhuman/.
[7] https://github.com/seethaaditi/Disruptive-Event-Identification.
[8] https://arxiv.org/pdf/2111.15629.pdf.
[9] https://www.tweepy.org/.

Table 4. Performance metrics used

Performance metrics	Formula	Definition
Normalized Mutual Information (NMI)	$NMI = \frac{\sum_{c,k} n_{c,k} log \frac{n_{c,k} \cdot N}{n_c \cdot n_k}}{\sqrt{(\sum_c n_c \log \frac{n_c}{N})(\sum_c n_c \log \frac{n_c}{N})}}$	This metric is used for the text clustering algorithm. It measures the reduction in uncertainty of class labels when cluster labels are given. The value of NMI ranges between 0 and 1
Homogeneity (H)	$H = 1 - \frac{\sum_{c,k} n_{c,k} log \frac{n_{c,k}}{n_k}}{\sum_c n_c \log \frac{n_c}{N}}$	Homogeneity (H) is maximized when samples within a cluster belong to the same class
Completeness (C)	$C = 1 - \frac{\sum_{c,k} n_{c,k} log \frac{n_{c,k}}{n_c}}{\sum_c n_c \log \frac{n_k}{N}}$	Completeness (C) is maximized when all class samples belong to the same cluster. Here, n_k is the number of samples in cluster k, $n_{c,k}$ is the number of samples in class c as well as cluster k, N is the number of total samples in the dataset
V measure	$V = \frac{2 * H * C}{H + C}$	V measure is calculated as the harmonic mean of Completeness and Homogeneity
Precision@k	$Precision@k = \frac{No.\ of\ correctly\ identified\ events}{No.\ of\ top\ k\ of\ identified\ events}$	It tells the proportion of correctly identified events(event clusters) in the top k selected events(event clusters)
Normalized Discounted Cumulative Gain (NDCG)	$NDCG = \frac{\sum_{i=1}^{n} \frac{r_i}{\log_2(i+1)}}{\sum_{i=1}^{\vert R \vert} \frac{r_i}{\log_2(i+1)}}$	NDCG ranks the top identified events relative to their ideal ranking. Its value ranges from 0 to 1

4.4 Experimental Results

This section discusses the experimental results to address aforementioned research queries.

Result Analysis for RQ1. We validated the effectiveness of the proposed DisEvD dataset by applying it to the disruptive event detection framework to identify the disruptive events. This RQ consists of two tasks. The first task includes classification of DETs and non-DETs (DET pre-processing module) by employing our DNN based Tweet classifier. The second task involves clustering of the resulted DETs to create close-knit communities (DET Identification Module). We evaluated both the tasks on the Tweets acquired between May 25, 2022 to July 25, 2022 from DisEvD dataset.

The total number of Tweets acquired over a 24-hour-time-window are around 30,000 to 40,000 Tweets. The DNN model results around 11000 to 15000 DETs in each 24-hour-time-window. The complete list of the Tweet count is present in the supplementary file[10]. We also compared the performance of our DNN based Tweet classifier with the standard ML based classifiers like Naive Bayes, Random Forest, Support Vector Machine (SVM), Decision Trees. The performance of this task is evaluated by computing the machine learning metrics like,

[10] https://github.com/seethaaditi/Disruptive-Event-Identification.

Accuracy, Precision, Recall and F1-score [23]. Figure 5 depicts that DNN outperforms all the other classification algorithms. The DNN model better represents the complex pattern present in the data. Furthermore, DNN based models offers greater tunability because of the high number of hyperparameters in comparison to ML-based models.

We applied five community detection algorithms, Louvain [24], Pycombo [25], Girvan Newman [26], Connected Components [27], Leidenalg [28] and a Online Tweet Clustering algorithm [2]. We evaluated the performance of the algorithms by computing Completeness [29], Homogeneity [29], V measure [29], NMI [30], and average Sihouette Coefficient [31].

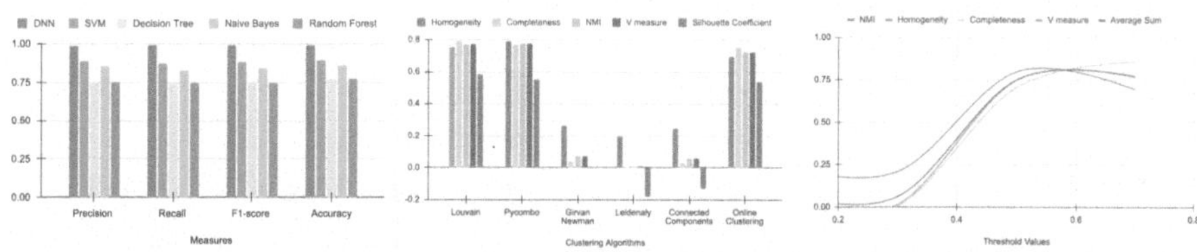

Fig. 5. a) Comparison of DNN based Tweet classifier with different ML-based classification algorithms, b) Clustering performance of various clustering algorithms, c) Clustering performance using different threshold values

Figure 5 represents that Louvain and Pycombo outperforms other clustering algorithms in terms of all the metrics in creating best DET clusters. The community detection algorithms are suitable for social network analysis where data can be represented as a graph of interconnected entities. We chose these methods for their ability to identify optimized communities and sub-communities within social networks. Unlike the k-means algorithm, these methods offer the advantage of not requiring the predetermined specification of the number of clusters. Notably, the Louvain and Pycombo methods emerged as top performers due to their efficient handling of large dynamic datasets commonly encountered in social media. What sets these methods apart is their small computational complexity, a feature derived from their optimized modularity-based solution, which ensures efficiency irrespective of the number of edges in the network. Compared to the online clustering algorithm, which processes data in a single pass and is unable to find sub-communities within larger communities, the community detection algorithm perform better.

All the performance metrics considered in the discussion were assigned equal importance. Therefore, to determine the optimum threshold value, we looked for the maximum average point across these metrics. As a result, the threshold value of 0.5 emerged as optimal, reflecting the peak performance across all measures (refer Fig. 5).

The maximum DET communities formed in a time window using the Louvain community detection algorithm was around 750–850. The DET communities with less than 200 Tweets were dropped. Next, these communities were blended

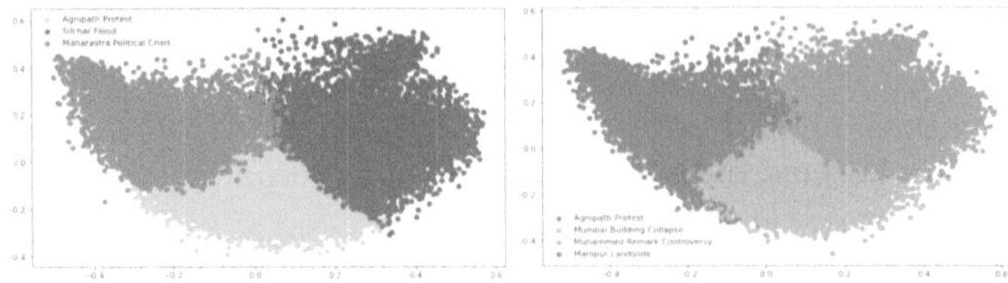

Fig. 6. a) DET clusters formed during 21st of June, b) DET clusters formed during 27th of June

using the threshold of 0.4 in 1st iteration and 0.5 in 2nd iteration. At last, around 1–10 clusters were formed in each time window. Cohen's kappa was used to determine the level of agreement between the annotators, which is set to 0.771. Only the clusters that were agreed upon by all annotators were used for testing.

Table 5. List of Disruptive Events found between 25th May and 25th July

Disruptive Events	Origin Date	End Date	No. of days this event last
Sidhu Moosewala Murder	29/05/2022	09/07/2022	42
Hapur Chemical Plant Explosion	04/06/2022	10/06/2022	7
National Herald Corruption Case	13/06/2022	22/06/2022	9
Agneepath Protest	16/06/2022	21/09/2022	96
Silchar Flood	19/06/2022	06/072022	18
Maharashtra in Political Crisis	21/06/2022	04/07/2022	14
Umesh Kolhe Murder	21/06/2022	08/07/2022	18
Mumbai Building Collapse	28/06/2022	03/07/2022	6
Manipur Landslides	30/06/2022	24/07/2022	25
Amarnath Cloud Burst	08/07/2022	15/07/2022	8
Unparliamentary Language Controversy	14/07/2022	20/07/2022	8
SSC Scam	23/07/2022	03/08/2022	11
Toxic Liquor Deaths	25/07/2022	29/07/2022	5

A total of 13 major disruptive events clusters were identified over a period of 2 months (May 25th, 2022 to June 25, 2022), show in Table 5. We have represented the DET clusters identified on 21st and 27th June. We can see from Fig. 6 that Maharashtra political crisis, the Silchar flood (Assam flood), and the Agnipath protest were the three major DET clusters that identified on 21*st* June. The largest cluster formed is on the Silchar flood, followed by Maharashtra political crisis and the Agnipath protest. Similarly, the Mumbai building collapse, Muhammed's remark controversy, the Manipur landslide, and the Agnipath protest were the four major DET clusters that occured on the 27th of June. The largest cluster formed is on Muhammed's remark controversy, followed by the Mumbai building collapse and Manipur landslide. The smallest cluster formed is of Agnipath protest.

4.5 Result Analysis of RQ2

Once the DET clusters are formed, we applied the DET information extraction module to extract disruptive event information. The goal is to extract the title, location, and datetime of the DETs. We applied a clustering algorithm to the Tweets acquired on the 16th of June. We took an excerpt from the tweet cluster, shown in Fig. 7. The tweets are given input to the keyword extraction function to generate keywords obtained using the spacy library. The confidence of the keywords is checked using the YAKE library to filter out the keywords with lower confidence. We gave high-confidence keywords to the keytotext model to generate a coherent sentence. The probable titles are shown in Fig. 7. All the titles of this cluster are related to the Agneepath event.

The second module, disruptive event location extraction, is used to extract the possible location of the disruptive event. It takes tweets sentence as input and results in possible locations of the Agneepath protest. The module outputs the location's longitude and latitude of the place using the OpenStreetMap (OSM) and Google Places offered by Google. The locations that resulted from our module are India, Bihar, UP, Haryana, and Uttarakhand. The third module, disruptive event time extraction, extracts the disruptive event's date and time. It takes keywords as input which are generated using the spacy library. The items whose labels equal DATE are added to the list. The list contains three items tomorrow, Thursday, and Friday. The Time of Human API is applied to extract the date and time of the items in the list.

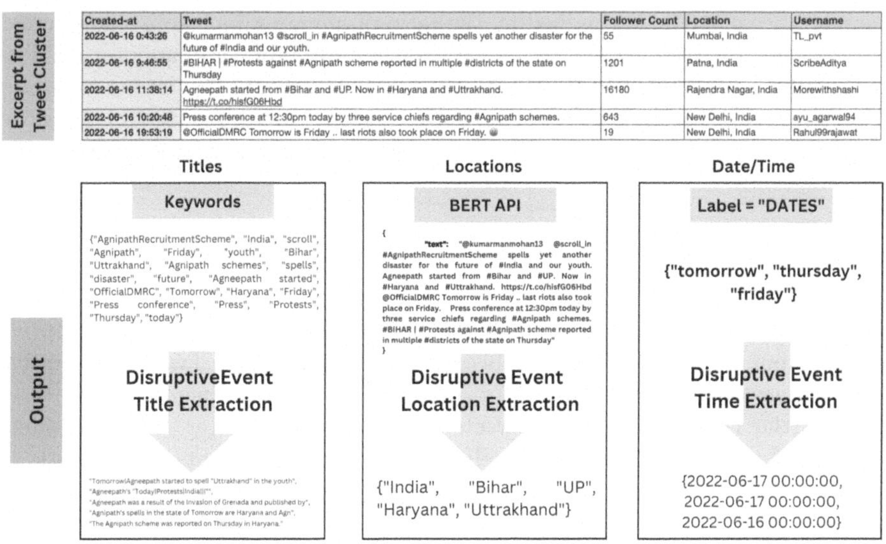

Fig. 7. DET Information Extraction

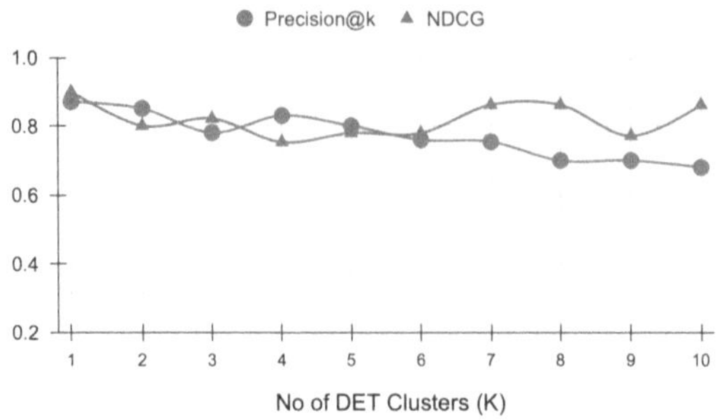

Fig. 8. Precision@K and NDCG at K of our proposed disruptive event detection framework

Result Analysis of RQ3. To address RQ3, our focus is on identifying the foremost disruptive events per day within the dynamic Twitter data streams. Our proposed framework consistently generates an average of 1 to 10 disruptive events daily (denoted by K). To evaluate the accuracy, three annotators manually assessed disruptive events by assigning a relevance score ranging from 0 to 3. Events with a relevance score exceeding 1.5 were marked correctly identified disruptive events. For instance, in Fig. 8, the precision@5 value is 0.8, indicating the accurate identification of four disruptive events out of the top 5 by our proposed framework.

The NDCG scores for various K values were computed by dividing the Discounted Cumulative Gain (DCG) of the ranking of K disruptive events (identified by our framework) by the DCG of their ideal ranking. In NDCG calculations, the relevance score is determined by the ratio of human annotators who marked the identified disruptive event as relevant to the total number of annotators. The ideal ranking is established by ordering the list of K disruptive events based on their relevance scores. The resulting NDCG scores for different K values are presented in Fig. 8, showcasing the framework's effectiveness in identifying disruptive events.

5 Conclusion

We presented D*is*E*v*D, a dataset for training and testing to identify disruptive events happening in the Indian subcontinent. It is the first Twitter disruptive event dataset containing the tweets of the Indian subcontinent. This dataset contains the 12.5 M geo-tagged tweets acquired from Twitter API. The effectiveness of the dataset has been validated by applying *DiEvD* framework to the tweets collected over a time window of one day. The performance measures' value shows that the clustering algorithms' clusters are significant. This proves that the dataset can help make analyses based on clustering. Additionally, the framework extracts disruptive event information like title, location, and time corresponding to each DET cluster. The result of this proposed methodology shows that it effectively extracts the information corresponding to a disruptive event.

Compliance with Ethical Standards

Conflict of Interest. The authors declare no potential competing interests with respect to the research, authorship, and/or publication of this article.

Ethical Approval. This article does not contain any studies with human participants or animals performed by any of the authors.

Informed Consent. Not Applicable.

References

1. Alsaedi, N., Burnap, P.: Feature extraction and analysis for identifying disruptive events from social media. In: Proceedings of the IEEE/ACM International Conference on Advances in Social Networks Analysis and Mining, pp. 1495–1502 (2015)
2. Alsaedi, N., Burnap, P., Rana, O.: Can we predict a riot? Disruptive event detection using twitter. ACM Trans. Internet Technol. (TOIT) **17**(2), 1–26 (2017)
3. Hananto, A.R., Rahayu, S.A., Hariguna, T.: Covid-19 vaccination: a retrospective observation and sentiment analysis of the twitter social media platform in Indonesia. Int. J. Inform. Inf. Syst. **5**(1), 56–68 (2022)
4. Das, M.: Social media posts trigger seven communal riots in a month in West Bengal (2020)
5. McMinn, A.J., Moshfeghi, Y., Jose, J.M.: Building a large-scale corpus for evaluating event detection on twitter. In: Proceedings of the 22nd ACM International Conference on Information & Knowledge Management, pp. 409–418 (2013)
6. Wang, X., et al.: Maven: a massive general domain event detection dataset. In: Proceedings of the Conference on Empirical Methods in Natural Language Processing (EMNLP), pp. 1652–1671 (2020)
7. Alnazzawi, N.: Using twitter to detect hate crimes and their motivations: the hatemotiv corpus. Data **7**(6), 69 (2022)
8. Giorgi, S., et al.: Twitter corpus of the# blacklivesmatter movement and counter protests: 2013 to 2021. In: Proceedings of the International AAAI Conference on Web and Social Media, vol. 16, pp. 1228–1235 (2022)
9. Wang, S., Giridhar, P., Kaplan, L., Abdelzaher, T.: Unsupervised event tracking by integrating twitter and instagram. In: Proceedings of the 2nd International Workshop on Social Sensing, pp. 81–86 (2017)
10. Huang, Y., Li, Y., Shan, J.: Spatial-temporal event detection from geo-tagged tweets. ISPRS Int. J. Geo Inf. **7**(4), 150 (2018)
11. Modha, S., Majumder, P., Mandl, T.: Filtering aggression from the multilingual social media feed. In: Proceedings of the First Workshop on Trolling, Aggression and Cyberbullying (TRAC-2018), pp. 199–207 (2018)
12. Morabia, K., Murthy, N.L.B., Malapati, A., Samant, S.: Sedtwik: segmentation-based event detection from tweets using wikipedia. In: Proceedings of the Conference of the North American Chapter of the Association for Computational Linguistics: Student Research Workshop, pp. 77–85 (2019)

13. Shah, Z., Dunn, A.G.: Event detection on twitter by mapping unexpected changes in streaming data into a spatiotemporal lattice. IEEE Trans. Big Data **8**(2), 508–522 (2019)
14. Liu, Y., Peng, H., Li, J., Song, Y., Li, X.: Event detection and evolution in multilingual social streams. Front. Comput. Sci. **14**(5), 1–15 (2020)
15. Alsmadi, I., O'Brien, M.: Event detection in twitter: a content and time-based analysis. arXiv preprint arXiv:2111.05274 (2021)
16. Chen, E., Ferrara, E.: Tweets in time of conflict: a public dataset tracking the twitter discourse on the war between Ukraine and Russia. In: Proceedings of the International AAAI Conference on Web and Social Media, vol. 17, pp. 1006–1013 (2023)
17. Suwaileh, R., Elsayed, T., Imran, M.: IDRISI-RE: a generalizable dataset with benchmarks for location mention recognition on disaster tweets. Inf. Process. Manag. **60**(3), 103340 (2023)
18. Alduailaj, A.M., Belghith, A.: Detecting Arabic cyberbullying tweets using machine learning. Mach. Learn. Knowl. Extr. **5**(1), 29–42 (2023)
19. Ahmad, K., Conci, N., Boato, G., De Natale, F.G.B.: USED: a large-scale social event detection dataset. In: Proceedings of the 7th International Conference on Multimedia Systems, pp. 1–6 (2016)
20. George, Y., Karunasekera, S., Harwood, A., Li, K.H.: Spatio-temporal event detection using poisson model and quad-tree on geotagged social media. In: Proceedings of IEEE International Conference on Big Data (Big Data), pp. 2247–2256. IEEE (2019)
21. Ristea, A., Al Boni, M., Resch, B., Gerber, M.S., Leitner, M.: Spatial crime distribution and prediction for sporting events using social media. Int. J. Geogr. Inf. Sci. **34**(9), 1708–1739 (2020)
22. Devlin, J., Chang, M.-W., Lee, K., Toutanova, K.: Bert: pre-training of deep bidirectional transformers for language understanding. arXiv preprint arXiv:1810.04805 (2018)
23. Tharwat, A.: Classification assessment methods. Appl. Comput. Inform. **17**(1), 168–192 (2020)
24. Blondel, V.D., Guillaume, J.-L., Lambiotte, R., Lefebvre, E.: Fast unfolding of communities in large networks. J. Stat. Mech. Theory Exp. **2008**(10), P10008 (2008)
25. Sobolevsky, S., Campari, R., Belyi, A., Ratti, C.: General optimization technique for high-quality community detection in complex networks. Phys. Rev. E **90**(1), 012811 (2014)
26. Girvan, M., Newman, M.E.J.: Community structure in social and biological networks. Proc. Natl. Acad. Sci. **99**(12), 7821–7826 (2002)
27. Fortunato, S.: Community detection in graphs. Phys. Rep. **486**(3–5), 75–174 (2010)
28. Traag, V.A., Waltman, L., Van Eck, N.J.: From Louvain to Leiden: guaranteeing well-connected communities. Sci. Rep. **9**(1), 1–12 (2019)
29. Rosenberg, A., Hirschberg, J.: V-measure: a conditional entropy-based external cluster evaluation measure. In: Proceedings of the Joint Conference on Empirical Methods in Natural Language Processing and Computational Natural Language Learning (EMNLP-CoNLL), pp. 410–420 (2007)
30. Estévez, P.A., Tesmer, M., Perez, C.A., Zurada, J.M.: Normalized mutual information feature selection. IEEE Trans. Neural Netw. **20**(2), 189–201 (2009)
31. Rousseeuw, P.J.: Silhouettes: a graphical aid to the interpretation and validation of cluster analysis. J. Comput. Appl. Math. **20**, 53–65 (1987)

Adaptive Temporal Random Walks for Graph Representation

Monachary Kammari(✉) and S. Durga Bhavani

School of Computer and Information Sciences, University of Hyderabad, Hyderabad 500046, Telangana, India
monaacharykammari@gmail.com, {sdbcs,19mcpc08}@uohyd.ac.in
https://scis.uohyd.ac.in/ sdbcs

Abstract. In recent times, Graph Representation Learning (GRL) has become popular due to the advancements in Deep Learning (DL) and Machine Learning (ML) algorithms. Most real-world networks are dynamic in nature, thus temporal information is intrinsic part of the network information. There are very few GRL schemes that take temporal aspect into consideration. In this work, we propose a GRL method, namely, *tNode2Vec*, in which the popular Node2Vec algorithm is extended by defining an adaptive temporal random walk that extracts the node neighborhood samples using degree-based and temporal distribution. We consider two downstream prediction tasks namely, link prediction and recommendation, to evaluate the quality of the proposed model. Extensive experimentation is performed on two real-world bibliographic datasets, Hep-Ph and DBLP V11. DBLP V11 is a large well-known dataset with a collection of 41, 07, 340 research papers and conference proceedings from computer science and related domains. We have implemented all the baseline methods to solve the link prediction and recommendation problems. The proposed tNode2Vec method achieved better results compared to all the baseline methods for the link prediction task as well as for the recommendation task. We have included the BERT model in the proposed tNode2vec model as well as baseline methods to extract textual node attributes and combine them with node representation to enhance the link prediction performance.

Keywords: Random Walks · Node2Vec · Temporal Aspects · Graph Representation Learning · Link Prediction · Citation Recommendation

1 Introduction

Most real-world complex multi-dimensional problems can be modeled conveniently as graphs and thus solved efficiently. Some of the examples are online social networks [7], biological networks [30], bibliographic networks [4] etc. Understanding and analyzing the structure of networks provides interesting solutions for popular problems like node classification, link prediction, community detection, recommendation etc. Analyzing these complex graphs is challenging

A. Dasgupta et al. (Eds.): BDA 2024, LNCS 15526, pp. 75–91, 2025.
https://doi.org/10.1007/978-3-031-81821-9_5

due to their high dimensionality and raw representation. Earlier works in the literature uses Laplacian Eigenmaps [5] to analyze the networks, which involve computationally expensive matrix operations.

Graph Representation Learning (GRL) [13,33] encodes the graph data into low-dimensional vectors without losing the topological properties. These low-dimensional vectors learned by GRL can be used as inputs to machine-learning models for any downstream prediction tasks. Representing graphs as feature vectors is a challenging problem. A straightforward way is to select rows of the adjacency matrix as feature vectors of the respective nodes. However, this can limit the feature representation to a one-hop neighborhood around the target node. Most real-world networks are very sparse thus these row vectors of size V are not useful representations. Also, this idea only represents the nodes from their respective direct neighbors which is very local. The challenge with higher-order neighborhood selection as feature vectors is their dynamic size. Graph Neural Network model [19,20,31] uses hop-wise topological information to encode node representations. These GNN models use the message-passing method that aggregates the information from neighborhood nodes to represent local structures. On very large real-world networks GNN models are computationally expensive.

Another way of effectively selecting neighborhoods is by computing random walks. Initial random walk-based GRL methods are DeepWalk [25] and Node2Vec [12]. The idea behind the random walk method is to select part of the network around the target node by following the link formation. The selection process of the next node from the current node follows the link formation between the current node and the next node. In networks with scale-free properties, link formation is based on preferential attachment [3] of degree. Recency or temporal information is another factor that impacts the link formation process. A few papers in the literature show that the inclusion of temporal information into GRL methods enhances the quality of representations [22,27]. In this work, we propose a tNode2Vec model that extends Node2Vec by effectively including temporal information. The contributions made in this work are listed below.

1.1 Contributions

- We propose an adaptive temporal random-walk-based graph representation method (tNode2Vec) that generates random walks using temporal and degree-based priorities and uses them to extract meaningful node representations.
- We use the proposed tNode2Vec model to solve two downstream prediction tasks: link prediction and recommendation problem.
- We perform experiments on two real-world Hep-Ph, and DBLP datasets and compare the results with baseline methods.
- We propose a method to enhance the link prediction quality by including textual representations using the BERT model.

The rest of the paper is organized as follows: Sect. 2 provides the review of the related literature. The proposed approach for adaptive temporal random walks is given in Sect. 3. Section 4 explains the experimental setup, pipelines of link prediction, and recommendation tasks followed by respective results and analysis. Conclusion and future work are presented in Sect. 5.

2 Literature Review

In recent times, Graph Representation Learning (GRL) has become popular as it enables the possibility of solving graph-based problems using Deep Learning (DL) and Machine Learning (ML) algorithms. Initial attempts to solve the GRL problem use Eigen matrices, Eigenvectors, and their properties. Graph Laplacian Eigenmaps [5] encode nodes by utilizing the spectral properties of the Laplacian matrix, which emphasizes the smoothness of node embeddings across the network. It is impractical to compute eigenvalues and eigenvectors of the Laplacian matrix, especially for large graphs as these operations are computationally expensive. Graph kernel methods [18,26] divide the graph into small substructures and use kernel functions to find the similarity among these substructures. These substructures or feature representations are extracted using handcrafted criteria which require domain expertise to set up.

LINE model proposed in [28] learns node representations by preserving higher-order proximities. This model considers first-order neighborhoods followed by second-order neighborhoods to generate edge sampling. The LINE model works using the Breadth First Search strategy that focuses more on the regional structures, thus losing information on global structures. Recent deep learning methods proposed in [19,20,31] learn node representations using graph neural networks. Authors in [20] proposed a semi-supervised Graph Convolution Networks (GCN) model to learn the node representations using convolution operations on the graph-structured data. An encoder-decoder architecture GAE model [19] uses the GCN layers as encoding and decoding layers to learn node representations. Velickovic et al. [31] proposed an attention mechanism into the convolution process to include prioritization.

The other effective way of generating neighborhood samples is the random walk approach to solve the GRL problem. The authors of [24] proposed a Skip-Gram model that is trained using a feed-forward neural network to learn representations of words such that similar words are closely represented in the embedding space. It is natural for similar words to be surrounded by many words in common. DeepWalk [25] and Node2Vec [12] use a random walk approach to generate sequences of nodes, by treating the graph as a corpus of sentences, and then apply techniques like Skip-Gram from natural language processing to learn node embeddings. Authors of [25] proposed an unsupervised feature learning technique DeepWalk that learns node representations using random walks on the network. The idea of this approach is to consider a network as a document and extract random walks on the network to form sentences. Further, the Skip-Gram model is trained with neighborhood samples as sentences to learn the node representations. Random walks used in this model do not preserve lower-order proximities as they follow the DFS approach. The random walks used by this model are not flexible enough to suit different scenarios. Node2Vec [12] by

Grover et al. is an approach that learns node representations by training the Skip-Gram model using random walks that are generated by balancing Breadth First Search and Depth First Search to preserve local and global structures. While generating the random walks in the Node2Vec model, the next node from the current node is selected randomly with equal priority. In reality, link formation is a priority-based process that depends on different features like high degree, recency, similarity etc.

Most real-world networks are dynamic thus temporal information is included naturally, for example, in e-commerce networks, bibliographic networks etc. Analyzing these networks by including the temporal information may produce more meaningful node representations. TDL2vec [22] is an extension of the Node2Vec model, that generates link embeddings using time factor and uses them to solve the problem of disappearing link prediction. Considering only the time factor to generate random walks may neglect the importance of degree. Authors of [27] proposed an embedding approach for dynamic networks using temporal snapshots of the network. Unlike Node2Vec which learns node embeddings by considering nodes as words, the tNodeEmbed model [27] approach learns node embeddings by considering nodes as sentences that are sequences of embeddings of snapshots over time.

In the current work, we extend the Node2Vec model for GRL by including the temporal aspect effectively to extract meaningful node representations. We propose degree and temporal distribution-based transition probabilities to include in the selection mechanism of the random walk method. In the literature, link prediction, node classification, clustering, and recommendation problems are generally used as downstream tasks to evaluate the proposed GRL methods. In this work, we consider link prediction and recommendation problems to evaluate the proposed tNode2Vec model.

3 Proposed Approach

Node2vec [12] approach extracts the node embeddings based on the selected neighborhood of nodes in a given network. The main idea of this algorithm is to control the random walks using two parameters p and q to cover meaningful neighborhoods. These parameters enable the selection of choices like moving closer, moving farther, and returning back to the target node. In all these choices, the next node from a possible set of nodes is chosen randomly i.e. with equal probability. In reality, this selection is not random. The next node is chosen based on factors like popularity, recency, relevance, etc. These edges represent meaningful interactions among nodes, for example, in citation networks, edge is a citation that depends on relevance, recency, and popularity; in social networks like Facebook, edge is a friendship that depends on common interests, proximity, etc. Hence it is more meaningful to consider these aspects.

We propose a model called *tNode2Vec* to incorporate these aspects into the Node2Vec model by defining adaptive random walks that choose the next node using preferential attachment-based probabilities.

3.1 Graph Representation Learning Framework

We adapt the feature learning formulation of Node2Vec [12] and define Graph Representation Learning (GRL) as a maximum likelihood optimization problem.

Assumptions

- Network follows homophily property, the nodes closer in the network tend to be related.
- Random walks extracted around the target node represent the network structure and properties of the underlying network.
- The network is assumed to be static while learning the node embeddings using the proposed model.
- A random walk in the network is similar to the sentence of a language that represents semantic and syntactic relationships of co-occurrences of words.

Given a graph $G(V, E)$ where V is a set of nodes and E is a set of edges, the GRL problem is to learn function $f : V \rightarrow \mathbf{R}^d$ mapping a node to its corresponding representation by training the Skip-Gram model using neighborhood samples N_S that can be used for any downstream task. The objective function is to maximize the log-likelihood of observing a network neighborhood of node u, $N_S(u)$, conditioned on its feature representation $\overrightarrow{f(u)}$ as given in Eq. 1.

$$\max_f \left(\sum_{u \in V} \log \Pr(N_S(u) | \overrightarrow{f(u)}) \right) \tag{1}$$

where

$$\Pr(N_S(u) | \overrightarrow{f(u)}) = \prod_{n_i \in N_S(u)} \Pr(n_i | \overrightarrow{f(u)}) \tag{2}$$

Equation 2 is based on the conditional independence assumption that observing any two neighboring nodes is independent.

$$\Pr(n_i | \overrightarrow{f(u)}) = \frac{\exp(\overrightarrow{f(n_i)} \cdot \overrightarrow{f(u)})}{\sum_{v \in V} \exp(\overrightarrow{f(v)} \cdot \overrightarrow{f(u)})} \tag{3}$$

Equation 3 is based on the symmetry in latent space assumption that any two connected nodes may have corresponding symmetric effect in the latent space. Finally, Eq. 1 can be simplified as shown in Eq. 4.

$$\max_f \sum_{u \in V} \left[-\log \left(\sum_{v \in V} \exp(f(u).f(v)) \right) + \sum_{n_i \in N_S(u)} (\overrightarrow{f(n_i)} \cdot \overrightarrow{f(u)}) \right] \tag{4}$$

The objective function given in Eq. 4 is optimized using Skip-Gram model [24] which is developed for natural language processing. The challenge in adapting the

Skip-Gram model to networks is the non-linear nature of the data. To overcome this challenge, Grover et al. [12] proposed a biased random walk method to generate node sequences to suit the Skip-Gram model. We update the random walk method by incorporating a selection mechanism that favors nodes having recency and high degree as defined in the next Sect. 3.2.

3.2 Adaptive Temporal Random Walks

The problem of neighborhood sampling can be modeled as a local search method around the target node. Any downstream prediction task on nodes depends on two kinds of similarities homophily and structural equivalance [11,15,16]. Node2Vec [12] model uses the biased random walk strategy for neighborhood sampling by balancing the BFS and DFS to control the coverage. In reality, the next node selection in the random walk is not a random process as it is considered in the existing approaches. Since many real-world networks are dynamic networks, the link formation in those networks depends on other aspects like recency, similarity etc. (Fig. 1).

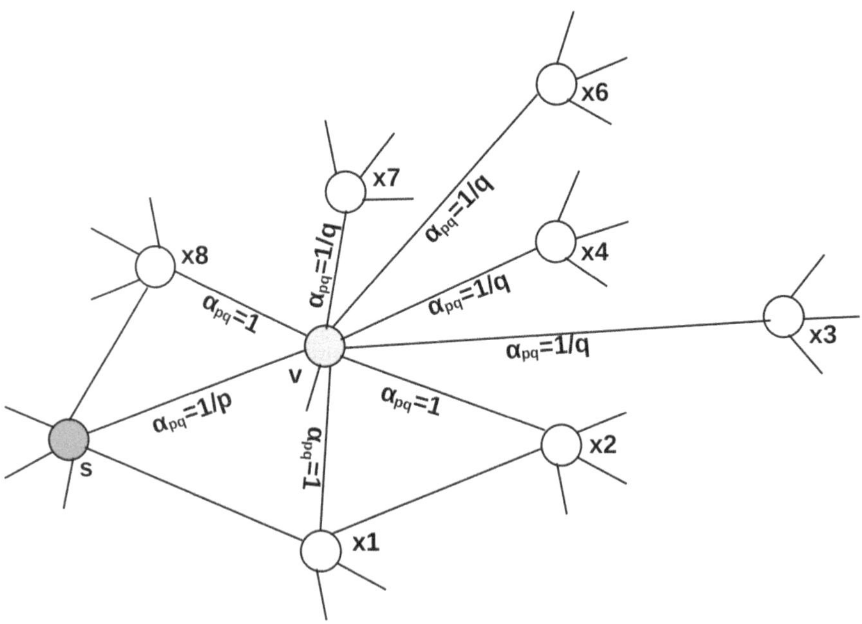

Fig. 1. Random walk procedure in tNode2Vec model

Given a graph $G(V, E)$ and a node u to start from, the idea of the proposed *adaptive temporal random walk* method is to generate a random walk using the underlying temporal and degree-based distribution.

We update the transition probability π_{vx} to include the temporal aspect and degree-based priority $tp_v(x)$ as defined in Eq. 5. The transition probability π_{vx} of moving from node v to x is defined as:

$$\pi_{vx} = \alpha_{pq}(s,x) \cdot tp_v(x) \cdot w_{vx} \tag{5}$$

where

$$\alpha_{pq} = \begin{cases} \frac{1}{p} & \text{if} \quad d_{sx} = 0 \\ 1 & \text{if} \quad d_{sx} = 1 \\ \frac{1}{q} & \text{if} \quad d_{sx} = 2 \end{cases} \tag{6}$$

$$tp_v(x) = \frac{deg(x) \cdot r_v(x)}{\sum_{y \in N(v)} (deg(y) \cdot r_v(y))} \tag{7}$$

The temporal aspect $tp_v(x)$ is defined as the normalized product of the degree and recency of node x. The idea behind this formulation is to select a node that is recent and well-connected. The denominator defined in Eq. 7 normalizes the product over all the neighboring nodes.

$$r_v(x) = \frac{1}{\log(t_v - t_x)} \tag{8}$$

The recency of the node x with respect to node v is $r_v(x)$ that is defined as given in Eq. 8 where t_v is the time at which node v is included into the network and t_x is the time at which node x is included into the network. In the current work, we follow the definition given to recency as reciprocal of age as in [17,22]. The log function is used in the denominator to modulate the diminishing effect of the recency.

For every node v in the walk, the above-mentioned transition probabilities are calculated to choose the next node.

Justification to the Model Choices. We modified the transition probability π_{vx} by including the temporal and degree aspects in the proposed tNode2vec model. The temporal and degree factor $tp_v(x)$ is defined as the normalized product of the product $deg(x) \cdot r_v(x)$ that favors the selection of a node with high degree as well as recency. This product penalizes the selection of an old node having a high degree and a recent node with a low degree thus ensuring the balance between recency and high degree.

The recency $r_v(x)$ of a node x is defined as the difference $t_v - t_x$ that gives the relative age of node x with respect to node v. That idea of relative age naturally favors the selection of recent research works which are important in understanding the recent advances and research gaps in the problem.

4 Experimental Setup and Results

We design extensive experimentation using two real-world bibliographic datasets to analyze the efficiency of the proposed tNode2Vec model. TDL2Vec model [22] is the latest model proposed to solve the disappearing link prediction problem that includes the temporal aspect given in Eq. 9. Higher values of t' result in very low values of τ in turn the transition probability will become very small. The proposed tNode2Vec model addresses this issue by modulating the denominator using log function as given in Eqs. 7 and 8 and also includes degree distribution to balance the recency and degree of a node. In this work, we implement the TDL2Vec, Node2Vec, and Deepwalk models along with the proposed tNode2Vec model to solve link prediction and recommendation tasks. The details of the datasets considered in the experimentation are given below.

4.1 Datasets

We have used two popular bibliographic datasets, Hep-Ph dataset [21], and DBLP dataset [29] for experimentation. The Hep-Ph dataset is a classical dataset used in most of the baseline methods that consists of research papers from the High-energy Physics domain. DBLP V11 is a large well-known dataset with a collection of $41,07,340$ research papers and conference proceedings from computer science and related domains. This dataset contains metadata of the research paper which covers the title, authors, year of publication, journal/ conference, keywords, abstract, etc. Table 1 gives the details of the datasets used. We construct a citation network by extracting papers, citations, and year of publication from the DBLP dataset. For experimentation, we extracted a connected sub-network from the DBLP V11 dataset that consists of 2,30,830 papers as nodes and 4,26,459 citation relationships as edges. We have extracted the titles of these 2,30,830 papers to perform additional experimentation later using the BERT model. We have also extracted a connected sub-network from the Hep-Ph dataset that consists of 34,546 nodes and 4,21,578 citations.

Table 1. Details of the datasets used for experimentation

Property	Hep-Ph	DBLP
Nodes (papers)	34,546	41,07,340
Edges (citations)	4,21,578	3,66,24,464
Node attributes (publication year)	yes	yes
Node attributes (titles)	no	yes

4.2 Baseline Methods

We consider the following models as baseline methods and compare the results against them.

- **DeepWalk** [25]**:** an encoding method that learns node representations using random walks on the network.
- **Node2Vec** [12]**:** extension of DeepWalk model by introducing biasing parameters to control the random walks to generate node representations.
- **TDL2vec** [22]**:** modified Node2Vec algorithm to incorporate temporal aspect. The temporal aspect used in this model is mathematically given as:

$$\tau = \frac{1}{t' - t}, \qquad t' > t \tag{9}$$

Parameter Setting. After extensive experimentation, we set the model parameters as follows: the number of walks per node r to 80, walk length l to 20, context window length w to 10, and number of epochs e to 5. To make the comparison of results meaningful, all the baseline models are implemented with the same parameter settings. The size of output embedding size is set to 64.

For experimentation, we considered two downstream tasks, namely, Link Prediction and Recommendation to evaluate the efficiency of the proposed model. We explain the implementation of link prediction and recommendation in this section.

4.3 Task: Link Prediction

We consider the widely known link prediction [14,23] problem to show the effectiveness of the proposed graph representation method. Given a graph $G(V, E)$ at time t, the link prediction problem is to predict the set of links to be formed by the time t' such that $t' > t$. Figure 2 depicts the link prediction problem.

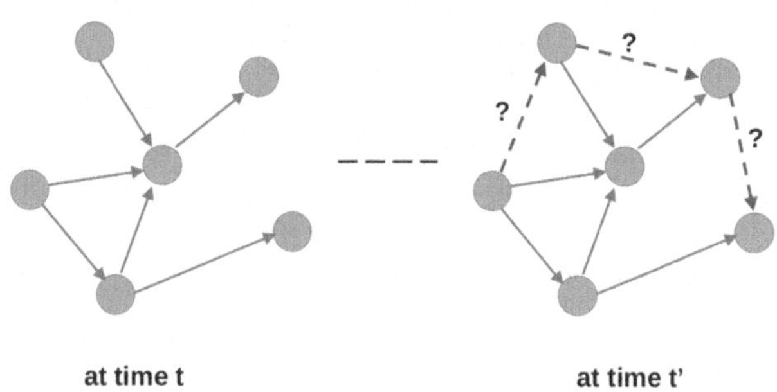

Fig. 2. Link prediction problem

The link prediction problem is generally treated as a binary classification problem. The pipeline used for link prediction is given in Fig. 3. The edge set E is considered as the positive samples and an equal number of non-edge node pairs NE are randomly selected as negative samples to form the data samples (where $|E| = |NE|$) such that the most common class imbalance problem is addressed. For every node pair (u, v), combine the node representations using $\overrightarrow{f(u)} + \overrightarrow{f(v)}$ (where $\overrightarrow{f(u)}$, $\overrightarrow{f(v)}$ are the node representations extracted from *tNode2Vec model*) to form edge embeddings as done in [12]. A classifier is trained on 70% of the data samples and validated on the remaining 30% of the data samples.

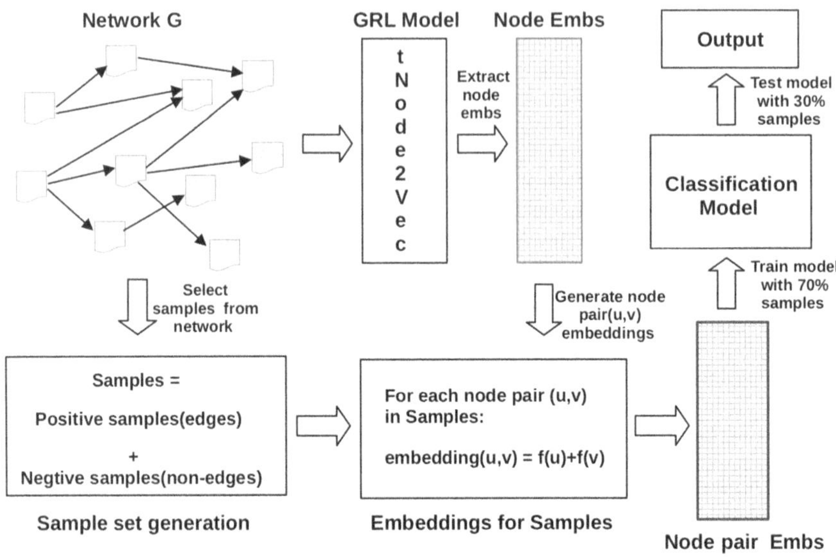

Fig. 3. Link prediction pipeline

4.4 Analysis of Link Prediction Results

We have constructed the paper citation network with nodes as papers and edges as the citations. We extracted a connected component from this network and given as input to the proposed tNode2Vec model to learn node representations. We train the simple logistic regression model as explained in Sect. 4.3.

Evaluation Metrics. We use the standard machine learning metrics AUC, precision, recall, MAP, and accuracy for the link prediction task [12,22,25].

The results obtained for the link prediction task on two datasets Hep-Ph and DBLP V11 are given in Table 2. It can be observed considering the temporal aspect improves the results as both TDL2Vec and tNode2Vec models perform better than Node2Vec and DeepWalk models. The temporal factor used in the TDL2Vec model diminishes the transition probabilities with higher values of t'. The temporal factor tp_v included in the proposed tNode2Vec model uses the product of degree and temporal factor that balances the importance

Table 2. Comparison of link prediction task results on different datasets

Dataset	Model	AUC	Precision	Recall	MAP	Accuracy
Hep-Ph	tNode2Vec	**0.724**	**0.671**	**0.674**	**0.615**	**0.699**
	TDL2vec [22]	0.683	0.634	0.648	0.601	0.653
	Node2Vec [12]	0.671	0.626	0.636	0.580	0.628
	DeepWalk [25]	0.669	0.628	0.627	0.581	0.627
DBLP V11	tNode2Vec	**0.968**	**0.948**	**0.910**	**0.907**	**0.930**
	TDL2vec [22]	0.936	0.914	0.901	0.882	0.912
	Node2Vec [12]	0.928	0.908	0.878	0.858	0.890
	DeepWalk [25]	0.915	0.889	0.863	0.847	0.881

of recency and degree. Further, the log function used in the temporal aspect to modulate the diminishing effect improves the TDL2Vec model. tNode2Vec model achieved 4.1%, 3.7%, 3.6%, 1.4%, and 4.6% improvements on AUC, precision, recall, MAP, and accuracy values over the best baseline TDL2vec model on the Hep-Ph dataset. The same trend of results can be seen in the DBLP dataset where tNode2Vec achieved 3.2%, 3.4%, 0.9%, 2.5%, 1.8% improvements on AUC, precision, recall, MAP and accuracy values over the best baseline TDL2vec model. These improvements confirm that the inclusion of temporal and degree-based distribution is meaningful and thus enhances the performance of the link prediction task.

Inclusion of the BERT Model. Devlin et al. introduced the BERT model [10], designed to pre-train deep bidirectional representations that are conditioned on both left and right context in all layers. In the literature on link prediction, the inclusion of textual representations aids in achieving better results [9,32]. We have used the BERT model to extract the representations for the titles of the papers. We have experimented with the proposed tNode2Vec by including the titles of the papers that are available in the DBLP V11 dataset. We have also reimplemented all the baseline methods by including the BERT model to extract textual representations. These title representations are concatenated with node representations extracted from tNode2Vec and baseline models. The results obtained from this experimentation are given in the Table 3.

Table 3. Additional experiments on the DBLP V11; Inclusion of BERT model to learn title representations

Dataset	Model	AUC	Precision	Recall	MAP	Accuracy
DBLP V11	tNode2Vec+BERT	**0.978**	**0.951**	**0.927**	**0.918**	**0.940**
	TDL2vec [22]+BERT	0.947	0.920	0.913	0.890	0.924
	Node2Vec [12]+BERT	0.938	0.909	0.891	0.864	0.901
	DeepWalk [25]+BERT	0.961	0.920	0.909	0.882	0.915

It can be observed from the results that including title representations extracted from the BERT model improves the results of all the models. We can also observe that the proposed tNode2Vec model without including the BERT model could achieve better results than the TDL2Vec and the Node2Vec models including the BERT model.

4.5 Task: Recommendation

Given the graph $G(V, E)$, a query q, and a set of seed nodes S, the recommendation problem is to extract top-k relevant nodes as recommendations. We consider the citation recommendation problem inspired by the recent literature [1,2,6,8,17]. CR-HITS [17] claims that the usage of 2-hop neighborhood is enough for effective representation of nodes. 2-hop neighbourhood of any node x is the set of all nodes in V which are at a hop distance less than or equal to 2 from x. We follow the approach of focusing on the 2-hop neighborhood instead of the entire network as given in CR-HITS model [17]. We apply the proposed tNode2Vec algorithm on the 2-hop neighborhood extracted around the set of seed nodes S to compute the node representations. These vectors are compared with the seed node representations to recommend relevant nodes. We describe the recommendation algorithm as given below.

Input: Graph $G(V, E)$, Query q and Seed set S
Output: Top-k recommendations

Step 1 Extract a 2-hop subnetwork $G^{'}$ around the seed set nodes S.
Step 2 Use the *tNode2Vec* model to generate the node embeddings of the subnetwork.
Step 3 For every node u, calculate the cosine similarities with each seed node $s \in S$ and consider maximum cosine similarity over S, as the similarity score $sim\_score(u)$ of node u.

$$sim\_score(u) = \max_{s \in S} \frac{\overrightarrow{f(s)} \cdot \overrightarrow{f(u)}}{\|\overrightarrow{f(s)}\| \|\overrightarrow{f(u)}\|}$$

Step 4 Rank the nodes in the decreasing order of sim_score and recommend the top-k nodes as recommendations.

Evaluation Metrics. We use the standard metrics used for recommendation algorithms [1,2,6,8,17], MAP, MRR, and Recall, for the recommendation task to evaluate the results, which are defined below.

MAP: Mean Average Precision (MAP) is a precision metric that emphasizes ranking relevant papers higher. Let Q be the set of query papers considered for evaluation. For a paper q in Q, let the actual bibliography set of q be $B(q)$, and suppose the algorithm predicts the reference papers to be R. We consider the top N recommended papers of the ranking list and call it $R@N$.

$$MAP = \frac{1}{\mid Q \mid} \sum_{q \in Q} \frac{1}{\mid B(q) \mid} \left(\sum_{\substack{p \in B(q) \\ rank(p) \neq 0}} \frac{c(p)+1}{c(p)} \right) \tag{10}$$

where $rank(p)$ is defined as the position of the paper p in the predicted list R if p is in R, otherwise $rank(p)$ is defined to be zero. $c(p)$ is defined to be the number of the correct reference papers obtained in $R@N$ which rank higher than p.

MRR: Mean Reciprocal Ranking(MRR) measures how far from the top does the first relevant reference paper appears. In the predicted reference list of $R@N$, let the position where the first correct reference appears from the top be denoted as $rank(p_f)$. Then MRR is given by

$$MRR = \frac{1}{\mid Q \mid} \sum_{p \in Q} \frac{1}{rank(p_f)} \tag{11}$$

where $rank(p_f)$ indicates the position of the first retrieved relevant reference papers in the reference list R.

Recall: The significance of the proposed model can be measured using recall which is the percentage of relevant recommendations from top N results calculated as recall@N. It is defined by calculating the percentage of the original bibliography of the given scientific paper that are present in the top N recommended papers.

$$recall@N = \frac{1}{\mid Q \mid} \sum_{q \in Q} \frac{\mid B(q) \cap R@N \mid}{|R@N|} \tag{12}$$

where R denotes the bibliography set of the paper and $R@N$ denotes top-N recommendations.

4.6 Analysis of Recommendation Results

We follow the experimental setup given in the recommendation literature [1, 2, 6, 8, 17] to solve the citation recommendation problem. We randomly select 150 query papers from the network to experiment with the proposed model and baseline methods. For each query paper, three seed papers are selected from its bibliography to extract the 2-hop neighborhood G' around them.

We perform experimentation by implementing the baseline methods Node2Vec, TDL2Vec, and DeepWalk along with the proposed method tNode2Vec for the recommendation task, on the two real-world datasets Hep-Ph and DBLP. The embeddings are computed for all the methods by taking only the subnetwork G' around the seed node set S. The citation recommendation results are tabulated in Table 4. It can be observed that the results obtained by all the random walk-based models are almost similar, with the proposed tNode2Vec achieving slightly better results on both the data sets Hep-Ph and DBLP V11. Further, in the case of Hep-Ph dataset, all the baseline methods along with the proposed method tNode2Vec achieve over 90% result for MAP, MRR and Recall, with 1–2% improvement shown by tNode2Vec.

Table 4. Comparison of recommendation task results on Hep-Ph and DBLP datasets

Dataset	Model	MAP	MRR	R@20	R@40	R@60	R@80	R@100
Hep-Ph	tNode2Vec	**0.93**	**0.92**	**0.94**	**0.94**	**0.94**	**0.94**	**0.94**
	TDL2Vec [22]	0.92	0.91	0.93	0.93	0.94	0.94	0.94
	Node2Vec [12]	0.92	0.91	0.93	0.93	0.93	0.94	0.94
	DeepWalk [25]	0.92	0.91	0.93	0.93	0.93	0.94	0.94
DBLP V11	tNode2Vec	**0.76**	**0.81**	**0.78**	**0.78**	**0.79**	**0.79**	**0.80**
	TDL2Vec [22]	0.75	0.79	0.77	0.78	0.79	0.79	0.80
	Node2Vec [12]	0.74	0.78	0.77	0.78	0.78	0.78	0.79
	DeepWalk [25]	0.73	0.76	0.75	0.75	0.76	0.77	0.77

4.7 Complexity Analysis

The proposed tNode2Vec model is an extension of the Node2Vec model that introduces a mechanism to generate random walks in an adaptive manner. So the proposed tNode2Vec model differs from the Node2Vec model only in the random walk generation.

The time complexity of the model grows linearly with the number of nodes V, the number of walks per node r, and the walk length l. Generating a random walk of length l includes a selection mechanism of the next node based on time and degree at every step. Assuming that the degree of nodes in the social network is of $O(1)$. Next node selection at each step takes constant time $O(1)$ hence each random walk requires $O(l)$ time. For every node($v \in V$) in the graph, generating r random walks requires $O(|V|.r.l)$. So even with the inclusion of temporal aspects into the random walk generation process, the time complexity remains the same as that of the Node2Vec model. The overall time complexity of the proposed tNode2Vec model is $O(|V|.r.l) + O(d.|V|.r.l)$ where $O(d.|V|.r.l)$ is the time complexity of the Skip-Gram model that produces the embeddings of size d.

5 Conclusion and Future Work

In this work, we have proposed a Graph Representation Learning(GRL) method *tNode2Vec* by extending the popular Node2Vec model by incorporating temporal aspects. We modified the transition probabilities used in random walks using temporal and degree-based distributions. We evaluated the model performance using link prediction and recommendation tasks and analyzed the results. We performed extensive experimentation on two real-world datasets Hep-Ph and DBLP V11. In case of the link prediction task, the proposed tNode2Vec model showed better performance than all the baseline methods, in particular showed more than 7% over Node2Vec model. It can be observed that the inclusion of title representations extracted using the BERT model into feature vectors improved the link prediction performance. The trend of improvement in these results is consistent among the two datasets. In the case of the recommendation task, the proposed tNode2Vec model performs marginally better than the baseline methods. The results are consistent over both Hep-Ph and DBLP datasets.

It is curious to note that in the citation recommendation literature [1,2,6, 8,17], the results are significantly lower than that of the random walk based methods used in the current work. It is possible that the results are lower since all the existing models have been proposed for heterogeneous bibliographic networks whereas the current work is restricted to homogeneous citation networks. This aspect needs to be investigated further. In future, extending the proposed random walks to heterogeneous networks could help in learning meaningful heterogeneous node representations.

References

1. Ali, Z., Qi, G., Muhammad, K., Ali, B., Abro, W.A.: Paper recommendation based on heterogeneous network embedding. Knowl.-Based Syst. **210** (2020)
2. Ali, Z., Qi, G., Muhammad, K., Bhattacharyya, S., Ullah, I., Abro, W.: Citation recommendation employing heterogeneous bibliographic network embedding. Neural Comput. Appl. **34**(13), 10229–10242 (2022)
3. Barabási, A.L., Albert, R.: Emergence of scaling in random networks. Science **286**(5439), 509–512 (1999)
4. Batagelj, V., Cerinšek, M.: On bibliographic networks. Scientometrics **96**(3), 845–864 (2013)
5. Belkin, M., Niyogi, P.: Laplacian eigenmaps for dimensionality reduction and data representation. Neural Comput. **15**(6), 1373–1396 (2003)
6. Bhagavatula, C., Feldman, S., Power, R., Ammar, W.: Content-based citation recommendation. In: Proceedings of the Conference of the North American Chapter of the Association for Computational Linguistics: Human Language Technologies, pp. 238–251 (2018)
7. Bourigault, S., Lagnier, C., Lamprier, S., Denoyer, L., Gallinari, P.: Learning social network embeddings for predicting information diffusion. In: Proceedings of the 7th ACM International Conference on Web Search and Data Mining, pp. 393–402 (2014)
8. Cai, X., Han, J., Li, W., Zhang, R., Pan, S., Yang, L.: A three-layered mutually reinforced model for personalized citation recommendation. IEEE Trans. Neural Netw. Learn. Syst. **29**(12), 6026–6037 (2018)
9. Daza, D., Cochez, M., Groth, P.: Inductive entity representations from text via link prediction. In: Proceedings of the Web Conference 2021, pp. 798–808 (2021)

10. Devlin, J., Chang, M.W., Lee, K., Toutanova, K.: Bert: pre-training of deep bidirectional transformers for language understanding. arXiv preprint arXiv:1810.04805 (2018)
11. Fortunato, S.: Community detection in graphs. Phys. Rep. **486**(3–5), 75–174 (2010)
12. Grover, A., Leskovec, J.: node2vec: scalable feature learning for networks. In: Proceedings of the 22nd ACM SIGKDD International Conference on Knowledge Discovery and Data Mining, pp. 855–864 (2016)
13. Hamilton, W.L.: Graph Representation Learning. Morgan & Claypool Publishers (2020)
14. Hasan, M.A., Zaki, M.J.: A survey of link prediction in social networks. Soc. Netw. Data Anal. 243–275 (2011)
15. Henderson, K., et al.: Rolx: structural role extraction & mining in large graphs. In: Proceedings of the 18th ACM SIGKDD International Conference on Knowledge Discovery and Data Mining, pp. 1231–1239 (2012)
16. Hoff, P.D., Raftery, A.E., Handcock, M.S.: Latent space approaches to social network analysis. J. Am. Stat. Assoc. **97**(460), 1090–1098 (2002)
17. Kammari, M., Bhavani, S.D.: Citation recommendation using modified hits algorithm. Computing 1–21 (2023)
18. Kashima, H., Tsuda, K., Inokuchi, A.: Marginalized kernels between labeled graphs. In: Proceedings of the 20th International Conference on Machine Learning (ICML 2003), pp. 321–328 (2003)
19. Kipf, T.N., Welling, M.: Variational graph auto-encoders. In: NIPS Workshop on Bayesian Deep Learning (2016). https://arxiv.org/abs/1611.07308
20. Kipf, T.N., Welling, M.: Semi-supervised classification with graph convolutional networks. In: International Conference on Learning Representations (ICLR) (2017). https://arxiv.org/abs/1609.02907
21. Leskovec, J., Krevl, A.: SNAP Datasets: Stanford large network dataset collection (2014). http://snap.stanford.edu/data
22. Li, L., Wang, W., Yu, S., Wan, L., Xu, Z., Kong, X.: A modified node2vec method for disappearing link prediction. In: 2017 IEEE 15th International Conference on Dependable, Autonomic and Secure Computing, 15th International Conference on Pervasive Intelligence and Computing, 3rd International Conference on Big Data Intelligence and Computing and Cyber Science and Technology Congress (DASC/PiCom/DataCom/CyberSciTech), pp. 1232–1235. IEEE (2017)
23. Liben-Nowell, D., Kleinberg, J.: The link prediction problem for social networks. In: Proceedings of the Twelfth International Conference on Information and Knowledge Management, pp. 556–559 (2003)
24. Mikolov, T., Chen, K., Corrado, G., Dean, J.: Efficient estimation of word representations in vector space. arXiv preprint arXiv:1301.3781 (2013)
25. Perozzi, B., Al-Rfou, R., Skiena, S.: Deepwalk: online learning of social representations. In: Proceedings of the 20th ACM SIGKDD International Conference on Knowledge Discovery and Data Mining, pp. 701–710 (2014)
26. Shervashidze, N., Schweitzer, P., Van Leeuwen, E.J., Mehlhorn, K., Borgwardt, K.M.: Weisfeiler-Lehman graph kernels. J. Mach. Learn. Res. **12**(9) (2011)
27. Singer, U., Guy, I., Radinsky, K.: Node embedding over temporal graphs. In: Proceedings of the Twenty-Eighth International Joint Conference on Artificial Intelligence, IJCAI 2019, pp. 4605–4612 (2019). https://doi.org/10.24963/ijcai.2019/640
28. Tang, J., Qu, M., Wang, M., Zhang, M., Yan, J., Mei, Q.: Line: large-scale information network embedding. In: Proceedings of the 24th International Conference on World Wide Web, pp. 1067–1077 (2015)

29. Tang, J., Zhang, J., Yao, L., Li, J., Zhang, L., Su, Z.: Arnetminer: extraction and mining of academic social networks. In: KDD 2008, pp. 990–998 (2008)
30. Theocharidis, A., Van Dongen, S., Enright, A.J., Freeman, T.C.: Network visualization and analysis of gene expression data using biolayout express 3D. Nat. Protoc. **4**(10), 1535–1550 (2009)
31. Velickovic, P., et al.: Graph attention networks. Stat **1050**(20), 10-48550 (2017)
32. Wang, X., van Harmelen, F., Cochez, M., Huang, Z.: Scientific item recommendation using a citation network. In: International Conference on Knowledge Science, Engineering and Management, pp. 469–484. Springer (2022)
33. Zhang, Z., Cui, P., Zhu, W.: Deep learning on graphs: a survey. IEEE Trans. Knowl. Data Eng. **34**(1), 249–270 (2020)

SDG_HHAG Framework: Homogeneous and Heterogeneous Attributed Graphs

A. Sandhya Rani(✉), G. Ashwitha, and K. Swarupa Rani

University of Hyderabad, Hyderabad, Telangana, India
{21mcpc18,20mcpc14,swarupacs}@uohyd.ac.in

Abstract. Graphs, comprising nodes, edges, and additional information known as attributed graphs, are essential for many applications which represents the complex data structures. Accessing these types of graphs is a significant challenge due to privacy concerns and the sensitivity of information. In addition, finding the attributed graphs with different node types and edge types is another challenge. Lack of attributed graph datasets with its heterogeneity factors hinders the powerful graph analysis algorithms across different fields. To address these challenges, we propose the SDG_HHAG Framework, a Synthetic Data Generation Framework comprising two approaches: 1) $Synt\_H_oAG$ (Synthetic Homogeneous Attributed Graphs) to enrich existing graphs with node and edge attributes, and 2) $Synt\_H_eAG$ (Synthetic Heterogeneous Attributed Graphs) to create and generate synthetic heterogeneous attributed graphs. We conducted experiments using the proposed approaches, and results reveal that these approaches are flexible to generate the synthetic graphs with attributes which are useful for graph analytics. Further, generated synthetic datasets are used for applications of social networks such as community detection and for the applications of routing problem such as query reachability.

Keywords: Attributed graphs · Heterogeneous graph · Homogeneous graph

1 Introduction

Data can be in various forms such as text, tables, images, audio or video. While some forms can be used directly, other forms required to be converted into graphs which is a challenging. To get more insights, the data can be represented in the forms of graphs (consists of nodes and edges) which captures the relational and structural information between nodes and edges, especially attributed graphs represents the complex relationships that exists in the data and provides the additional context and detailed information. Regardless of form, every information consists in various forms contributes to the field of big data [26]. However, these graphs manage massive datasets (Volume) by organizing data into nodes and edges with attributes. They efficiently process high-speed incoming data

A. Dasgupta et al. (Eds.): BDA 2024, LNCS 15526, pp. 92–112, 2025.
https://doi.org/10.1007/978-3-031-81821-9_6

(Velocity) through dynamic updates, accommodating real-time data changes. Attributed graphs handle diverse data types like text, images, and videos (Variety). Additionally, they improve data accuracy and reliability (Veracity) by capturing detailed attributes and relationships, thereby enhancing context and clarity. Representing attributed graphs into the domain of big data enhances the effectiveness of data management, processing, and analysis. Attributed graphs, in particular, provide enriched context by incorporating additional attributes for nodes and edges, enabling more sophisticated analysis and insights into the relationships within the data, mainly in domains like social networks, bibliographic networks [30], and recommended systems [6] as described in Table 1. Attributed Graphs are useful for representing the complex networks where the nodes and edges are of single or multiple types. Researchers and developers frequently use many graphs with different properties to validate the efficiency of the discoveries in order to clarify the applicability and limitations of graph analytic methods. However, there is a lack of real-data in both homogeneous and heterogeneous attributed graphs. Due to unavailability and limited access of data, there is a demand to generate synthetic data. Hence, these aspects motivate us to generate synthetic attributed graph datasets to help researchers, to analyze their proposed algorithms and discoveries in various applications. Generation of synthetic data based on state-of-the-art algorithms for simple graphs by using existing tools such as NetworkX [1], SNAP [10], igraph [9] or using the models such as SYNTH [6], DATAGEN [3], these models are trained from the limited available data. Synthetic generation frameworks are popular because of cost-effective, readily accessible, and progressing in data science field without any issues of privacy concerns.

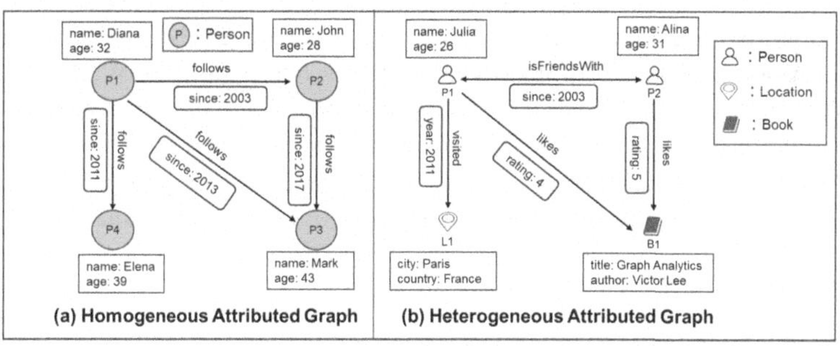

Fig. 1. Example of a homogeneous and heterogeneous attributed graph

Figure 1 illustrates two types of attributed graphs: homogeneous and heterogeneous with different vertex attributes, and edge attributes. Figure 1a shows a homogeneous graph with a single type of node such as Person connected by single edge type 'follows'. And, Fig. 1b represents a heterogeneous graph with different node types such as Person, Location, Book and with multiple edge types 'isFriendsWith', 'likes', 'visited' with directional and bidirectional relationship.

In the literature, some works related to homogeneous and heterogeneous graphs exists but with limited relational and structural information. A homogeneous graph [14] consists of a single type of node and edge. For example, in Co-authorship network [13] node represents author, the relation between two nodes is 'co-author'. The node attributes considered in this network as *research area, country*. Communities can be detected considering these structural and node attribute information. By adding additional node attributes, the analysis process provides a better context for understanding the nodes in a graph, which improves the overall quality and depth of the analysis. In contrast, heterogeneous graphs [14] incorporates multiple types of nodes and edges. For research collaboration, the bibliographic networks [12] consists of different node types author, paper, and venue. The relation between node author and node paper is {author 'writing' paper}, the relation between node paper and node venue is {paper 'publishedIn' venue}, the relation between the two nodes paper is {paper 'cited' paper}. This network identifies similar objects based on these relationships without considering attribute information. By incorporating attribute information, it enables in-depth knowledge and insights in heterogeneous bibliographic networks [30]. In the real world, there exist different set of problems with attributed graphs in query reachability [25], routing [19], link prediction [23], community detection [27] etc. Analysing these problems by considering the attributes plays a significant role in the field of graph analytics. But due to the limited availability of data, unable to analyse the graphs efficiently. Hence, there is demand and it is essential to generate the synthetic attributed graph data.

NetworkX [1] proposed various classical graph model types to generate basic graphs but does not directly supports the heterogeneous attributed graphs. SYNTH [6], DataGen [3], and SyntHING [5] approaches are to generate synthetic graphs, but it is having own limitations of training bias, not simulating the original behaviour as real graphs, computationally expensive, specific to domain. Due to the limitations of existing datasets and data generation approaches which is not suitable for various applications and this could be one of the significant challenges. In order to handle these applications, homogeneous and heterogeneous graphs data are necessary. Furthermore, we observed that there is a scope to generate synthetic datasets and there is a demand in the field of graph analytics. Despite extensive interest in synthetic attributed graph data, to encompass the diverse applications and approaches while addressing privacy concerns and sensitivity, we propose the following definition.

Definition: Generation of Synthetic Attributed Graph Data creates artificial graphs with informative attributes for both nodes and edges. It involves generation of values to the attributes using statistical measures based on the respective attribute type, incorporating relational and structural information to mimic real-world graphs. The synthetic data is designed to be free from privacy concerns and sensitivity issues, as it does not directly replicate any individual's real data.

1.1 Contributions

We proposed a framework for generating the synthetic data. Our main contributions includes

1. We proposed an approach $Synt\_H_oAG$ to enrich existing homogeneous datasets with node and edge attributes.
2. We proposed an approach $Synt\_H_eAG$ to create and generate synthetic heterogeneous attributed graph data.

Section 2 explains the literature on synthetic data generation for graphs. Section 3 presents our proposed approach with examples. Section 4 presents the experiments of our proposed techniques. Section 5 describes the applications of our proposed approach. Section 6 presents the conclusions and scope for further research.

Table 1. Different Graph Applications

Domain	Networks	Applications
Social Networks	Citation Networks	Research Collaborations [29]
	Twitter	Targeting advertisements, topic detection, detecting events [16]
Geographic Information Systems (GIS)	Transportation Network	• Mapping accidents, traffic stops, and traffic counts along streets • Maintaining fleet vehicle locations and asset management functions [17]
E-Commerce business	Online Shopping Networks	• Explore similar customer preferences in shopping networks for personalized recommendations • Identify influential customers for better recommendations in E-commerce [18]
Biological Networks	Metabolic networks	• Analyzing metabolic flux changes and rerouting pathways post genetic perturbations • Identifying short alternative pathways for rerouting metabolic flux [19]
	Genomic networks	• Protein function prediction, disease gene prioritization, network-based genome-wide association study • Analyzing and visualizing complex biological activities using network representations [20]
	Inter-cellular networks	• Investigating functional connectivity patterns in multicellular systems using cellular networks • Determining molecular interaction networks within cells through cellular network analysis [21]
Ecological systems	Climate network	Analyzing climate subsystem interconnections and network influences on stability [22]

2 Related Work

To handle various applications and to discover algorithms, we must rely on data; if unavailable, synthetic data generation becomes necessary. The generation of synthetic data plays an important role where real benchmark datasets are not available. There exists the traditional models for generating synthetic datasets like Erdős-Rényi [8], Barabási-Albert [7] for simple graphs. These models generates the random networks which are not applicable to generate single or multiple node types along with the node and edge attributes for homogeneous and heterogeneous attributed graphs.

There are various tools for generating the graphs, such as NetworkX [1], SNAP [10], igraph [9]. NetworkX [1] derived classical Graph, DiGraph, MultiGraph, MultiDiGraph model types to generate basic graphs but does not directly support heterogeneous attributed graphs, which consist of graphs with different node and edge types. Darabi et al. [6] proposed Synthetic Graph Generation (SYNTH) framework for generating the synthetic dataset by preserving the properties of original graph (nodes and edge attributes, link information). This approach generates the homogeneous graph with node and edge attributes. However, there is a lack of original graph with all characteristics. Ioannis et al. proposed GenCAT [4], which is used for generating the synthetic data for homogeneous attributed graphs with only node attribute information. This approach generates the graph with class labels. Nettleton [11] proposed an approach for generating the synthetic datasets exclusively for an application of Online Social Networks to address the issues on data availability and privacy. The approach handles the homogeneous graph with only node attributes and edge weights. Here, the edge weight indicates the strength of the relation.

Erling et al. [3] proposed LDBC SNB Data Generator (DATAGEN) for generating synthetic social network graphs that mimic real-world data. It generates heterogeneous graphs for social networks, which will not support for other domain applications. Ming-Yi et al. [5] proposed Synthetic Heterogeneous Information Network Generation (SyntHING) approach that uses the real-world heterogeneous information network as a reference to understand the patterns and generates the synthetic dataset. It generates a heterogeneous graph with only node attribute information.

In summary, there are various methods that exist for generating synthetic datasets. Most existing approaches have few limitations in addressing homogeneous and heterogeneous graphs. Therefore, to overcome the limitations, we are proposing $Synt\_H_oAG$, and $Synt\_H_eAG$ approaches for generating the synthetic dataset for homogeneous and heterogeneous attributed graphs, which can be applicable to various domains.

3 Proposed Approaches

The proposed techniques Synt_H_oAG, and Synt_H_eAG are described in this section to generate the synthetic datasets for homogeneous and heterogeneous attributed graphs to solve the data scarcity issues in the domain of graph analytics.

3.1 Synthetic Data Generation Framework

The proposed framework consists of three components: i) Structure extractor for homogeneous graph, ii) Structure generator for heterogeneous graph and iii) Attribute value generator as depicted in Fig. 2. The proposed three components will work by using NetworkX [1]. The motivation for using NetworkX [1] is, its comprehensive, user-friendly, and flexible platform that supports for creating, manipulating, analyzing and visualizing the structure of the networks. It is an open source and supports the creation of different types of graphs such as Directed, Undirected, MultiGraphs (Undirected graph with multiple edges), MultiDiGraphs (directed graph with multiple edges), which allows it to create synthetic attributed graph datasets for various applications. The purpose of using MultiDiGraph in NetworkX is to handle multiple directed edges between nodes, each with its own set of attributes for generating heterogeneous synthetic data. The proposed components are explained in detail below.

Structure Extractor: The homogeneous attributed graph will be generated based on the existing graph data by adding new nodes and edge attributes to the graph. The input for Synt_H_oAG approach is an existing dataset, which may contain the single type of node with few node and edge attributes, or it can be a simple graph (consists of nodes and edges without any nodes and edge attributes). The module *Structure Extractor* shown in Fig. 2 is to extract structure. To extract structure information ($g_\theta(S)$) for a given existing dataset, the *Graph* class object in NetworkX reads the edge list. If the dataset contains attributes, NetworkX will parse and extract node and edge attributes ($g_\phi(A)$). Once the information is extracted, the homogeneous attributed graph is enriched through *Enriching Attribute* module as shown in Fig. 2 for additional node and edge attributes information.

Structure Generator: As stated in Sect. 1, the heterogeneous attributed graph with all characteristics (different types of nodes and edges and their attributes) is not available in the literature. Therefore, for our proposed framework, giving an User Input Configuration file which consists of all characteristics (different node and edge types, number of nodes and edges, node and edge attributes). For this approach, the *Structure Generator* ($g_\theta(S)$) module uses MultiDiGraph by taking the user Input Configuration file and creates the heterogeneous attributes graph with multiple directed edges between different types of nodes.

Attribute Value Generator: Assigning attribute values to the obtained homogeneous and heterogeneous graph using the measures (Random range, Predefined set, and Gaussian Distribution [15]) to the node and edge attributes. After exe-

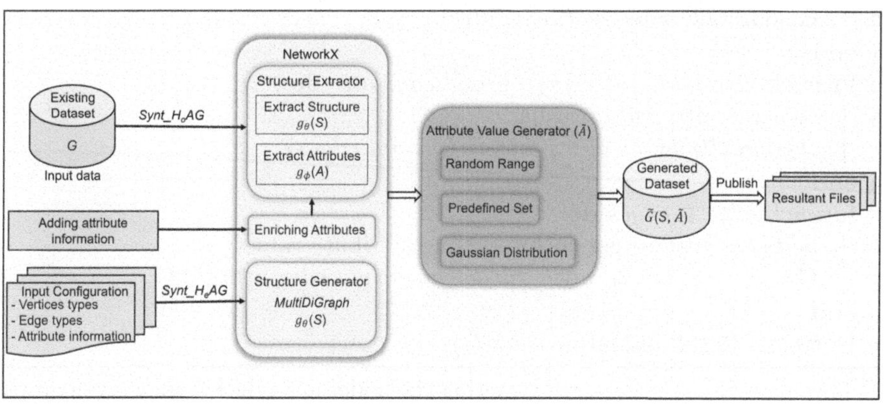

Fig. 2. Synthetic Data Generation Framework

cution of *Attribute Value Generator*, the resultant dataset is derived and subsequently published in output data as Resultant Files. The proposed framework is suitable for a wide range of applications for further analysis in the field of graph analytics.

3.2 Synthetic Homogeneous Attributed Graph (Synt_H_oAG) Generation Approach

The proposed approach is a novel method to enrich the existing graphs, which contains single type of nodes by adding node and edge attributes.

3.2.1 Procedure to Generate Homogeneous Attributed Graphs

The proposed Synt_H_oAG takes the existing graph and the additional attributes information as an input to generate homogeneous attributed graph. The existing graph could be a partial graph, which may contain single type of node with few node and edge attributes, or it may be a simple graph. Our approach enriches the existing graph by adding additional attribute information.

Algorithm 1: Synthetic Homogeneous Attributed Graph (Synt_H_oAG) Generation

Input: `Existing Graph G`, *add_node_attr*, *add_edge_attr*
Output: Synthetic Attributed Graph Data

1 $nodes \leftarrow G.nodes()$
2 $edges \leftarrow G.edges()$
3 $node\_attr\_info \leftarrow \{node : G.nodes[node]\}$
4 $edge\_attr\_info \leftarrow \{edge : G.edges[edge]\}$
5 **foreach** $attr\_name, attr\_info \in add\_node\_attr$ **do**
6   Generate $attr\_values$ using $AssignAttributeValues(attr\_info, num\_nodes)$;
7 **end**
8 **foreach** $node \in nodes$ **do**
9     $add\_node\_attr \leftarrow$ get corresponding attribute values;
10   G.add_node(*node*, *add_node_attr*);
11 **end**
12 **foreach** $attr\_name, attr\_info \in add\_edge\_attr$ **do**
13   Generate $attr\_values$ using $AssignAttributeValues(attr\_info, num\_edges)$;
14 **end**
15 **foreach** $edge \in edges$ **do**
16     $add\_edge\_attr \leftarrow$ get corresponding attribute values;
17   G.add_edge(*source_node*, *target_node*, *add_edge_attr*)
18 **end**
19 Output$\leftarrow$ ResultantOutputData(G)

The Algorithm 1 begins by extracting the nodes and edges from a given graph G (lines 1–2). It then collects the attribute information from given graph for both nodes and edges (lines 3–4). To enrich the existing graph, the additional node *add_node_attr*, and edge *add_edge_attr* attributes are provided in input. For the additional node attributes, the attribute values are generated using the measures as described in Sect. 3.2.2 (lines 5–7). These generated values are assigned to the additional node attributes and added to the graph (lines 8–11). Subsequently, the algorithm repeats the same process as node attributes for edge attributes. For each additional edge attributes, the attribute values are generated using the Algorithm 2 as described in Sect. 3.2.2 (lines 12–14). These generated values are assigned to edges with their corresponding source and target node and added to the graph (lines 15–18). The resultant data of homogeneous attributed graph will be published in output files, in the format of text as Resultant Files.

3.2.2 Statistical Measures: Assigning Attribute Values

The Algorithm 2 focuses on assigning values to the node and edge attributes using statistical measures. This section describes the details of how to generate the attribute values. In the literature several attribute statistical measures such as Random [24], Bernoulli and Gaussian distributions [4] are described. In this work, we considered three attribute statistical measures, two from the literature Random range (continuous numerical values), Gaussian distribution (values falls under bell curve [15]) and one measure we defined and termed as Predefined set (fixed set of values). These measures are most suitable and covers all type of attributes commonly encountered in various fields. The details of statistical measures are explained with illustrations.

Algorithm 2: AssignAttributeValues

```
Input: attr_info, k as num_values
Output: List of attribute values
if attr_info['measure'] == 'random_range' then
    return k values within attr_info['range']
end
else if attr_info['measure'] == 'predefined_set' then
    return k values from attr_info['values']
end
else if attr_info['measure'] == 'gaussian_distribution' then
    return k values from Gaussian distribution with mean and std
end
```

1. **Random range:** Generate random values within the specified range provided in the attribute information.
2. **Predefined set:** It selects the elements such as constant, domain values from a predefined set of values listed in attribute information.
3. **Gaussian distribution:** It is a bell-shaped curve, also known as normal distribution [15]. It generates the sample values from a Gaussian distribution

with the mean and standard deviation specified in attribute information. This statistical measure is used for assigning continuous numerical values to the attributes.

Illustrations: Let us consider the input as simple graph (Fig. 3(i)a) or partial homogeneous attributed graph (Fig. 3(ii)a). Using the proposed $Synt\_H_oAG$ approach, the resultant graph is enriched with additional node and edge attributes as shown in Fig. 3(i)b and 3(ii)b. For example, in Fig. 3(i)b, *Age* attribute of node P1, and *since* attribute of the edge "follows" are specified using the measure (Random range), within a provided minimum and maximum range. The *Name* attribute of node P2 is assigned from a measure (predefined set). Similarly, the *Height* attribute of node P3 is assigned using the measure (Gaussian distribution) with values based on the provided mean and standard deviation. Similarly, for the partial homogeneous attributed graph (Fig. 3(ii)a), we can enrich by assigning attribute values to the additional attribute information using the statistical measures as shown in Fig. 3(ii)b.

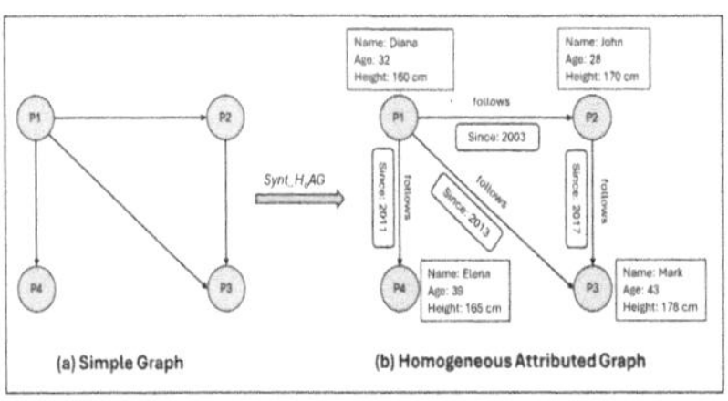

(i) Attribute assignment to simple graph

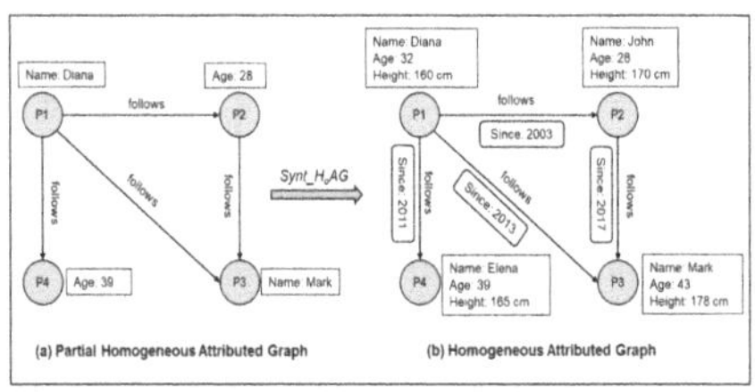

(ii) Attribute assignment to partial homogeneous graph

Fig. 3. Generating the attribute values through statistical measures

3.3 Synthetic Heterogeneous Attributed Graph (Synt_H_e*AG*) Generation Approach

We proposed a novel approach to create and generate synthetic data for heterogeneous attributed graphs without using complex machine learning and deep learning models. The proposed approach consist of three phases as explained in this section.

Phase I: Read user input configuration file
Phase II: Heterogeneous attributed graph generation
Phase III: Resultant files

Algorithm 3: Synthetic Heterogeneous Attributed Graph (Synt_H_eAG) Generation

Input: User input file
Output: Synthetic Graph Data
node_types, *edge_types*, *num_nodes*, *num_edges*, *node_attributes*, *edge_attributes* ← Read_input_file(config)
G ← GenerateHeteroGraph(*node_types*, *edge_types*, *num_nodes*, *num_edges*,*node_attributes*, *edge_attribute*)
Output ← ResultantOutputData(G)

3.3.1 Read User Input Configuration

User will provide sample input configuration file based on the application for creating synthetic heterogeneous attributed graph. The *Synt_H_eAG* approach will parses the configuration file to obtain node and edge information for graph generation process. The file includes details of the different types of nodes and edges, number of nodes and edges, and attribute information for both nodes and edges. After processing the entire file along with specified information one can build the heterogeneous attributed graph. For instance, the input configuration file (Fig. 4a) includes the details of graph with different node types (Person, Company, Technology), and edge types ('worksAt', 'knows', 'friendOf') along with number of nodes and edges to create graph. Node attributes *age, name* for Person node, *revenue* for Company node, *name* for Technology node are specified along with their measures (Random range, Predefined set, Gaussian distribution). Similarly, edge attributes, *since* for 'worksAt' edge, *level* for 'knows' edge, and *mode* for 'friendOf' edge are specified with the measure (Predefined set) which is shown in Fig. 4a.

3.3.2 Procedure to Generate Heterogeneous Attributed Graphs

Algorithm 4 constructs a graph G with different node and edge types corresponding to node and edge attribute information. Initially, a graph with different node and edge types is generated using the MultiDiGraph [1] model. It then iterates through each node type (lines 3–14) retrieves its corresponding attributes from the User Input Configuration file, and uses the *AssignAttributeValues* (Algorithm 2) to generate attribute values for the respective nodes (lines 5–7). Subsequently, the algorithm iterates through each node, assigns a unique ID and adds the respective attributes with the assigned values to the graph G (lines 8–13).

Subsequently, the algorithm iterates through each edge type (lines 15–26), repeating the same process as node attributes, for edge attributes. For each edge type, it extracts source node, target node types along with the relation (line 16). The algorithm retrieves its corresponding edge attributes from the User Input Configuration file and generate the attribute values through Algorithm 2 for the edge attributes (lines 17–20). These generated values are assigned to edges with their corresponding source and target node type and added to the graph (lines 21–25). Therefore, heterogeneous attributed graph G is generated with all characteristics which includes different type of nodes and edge information along with their attributes.

3.3.3 Resultant Files

The generated synthetic heterogeneous attributed graph (G) can be published based on User Input Configuration file, which is shown in Fig. 4a. The synthetic heterogeneous attributed graph 'G' is generated and published in resultant files (text or cypher files). The Cypher files are used in the Neo4j database environment to support query processing.

Algorithm 4: GenerateHeteroGraph

```
   Input: node_types, edge_types, num_nodes, num_edges,
          node_attributes, edge_attributes
   Output: Graph G
1  Initialize G as an empty MultiDiGraph/MultiGraph;
2  Initialize node_count for each node_type;
3  foreach node_type ∈ node_types do
4      Get attributes for node_type from node_attributes;
5      foreach attr_name, attr_info ∈ attributes do
6          Generate attr_values using
           AssignAttributeValues(attr_info, num_nodes);
7      end
8      for i ← 0 to num_nodes do
9          node_id ← node_type_i;
10         node_attr ← get corresponding attribute values;
11         G.add_node(node_id, node_attr);
12         Increment node_count[node_type];
13     end
14 end
15 foreach edge_type ∈ edge_types do
16     Parse source_type, relation, target_type from edge_type;
17     Get attributes for edge_type from edge_attributes;
18     foreach attr_name, attr_info ∈ attributes do
19         Generate attr_values using
           AssignAttributeValues(attr_info, num_edges);
20     end
21     for i ← 0 to num_edges do
22         Generate source_node and target_node;
23         edge_attr ← get corresponding attribute values;
24         G.add_edge(source_node, target_node, edge_attr)
25     end
26 end
27 return G
```

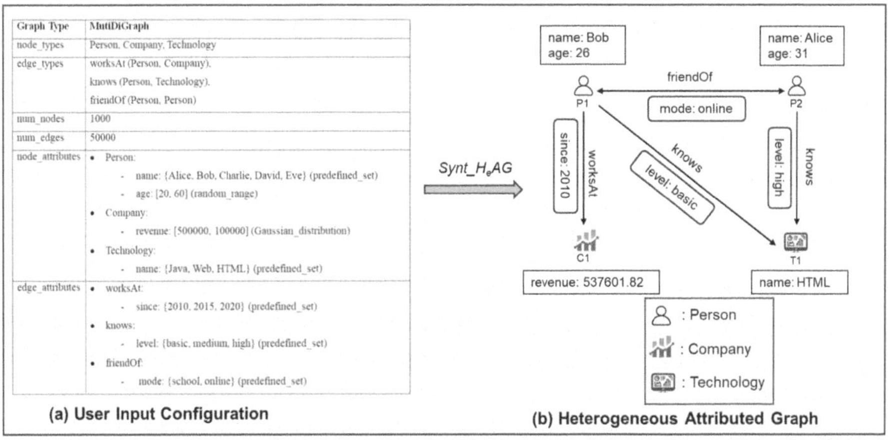

Fig. 4. Synthetic Heterogeneous Attributed Graph Generation

Illustrations: The proposed Synt_H_eAG approach takes User Input Configuration file (Fig. 4a) to generate synthetic heterogeneous attributed graph (Fig. 4b). For example, in the given Fig. 4b node Person P2 has an attribute *Age* is assigned from a measure (Random range), within a provided minimum and maximum range. The node Technology T1 has a node attribute 'Name', edge relation 'knows' has an edge attribute 'level' is assigned with values from a measure (predefined set). Similarly, *Revenue* attribute of node Company C1 is assigned using the measure (Gaussian distribution) with values based on the provided mean and standard deviation.

3.4 Complexity Analysis

For the $Synt\_H_oAG$ approach, extracting the structural and relational information from the given graph takes $\mathcal{O}(1)$. Then enriching the existing graph with additional node and edge attribute information and assigning the values has a complexity of $\mathcal{O}(na_n)$ and $\mathcal{O}(ma_e)$ respectively, where n and m are the number of nodes and edges, and correspondingly a_n, a_e are the number of node and edge attributes. Therefore, the overall time complexity for homogeneous attributed graph is given in Eq. (1)

$$T_{Synt\_H_oAG} = \mathcal{O}(na_n + ma_e) \tag{1}$$

For the $Synt\_H_eAG$ approach, initializing the graph takes $\mathcal{O}(1)$. For node and edge processing, assigning attribute values for nodes and edges, has a time complexity of $\mathcal{O}(t_n a_n)$ and $\mathcal{O}(t_e a_e)$, where t_n and t_e are the number of node and edge types respectively. Therefore, the Eq. (2) represents the overall time complexity for constructing a heterogeneous attributed graph.

$$T_{Synt\_H_eAG} = \mathcal{O}(t_n a_n + t_e a_e) \tag{2}$$

4 Experimental Results

This paper focuses on two types of synthetic attributed graph generation. Due to unavailability and limited access of attributed graphs, we proposed framework to generate synthetic attributed graphs. The following are the three strategies to generate both synthetic homogeneous and heterogeneous attributed graphs.

Strategy 1: Attribute Information Enrichment- Based on the suitability of the application, the available datasets can be enhanced to homogeneous attributed graphs

Strategy 2: Creation of Homogeneous Attributed Graphs- For the desired application one can create and generate dataset with our proposed framework

Strategy 3: Creation of Heterogeneous Attributed Graphs- Creating and generating different types of node and edge attributes information

4.1 Experimental Setup

Datasets generation experiments are conducted on laptop with 12th Gen Intel®Core^{TM} i7-9750H 2.30GHz, 32 GB RAM in Windows OS. We have implemented our proposed approaches in Python programming. We have used the NetworkX API to construct and manage the graph datasets [1].

4.2 Generating Synthetic Dataset

In this study, we have proposed two approaches ($Synt\_H_oAG$, $Synt\_H_eAG$) for synthetic graph data generation. The $Synt\_H_oAG$ approach uses available benchmark datasets and $Synt\_H_eAG$ approach creates the synthetic heterogeneous attributed graph datasets from scratch based on user input configuration file. Experiments are conducted for three strategies explained in this section through the proposed algorithms $Synt\_H_oAG$, $Synt\_H_eAG$ and provided in Table 2, serial number 5 and 6.

4.2.1 Homogeneous Attributed Graphs

We conducted experiments on available benchmark datasets by enriching node and edge attributes shown in Table 3 in columns number 6 and 8, and the graph generation time in seconds (column number 9). While the Lazega and Twitter datasets already have some node attributes, we enhanced these datasets by adding additional node and edge attributes for more analysis and to know the behaviour of proposed algorithms in various fields. Even though node attribute information is available to benchmark datasets European Deezer, Lastfm Asia, and Facebook datasets, which cannot be used due to the preprocessing of information. Therefore, with our proposed approach $Synt\_H_oAG$ the node and edge attribute information is generated from scratch for those datasets (Strategy 1).

Table 2. Experimental Results on properties of graph generation

SNo.	Graph Generator	Graph Type	Attributes Information		Graph Size	Techniques and Tool/Model
			Node Attributes	Edge Attributes		
1	NetworkX [1]	Homogeneous Attributed Graph	✗	✗	Large	Standard algorithms (e.g. Erdős-Rényi etc.) and Tool
2	SYNTH (Synthetic Graph Generation) [6]	Homogeneous Attributed Graph	✓	✓	Medium	Standard algorithms (e.g. Erdős-Rényi, Deep Learning models etc.) and Model *
3	LDBC SNB Data Generator (DataGen) [3]	Heterogeneous Attributed Graph	✓	✓	Large	Hadoop/Spark based and Model *
4	SyntHING (Synthetic Heterogeneous Information Network Generation) [5]	Heterogeneous Attributed Graph (only node attributes)	✓	✗	Large	Clustering, Node embedding and Model *
5	***Synt_H$_o$AG* (Proposed approach)**	**Homogeneous Attributed Graph (Single type of node with both node and edge attributes)**	✓	✓	**Large**	***Synt_H$_o$AG* and Tool**
6	***Synt_H$_e$AG* (Proposed approach)**	**Heterogeneous Attributed Graph (Different type of nodes with both node and edge attributes)**	✓	✓	**Large**	***Synt_H$_e$AG* and Tool**

* indicates available benchmark dataset applied to the model to mimic as real dataset, but there is a lack of simulating like benchmark datasets with all characteristics

For some applications, available benchmark datasets may not be useful and possible to enhance, in that scenario, our proposed framework will support to create synthetic attributed graph dataset by taking user input configuration file with single type of node (Strategy 2).

Table 3. Enriching Homogeneous Datasets with Attributes

SNo.	Datasets	Nodes	Edges	Node Attributes			Edge Attributes	Graph Generation Time (in Seconds)
1	2	3	4	5	6	7	8	9
				Existing	**Inserted**	**Total**	**Inserted**	
1	Lazega	71	575	7	4	11	6	0.03
2	Lastfm_asia	1150	27806	-	11	11	7	0.81
3	Twitter	2511	37154	1	4	5	2	0.15
4	Facebook	4039	88234	-	15	15	6	2.28
5	European Deezer	28281	92752	-	14	14	7	2.98

For example, considering available benchmark Twitter dataset [2] contains 2511 nodes, 37154 edges, and *Tag* is the only available node attribute. In this dataset, all the characteristics of structural and attribute information is not available. But in the real time scenario, some applications requires both node and edge attribute information for analysis. To support real time applications, in this study, we are enriching the available benchmark datasets by adding the additional node and edge attribute information, which is shown in framework (Fig. 2) as *Enriching Attributes* and the values are assigned through *Attribute Value Generator*. In this study, Table 4 is the available benchmark twitter dataset with only one node attribute *tag*, we synthetically populated four node attributes that denotes *Visibility*, *Gender*, *Age*, and *Followers* which are underlined and the two edge attributes that denotes *Disclose*, *Status* which are underlined and assigning attribute values through the component *Attribute Value Generator* in the framework. For instance, node attributes *Age*, and *Followers* attribute values are generated using the measure (Random range), and *Gender* node attribute used a measure (Predefined set) {Male, Female}. The edge attributes were assigned by using the measure (Predefined set) i.e., 'Family', 'Friend', 'Blocked', 'Muted' for *Status* attribute, 'Public', 'Secret' for *Disclose* attribute. Hence, the homogeneous attributed graph is generated and published the output as resultant files which is shown in Fig. 5.

Table 4. Addition of node and edge attributes to Twitter dataset

Node Attributes	Domain Size, Measure	Node Attribute Values
• Tag	• n, real	• $[@sunterryjones, @ebs\_14, ...n]$
• **Visibility**	• 2, predefined set	• $[True, False]$
• **Followers**	• n, random	• $[1...n]$
• **Gender**	• 2, predefined set	• $[20....60]$
• **Age**	• 40, random	• $[Male, Female]$
Edge Attributes {Link_info}	Domain Size, Measure	Edge Attribute Values
• **Status** (Follows)	• 4, predefined set	• $[Family, Friend, Blocked, Muted]$
• **Disclose** (Follows)	• 2, predefined set	• $[Public, Secret]$

4.2.2 Heterogeneous Attributed Graphs

As there are no benchmark heterogeneous attributed graph datasets are available, we are creating and generating these types of graphs from scratch as mentioned in Strategy 3 of Sect. 4. As stated in Sect. 1 heterogeneous graph requires different types of nodes and edges, which are necessary for certain applications such as Research Collaborations [29], Recommendations in E-commerce [18], etc. To handle these types of applications, the proposed technique $Synt\_H_oAG$ is also not sufficient. Hence, the other approach $Synt\_H_eAG$ is proposed to generate heterogeneous attributed graphs.

Table 5. Generating Heterogeneous Datasets with Attributes

Scale	Nodes	Edges	Node Types	Edge Types	Node Attributes	Edge Attributes	Graph Construction Time (in Seconds)
1x	856	33793	4	3	6	3	0.369
2x	1712	270336	3	3	5	3	2.796
3x	2568	912384	3	2	4	3	9.332
4x	3424	2162688	3	2	4	3	22.060
5x	4280	4224000	2	1	2	2	40.921

This approach depends on user input configuration file (sample given in Fig. 4a) to generate synthetic heterogeneous attributed graphs. The process involves three phases (i) Input user configuration file, (ii) Graph generation, and (iii) Resultant files. The algorithmic steps are listed in proposed approaches of Sect. 3. The details of constructing the graph by three phases are given. As part of the first phase, parses the user input configuration file as given in Fig. 4 and obtains the related information of node and edge types, number of nodes and edges, node and edge attributes. As part of second phase, as stated in the framework (Fig. 2), the component *Structure Generator* executes by using MultiDiGraph model and through the component *Attribute Value Generator* in the framework, the values are assigned to node and edge attributes. For instance in Fig. 4a, Person node, the node attributes *Age* is assigned using the measure (Random range) between 20, and 60, for Company node, the attribute *Revenue* is assigned using the measure (Gaussian distribution) and for Technology node the *Name* attribute is assigned using the measure (Predefined set) {Java, Web, HTML}. Similarly, edge attributes are assigned as: the edge attribute *Mode* for 'friendOf' relation, from node Person P1 to Person P2 is assigned using the measure (Predefined set) {school, online}, the edge attribute *level* for 'knows' relation from node Person P1 to node Technology T1 is assigned using the measure (Predefined set) {basic, medium, high}, the edge attribute *since* for 'worksAt' relation from node Person P1 to node Company C1 is assigned using the measure (Predefined set) 2010, 2015, 2020. Further in third phase, the resultant files of synthetic graph is generated for each node and edge type. The Fig. 6 refers to the node Technology which contains the node attribute *Name* and the edge attribute *level* for 'knows' relation is assigned between nodes Person and Technology.

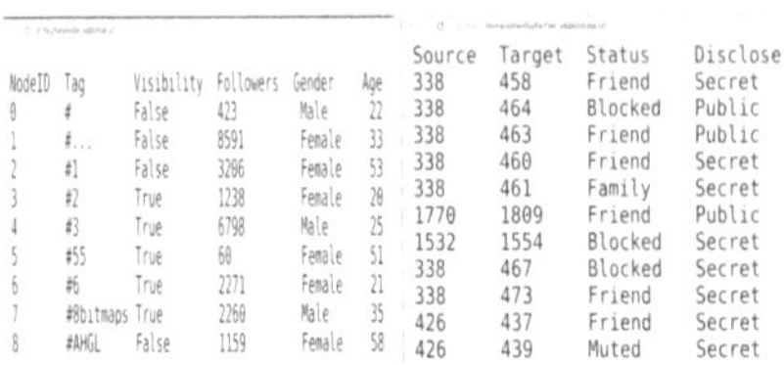

NodeID	Tag	Visibility	Followers	Gender	Age
0	#	False	423	Male	22
1	#...	False	8591	Female	33
2	#1	False	3206	Female	53
3	#2	True	1238	Female	20
4	#3	True	6798	Male	25
5	#55	True	60	Female	51
6	#6	True	2271	Female	21
7	#8bitmaps	True	2260	Male	35
8	#AHGL	False	1159	Female	58

Source	Target	Status	Disclose
338	458	Friend	Secret
338	464	Blocked	Public
338	463	Friend	Public
338	460	Friend	Secret
338	461	Family	Secret
1770	1809	Friend	Public
1532	1554	Blocked	Secret
338	467	Blocked	Secret
338	473	Friend	Secret
426	437	Friend	Secret
426	439	Muted	Secret

Fig. 5. Resultant files of *Synt_H_oAG*

Node_Technology.txt

```
T0 { name:Java }
T1 { name:Java }
T2 { name:HTML }
T3 { name:Web }
T4 { name:HTML }
T5 { name:Java }
T6 { name:Web }
```

Edge_Knows.txt

```
P0 : T15 {  level:medium }
P0 : T16 {  level:basic }
P0 : T9 {  level:basic }
P0 : T8 {  level:medium }
P0 : T2 {  level:high }
P1 : T15 {  level:high }
P1 : T14 {  level:basic }
```

Fig. 6. Resultant files of *Synt_H_eAG*

We conducted experiments to generate the heterogeneous attributed graph datasets by scaling size of the nodes linearly and size of the edges cubically by varying the types of nodes and edges along with their respective attributes and graph construction time as shown in Table 5. The proposed approach *Synt_H_eAG* demonstrate that it can handle any size of large dataset to generate synthetic graph.

We observed that to enrich homogeneous attributed graph and to create synthetic heterogeneous attributed graph our proposed *Synt_H_oAG* and *Synt_H_eAG* approaches are taking considerably less amount of time as shown in Table 3 and Table 5. These generated synthetic homogeneous and heterogeneous attributed graph datasets can be used in the domain of graph analytics specifically to solve the problems of query reachability, routing, link prediction, etc. The generated graphs useful for various applications and given in section below.

5 Applications

Attributed graphs plays an important role in real-world applications. The available benchmark datasets are not suitable for various applications, hence there is a demand to generate synthetic graphs. In this study, the generated synthetic homogeneous and heterogeneous datasets are designed to mimic real-world data. To evaluate the quality of these generated datasets, we applied to two different applications. Further, it can also be used for the part of graph analysis. In this section, the generated homogeneous datasets is used for one of the applications of community detection, and the generated heterogeneous dataset is used to solve the problem of query reachability. Below are the applications demonstrated through generated datasets.

5.1 Community Detection

Homogeneous Attributed Graphs: Using *Synt_H_oAG* approach, for the Political Books dataset [31], Structure Extracter component is used to extract the nodes and edges. Further, we enriched the dataset with four node attributes and applied to community detection algorithm such as SAC (Structure-Attribute Clustering) [27]. The SAC algorithm requires both structural and node attribute information. Hence, the generated homogeneous dataset has such attribute information and found suitable to work with these type of algorithms and carried the experiments by using SAC algorithm.

Fig. 7. Community Detection on Political Books

Figure 7 represents the detected communities for the generated dataset. Each node is represented by a color, indicating the community to which the node belongs and the edges represent the connections between nodes. Figure 7 shows the formation of groups by the SAC algorithm with the modularity of 0.483. The good community structure obtains from the modularity measure ranging from −1 to 1. Hence, the communities formed and shown in Fig. 7 with the modularity value implies $Synt\_H_oAG$ approach is suitable and performing better. Also we observed that the generated homogeneous data is suitable for various kinds of applications.

5.2 Query Reachability

Heterogeneous Attributed Graphs: Using $Synt\_H_eAG$ approach, synthetic heterogeneous attributed graph is generated. The generated graph dataset consists of multiple cypher files, each file representing different node and edge types. To generated cypher files are uploaded to Neo4j desktop browser to establish the nodes and relationships. The attributed graph consists of Course, Instructor, and University are the node types and 'taughtBy', 'offerIn', and 'worksIn' are the edge types. Node attributes *level*, *mode* for Course node, *service*, *domain* for Instructor node, *type*, *rank* for University node are specified along with their measures (Random range, Predefined set). Similarly, edge attributes, *numberoftimes* for 'taughtBy' edge, *certification* for 'offerIn' edge, and *years* for 'worksIn' edge are specified with the measure (Random range, Predefined set). The resultant files of synthetic attributed graph is given to Neo4j to extract both structure and relational attribute information. Also Neo4j [28] uses Cypher Query Language (CQL) which is designed for query processing.

For instance, the query is that students want to "Identify an Instructor who taught a particular Course, C1 (source), at reputed Universities {U2 or U15 or U21} (destination), with the constraints that the Instructor must have taught the Course more than *eight* times and is currently offering it *online*". This query can be translated into the CQL to execute in Neo4j database environment. This Cypher query ensures that the resultant paths follow the specified relationship types with given attribute constraints. Figure 8a is the sample generated heterogeneous attributed graph which is used to execute the given query, further the query translates to the CQL (shown in Fig. 8b) to find the resultant paths (shown in Fig. 8c).

This process demonstrates the effective utilization of the generated dataset within the Neo4J environment, providing valuable insights. It proves that the generated dataset is well-suitable for the query reachability to identify specific relationships and attributes, thereby facilitating the discovery of resultant paths for a given query.

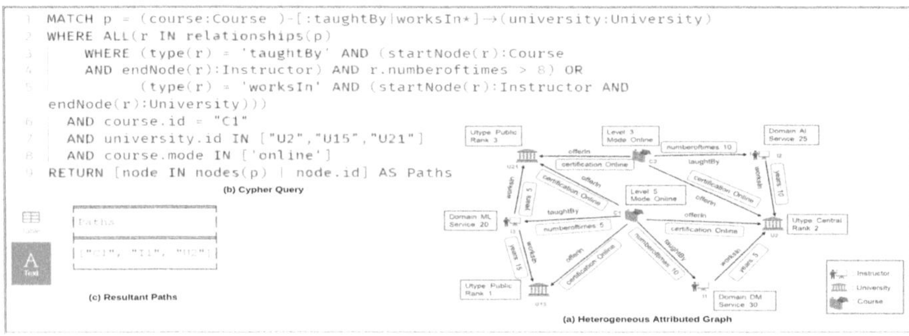

Fig. 8. Demonstration on the application of query reachability

Due to the scarcity of both homogeneous and heterogeneous datasets with specific attributes, the experimental results are proving that our proposed approaches $Synt\_H_oAG$ and $Synt\_H_eAG$ which is shown in Table 2 and listed in S.No. 5 and 6 are having all types of characteristics to generate synthetic attributed graph dataset which are compatible with different algorithms and applications.

6 Conclusions and Future Scope

In this study, we proposed a framework, which includes $Synt\_H_oAG$ approach to enrich the existing graph and $Synt\_H_eAG$ to create and generate synthetic heterogeneous attributed graph datasets in the domain of graph analytics. Experiments are conducted by enriching the datasets with few characteristics to generate homogeneous attributed graphs and this dataset used by community detection algorithm SAC [27]. Based on the User Input Configuration file, our approach created and generated heterogeneous attributed graph datasets and demonstrated for query reachability application. There is a scope to extend these approaches not only by using the tools but also by considering the other models of machine learning algorithms to generate the synthetic graph dataset. Further, the results of our proposed approach can pass to the SYNTH [6] tool to create an arbitrary size graph which follows the characteristics that includes structure with associated node and edge attributes.

References

1. Hagberg, A., Swart, P.J., Schult, D.A.: Exploring network structure, dynamics, and function using NetworkX (2008)
2. He, T., Chan, K.C.C.: Discovering fuzzy structural patterns for graph analytics. IEEE Trans. Fuzzy Syst. **26**(5), 2785–2796 (2018)
3. Erling, O., et al.: The LDBC social network benchmark: interactive workload. In: International Conference on Management of Data (2015)
4. Maekawa, S., Sasaki, Y., Fletcher, G., Onizuka, M.: GenCAT: generating attributed graphs with controlled relationships between classes, attributes, and topology. Inf. Syst. **115**, 102195 (2023)
5. Hong, M.-Y., Huang, Y.-H., Lin, S.-E., Teng, Y.-C., Wang, C.-Y., Lin, C.: SynHING: Synthetic Heterogeneous Information Network Generation for Graph Learning and Explanation (2024)
6. Darabi, S., Bigaj, P., Majchrowski, D., Kasymov, A., Morkisz, P., Fit-Florea, A.: A Framework for Large Scale Synthetic Graph Dataset Generation, arXiv preprint arXiv:2210.01944 (2022)
7. Barabási, A.L., Pósfai, M.: Network Science. Cambridge University Press, Cambridge (2016)
8. Erdős, P., Rényi, A.: On the Evolution of Random Graphs. Publications of the Mathematical Institute of the Hungarian Academy of Sciences (1960)
9. Csardi, G., Nepusz, T.: The igraph software package for complex network research. Int. J. Complex Syst. **1695**(5), 1–9 (2006)
10. Leskovec, J., Sosic, R.: SNAP: a general purpose network anlaysis and graph mining library. ACM Trans. Intell. Syst. Technol. **8**(1), 1–20 (2016)
11. Nettleton, D.F.: A synthetic data generator for online social network graphs. Soc. Netw. Anal. Mining **6**(44), 1–33 (2016)
12. Sun, Y., Han, J., Yan, X., Yu, P., Wu, T.: PathSim: meta pathbased top-k similarity search in heterogeneous information networks. Proc. VLDB Endow. **4**, 99–1003 (2011)
13. Yang, Y., Liu, H., Guan, Z., He, X., Liu, G.: CoHomo: a cluster-attribute correlation aware graph clustering framework. Neurocomputing **412**, 327–338 (2020)
14. Shi, C., Li, Y., Zhang, J., Sun, Y., Philip, S.Y.: A survey of heterogeneous information network analysis. IEEE Trans. Knowl. Data Eng. **29**, 17–37 (2016)
15. Gauss, C.F.: Theoria Motus Corporum Coelestium in Sectionibus Conicis Solem Ambientium (1809)
16. Alnajran, N., Crockett, K., McLean, D., Latham, A.: Cluster analysis of twitter data: a review of algorithms. In: International Conference on Agents and Artificial Intelligence, pp. 239–249 (2017)

17. Curtin, K.M.: Network analysis in geographic information science: review, assessment, and projections. Cartogr. Geogr. Inf. Sci. **34**, 103–111 (2007)
18. Xu, Y.: Using social network analysis as a strategy for e-commerce recommendation. In: 13th Pacific Asia Conference on Information Systems (2009)
19. Shlomi, T., Berkman, O., Ruppin, E.: Regulatory on/off minimization of metabolic flux changes after genetic perturbations. Proc. Natl. Acad. Sci. **102**, 7695–7700 (2005)
20. Yu, D., Kim, M., Xiao, G., Hwang, T.H.: Review of biological network data and its applications. Genomics Inform. **11**, 200–210 (2013)
21. Gosak, M., et al.: Network science of biological systems at different scales: a review. Phys. Life Rev. **24**, 118–135 (2018)
22. Holme, P., Rocha, J.C.: Networks of climate change: connecting causes and consequences. Appl. Netw. Sci. **8**(10), 1–20 (2023)
23. Nasiri, E., Berahmand, K., Rostami, M., Dabiri, M.: A novel link prediction algorithm for protein-protein interaction networks by attributed graph embedding. Comput. Biol. Med. **137**, 104772 (2021)
24. Yung, D., Chang, S.-K.: Fast reachability query computation on big attributed graphs. In: IEEE International Conference on Big Data, pp. 3370–3380 (2016)
25. Bhargavi, B., Rani, K.S., Neog, A.: Finding multidimensional constraint reachable paths for attributed graphs. EAI Endorsed Trans. Scalable Inf. Syst. **10**, e8 (2023)
26. Wang, L.: Heterogeneous data and big data analytics. Autom. Control Inf. Sci. **3**(1), 8–15 (2017)
27. Dang, T.A., Viennet, E.: Community detection based on structural and attribute similarities. In: International Conference on Digital Society (2012)
28. Aleksa, V., Nicki, W., Tareq, A., Dominic, F., Jonas, P.: Neo4j in action. Shelter Island: Manning, vol. 22 (2015)
29. Wang, X., Sukthankar, G.: Link prediction in heterogeneous collaborations networks. In: Social Network Analysis-Community Detection and Evolution, pp. 165–192 (2015)
30. Liu, X., Yu, Y., Guo, C., Sun, Y.: Meta-path-based ranking with pseudo relevance feedback on heterogeneous graph for citation recommendation. In: 23rd ACM International Conference on Conference on Information and Knowledge Management, pp. 121–130 (2014)
31. Krebs, V.: (2004). http://www.orgnet.com/

Big Data Analytics

Dynamic Deployment of Serverless Data Pipeline on Fog Nodes

Sreenivasu Mirampalli, Rajeev Wankar, and Satish Narayana Srirama(✉)

School of Computer and Information Sciences, University of Hyderabad, Telangana, India
satish.srirama@uohyd.ac.in

Abstract. Currently, the software industry's view of application development and deployment is changing frequently. It is common for the developers to expect software modifications/updation suggestions from the clients frequently. It is challenging for the developers to reflect those changes in the production environment in time. Moreover, prominent cloud providers follow their own application deployment mechanism to access their cloud resources. In this scenario, Topology and Orchestration Specification for Cloud Applications (TOSCA) offers a standardized mechanism for modeling the cloud and fog-based applications and deploying them on any type of hardware infrastructure. In this work, we modeled and dynamically deployed a Serverless Data Pipeline (SDP) application on fog devices with the adoption of the TOSCA modeling language. The required SDP configuration setup and relationships are defined in the modeling template. The xOpera orchestrator uses Ansible deployment tools internally to deploy the TOSCA service template on the defined nodes with the specified node relationships.

Keywords: Dynamic Deployment · Serverless Data Pipeline (SDP) · Fog computing · Internet of Things · TOSCA

1 Introduction

In the current technological world, business requirements frequently change, including service scaling, service interoperability, continuous integration, version updation, and micro-service deployments. These operations must be done quickly enough to meet the business requirements in time. In particular, in setting up large-scale complex IT infrastructure or deploying complex distributed system services, the system engineers perform a lot of repetitive work. Dynamic deployment is a concept of deploying/undeploying the software components on the remote host on demand. The repetitive overhead of tasks would be abstracted with the adoption of dynamic deployment concepts, which can automate the deployment process and minimize the configuration efforts [1]. Additionally, dynamic deployments could provide reliability and consistency across the network [17].

A. Dasgupta et al. (Eds.): BDA 2024, LNCS 15526, pp. 115–132, 2025.
https://doi.org/10.1007/978-3-031-81821-9_7

Some of the automation deployment tools like Ansible[1], Puppet[2], Terraform[3] provide the Infrastructure as code (IaC) to configure the cloud applications. However, these tools are not mature enough to simplify the design of complex system configurations [7]. Major cloud providers like Amazon, Google, and Microsoft designed their own CLI tools and frameworks to deploy the applications in their cloud systems. For example the proprietary deployment tools like Azure DevOps[4], Google Cloud Deployment Manager[5], AWS CodePipeline[6] allows deploying the services on their respective targeted platform. Most of the cloud providers follow their own deployment mechanisms, which makes it difficult to span the services across multiple cloud platforms [18].

To avoid the vendor lock-in nature and providing the standardization, a Topology and Orchestration Specification for Cloud Applications (TOSCA) is approved by OASIS. TOSCA [3] is a modeling language for describing the application components, requirements, capabilities, dependencies and relationships in the form of TOSCA YAMl file. With the TOSCA file description, the entire life cycle of cloud services can be controlled and automated. Moreover, TOSCA automates this process across multiple cloud platforms by specifying the node types; TOSCA is platform agnostic [22].

After the evolution of fog computing [10], some of the cloud computing capabilities are brought nearer to the data sources; in addition, the fog computing paradigm addresses some of the cloud challenges, i.e., latency issues, privacy concerns, storage cost, bandwidth utilization [6]. Fog environment may have heterogeneous nodes with different computing capabilities. So, deploying and automating the services in such a diversified environment is challenging. The cloud based TOSCA framework extended to fog nodes and achieved service portability and interoperability in FogDEFT Framework [21].

On the other hand, the usage of sensors has increased nowadays in the IoT domain [19]; most of these IoT applications are in real time and operate on different hardware systems [9]. Centrally processing IoT-generated data is not feasible to meet business requirements. In this scenario, the Data pipeline concept allows the processing of the data in a distributed manner across the network, where the output of one service is an input to the next service in the pipeline operations. The basic idea behind pipelines is to move data between systems while performing operations on it in transit from source to destination. Researchers recently established Serverless Data Pipelines (SDP) [13] across the edge, fog and cloud.

These SDPs carry out the data through multiple fog devices and orchestrate the flow control between the devices. In SDP's pipeline, tasks are triggered based on event generation. These events can be generated in the form of data arrival into the system, application log information, external API calls, IoT device data,

---

[1] https://www.ansible.com/.
[2] https://www.puppet.com/.
[3] https://www.terraform.io/.
[4] https://azure.microsoft.com/en-in/products/devops.
[5] https://cloud.google.com/deployment-manager/docs.
[6] https://aws.amazon.com/codepipeline/.

and user interventions. The built-in characteristics of serverless computing leverage the event-driven application development, which is more suitable in IoT applications development [16].

In addition, server provisioning responsibilities are taken away from the developer, which opens the scope of better focus on building the pipeline and flow orchestration. Moreover, the fine-grained task scalability will automatically scale up/down the bottleneck tasks to avoid the pipeline throughput performance issues and improve resource utilization. Even though the pipeline tasks are independently executable, they depend on other tasks that should finish their execution first in the pipeline. So deploying such inter related SDP tasks must be configured properly to avoid pipeline interruptions.

Moreover, the stateless properties of serverless systems do not maintain any queuing mechanism to store the intermediate data. However, to store these intermediate task results, SDPs rely on either external object storage systems like Amazon S3[7], Google Cloud Storage[8], Azure Blob Storage[9] or message queuing systems like MQTT[10], Apache Kafka[11], Rabit MQ[12]. The relationships between serverless functions and storage systems required proper configuration setup for the data in/out among the systems.

All the above mentioned challenges motivated us to attempt the dynamic deployment of SDP on the edge and fog nodes. The deployment process is more complicated to automate. Specially in heterogeneous resource-diversified environments deploying/undeploying the SDP tasks dynamically is challenging. In this paper, we tried to address all these challenges and dynamically deployed a QoE-based dynamic data path SDP on edge/fog nodes. In this work, the TOSCA standardized framework is used in our deployments due to its functionalities of interoperable component deployments across N-tier, allowing simple YAML file format to specify the service requirements, service portability, component relationships, and automation.

This paper's contributions are listed below.

- Discussed building of QoE based dynamic path SDP on edge and fog nodes. Furthermore, demonstrated how QoE based SDP dynamically reroutes the data packets on to node which provides better experience to the users.
- Dynamically deployed and undeployed image processing SDP application on fog nodes using TOSCA template with the assistance of xOpera.

The current paper is organised as follows. Section 2 discuss the current state of the art in dynamic deployment and building SDP in edge/fog ecosystem. Section 3 elaborates our developed image processing SDP system and it's dynamic path flow control mechanism. Section 4 mentions technical details of

[7] https://aws.amazon.com/s3/.
[8] https://cloud.google.com/storage/?hl=en.
[9] https://azure.microsoft.com/en-in/products/storage/blobs.
[10] https://mqtt.org/.
[11] https://kafka.apache.org/.
[12] https://www.rabbitmq.com/.

TOSCA and Opera orchestrator. Section 5 explains the dynamic deployment of SDP on edge/fog nodes wiht the help of TOSCA service template. Section 6 Finally concludes the paper.

2 Related Work

Tomarchio et al. [22] proposed a TORCH, a TOSCA-based framework for IT customers to select the appropriate cloud providers based on application specifications. The developed framework allows application deployment and orchestration on multi-cloud platforms in the classical or containerized way. With the low implementation cost and slight modification in specification models, the TORCH framework supports any cloud platform to deploy cloud applications.

Yussupov et al. [24] introduced a method for modeling and deploying serverless function orchestrations that is independent of vendors and technologies. The suggested approach for function orchestration and deployments is based on the business process model and notation (BPMN) and TOSCA modeling standards. The developed toolchain allows serverless function orchestration in BPMN form, which generates different function orchestration vendor-specific models for deployment.

Srirama et al. [21] extended the TOSCA framework to deploy the services onto the constraint fog devices. In this work, authors achieved platform independence with the help of docker containers. The primary work focuses on deploying the services on the fly-on to the remote systems and achieving service coordination between the fog nodes. Their considered use case, "climate control service", demonstrates the work in real time.

Challita et al. [4] worked on integrating Open Cloud Computing Interface (OCCI) and TOSCA. OCCI provides standardized cloud infrastructure provisioning through a common API. TOSCA is a standardized template language that defines cloud topology, service deployment, and orchestration. With the integration of these two, the authors studied how to provide complete standardization in managing both cloud infrastructure and application layers.

Wurster et al. [23] tried to address the gap between service deployment state of the art in research and the state of practice in the industry. The specified TOSCA Light toolchain transforms the TOSCA deployment models into different technology-specific deployment models that industry practitioners commonly use.

Adopting cost-effective service providers for a particular serverless pipeline is challenging. Mathew et al. [11] addresses this issue by studying a set of patterns in proprietary modelling languages and identifying the specific orchestration provider.

Paraskevoulakou [15] designed an ML-based serverless data pipeline system on the Apache OpenWhisk serverless framework. Adopting serverless concepts, they abstract the server maintenance responsibilities to focus only on the ML operation workflows. The effectiveness of the pipeline is illustrated through an anti-fraud detection system.

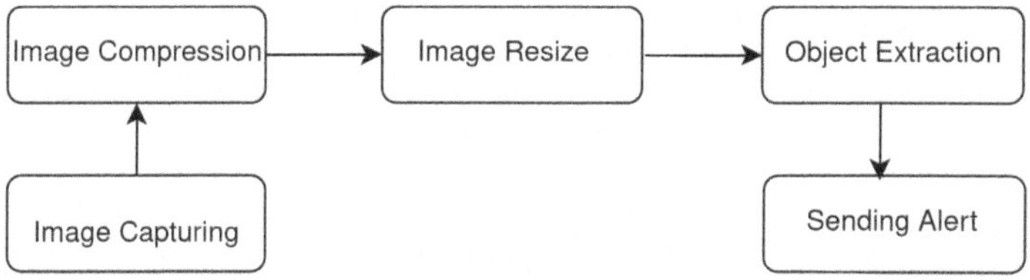

Fig. 1. Pipeline Tasks

In order to model the data pipeline, Dehury et al. [7] extend TOSCA standards, the designed TOSCAdata framework, modelling the pipeline in YAML file format and deploy the services on the cloud. The work focuses on mentioning the pipeline requirements in TOSCA and deploying the components independently on any cloud platform by maintaining effective data flow between the pipeline tasks.

Cordingly et al. [5] investigated the impact of programming languages in designing SDP. The study focuses on data processing implications by varying the data size, function memory, and scaling parameters. The outcome showed there is no single programming language that is perfect for every stage of data processing in the pipeline. Only the hybrid mix languages produced optimized performance in their experiments.

Mirampalli et al. [13] proposed two approaches to designing SDP on edge and fog networks. A real time image processing SDP was implemented on heterogeneous devices using two proposed approaches. The study evaluates the pipeline performance in terms of CPU and memory. Finally, the authors recommended a suitable approach for the considered use case.

To the best of our knowledge, no research study has attempted to use the TOSCA modeling language to dynamically deploy the real-time SDP across the fog nodes.

3 Design of Image Processing Dynamic SDP

In computer vision, object detection is a primary challenging task, which is all about locating the instance objects in an image or video. These applications are broadly used in domains of security systems, self-driving cars, medical imaging, etc. This section specifies the design and implementation details of the constructed real time image processing application. The designed SDP tasks are distributively spread over edge fog and cloud layers. These tasks communicate with each other and perform the operations on the captured real time image data. The designed system consists of five pipeline tasks which are shown in Fig. 1. All these tasks are executed sequentially inside the pipeline.

The deployed SDP system is primarily divided into three parts; these details are described in the following subsection.

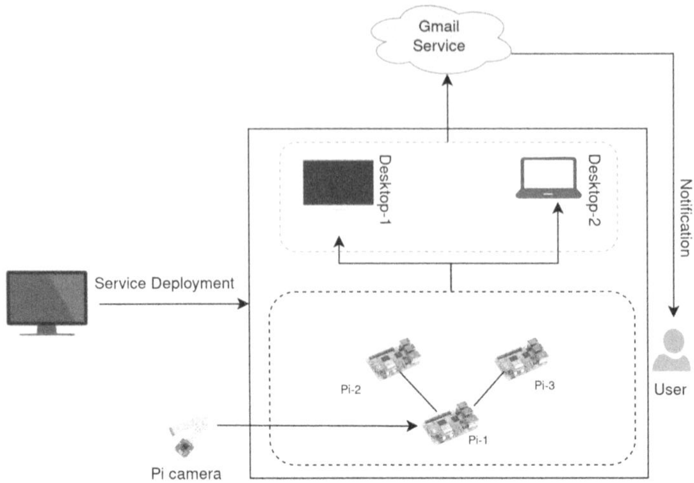

Fig. 2. Hardware Design

3.1 Hardware Design

Establishing a Pi camera setup, Raspberry Pis, and desktop connections are all part of hardware design. Figure 2 depicts the hardware setup of the considered SDP application. One of the Raspberry Pi's camera slots is where the Pi camera is mounted. SSH connections allow Pi devices to communicate and exchange staged data. After obtaining the pipeline staged data from the Raspberry Pi, desktop devices perform the subsequent pipeline operations. Finally, cloud services are triggered by event-based API calls based on the desired conditions.

3.2 Software Design

The prerequisite tools and technology stacks used to construct the image processing SDP are described in this subsection. The staging data processed at each stage of the SDP process must be stored somewhere. Either the subsequent tasks of the pipeline receives this staged data as input, or another system collects it for monitoring. In our experiments, we collected the intermediate pipeline data using the MQTT message publish subscriber model. Due to its features, which include a lightweight protocol, low bandwidth usage, options for configuring QoS levels, and retained message flag configurations [8], MQTT is a highly adopted technology in environments with limited resources, such as Raspberry Pis. We configured Mosquitto MQTT broker on Raspberry Pi-1 to serve as a central broker for several MQTT clients. We have configured the broker to accept anonymous connection requests from other MQTT clients. Every Raspberry Pi device has a MQTT client installed.

A single node local Kubernetes cluster, "minikube", is installed on two desktops. Minikube acts as both a master and worker node. Its primary purpose is application development and testing in a developer environment. To interact with the minikube, developers need an interface tool called Kubectl. kubectl allows users to interact with the Kubernetes API server with a command line and manages the cluster resources.

In addition, we set up Arkade, a command-line interface tool that makes installing additional Kubernetes tools and programs easier. We set up an open-source serverless framework called "OpenFaaS" on top of minikube with the assistance of Arkade. On successful installation of OpenFaaS, two namespaces, openfaas and openfaas-fn, will be created inside minikube. The name space "openfaas" contains the default FaaS architecture components: those are alert-manager, gateway, nats, prometheus, queue-worker. The Gateway acts as the OpenFaaS platform's entry point for HTTP requests. For controlling functions and triggering function executions, it offers a REST API. Requests from outside the cluster must be directed via the Gateway to the appropriate function pods.

The openfaas-fn namespace contains user-defined FaaS functions. OpenFaaS framework allows the design and development of serverless functions. The designed FaaS functions are deployable on any container orchestration platform. The OpenFaaS community designed the FaaS-cli tool and provides scaffold templates for Python, node.js, Go language, Ruby, etc. These built in boilerplate templates speed up the function development process.

Once the FaaS function is created, it is converted into a docker image and pushed to the docker hub repository with the help of FaaS-cli tool. The command `FaaS-cli deploy function-name` pulls the image from the docker repository and deploys it to any device which is already configured with the docker swarm or Kubernetes. The user must authenticate to the FaaS Gateway for successful deployments by providing valid credentials. The user can access the gateway services or run the FaaS functions in the minikube by port forwarding. Finally, we enable less secure apps to set up a Gmail account to accept calls from third-party APIs and send email alerts. The hardware configurations and installed software components are mentioned in Table 1.

Table 1. Hardware and Software configurations

Device Name	Hardware Configuration	Operating System	Installed Technologies
Pi-1 (Raspberry Pi Model4)	Quad core Cortex-A72, 8GB RAM	Raspbian v11	Mosquitto MQTT Broker
Pi-2 (Raspberry Pi Model4)	Quad core Cortex-A72, 4GB RAM	Raspbian v11	MQTT
Pi-3 (Raspberry Pi Model4)	Quad core Cortex-A72, 4GB RAM	Raspbian v11	MQTT
Desktop-1	Intel® Xeon(R) CPU E5-2620 v3 @ 2.40GHz × 24, 64GB RAM	Ubuntu 22.04.3 LTS	Minikube, OpenFaaS, MQTT
Desktop-2	11th Gen Intel® Core™ i5-1135G7 @ 2.40GHz × 8, 8GB RAM	Ubuntu 22.04.3 LTS	Minikube, OpenFaaS, MQTT

3.3 Data Flow Design

Typically, the data flow is fixed at the design phase in standard pipeline systems [20]. In such systems, if the data source and operations of pipelines are the same, then every data packet follows its predecessor route. However, in dynamic shared environments like fog, we can not expect all the resources to be reserved and able to provide sufficient resources as per pipeline demands. In such scenarios, we need a pipeline system which can dynamically adapt real time adjustments and reroute its data path onto the nodes, which can offer better performance.

In the designed SDP, we included the dynamic nature of path rerouting. The system can self-reroute onto the nodes, which can offer better performance. For this, the system considers the task expectation parameters such as access rate, required resource, processing time and dynamically fetches the current status metrics such as round trip time, resource availability, processing speed of available nodes. By using our earlier fuzzy-based task scheduling strategy [12], we assigned the priorities to the task as well as nodes. Furthermore, by applying the Hungarian maximization algorithm, we mapped the task with a suitable node in such a way as to maximize the QoE. In the designed SDP [14], once the node gets selected, the system transfers the staging data onto the selected node to perform the next pipeline execution step.

However, pipeline tasks are diversified, and each task is responsible for performing certain operations. These task resource requirements vary from task to task. The tasks that demand higher resources should be separately scheduled from those that are not resource-hungry. Our hardware setup is formed with three Raspberry pi's and two desktops. In the pipeline system, object extraction is a resource-intensive operation which can only be supported by desktops. The other reaming operations, image capturing, compression, and resizing, can be performed on any of the selected Raspberry Pis.

The developed dynamic SDP is shown in the Fig. 3. SDP pipeline execution starts by capturing the images through the Pi camera, which is connected to one of the Raspberry Pi. These images are published to the MQTT broker. Our scheduling algorithm selects the best node to perform the compression operation on the published image from the Fig. 3 orange lines indicating the flow of an Img1. Img1 is compressed on pi-2 and then published this staging image data onto the broker. At pi-2, again, we select the best node to perform the resize operation; this time, pi-3 is selected and pushed the resize data onto the broker. Finally, one of the nodes is selected for object extraction from the resized image. If the original image contains any human objects, those objects are separated and pushed onto the MQTT. One subscriber scripts listen to that MQTT topic for any state changes. As a state change response, the systems is configured to trigger an API call to send an alert notification to the user. The same process is continuous for the rest of the images. The designed pipeline maintains high availability; even if any devices go down, the system reroutes its path and continues its pipeline operations.

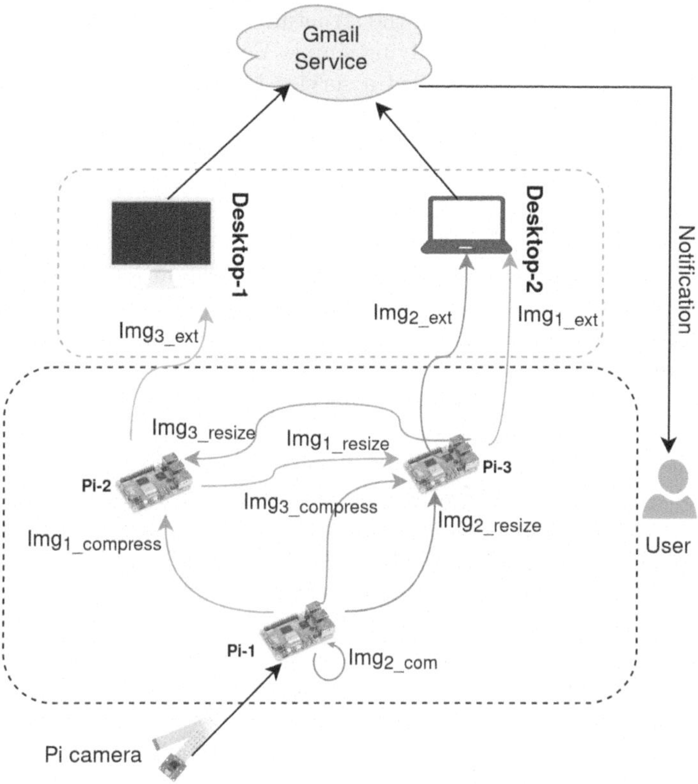

Fig. 3. Dynamic Flow Design

We tested our QoE SDP in real time under a significant workload. Our test results concluded that, instead of using the predefined path SDP, QoE-based SDP showed significant improvement in the system performance. The designed QoE based SDP is almost free from the bottleneck effect with the loss of 2% packet loss, where as predetermined showed 74% packet drop during 100 images streaming from end to end pipeline execution. The detailed evaluation metrics are presented in the paper [14].

4 TOSCA Literature

In 2011, OASIS introduced the Topology and Orchestration Specification for Cloud Applications (TOSCA). TOSCA addresses the issue of application portability among cloud service providers. TOSCA language is available for specifying lifecycle management procedures for the creation or modification of services through orchestration processes, as well as for describing application components and their relationships through a service topology. Topology and orchestration work together to allow for the automation of the entire service lifecycle management process in addition to the deployment.

TOSCA allows the modeling of the structure of cloud applications and their functionalities in the form of specifying simple YAML files in an easily readable format. This file is commonly called a service template. The service template contains multiple sections, primarily two basic sections: node type and relationship. Node type describes the infrastructure components such as compute instances, servers, load balancers, network components, and databases. The node section contains the properties, capabilities and interfaces of every component. Capability represents the ability of what a node can do. Properties explain the characteristics of a node. Finally, the interface defines how a node interacts with other components.

Connections between the nodes are defined in the relationship section. Usually, this describes how nodes interact with other nodes, dependency requirements, hosted on relations, and communication mediums. For example, in the installation of the OpenFaaS framework, the TOSCA orchestration engine first needs to install the FaaS framework requirements, either Docker Swarm or Kubernetes. After successful installation and configuration setup then only the FaaS framework components can be deployed inside the Kubernetes. Here, the Hosted on relationship is formed between the OpenFaaS and Kubernetes.

In addition to the basic types (node type and relationship) provided in the TOSCA specifications, it is anticipated that a TOSCA community will define many more types for use in service templates in the future. Moreover, template authors can often start writing service templates that use pre-existing types without defining any types. There are several key components associated with the basic normative types to organize the service template. Those are:

- **Type**: Type specifies the type of a node template or relationships template. If it is a node template, it describes the capabilities of a node and its characteristics. In the relationship template, it describes the interactions among the nodes.
- **Properties**: Defines the TOSCA instance characteristics and its parameter configurations.
- **Interfaces**: Defines the custom actions performed on the template instance. Such actions could be deployments, create, start, stop, or delete.
- **Artifact**: These are the necessary binaries or files to deploy or configure the nodes.
- **Requirements**: Specifies the dependencies among the nodes. For example, a serverless frameworks depends on a container orchestration tool for the infrastructure provisioning.

A TOSCA standard is referenced in the service template. The service template must be parsed after it has been created. To do this, the "TOSCA processor" serves as an engine for parsing the service template. The roles of the TOSCA processor are translation, rendering, and validation. The service template must be deployed along with its component parts on the fog nodes after it has successfully parsed. For this, An open-source orchestration tool for deploying TOSCA templates called xOpera[13] is used, and it is developed in Python. Opera can work with a wide range of operating systems. Additionally, it is lightweight, and to implement the TOSCA standards, it internally uses Ansible automation tools. The TOSCA service template specifies the precise order in which the Ansible playbook execution occurs during deployment. XOpera is responsible for following the topological order when deploying or undeploying components from remote hosts.

In TOSCA, to scale the services or to distribute the load among the fog devices, the services need to be modelled accordingly in the TOSCA template. By setting up the scalability policies (min_instances: number, max_instances: number) we can scale the workloads dynamically in the fog network based on resource usage (CPU, memory) or incoming traffic. With the help of node type "tosca.nodes.LoadBalancer" it is used to route the network traffic inside the fog network. However, TOSCA does not directly inherit the load distribution in the network; it uses the orchestration frameworks to do it. For example, OpenStack, Kubernetes, FogFlow, and OpenFog are the actual carriers that do load distribution based on the TOSCA modeling. To be precise, TOSCA offers a blueprint for service orchestration and it is the orchestration framework responsibility to distribute the load based on the TOSCA template.

5 Dynamic Deployment of SDP on FOG Nodes

The SDP application for image processing is dynamically deployed on the fog nodes, as described in this section. The TOSCA Service Template from Listing 1.1 was successfully deployed on the hardware configuration shown in Table 1 with the assistance of the xOpera orchestrator running remotely on a workstation. Our deployed TOSCA service template is displayed in Listing 1.1. The TOSCA template includes the relationship and node types as well as the stack of technologies required to support the pipeline deployment. The deployment and undeployment of nodes are organized in a topological sorting of the application blueprint to ensure a proper execution sequence.

Regarding the SDP framework, five generic node types are needed to specify the behaviour and topology of our developed image processing applications. Table 2 provide specific information about the node types that are required for the SDP. These customized types enable the deployment of fog-based applications and the establishment of appropriate relationships within the SDP.

Following the TOSCA service template deployment, the installed services are described in Fig. 4. The Raspberry Pis are represented by the orange boxes at the bottom. The MQTT publisher and subscriber components hosted on each of the three Raspberry Pis are indicated by the pink boxes. The connections between the nodes and services are represented by the edges. The publisher and subscriber services in this case rely on the deployment of MQTT brokers. The MQTT Broker is hosted on Pi-1. Publisher clients can post data to topics and subscribers can receive the data by subscribing to the corresponding topic. The

[13] https://xlab-si.github.io/xopera-docs/02-cli.html.

Table 2. List of Custom Node Types

Nodes Types	Description
`Pi-1`	Connected with Raspberry Pi-camera. MQTT Broker acts as a central hub to store the intermediate pipeline data among Pi devices. Execute QoE node selection algorithm to find the suitable node to perform the Pipeline operations. Image Capturing, Compression, Resizing operations performed on it
`Pi-2`	Execute the QoE node selection algorithm to find the suitable node to perform the Pipeline operations. Retrieves the intermediate data from the MQTT Broker. Performs Compression and Resize Operations
`Pi-3`	Execute the QoE node selection algorithm to find the suitable node to perform the Pipeline operations. Retrieves the intermediate data from the MQTT Broker. Performs Compression and Resize Operations
`Desktop-1`	Accepts the resized image data from the Raspberry Pis. Setup the minikube cluster on the host node. Deploy the OpenFaaS components inside the minikube. Pull the FaaS function from the docker repository and deploy it on to the minikube. Send an API call to trigger the FaaS functions (Object Extraction, sendAlert) deployed on the OpenFaaS
`Desktop-2`	Accepts the resized image data from the Raspberry Pis. Setup the minikube cluster on the host node. Deploy the OpenFaaS components inside the minikube. Pull the FaaS function from the docker repository and deploy it on to the minikube. Send an API call to trigger the FaaS functions (Object Extraction, sendAlert) deployed on the OpenFaaS

required necessary pipeline code is deployed and configured on the Raspberry Pis. By running the scripts mentioned in Listing 1.1 the system start SDP execution.

The desktop nodes are represented by the two yellow boxes in the center. Single node Kubernetes cluster minikube is hosted on these two desktops. The open source OpenFaaS components installed inside the minikube. MQTT Broker is hosted on each desktops. These Brokers allows the connections from all other MQTT publish&subscribers shown in Fig. 4 with the edges labeled by depends on.

The deployment/undeployment topological order of installing and uninstalling components is mentioned in Fig. 5. The fog services uninstalled order is determined by their dependencies, as specified in the Topology Template. Our experiments showed that SDP deployment took around 1200 s, and undeployment took approximately 600 s. The internet speed and device resource capacities affect the deployment and undeployment times.

```yaml
---
tosca_definitions_version: tosca_simple_yaml_1_3

description: Service Template for Human Detection Services

metadata:
  template_name: "Service template"
  template_author: "srinivas & anant"
  template_version: "1.0"

imports:
  - nodetypes/desktop/desktop-service.yaml
  - nodetypes/rasp-pi/pie-service.yaml

topology_template:
  inputs:
    node_1:
      type: string
      description: Input IP address of node_1

    node_2:
      type: string
      description: Input IP address of node_2

    node_3:
      type: string
      description: Input IP address of node_3

    node_4:
      type: string
      description: Input IP address of node_4

    node_5:
      type: string
      description: Input IP address of node_5

  node_templates:
    rasp-pie-1:
      type: tosca.nodes.Compute
      attributes:
        private_address: { get_input: node_1 }
        public_address: { get_input: node_1 }

    rasp-pie-2:
      type: tosca.nodes.Compute
      attributes:
        private_address: { get_input: node_2 }
        public_address: { get_input: node_2 }

    rasp-pie-3:
      type: tosca.nodes.Compute
      attributes:
        private_address: { get_input: node_3 }
        public_address: { get_input: node_3 }
```

```
    desktop-1:
      type: tosca.nodes.Compute
      attributes:
        private_address: { get_input: node_4 }
        public_address: { get_input: node_4 }

    desktop-2:
      type: tosca.nodes.Compute
      attributes:
        private_address: { get_input: node_5 }
        public_address: { get_input: node_5 }
    pie-service-3:
      type: pie.service
      properties:
        name: pi-2
        script_url: /home/srinivas/srinivas/pi-1/scripts/
        apt_packages:
          - openjdk-8-jdk
          - mosquitto
          - mosquitto-clients


        pip_packages:
          - picamera
          - paho-mqtt==1.6.1
          - paramiko

        run_scripts:
              - "read.py"
              - "read2.py"
              - "listenforresize.py"
              - "decision.py"
      requirements:
        - host: rasp-pie-2

pie-service-2:
  type: pie.service
  properties:
    name: pi-3
    script_url: /home/srinivas/srinivas/pi-1/scripts/
    apt_packages:
      - openjdk-8-jdk
      - mosquitto
      - mosquitto-clients
    pip_packages:
      - picamera
      - paho-mqtt==1.6.1
      - paramiko
    run_scripts:
      - "read.py"
      - "read2.py"
      - "listenforresize.py"
      - "decision.py"
  requirements:
    - host: rasp-pie-3
```

```
desktop-service-1:
    type: desktop.service
    requirements:
      - host: desktop-1

    properties:
      apt_packages:
        - mosquitto
      pip_packages:
        - paho-mqtt==1.6.1
        - requests
      script_url: /home/srinivas/srinivas/systems/
      run_scripts:
      - "filereceiver_humanobject.py"
      - "filesreceiver.py"

desktop-service-2:
    type: desktop.service
    requirements:
      - host: desktop-2
    properties:
      apt_packages:
        - mosquitto
      pip_packages:
        - paho-mqtt==1.6.1
        - requests
      script_url: /home/srinivas/srinivas/systems/
      run_scripts:
      - "filereceiver_humanobject.py"
      - "filesreceiver.py"

pie-service-1:
  type: pie.service
  properties:
    name: pi-1
    script_url: /home/srinivas/srinivas/pi-1/scripts/
    apt_packages:
      - openjdk-8-jdk
      - mosquitto
      - mosquitto-clients
    pip_packages:
      - picamera
      - paho-mqtt==1.6.1
      - paramiko
    run_scripts:
      - "read.py"
      - "read2.py"
      - "listenforresize.py"
      - "decision.py"
      - "pipe2_captureimage.py"
  requirements:
    - host: rasp-pie-1
```

Listing 1.1. TOSCA Service Template for Image Processing SDP

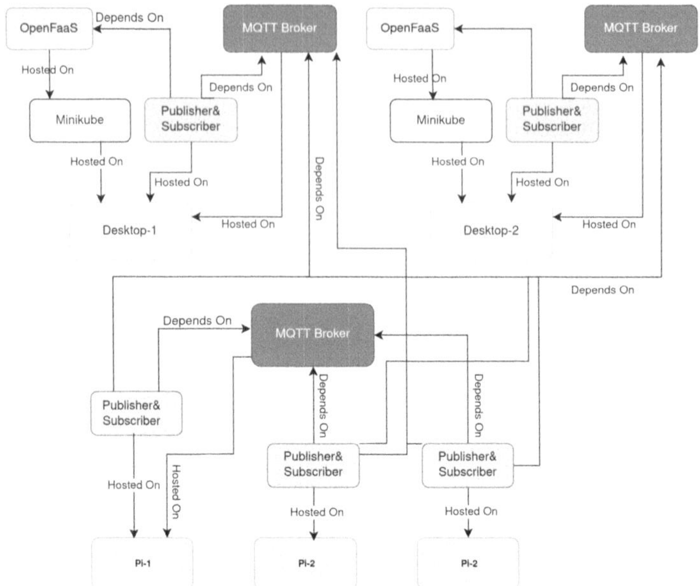

Fig. 4. Service Design

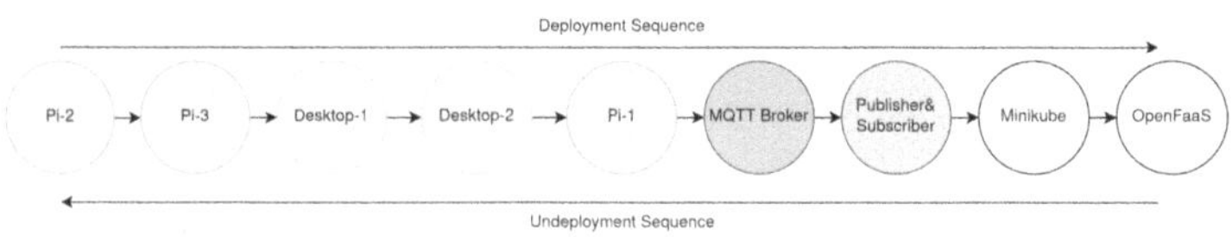

Fig. 5. Service Deployment Sequence

6 Conclusion

In this work, we tried to dynamically deploy and undeploy the SDP on constraint fog resources. The proposed approach uses the TOSCA standards for structuring and deploying the SDP and its lifecycle management. The portability and interoperability features of TOSCA standards allow us to deploy the SDP on any hardware infrastructure, which avoids vendor lock-in. xOpera orchestrator assisted in the dynamic deployment of the TOSCA templates on the remote hosts through SSH connections. In real time event driven IoT based SDP applications, the tasks need to be communicated to each other. Such applications follow some specific order of task executions and may interdependent on each other. TOSCA relationship types allow us to define the appropriate SDP task dependency relationships in the TOSCA service template. Finally, from our work, we concluded that with the TOSCA modelling language, we modelled and dynamically deployed a complex SDP application across the edge, fog and cloud. As future work, we will be adapting the final TOSCA node types and relationship types developed during the design of FogDEFT framework, for defining the service templates used for the dynamic SDP [2].

Acknowledgments. This work is supported by SERB, India, through grant CRG/2021/003888. We also thank financial support to UoH-IoE by MHRD, India (F11/9/2019-U3(A)).

References

1. Aksakalli, I.K., Çelik, T., Can, A.B., Tekinerdoğan, B.: Deployment and communication patterns in microservice architectures: a systematic literature review. J. Syst. Softw. **180**, 111014 (2021)
2. Basak, S., Srirama, S.N.: Fog computing out of the box: dynamic deployment of fog service containers with TOSCA. Int. J. Network Manage **34**, e2246 (2024)
3. Caballer, M., Moltó, G., Calatrava, A., Blanquer, I.: Infrastructure manager: a TOSCA-based orchestrator for the computing continuum. J. Grid Comput. **21**(3), 51 (2023)
4. Challita, S., Korte, F., Erbel, J., Zalila, F., Grabowski, J., Merle, P.: Model-based cloud resource management with TOSCA and OCCI. Softw. Syst. Model. 1–23 (2021)
5. Cordingly, R., et al.: Implications of programming language selection for serverless data processing pipelines. In: 2020 IEEE International Conference on Dependable, Autonomic and Secure Computing, International Conference on Pervasive Intelligence and Computing, International Conference on Cloud and Big Data Computing, International Conference on Cyber Science and Technology Congress (DASC/PiCom/CBDCom/CyberSciTech), pp. 704–711 (2020). https://doi.org/10.1109/DASC-PICom-CBDCom-CyberSciTech49142.2020.00120
6. Das, R., Inuwa, M.M.: A review on fog computing: issues, characteristics, challenges, and potential applications. Telemat. Inform. Rep. **10**, 100049 (2023)
7. Dehury, C.K., Jakovits, P., Srirama, S.N., Giotis, G., Garg, G.: TOSCAdata: modeling data pipeline applications in TOSCA. J. Syst. Softw. **186**, 111164 (2022)
8. Gamess, E., Ford, T.N., Trifas, M.: Performance evaluation of a widely used implementation of the MQTT protocol with large payloads in normal operation and under a dos attack. In: Proceedings of the 2021 ACM Southeast Conference, ACM SE 2021, pp. 154–162. Association for Computing Machinery, New York (2021). https://doi.org/10.1145/3409334.3452067
9. Jamshed, M.A., Ali, K., Abbasi, Q.H., Imran, M.A., Ur-Rehman, M.: Challenges, applications, and future of wireless sensors in internet of things: a review. IEEE Sens. J. **22**(6), 5482–5494 (2022). https://doi.org/10.1109/JSEN.2022.3148128
10. Mahmud, R., Kotagiri, R., Buyya, R.: Fog computing: a taxonomy, survey and future directions. In: Internet of Everything: Algorithms, Methodologies, Technologies and Perspectives, pp. 103–130 (2018)
11. Mathew, A., Andrikopoulos, V., Blaauw, F.J., Karastoyanova, D.: Pattern-based serverless data processing pipelines for function-as-a-service orchestration systems. Futur. Gener. Comput. Syst. **154**, 87–100 (2024)
12. Mirampalli, S., Srirama, S.N., Wankar, R., Chillarige, R.R.: Hierarchical fuzzy-based quality of experience (QoE)-aware application placement in fog nodes. Softw. Pract. Exp. **53**(2), 263–282 (2023)

13. Mirampalli, S., Wankar, R., Srirama, S.N.: Evaluating NIFI and MQTT based serverless data pipelines in fog computing environments. Futur. Gener. Comput. Syst. **150**, 341–353 (2024)
14. Mirampalli, S., Wankar, R., Srirama, S.N.: Quality of experience based dynamic path serverless data pipelines in edge/fog computing. Commun. Softw. Pract. Exp. (2024, under review)
15. Paraskevoulakou, E., Kyriazis, D.: Leveraging the serverless paradigm for realizing machine learning pipelines across the edge-cloud continuum. In: 2021 24th Conference on Innovation in Clouds, Internet and Networks and Workshops (ICIN), pp. 110–117 (2021). https://doi.org/10.1109/ICIN51074.2021.9385525
16. Pogiatzis, A., Samakovitis, G.: An event-driven serverless ETL pipeline on AWS. Appl. Sci. **11**(1), 191 (2020)
17. Santos, Á., Bernardino, J., Correia, N.: Automated application deployment on multi-access edge computing: a survey. IEEE Access **11**, 89393–89408 (2023). https://doi.org/10.1109/ACCESS.2023.3307023
18. Shukla, P.R., Patil, V.M.: A comprehensive review of frameworks for achieving interoperability in multi-cloud environments. In: 2023 Second International Conference on Informatics (ICI), pp. 1–6 (2023). https://doi.org/10.1109/ICI60088.2023.10421703
19. Srirama, S.N.: A decade of research in fog computing: relevance, challenges, and future directions. Softw. Pract. Exp. **54**(1), 3–23 (2024)
20. Srirama, S.N.: Distributed edge analytics in edge-fog-cloud continuum. Internet Technol. Lett. e562 (2024)
21. Srirama, S.N., Basak, S.: Fog computing out of the box with fogdeft framework: a case study. In: 2022 IEEE 15th International Conference on Cloud Computing (CLOUD), pp. 342–350 (2022). https://doi.org/10.1109/CLOUD55607.2022.00057
22. Tomarchio, O., Calcaterra, D., Di Modica, G., Mazzaglia, P.: Torch: a TOSCA-based orchestrator of multi-cloud containerised applications. J. Grid Comput. **19**(1), 5 (2021)
23. Wurster, M., Breitenbücher, U., Harzenetter, L., Leymann, F., Soldani, J.: TOSCA lightning: an integrated toolchain for transforming TOSCA light into production-ready deployment technologies. In: Herbaut, N., La Rosa, M. (eds.) CAiSE 2020. LNBIP, vol. 386, pp. 138–146. Springer, Cham (2020). https://doi.org/10.1007/978-3-030-58135-0_12
24. Yussupov, V., Soldani, J., Breitenbücher, U., Leymann, F.: Standards-based modeling and deployment of serverless function orchestrations using BPMN and TOSCA. Softw. Pract. Exp. **52**(6), 1454–1495 (2022)

Matrix Factorization of Large Scale Data Using Block Based Approach and GPU Acceleration

Prasad Bhavana(✉), Vineet Padmanabhan, and Wilson Naik Bhukya

Artificial Intelligence Lab, School of Computer and Information Sciences, University of Hyderabad, Hyderabad, India
{17mcpc14,vineetnair,rathore}@uohyd.ac.in

Abstract. Matrix Factorization (MF) on large scale data takes substantial time on a Central Processing Unit (CPU). While Graphics Processing Unit (GPU)s could expedite the computation of MF, the available memory on a GPU is finite. Leveraging GPUs require alternative techniques that allow not only parallelism but also address memory limitations. Synchronization of data between CPU and GPUs is one challenge that arises while leveraging GPUs for MF. Similalry, isolation of data related to a specific computational unit so as to avoid `race` conditions while ensuring sharing of data between computational units is yet another challenge. Currently available matrix factorization models only take advantage of the GPU(s) for independent vector operations which require multiple back and forth data transfers between computational units thereby leading to only fractional utilization of GPUs. The existing works are also limited to approximate computation due to `race` conditions while updating the matrices. We propose a novel approach to matrix factorization using a block based technique that leverages GPUs. The proposed method addresses factorization of large scale data by identifying independent blocks, each of which are factorized in parallel using multiple computational units. The approach can be extended to one or more GPUs and even to distributed systems. The RMSE results of the block based approach are within acceptable delta in comparison to the results of CPU based variant and multi-threaded CPU variant of similar Stochastic Gradient Descent (SGD) kernel implementation. The advantage, of the block based variant, in-terms of speed is significant in comparison to other variants.

1 Introduction

Recommender systems are a critical component of many modern online platforms, playing a key role in enhancing user experience by providing personalized suggestions. The core objective of these systems is to predict the preferences of users for various items based on past interactions, thereby delivering relevant and engaging content. Over the past decades, significant advancements in machine learning (ML) have driven the evolution of recommender systems,

A. Dasgupta et al. (Eds.): BDA 2024, LNCS 15526, pp. 133–147, 2025.
https://doi.org/10.1007/978-3-031-81821-9_8

introducing sophisticated techniques to better model user-item relationships. Traditionally, recommendation techniques are categorized into three main types: content-based, collaborative filtering, and hybrid approaches. Content-based recommendation relies on the features of items and user profiles to make suggestions. For instance, in movie recommendation, this approach uses the genre, director, and cast information to recommend movies similar to those a user has liked previously. Collaborative filtering, on the other hand, leverages the preferences of similar users to recommend items. This method can be divided into user-based and item-based collaborative filtering. Hybrid recommendation systems combine both content-based and collaborative filtering techniques to mitigate their individual limitations and enhance recommendation accuracy.

While improving the efficiency of recommendation algorithms is one direction of research, the scalability and performance of various applications, especially in recommender systems is also an important direction to focus. One of the core challenges in these systems is the computation-intensive nature of matrix factorization techniques, such as Singular Value Decomposition (SVD) [2] and Non-Negative Matrix Factorization (NNMF) [3]. These techniques are pivotal for decomposing large user-item interaction matrices into latent factors that capture the underlying relationships between users and items. However, the computational demands of these algorithms can be prohibitive, particularly when dealing with large-scale datasets. Recent advancements in parallel computing, particularly through the utilization of Graphics Processing Units (GPUs) [4], have revolutionized the way we approach these computational challenges. GPUs, originally designed for rendering graphics, are highly efficient at performing parallel operations, making them ideal for the intensive computations required in machine learning and scientific computing. By leveraging the parallel processing capabilities of GPUs, researchers and practitioners have significantly accelerated the training and inference phases of matrix factorization algorithms, reducing computational time from hours or days to mere minutes or seconds.

The shift towards GPU acceleration in matrix factorization is underscored by several key benefits. Firstly, GPUs offer a massive number of cores that can handle thousands of threads concurrently, enabling them to perform parallel computations at unprecedented speeds. This parallelism is particularly advantageous for matrix factorization, where operations on large matrices can be decomposed into smaller, independent tasks that can be executed simultaneously. Secondly, GPUs are equipped with high-bandwidth memory, allowing for the rapid transfer of data between the CPU and GPU, thereby minimizing bottlenecks and enhancing computational throughput. In addition to hardware advancements, software frameworks and libraries have evolved to harness the power of GPUs effectively. Libraries such as CUDA[1] (Compute Unified Device Architecture) [4] and OpenCL[2] (Open Computing Language) provide the necessary tools and abstractions to develop and optimize GPU-accelerated algorithms. These frameworks simplify the process of writing parallel code, enabling researchers to focus

[1] https://developer.nvidia.com/cuda-zone.
[2] https://developer.nvidia.com/opencl.

on algorithm development rather than low-level hardware management. The advantages of GPU acceleration extend beyond computational speed. GPUs also enhance the scalability of matrix factorization algorithms, allowing them to handle larger datasets. This scalability is crucial for modern recommender systems, which must process vast amounts of data from millions of users and items. Moreover, GPUs enable the deployment of more sophisticated models, such as deep learning-based collaborative filtering, which require substantial computational resources. By leveraging GPUs, researchers can explore new algorithmic innovations and develop efficient recommender systems.

In this paper, by leveraging on our prior work [1] related to block based approach to matrix factorization and adapting the algorithm for GPU we propose a new technique BGMF (Block based GPU matrix factorization) that details an alternative method that is effective in terms of memory management and provides a way to parallelism in computation.

2 Related Work

Beyond traditional matrix factorization techniques, recent research has explored GPU based approaches for matrix factorization. GPU accelerated matrix factorization based on SGD is proposed in [8] and [9]. However, both of these are approximate techniques. In [9], an optimization based approach for memory management within GPU is proposed. While this approach is beneficial for leveraging GPU, our approach avoids the need for cumbersome memory management altogether. GPU accelerated Non-Negative Matrix Factorization (NNMF) for CUDA (Compute Unified Device Architecture) capable hardware has been proposed in [4]. Similarly [11] proposed NNMF with GPU acceleration for text mining purposes. In [10] a unified framework for achieving MF on multiple CPUs as well as multiple GPUs is proposed.

From the literature it can be seen that various matrix factorization based approaches have been proposed which includes parallel as well distributed frameworks for scaling up the factorization process. In [14] and in [17], techniques based on divide and conquer have been proposed, whereas in [15,20,22], distributed methods are outlined. A parallel framework for matrix factorisation based on Stochastic Gradient Descent is outlined in [18,19,24]. Another parallel framework based approach to matrix factorization has been proposed in [21] and a large scale matrix factorization approach is proposed in [16]. It is needless to say that all these techniques need further adaptations for GPU acceleration. BGMF, as proposed in this work, provides a framework that can be used by other matrix factorization methods for scaling up as well as providing parallelism and distributed computation along with GPU acceleration. BGMF relies on kernel implementation that takes partial data elements to achieve matrix factorisation. Even though it is currently implemented using Stochastic Gradient Descent (SGD), the approach can be extended to approximate techniques based on Alternating Least Squares (ALS) approach that uses partial data.

3 Our Approach

Let $X \in \mathbb{R}^{\{n \times m\}}$ be a matrix with n users and m observations per user. Typically, in a regular MF approach, the data matrix X can be approximated as $X \approx UV^T$; where user latent factor matrix $U \in \mathbb{R}^{\{n \times k\}}$, for some given k dimensional latent feature vectors, is derived from the given observations. Similarly, the item latent factor matrix $V \in \mathbb{R}^{\{m \times k\}}$, for k latent features, is derived from given observations. The optimization problem of MF can be represented as:

$$\min_{U,V} \sum_{x_{ij}} ((x_{ij} - u_i^T v_j)^2 + \beta(||u_i||^2 + ||v_j||^2)) \tag{1}$$

where β is the regularization parameter. The update equation for i^{th} row of U for SGD convergence is:

$$u'_{ik} = u_{ik} + \alpha\big(2e_{ij}v_{kj} - \beta u_{ik}\big) \tag{2}$$

where α is the learning parameter. Similarly the update equation for j^{th} column of V for SGD convergence is given as:

$$v'_{kj} = v_{kj} + \alpha\big(2e_{ij}u_{ik} - \beta v_{jk}\big) \tag{3}$$

The block matrix representation as show in Eq. (4) for the data matrix X is based on concatenation of blocks/sub-matrices of equal dimension as shown below:

$$X = \begin{bmatrix} \underline{X}_{11} & \underline{X}_{12} & \cdots & \underline{X}_{1j} & \cdots & \underline{X}_{1J} \\ \vdots & \vdots & \cdots & \vdots & \ddots & \vdots \\ \underline{X}_{i1} & \underline{X}_{i2} & \cdots & \underline{X}_{ij} & \cdots & \underline{X}_{iJ} \\ \vdots & \vdots & \cdots & \vdots & \ddots & \vdots \\ \underline{X}_{I1} & \underline{X}_{I2} & \cdots & \underline{X}_{Ij} & \cdots & \underline{X}_{IJ} \end{bmatrix} \tag{4}$$

Since the data is assumed to be independent and identically distributed (i.i.d) for each user and respective item, Matrix Factorization does not constrain the sequence of processing of ratings to derive latent features. The technique results in optimized results when each of the data elements are processed once per iteration. Block based matrix factorization approach decomposes any given matrix into a block matrix and each sub-matrix is independently factorized, while the resultant latent feature sub-matrices are reused for factorization of other relevant sub-matrices. The overall process is repeated till convergence. The GPU acceleration is achieved by dedicating different independent blocks to different GPU cores and each block is factorized in parallel. Approximate, but comparable to global optima, results can also be achieved with factorization of each block for more than one iteration as detailed in Sect. 4.2.

The objective of block based MF is to find $\underline{U_i}, \underline{V_j}$ latent blocks for $\underline{X_{ij}}$ data block such that $\underline{X_{ij}} \approx \underline{U_i}.\underline{V_j}^T$. Let $\underline{E_{ij}}$ represent the deviation of estimate $\underline{X_{ij}}'$

from actual $(\underline{X_{ij}})$. Hence sum of squared deviations $(\mathcal{E})$, for a block, can be represented as:

$$\mathcal{E} = ||\underline{X_{ij}} - \underline{X_{ij}}'||^2 = ||\underline{X_{ij}} - \underline{U_i}\underline{V_j}^T||^2 \tag{5}$$

Our objective now is to find $\underline{U_i}, \underline{V_j}$ such that the sum of squared deviations is minimal. The optimum value for g^{th} latent feature of $\underline{U_i}$ block can be arrived by minimizing Eq. (5) w.r.t $\underline{U_i}$ as:

$$\frac{\partial}{\partial \underline{U_i}_{*g}} \mathcal{E} = -2(\underline{X_{ij}} - \underline{X_{ij}}').\underline{V_j}_{*g} \tag{6}$$

Similarly, optimal value for for g^{th} latent feature of $\underline{V_j}$ can be arrived by minimizing Eq. (5) w.r.t $\underline{V_j}$ as:

$$\frac{\partial}{\partial \underline{V_j}_{*g}} \mathcal{E} = -2(\underline{X_{ij}} - \underline{X_{ij}}')^T \underline{U_i}_{*g} \tag{7}$$

Using the block level gradients for a latent feature, the update equation for g^{th} column of $\underline{U_i}$ can be arrived as:

$$\underline{U_i}'_{*g} = \underline{U_i}_{*g} + \alpha \frac{\partial}{\partial \underline{U_i}_{*g}} \mathcal{E} = \underline{U_i}_{*g} + 2\alpha \underline{E_{ij}} \underline{V_j}_{*g} \tag{8}$$

Similarly, the update equation for g^{th} column of $\underline{V_j}$:

$$\underline{V_j}'_{*g} = \underline{V_j}_{*g} + \alpha \frac{\partial}{\partial \underline{V_j}_{*g}} \mathcal{E} = \underline{V_j}_{*g} + 2\alpha \underline{E_{ij}}^T \underline{U_i}_{*g} \tag{9}$$

Factoring in regularization term into the optimization Eq. (5) gives us:

$$\mathcal{E} = ||\underline{X_{ij}} - \underline{X_{ij}}'||^2 + \frac{\beta}{2}(\sum ||\underline{U_i}||^2 + ||\underline{V_j}||^2) \tag{10}$$

Now, the update Eq. (8) changes accordingly with minimization of Eq. (10) w.r.t to $\underline{U_i}$ to include regularization term:

$$\underline{U_i}'_{*g} = \underline{U_i}_{*g} + \alpha(2\underline{E_{ij}\underline{V_j}_{*g}} - \beta \underline{U_i}_{*g}) \tag{11}$$

Similarly, the Eq. (9) for update of gth column of $\underline{V_j}$ changes accordingly to include regularization term:

$$\underline{V_i}'_{*g} = \underline{V_j}_{*g} + \alpha(2\underline{E_{ij}}^T \underline{U_i}_{*g} - \beta \underline{V_j}_{*g}) \tag{12}$$

4 Parallelism

U_i, V_j matrices can be derived simultaneously for multiple blocks by identifying the data blocks whose latent factors do not depend on each other and dedicating a GPU core (Computation unit) for each such block. Figure 1 shows one such example with a 6×6 block matrix wherein blocks are represented with the same color (and a number placed at lower right corner of the cell) can be simultaneously factorized; i.e. blocks on the diagonal are processed in first parallel step and then the factorization is moved on to the blocks below them while considering the entire column as a loop starting from the diagonal block.

X_{11} 1	X_{12} 6	X_{13} 5	X_{14} 4	X_{15} 3	X_{16} 2
X_{21} 2	X_{22} 1	X_{23} 6	X_{24} 5	X_{25} 4	X_{26} 3
X_{31} 3	X_{32} 2	X_{33} 1	X_{34} 6	X_{35} 5	X_{36} 4
X_{41} 4	X_{42} 3	X_{43} 2	X_{44} 1	X_{45} 6	X_{46} 5
X_{51} 5	X_{52} 4	X_{53} 3	X_{54} 2	X_{55} 1	X_{56} 6
X_{61} 6	X_{62} 5	X_{63} 4	X_{64} 3	X_{65} 2	X_{66} 1

Fig. 1. Example of parallel BMF

4.1 Algorithm

A broad outline of our approach is given in Algorithm 1. The outer **for** loop in Algorithm 1 runs precisely CPU_STEPS number of times to execute MF on the entire dataset. The inner **for** loop in Algorithm 1 iterates through each set of parallel blocks and invokes $gpumf()$ function for each such block. In the given algorithm the blocks are identified diagonally from top left to bottom right. The variables l, m denote the indexes of each such block. Each set of parallel blocks are executed simultaneously in a different computational unit. Assigning a computational unit for each such calls is either automatically done by the GPU or can be specified using firmware specific programming. The $gpumf()$ is a kernel function of the underlying MF variant. The function encapsulates GPU firmware specific implementation of MF and is implemented to factorize the given block for GPU_STEPS number of times. The two **for** loops, iterating CPU_STEPS, GPU_STEPS (number of times for each sub-matrix), together contribute to I (number of iterations) which is independent of data elements and hence can be

Algorithm 1. Block based GPU accelerated MF

Require: Input: Data matrix $X \in \mathbb{R}^{\{n \times m\}}$, number of features k

Initialize: latent feature matrices with random values $U \in \mathbb{R}^{\{n \times k\}}, V \in \mathbb{R}^{\{m \times k\}}$
Let I, J be two constants such that X is represented by $I \times J$ number of sub-matrices

Represent data matrix X as block matrix with sub-matrices X_{ij} where $i \in 1..I$ and $j \in 1..J$

Similarly, represent feature matrices U, V as block matrix with sub-matrices U_i, V_j where $i \in 1..I$ and $j \in 1..J$

Let CPU_STEPS, GPU_STEPS be constants representing maximum iterations of MF in CPU and GPU. Let α be the learning rate, β be the regularization factor and δ the minimum deviation of error between iterations

for step 1 to CPU_STEPS **do**
  for each column j in block matrix X **do**

    $l \leftarrow mod((i + step), I)$
    $m \leftarrow mod((j + step), J)$

    Wait if there is a thread running on same row as l

    Invoke thread: $U_l, V_m \leftarrow bgmf(X_{lm}, U_l, V_m, k, \alpha, \beta,$ GPU_STEPS)
    $\{bgmf()$ function is SGD based MF algorithm implementation, at block level, for GPU_STEPS and derives k latent features. The function is executed on one GPU thread per call executed in a multi-threaded mode$\}$

  end for
  Terminate convergence if RMSE improvement is $< \delta$ for entire matrix
end for

return Latent feature matrices $U \in \mathbb{R}^{\{n \times k\}}, V \in \mathbb{R}^{\{m \times k\}}$

considered as constant. Hence the time complexity of the algorithm remains same as MF. Considering the block size as constant, the space required for factorizing each block is $\approx (n \times m + n \times k + k \times m) \times c$ for some constant c. The space required for factorization of block can be explained as the space taken up by the data block and by the two latent factor blocks. As the blocks are of fixed dimension, the space complexity at a block level is constant. As the total computational units on a GPU are constant, and as the number of blocks factorized in parallel are constant, the space complexity of the algorithm remains $\approx O(c)$ (constant).

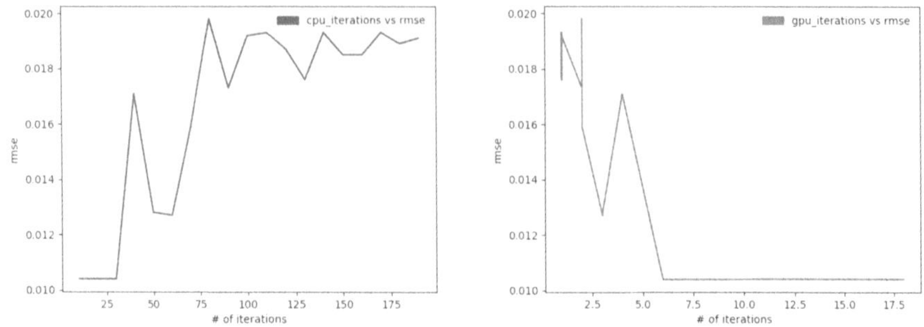

Fig. 2. Comparison of different combinations of GPU, CPU iterations

The implementation of the BGMF algorithm is made available for public on Github at[3] along with code for the experiments presented in Sect. 5.

4.2 CPU Vs GPU Iterations

Frequent transfer of blocks to and from GPU is a major cost in GPU computation. Factorizing each block for more than one iteration reduces the number of data transfers. Factorization is optimal when each block is factorized once and the respective $\underline{U}_i$, $\underline{V}_j$ latent features are passed on to factorize the following sub-matrices. It is also the case that approximate but nearly equivalent to optimal results are observed when each block is factorized more than once per step.

Figure 2 demonstrates differences in convergence at different combinations of CPU and GPU iterations while keeping the overall iterations (CPU_STEPS $\times$ GPU_STEPS) constant at 200 iterations. With the increase in GPU iterations, the CPU iterations come down and vice-versa; i.e. the CPU_STEPS variable is incremented from 1 to 100 and simultaneously GPU_STEPS variable is decreased from 200 to 1 so that the total iterations (CPU_STEPS $\times$ GPU_STEPS) is kept constant at 200 for this experimental setup. The data matrix used for the experiment is a dense matrix with size 1000 $\times$ 1000 with k value taken as 30.

As the GPU iterations improve the convergence at an individual block level, the RMSE also improves at an individual block level. The CPU iterations improve the transfer of latent features between the blocks thereby resulting in improvements in RMSE convergence across the entire matrix. In the above case 25 GPU iterations and 8 CPU iterations are optimal combination.

Hence, under constrained (overall number of iterations) scenario, identifying the right combination of CPU, GPU iterations is the key to achieving quicker convergence, without compromising on the outcome. While the combination of CPU, GPU iterations change with the overall number of iterations, in our observation the combination also differs with the nature of data. When the block sizes are relatively large and when the blocks are true representative sample of the overall data, the interdependence between the blocks is reduced. The latent

[3] https://github.com/17mcpc14/blockgmf.

features for each block are nearly equal to the latent features corresponding to other blocks and thus the results of factorization of just few blocks is nearly equivalent to the factorization of the entire matrix.

Clearly there exists an alternative approach that gives better performance than that of the regular approach of considering each element, block once per iteration for factorization. The experiments detailed in Sect. 4.2 (Block level factorization) dwells further in this direction.

5 Experiments

The experiments mentioned below are conducted on a shared hardware with Intel(R) Xeon(R) CPU E5-2640 v3 @ 2.60GHz and with Nvidia Tesla M40 GPU with 24GB of CPU memory and 8GB of GPU memory. The programming environment is Python 2.7.5 that leverages NVIDIA CUDA with driver version 8.0 and with PyCUDA application programming interface version 1.8. The data sets used are MovieLense 100K[4], MovieLense 1M[5], MovieLense 20M[6] and Jester[7]. All experiments are conducted with configuration of $\alpha = 0.0002$, $\beta = 0.0225$ as the learning, regularization parameters while $\delta = 0.01$ as the minimum difference of error between steps considered for early termination. Several experiments were conducted with different number of latent features. It is observed that a value of $k = 12$ latent features is generating optimal results, hence we choose to use the same configuration across all the experiments below.

5.1 Performance Analysis of Block Based MF

In order to compare the benefits of BMF, we took a basic Probabilistic Matrix Factorization (PMF) [25] implementation based on SGD. We also implemented a Block based implementation of PMF (BMF). In order to achieve parallelism, we have implemented multiple CPU threads to work in tandem. Each thread is dedicated to processing one block at a time, while splitting the data matrix into square block matrix. Several combination of number of blocks have been experimented. While increasing the number of blocks increases parallelism, however, we observed performance drops beyond a critical point. We hypothesize that this observation is due to increased pagination as the threads compete for resources. As our objective is to implement and analyze BMF, we limited the scope to parallelism through BMF. Hence, we configured just 8 parallel threads in accordance with the number of CPU cores available on our hardware. Accordingly we split the data matrix into 8×8 block matrix. With MovieLense 1 million data set (6040 users and 3900 movies), each block is taken as 755×486 dimension. We used 64 bit float data types for U, V matrices and 32 bit integer data type for ratings. Hence, the maximum memory required for each block including the rating

[4] http://grouplens.org/datasets/movielens/100k/.

[5] http://grouplens.org/datasets/movielens/1\protect\kern+.1667em\relaxm/.

[6] https://grouplens.org/datasets/movielens/20\protect\kern+.1667em\relaxm/.

[7] http://www.ieor.berkeley.edu/goldberg/jester-data/.

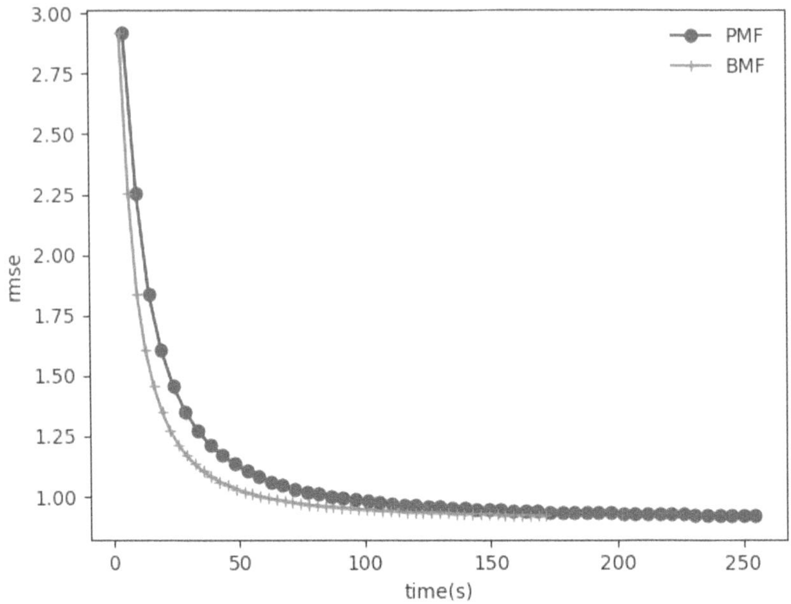

Fig. 3. Comparison of PMF with Block based PMF (BMF) against MovieLens 1M data set

data and latent feature vectors, is estimated to be of 1.513 Mega Bytes (MB). It is observed that memory needed per block is much lower (approximately 0.3 MB per block) due to sparse nature of the ratings. The memory needed for MF at a block level is estimated to be constant and is a maximum of 12.1 MB. This estimate excludes the memory needed for the code, program variables etc.

Figure 3 shows the comparative performance advantage of BMF. The time per iteration and RMSE values of the algorithms are detailed in Table 1. The RMSE convergence of both algorithms are nearly equal as the underlying kernel is the same. However, the convergence time is clearly an indicator of performance advantage of BMF over basic implementation.

Remark 1. It may be noted that the RMSE values reported in Table 1 may not be the best values in comparison to the state-of-the-art frameworks. The objective of the present work is to study the relative performance of BGMF over the basic PMF implementation. Hence, the rigorous parameter tuning exercise is not taken up. Nevertheless, our experiments show that BGMF is better in terms of performance and memory. Such state-of-the-art implementations could be leveraged in place of $mf()$ function within Algorithm 1 to replicate the same results.

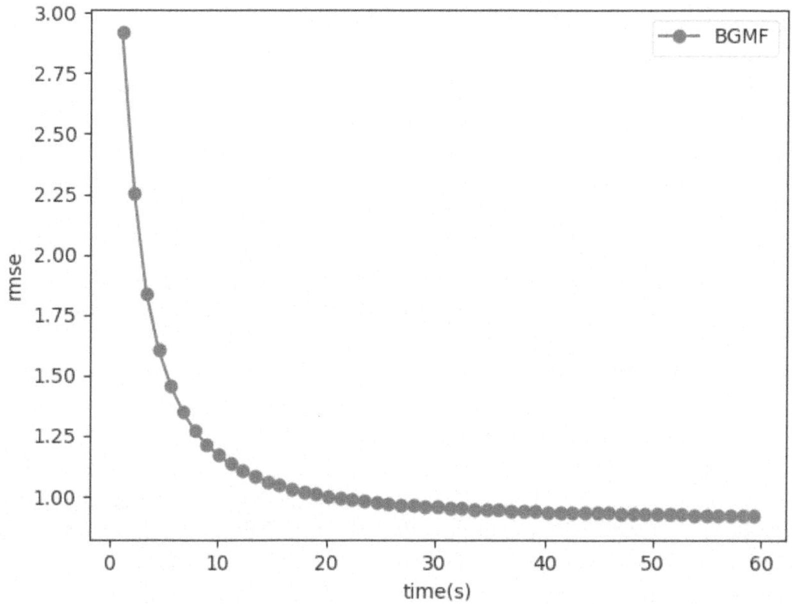

Fig. 4. RMSE vs time taken by Block based GPU implementation of PMF (BGMF) against MovieLens 1M data set

Table 1. Comparison of CPU based PMF, Block based CPU implementation of PMF (BMF), Block based GPU implemetnation of PMF (BGMF) against standard data sets.

	Data set −>	MovieLense 100K	MovieLense 1M	Jester 4M	MovieLense 20M
	n (users)	1000	6000	73421	138000
	m (items)	1700	4000	100	27000
Training Time (sec/iteration)	PMF	4.80	44.14	126.98	967.840
	BMF	3.24	28.78	74.90	578.903
	BGMF	1.120	3.391	9.825	84.521
Test RMSE	PMF	0.955	0.921	2.163	0.862
	BMF	0.959	0.928	2.173	0.867
	BGMF	0.958	0.926	2.172	0.868

5.2 Performance of BGMF

The advantage of block based approach are realized better with the usage of GPUs. Hence we have implemented Block based GPU implementation of MF (BGMF) using CUDA capable hardware with a maximum of 3072 threads, with each thread working in parallel on a separate block. The underlying MF kernel is again PMF and configured with same model parameters as discussed in Sect. 5. The $GPU\_ITERATIONS$ is set to 1 for comparison purposes. The implementation is specific to CUDA capable hardware supported by Nvidia. We have leveraged 3072 threads which are the maximum supported by the hardware. Accordingly, we have split the MovieLens 1M data set into block matrix of 3072×3072 dimension. We have used 32 bit floating point data types for U, V and 32 bit integer data types for data matrix. The estimated total memory needed is 12.5 MB for processing the entire 1 million data matrix on the GPU. The estimate excludes memory taken by the binaries. It is important to note that no memory optimization techniques were used for GPU acceleration. We have not leveraged the shared memory, nor used different caching techniques to fine tune the performance. Figure 4 shows RMSE convergence with time for BGMF. It is observed from the values in Table 1 that the RMSE convergence of all variants are nearly equal (within the acceptable delta). The underlying kernel implementation of all variants are the same and the focus of these experiments is to evaluate performance of BGMF against other approaches. This confirms our hypothesis that the block based approach has no impact on the quality but only on the performance. The results also demonstrate the comparative performance advantage of GPU variants of MF over CPU variants. It is important to note that the per iteration time observed, for GPU variants, includes the time to transfer data from and to the GPU. Considering the fact that the hardware architecture of the CPU and the GPU are significantly different, the computational time on CPU is substantially more than that observed in GPU variants and the time taken by CPU variants are not necessarily to be taken as the basis for comparison against GPU variants.

6 Conclusions

BGMF fundamentally does not change the underlying MF kernel itself, but merely makes it possible to achieve matrixfactorisation with block based approach. As demonstrated in the experimental results, the proposed approach provides computational advantage, with respect to time and memory, without compromising on RMSE convergence. By identifying optimal combination of GPU_STEPS, CPU_STEPS, suitable for the specific data set, it is possible to achieve quicker convergence. Further performance gain is possible with variable block level iterations with approximate (but nearly optimal) convergence. By increasing the number of blocks, and by running in parallel exact number of blocks as the number of available cores, it is possible to factorize data sets of any scale. The technique can be adapted to run block level factorization on multiple GPUs as well as on distributed systems. As our approach is based on identifying a precise number of blocks that can be factorized in parallel, if the number of computational units is not exactly equivalent to the number of parallel blocks, it is possible that the approach results in idling of a few computational units.

Proposed technique is implemented and experimented using CUDA framework and supported hardware, however similar implementation is possible for OpenCL framework as well. The proposed approach is limited by the availability of GPU memory and when the memory needed to execute the approach.

When the available memory for specific hardware configuration is less than the memory needed to compute MF for a given dataset, it will lead to multiple memory transfers between CPU and GPU, thereby taking away the advantages of the proposed approach. The current GPU architectures, especially the hardware memory management aspects, are significantly different from one another. It requires firmware specific custom programming to leverage the hardware memory. While the programming needed to achieve our approach is different for each manufacturer, the approach remains the same and can be easily adapted to other platform specific programming.

GPU_ITERATIONS, CPU_ITERATIONS are two additional configurable parameters associated with our approach. Setting GPU_ITERATION closer to 1 will result in frequent transfer of data between CPU and GPU and thereby increasing the computational time. At the same time, configuring GPU_ITERATIONS to a higher number will result in approximate results. In future, as the available memory within GPU is increased, this limitation can be overcome by loading all the required matrices into GPU.

The technique does not put limitation on the size of the data set. The block size is only limited by the available memory on the computation unit. While there is no limitation on the total number of blocks, it is ideal to fix the number of blocks equivalent to the number of computational units. Although we focused mostly on BGMF with equivalent block sizes, further experimentation is possible with variable block sizes and variable CPU, GPU iteration combinations at different phases of convergence. BGMF can further be enhanced to make it a divide and conquer approach with block level convergence, by identification of error limits at block level and by combining the results from multiple blocks. Further research is also possible in the direction of re-ordering the data elements so as to improve the efficiency of BGMF.

References

1. Bhavana, P., Padmanabhan, V.: BMF: matrix factorization of large scale data using block based approach. In: Nayak, A.C., Sharma, A. (eds.) PRICAI 2019. LNCS (LNAI), vol. 11671, pp. 431–436. Springer, Cham (2019). https://doi.org/10.1007/978-3-030-29911-8_33
2. Bhavana, P., Kumar, V., Padmanabhan, V.: Block based singular value decomposition approach to matrix factorization for recommender systems. arXiv https://arxiv.org/abs/1907.07410 (2019)
3. Lee, D.D., Seung, H.S.: Algorithms for non-negative matrix factorization. In: Proceedings of the 13th International Conference on Neural Information Processing Systems, NIPS'00, pp. 535–541. MIT Press, Cambridge (2000)
4. Koitka, S., Friedrich, C.M.: nmfgpu4r: Gpu-accelerated computation of the non-negative matrix factorization (nmf) using cuda capable hardware. R J. **8**(2), 382–392 (2016)
5. Zhang, L., Luo, T., Zhang, F., Wu, Y.: A recommendation model based on deep neural network. IEEE Access **6**, 9454–9463 (2018). https://doi.org/10.1109/ACCESS.2018.2789866

6. Patel, K., Patel, H.B.: A state-of-the-art survey on recommendation system and prospective extensions. Comput. Electron. Agric. **178**, 105779 (2020)
7. Elahi, M., Kholgh, D.K., Kiarostami, M.S., Oussalah, M., Saghari, S.: Hybrid recommendation by incorporating the sentiment of product reviews. Inf. Sci. **625**, 738–756 (2023)
8. Kilitcioglu, D., Greenquist, N., Zahran, M., Bari, A.: GPU accelerated matrix factorization for recommender systems. In: 2021 IEEE 6th International Conference on Big Data Analytics (ICBDA), pp. 272–281 (2021)
9. Tan, W., Chang, S., Fong, L.L., Li, C., Wang, Z., Cao, L.: Matrix factorization on gpus with memory optimization and approximate computing. In: ICPP (2018)
10. Huang, Y., Liu, Y., Bai, Y., Chen, S., Li, R.: UMA-MF: a unified multi-CPU/GPU asynchronous computing framework for SGD-based matrix factorization. IEEE Trans. Parallel Distrib. Syst. **34**, 2978–2993 (2023). https://doi.org/10.1109/TPDS.2023.3317535
11. Kysenko, V., Rupp, K., Marchenko, O., Selberherr, S., Anisimov, A.: GPU-accelerated non-negative matrix factorization for text mining. In: Bouma, G., Ittoo, A., Métais, E., Wortmann, H. (eds.) NLDB 2012. LNCS, vol. 7337, pp. 158–163. Springer, Heidelberg (2012). https://doi.org/10.1007/978-3-642-31178-9_15
12. Chen, W., Li, Y., Pan, B., Xu, C.: Block kernel non-negative matrix factorization and its application to face recognition. In: 2016 International Joint Conference on Neural Networks (IJCNN), pp. 3446–3452 (2016)
13. Chin, W.-S., Zhuang, Y., Juan, Y.-C., Lin, C.-J.: A learning-rate schedule for stochastic gradient methods to matrix factorization. In: Cao, T., Lim, E.-P., Zhou, Z.-H., Ho, T.-B., Cheung, D., Motoda, H. (eds.) PAKDD 2015. LNCS (LNAI), vol. 9077, pp. 442–455. Springer, Cham (2015). https://doi.org/10.1007/978-3-319-18038-0_35
14. Rundong, D., Kuang, D., Drake, B., Park, H.: Dc-nmf: non-negative matrix factorization based on divide-and-conquer for fast clustering and topic modeling. J. Global Optim. **68**(4), 777–798 (2017)
15. Gemulla, R., Nijkamp, E., Haas, P.J., Sismanis, Y.: Large-scale matrix factorization with distributed stochastic gradient descent. In: Proceedings of the 17th ACM SIGKDD International Conference on Knowledge Discovery and Data Mining, San Diego, CA, USA, 21–24 August 2011, pp. 69–77 (2011)
16. Li, B., Tata, S., Sismanis, Y.: Sparkler: supporting large-scale matrix factorization. In: Joint 2013 EDBT/ICDT Conferences, EDBT '13 Proceedings, Genoa, Italy, 18–22 March 2013, pp. 625–636 (2013)
17. Mackey, L.W., Jordan, M.I., Talwalkar, A.: Divide-and-conquer matrix factorization. In: Shawe-Taylor, J., Zemel, R.S., Bartlett, P.L., Pereira, F., Weinberger, K.Q. (eds.) Advances in Neural Information Processing Systems, vol. 24, pp. 1134–1142. Curran Associates, Inc. (2011)
18. Oh, J., Han, W.S., Yu, H., Jiang, X.: Fast and robust parallel SGD matrix factorization. In: Proceedings of the 21th ACM SIGKDD International Conference on Knowledge Discovery and Data Mining, Sydney, NSW, Australia, 10–13 August 2015, pp. 865–874 (2015)
19. Recht, B., Ré, C., Wright, S.J., Niu, F.: Hogwild: a lock-free approach to parallelizing stochastic gradient descent. In: Advances in Neural Information Processing Systems 24: 25th Annual Conference on Neural Information Processing Systems 2011. Proceedings of a meeting held 12-14 December 2011, Granada, Spain, pp. 693–701 (2011)
20. Schelter, S., Satuluri, V., Zadeh, R.: Factorbird - a parameter server approach to distributed matrix factorization. CoRR arxiv:1411.0602 (2014)

21. Yu, H.F., Hsieh, C.J., Si, S., Dhillon I.S.: Scalable coordinate descent approaches to parallel matrix factorization for recommender systems. In: 12th IEEE International Conference on Data Mining, ICDM 2012, Brussels, Belgium, 10–13 December 2012, pp. 765–774 (2012)
22. Yun, H., et al.: Nomad: non-locking, stochastic multi-machine algorithm for asynchronous and decentralized matrix completion. Proc. VLDB Endow. **7**(11), 975–986 (2014)
23. Zhang, Y., Zhang, M., Liu, Y., Ma, S., Feng, S.: Localized matrix factorization for recommendation based on matrix block diagonal forms. In: Proceedings of the 22Nd International Conference on World Wide Web, WWW '13, pp. 1511–1520. ACM, New York (2013)
24. Zhuang, Y., Chin, W.S., Juan, Y.C., Lin, C.J.: A fast parallel sgd for matrix factorization in shared memory systems. In: Proceedings of the 7th ACM Conference on Recommender Systems, RecSys 2013, pp. 249–256. ACM, New York (2013)
25. Salakhutdinov, R., Mnih, A.: Probabilistic matrix factorization. In: Proceedings of the 20th International Conference on Neural Information Processing Systems, NIPS'07, pp. 1257–1264. Curran Associates Inc. (2007)
26. Srebro, N., Rennie, J.D.M., Jaakkola, T.S.: Maximum-margin matrix factorization. In: Proceedings of the 17th International Conference on Neural Information Processing Systems, NIPS'04, pp. 1329–1336. MIT Press, Cambridge (2004)
27. Sowmini Devi, V., Padmanabhan, V., Pujari, A.K.: A matrix factorization & clustering based approach for transfer learning. In: Shankar, B.U., Ghosh, K., Mandal, D.P., Ray, S.S., Zhang, D., Pal, S.K. (eds.) PReMI 2017. LNCS, vol. 10597, pp. 77–83. Springer, Cham (2017). https://doi.org/10.1007/978-3-319-69900-4_10
28. Koren, Y., Bell, R., Volinsky, C.: Matrix factorization techniques for recommender systems. Computer **42**(8), 30–37 (2009)

Advanced HLS for Reconfigurable Image Processing in Big Data Systems

K. Bhagirath(✉), Dheeraj Pant, Abhishek Tiwari, and Vivek Khaneja

Embedded Systems Division, Centre for Development of Advanced Computing (CDAC), Noida 201309, India
{kbhagirath,dheerajpant,abhishek,vivekkhaneja}@cdac.in

Abstract. This paper introduces novel HLS techniques for reconfigurable and memory-efficient image processing within deep learning frameworks, addressing inherent limitations of current deep learning accelerators (DLAs) due to their dependency on the host processor, leading to pipeline delays and reduced throughput. The proposed approach emphasizes parallelization, on-chip buffering, and scalability, enhancing computational efficiency and throughput for AI applications in big data and edge computing environments. Methodologically, the paper emphasizes mathematical analysis, hardware architecture specification, HLS modeling, Verilog RTL implementation. By addressing these challenges, this work contributes to advancing the performance capabilities of DLAs in deep learning inference tasks. The algorithms integrates seamlessly with DLA IP subsystems, such as AXI interfaces, external (DRAM) and internal (SRAM) memory interfaces, and activation/pooling engines. Key contributions include advanced algorithm design to enhance memory efficiency and reconfigurability, essential for adapting to diverse hardware architectures. Experimental validation demonstrates substantial improvements in computational speed and resource utilization compared to traditional methods, making it suitable for intensive AI tasks in big data analytics. In conclusion, this research presents a framework for integrating HLS techniques into hardware accelerators, facilitating efficient image processing across diverse applications in big data and edge computing. These advancements aim to bolster DLAs' capabilities in handling complex AI workloads effectively, promising enhanced solutions for AI-driven domains.

Keywords: HLS Techniques · AI Accelerators · Big Data

1 Introduction

1.1 Background

The landscape of deep learning and artificial intelligence (AI) is rapidly evolving, with significant advancements in hardware accelerating complex computational tasks. Image processing, a crucial component of these systems, plays a vital role

A. Dasgupta et al. (Eds.): BDA 2024, LNCS 15526, pp. 148–162, 2025.
https://doi.org/10.1007/978-3-031-81821-9_9

in preparing data for neural network training and inference, as well as interpreting results. Traditionally, image processing tasks have been handled by general-purpose CPUs or GPUs, which, while powerful, are not always optimized for the specific demands of image processing. These processors are designed to handle a wide range of computational tasks, but the efficiency of pre-processing and post-processing stages often remains suboptimal.

As data volumes continue to grow and real-time processing becomes more critical, there is an increasing need for specialized hardware that can efficiently manage image processing tasks. High-Level Synthesis (HLS) techniques present a promising solution by enabling the design of custom hardware accelerators that can reduce design time, complexity and can be suited to specific processing needs [1]. These techniques allow for the creation of hardware that integrates image processing functions directly, addressing the inefficiencies associated with traditional methods [2]. The ability to design hardware that is both reconfigurable and memory-efficient is particularly valuable in the context of big data and edge computing, where performance and scalability are crucial.

The integration of specialized image processing hardware into deep learning frameworks has the potential to enhance overall system performance by reducing latency and improving throughput. This approach represents a significant advancement over conventional methods, offering a more efficient solution for managing image processing tasks within AI applications.

1.2 Problem Statement

Deep learning systems, particularly those dealing with big data and edge computing, face a significant challenge in efficiently managing image processing tasks. Traditional approaches often involve using general-purpose CPUs or GPUs to handle these tasks, which can lead to substantial inefficiencies. These processors, while effective for neural network computations, are not optimized for the specific requirements of image processing, such as pre-processing and post-processing. As a result, the performance of the overall system can be bottle-necked by these stages, leading to increased latency and reduced throughput(transferring the data to the host processor to perform the processing and again transferring back the processed data to the DLA IP).

The core issue is that existing hardware solutions do not adequately address the need for efficient and flexible image processing within deep learning pipelines. Host processors, which are responsible for these tasks, often introduce delays that can undermine the performance benefits of specialized neural network accelerators. This inefficiency is especially problematic in scenarios where real-time processing is required or where large volumes of data must be managed quickly and effectively.

To address these challenges, there is a need for a hardware accelerator that integrates image processing functions directly into the design. Such a solution would need to be reconfigurable to handle a variety of image processing tasks and optimized for memory efficiency to improve performance across different applications. The goal is to develop a hardware accelerator that enhances the overall efficiency of deep learning systems by addressing these specific processing bottlenecks.

1.3 Contribution/Novelty

This paper introduces a novel hardware accelerator designed to improve the efficiency of image processing tasks within deep learning frameworks through the use of advanced High-Level Synthesis (HLS) techniques. The primary contribution is the development of a custom hardware solution that integrates image processing functions directly into the accelerator, addressing the limitations of traditional host processor-based approaches.

The novelty of this work lies in the application of HLS techniques to create a versatile and memory-efficient accelerator that can be adjusted to a wide range of image processing tasks. Unlike existing solutions, which often focus solely on computational performance, our approach provides a comprehensive solution that enhances both pre-processing and post-processing stages. The hardware design allows for high levels of parallelism and reconfigurability, making it adaptable to various deep learning applications and scalable for different data volumes.

By exploiting HLS techniques, the proposed accelerator achieves significant improvements in processing speed and overall efficiency. It is compatible with both FPGA and ASIC implementations, offering flexibility and adaptability across different hardware platforms. This work represents a significant advancement in hardware design for deep learning, providing a more effective solution for managing image processing tasks and improving the performance of AI systems in big data and edge computing environments.

2 Related Work

In the context of big data and artificial intelligence (AI), hardware acceleration has become pivotal for managing and processing large volumes of data efficiently. Specialized processors, such as Graphics Processing Units (GPUs), Tensor Processing Units (TPUs), and Field-Programmable Gate Arrays (FPGAs), have been extensively used to enhance the performance of deep learning tasks. GPUs, for instance, are renowned for their ability to perform parallel computations, making them well-suited for training large neural networks. Google's TPUs are specifically designed to accelerate tensor operations, offering substantial speed improvements for AI workloads. Similarly, FPGAs provide customizable hardware solutions that can be suited to specific deep learning and big data applications.

However, while these specialized processors excel in computational tasks related to neural networks, they often do not address the inefficiencies in image processing stages such as pre-processing and post-processing. Companies like Intel and AMD have developed custom accelerators aimed at optimizing AI tasks, but these solutions typically focus on the computational aspects of deep learning rather than the end-to-end processing pipeline required for big data applications.

Recent advancements are beginning to bridge this gap by integrating image processing capabilities directly into hardware accelerators. For example, Intel's Movidius and Google's Edge TPU target edge computing scenarios, providing specialized solutions for real-time image processing. Despite these innovations, there is a continued need for hardware that efficiently handles diverse image processing tasks within big data and AI frameworks. The proposed High-Level Synthesis (HLS) techniques in this paper aim to address these challenges by offering a more adaptable and memory-efficient solution, enhancing the overall performance and throughput of image processing in deep learning systems.

3 Proposed Method

3.1 Algorithm Design

Initially, the task is to transform hls ap types [3] to `hlslib`'s ac types [4], since hls ap types are synthesizable only on AMD fpgas. To do this, the `hlslib`'s ac types are used to design the algorithm. The `hlslib`'s are open-source libraries which can be synthesized into both FPGA and ASIC.

HLS Arbitrary Precision Type Library provides data types and operations for high-level synthesis (HLS) tools. It is designed to allow the use of arbitrary precision arithmetic in hardware design, which is particularly useful when dealing with computations that require more precision than standard hardware types can offer. `hlslib`'s `ac` types provide arbitrary-precision arithmetic for HLS designs. They include `ac_int` for high-precision integers, `ac_fixed` for configurable fixed-point numbers, and `ac_float` for customizable floating-point types. These types allow precise and efficient hardware implementations, suited for specific computation needs.

Tools Used

- HLS: Cadence Stratus HLS
- Synthesis: Synopsys Design Compiler
- PDKs: Skywater PDK
- Power Analysis: Cadence Innovus

Figure 1 is the DLA IP, in this DLA IP, there is no specific functionality with respect to pre-processing and post-processing operations to map on the hardware logic, which are present in the neural network architecture. To perform these operations, the data is transferred to the processor(which the DLA IP is connected to) and the sent to the DLA IP. To mitigate this latency prone issue, the image processing operations are directly integrated in the IP itselt with enough buffers to process the data with much efficiency.

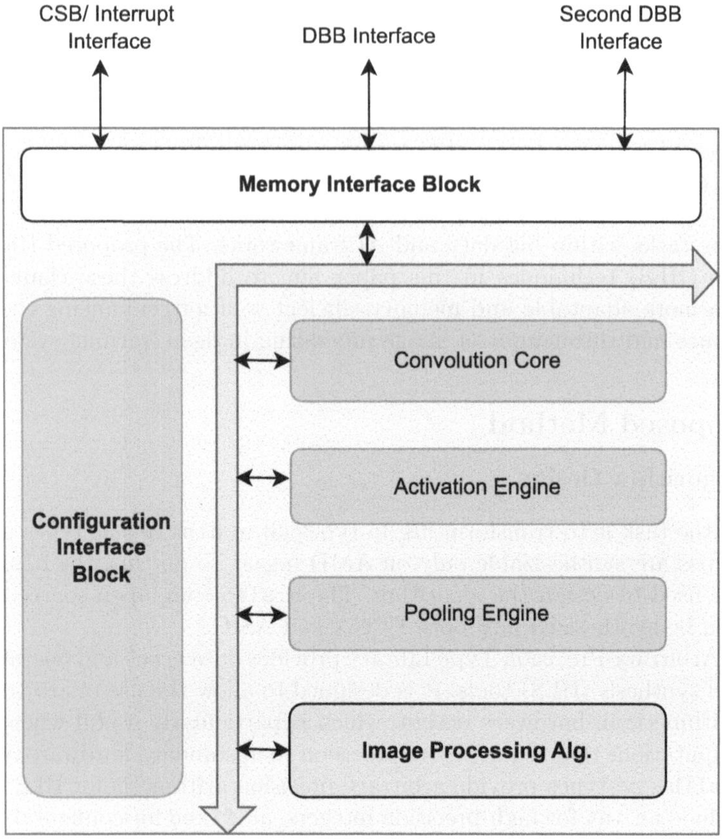

Fig. 1. DLA IP integrated with image processing algorithm

3.2 Reconfigurability Feature

Reconfigurable hardware approach. Generally, while developing image related algorithms, row and column loops will be hardcoded, this means arbitrary image of smaller size could not processed by our hardware. To make hardware configurable, we need to make some changes in the code such as computing minimum bit-width needed for counters(loops) and other type of operations in the design, there are several helper function, which help us to obtain the above discussed information that are included in the ac datatype library [5]. The function we use the most are ac::log2_ceil which returns number of bits required in the x elements and it is statically determinable at the compile time and ac::nbits (these are mention in the provided reference manual at pp. 76 [5]) which returns number of bits required to represent the value of x which is also statically determinable at the compile time.

Making a design configurable typically means restructuring the loops in design to exit based on some condition, when re-coding the loop the first thing to do is to remove the exit loop condition, this is because the exit condition will never hit if it is based on external variable, this will end up being a coverage hole in RTL code coverage. The next thing to use unsigned bit accurate datatype for loop counter, by removing the loop exit condition, tool is not able to automatically determine the optimal number of bits for loop counter, lastly conditional break used to exit the loop should be cast to unsigned bit accurate datatype, this is also required to ensure that the loop exit could not possibly have negative values which would show up as coverage hole in RTL code coverage (Fig. 2).

```
// Template parameters to make the code reusable and  allow configuring size of the image
template <int maxImageWidth, int maxImageHeight>
class NNInterpol

public:
  // to statically compute minimum bits required for both image width and height
  typedef ac_int<ac::nbits<maxImageWidth>::val,false> width;
  typedef ac_int<ac::nbits<maxImageHeight>::val,false> height;

void convert {
        width       &widthIn,
        height      &heightIn,
VROW: for (height y = 0; ; y++) { // loop exit condition has been removed
      VCOL: for (width x = 0; ; x++) {
        if (y <= maxImageHeight) {
        if (x == maxImageWidth)

        // programmable width exit condition
        if (x == height(widthIn)) { // cast to maxWType for RTL code coverage
          break;
        }
      }
      // programmable height exit condition
      if (y == heightIn) {
        break;
      }
```

Fig. 2. Reconfigurable c++

3.3 Memory Efficiency and Power Efficiency Techniques

Our approach includes advanced memory management strategies to enhance the efficiency of data handling, reducing latency and improving overall throughput.

When dealing with memories in the design, the best way is to use the single port memory since dual port memory is costly for ASIC phases. While using the single port memory, resource conflict appear in the design. Resource conflict occur when trying to simultaneously read/write from a memory more times than there are read/write ports. Pipelining the design overlapped read/writes in Fig. 4 (Fig. 3).

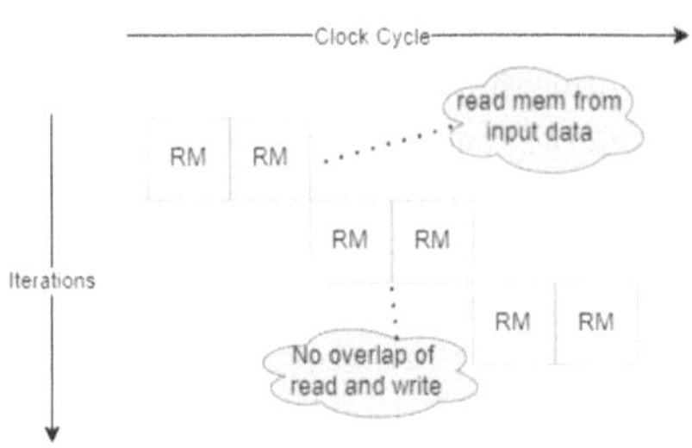

Fig. 3. No Pipelining

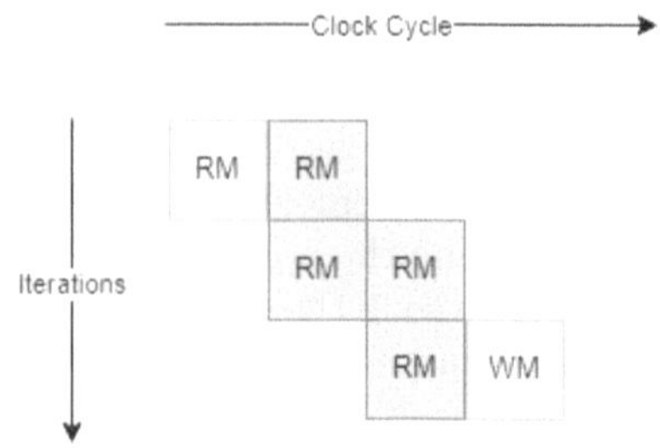

Fig. 4. Pipelined Conflict

Typically, pure algorithmic descriptions have performance bottlenecks due to no explicit memory architecture coded in the design. So to create a memory architecture the algorithm needs to be analysed. Efficient memory architectures reduce the reading of a memory or an I/O to once per clock cycle. The goal is to reducing the reading of memory or I/O to once per clock cycle. Which allows pipelining the design with II = 1 (Fig. 5).

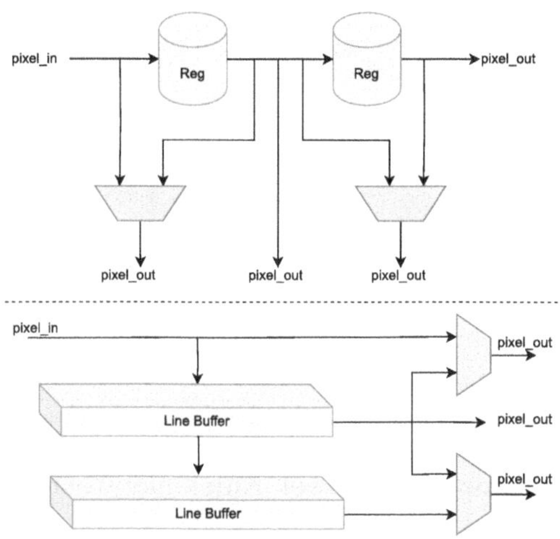

Fig. 5. Dataflow architecture with 1PPC(pixel per cycle)

In the context of ASIC (Application-Specific Integrated Circuit) design, this research paper proposes substituting dual-port memory implementations with single-port memory alternatives. Dual-port memory, while providing simultaneous read and write access, significantly increases the area required on the ASIC chip, leading to higher manufacturing costs. In contrast, single-port memory implementations are more area-efficient and cost-effective. By focusing on single-port memory, the overall chip area can be reduced, which not only decreases the production expenses but also enhances the scalability of the design. This shift to single-port memory solutions aims to optimize both the economic and spatial efficiency of ASICs, aligning with industry needs for cost-effective and high-performance designs (Fig. 6).

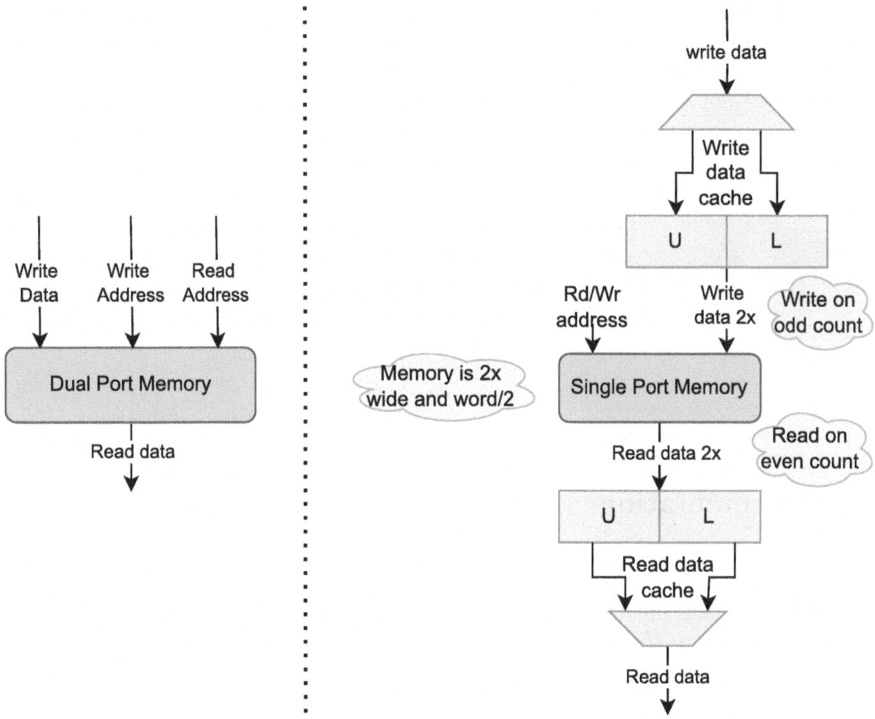

Fig. 6. Dual Port(DP) Memory and Pseudo Dual Port(PDP) Memory

Upon analyzing the fully optimized code, it was noted that power consumption peaks at specific points due to inefficiencies in the line buffer memory architecture previously utilized. This architecture proves to be power-inefficient because both line buffer memories are written to for every line of data. Specifically, pixel data streams into the first line buffer, which then shifts the existing data into the second line buffer, and so on. This continuous shifting and writing process for every line of data results in increased power consumption. The constant read and write operations across multiple buffers create unnecessary power overhead. An alternative to this approach is the circular buffer memory architecture. Unlike the line buffer, the circular buffer design minimizes unnecessary data movement by reusing memory locations in a circular manner. This reduces the frequency of write operations and, consequently, lowers power consumption, making it a more efficient choice for power-sensitive applications (Fig. 7).

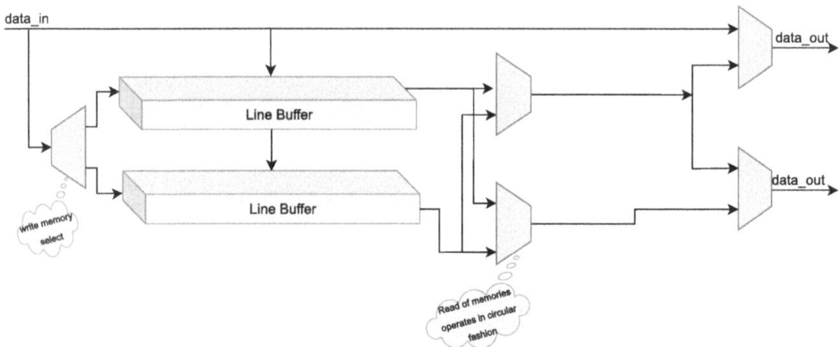

Fig. 7. Circular Buffer

4 Implementation

4.1 Software Integration

The software stack has been meticulously developed by our software team to facilitate seamless integration with the hardware accelerator. This stack includes user module drivers and kernel module drivers, which are essential for enabling effective communication between the accelerator and the host system. By optimizing this interaction, we enhance the management of deep learning tasks specifically within big data contexts (Fig. 8).

Designed to support a wide range of image processing functions, the software stack enables straightforward configuration of the hardware, which is crucial for adapting to the dynamic nature of big data workloads. We have created high-level abstractions for image processing operations, simplifying application development and ensuring that the hardware accelerator can efficiently handle large-scale data sets while minimizing latency and maximizing throughput.

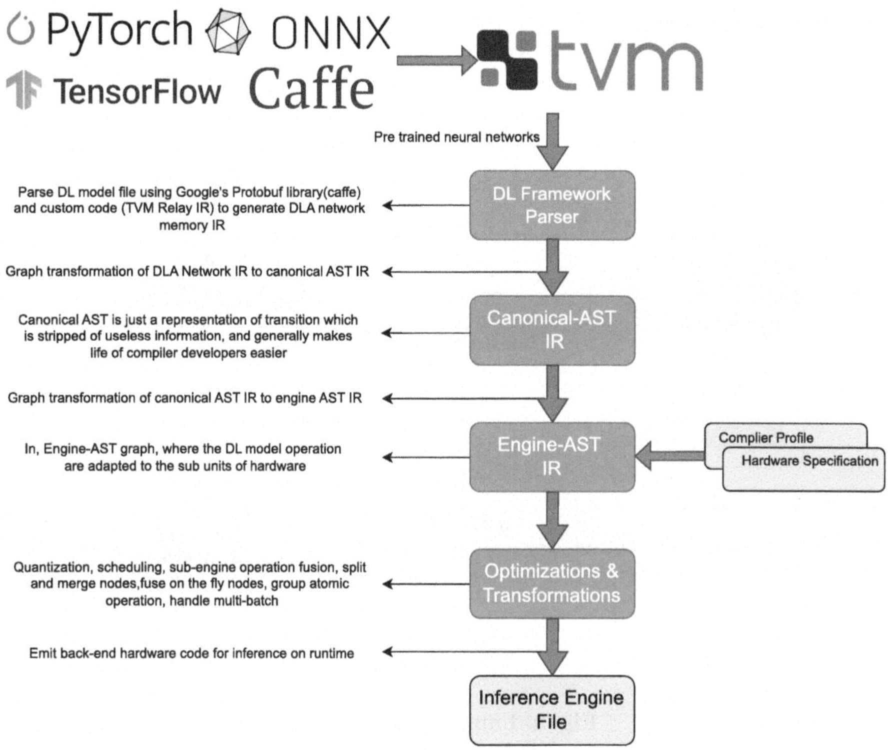

Fig. 8. Different stages of software stack

The integration process incorporates support for various memory interfaces, including DRAM and SRAM, allowing the system to efficiently accommodate diverse data sizes and types commonly encountered in big data analytics.

To validate the effectiveness of this integration, extensive testing has been conducted to ensure compatibility with leading deep learning frameworks and tools. This benchmarking process evaluates the performance of the accelerator in real-world big data scenarios, confirming its significant impact on overall system efficiency. Our goal is to provide a robust and user-friendly interface that empowers developers to fully leverage the capabilities of the hardware accelerator in their AI-driven big data applications.

4.2 Hardware Flow

The explanation of this flow-chart is given at Sect. 5.1 (Fig. 9).

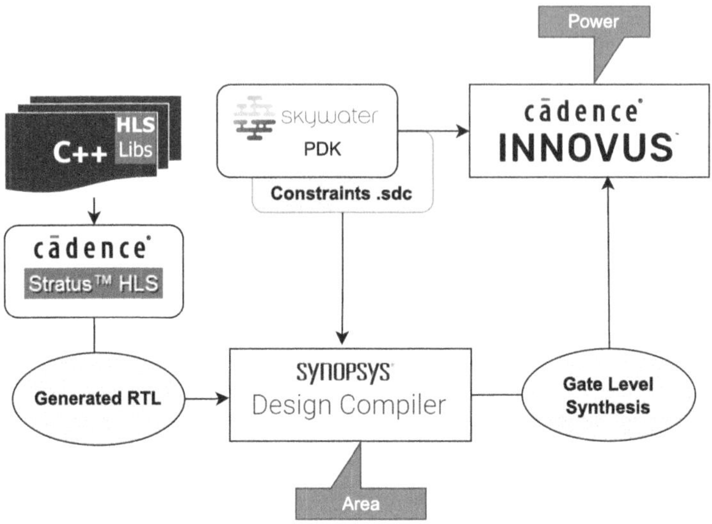

Fig. 9. Implementation flow

5 Experimental Results

5.1 Setup

The code is designed in hls/c++, then the code is made to go through the verification process which is a framework originally pre-built in the tool. After the verification, the code is transformed into synthesizable RTL. The generated RTL is given as input to the Synopsys DC tool which is then, along with the pdk libraries and constraint file (.sdc), is synthesized and gate-level-netlist(GLS) is generated along with the area report. The generated gate-level-netlist(GLS) is given to Cadence Innovus tool, which outputs the power report.

5.2 Benchmarking

(See Tables 1 and 2).

Table 1. Comparison of Integrated DLA with State-of-the-Art Accelerator Designs

Parameters	Eyeriss	Shi-diannao	Existing Work	This Work	Performance Analysis
Latency	1.00	1.20	1.28	1.30	Slightly higher than Eyeriss, competitive with others
Throughput	1.00	1.25	1.30	1.32	Improved compared to all previous works
Area	1.00	1.15	1.20	1.28	Increased area usage;trade off for performance gains
Power	1.00	1.14	1.16	1.20	Reasonable power consumption comparable to existing designs

Table 2. Resource Utilization of Various Accelerators on Different FPGA Boards

Accelerator	FPGA Board	Frequency (MHz)	LUTs Utilized	DSPs Utilized	BRAM Utilized (kb)
Eyeriss [6]	VCK190	250	6500	45	35
	ZCU104	200	7000	50	32
	ZCU102	180	7200	55	36
	Ultra96	150	7500	60	40
Shi-Diannao [7]	VCK190	250	7000	50	40
	ZCU104	200	7200	55	38
	ZCU102	180	7400	60	42
	Ultra96	150	7600	65	45
Existing Work [8]	VCK190	250	6800	40	30
	ZCU104	200	6900	45	32
	ZCU102	180	7100	50	34
	Ultra96	150	7400	55	39
This Work	VCK190	250	6600	45	33
	ZCU104	200	6800	50	31
	ZCU102	180	7000	55	35
	Ultra96	150	7300	60	38

5.3 Performance Analysis

The results highlight the advantages of our HLS techniques in enhancing the performance of image processing tasks, with notable gains in processing speed and efficiency (Fig. 10).

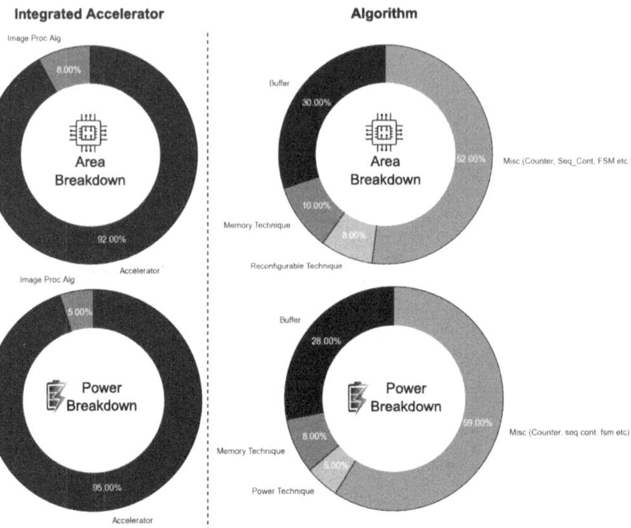

Fig. 10. Area and Power Breakdown

6 Discussion

6.1 Comparision with Existing Work

The proposed hardware accelerator significantly advances the field of image processing within deep learning frameworks, particularly when compared to existing technologies. Traditional accelerators, such as Graphics Processing Units (GPUs) and Tensor Processing Units (TPUs), have proven highly effective for neural network computations due to their parallel processing capabilities and specialized architectures. However, their focus is predominantly on accelerating the core computations of deep learning models, often leaving the image processing stages—such as pre-processing and post-processing—less optimized. These stages are crucial for preparing data for neural networks and interpreting outputs, yet they frequently suffer from inefficiencies when handled by general-purpose processors or even specialized accelerators.

In contrast, our proposed solution integrates image processing capabilities directly into the hardware accelerator, addressing this gap. The advanced High-Level Synthesis (HLS) techniques employed allow for a design that is not only more efficient but also highly adaptable. Unlike traditional methods, which may require complex integration with host processors, our hardware provides a solution that significantly reduces latency and improves throughput. This integration results in faster data handling and enhanced performance in big data applications and edge computing scenarios.

The design's ability to be synthesized for both FPGA and ASIC implementations further distinguishes it from existing solutions, providing a versatile and scalable approach. This flexibility ensures that the accelerator can be suited to various applications, from real-time image processing in edge devices to large-scale data processing in data centers. Overall, the proposed hardware represents a notable improvement over current methods by offering a more comprehensive solution that integrates image processing seamlessly into the deep learning pipeline.

6.2 Limitations and Future Work

Reconfigurable processors face inherent limitations due to their static nature, which restricts adaptability and efficiency in varying computational tasks. These processors rely on predetermined configurations that must be reconfigured and re-synthesized manually, leading to latency and increased power consumption during transitions between different applications or computing demands. This rigidity impedes their optimal performance in dynamic computing environments where tasks vary significantly in complexity and type.

However, the evolution towards dynamically reconfigurable processors represents a promising future direction in computing architecture. By enabling real-time adaptation of hardware configurations based on workload demands, dynamically reconfigurable processors can optimize resource allocation, enhance performance efficiency, and reduce energy consumption. This flexibility not only promises to improve overall system performance but also opens up new possibilities for adaptive and responsive computing systems capable of meeting the evolving demands of modern applications.

7 Conclusion

The development of the proposed hardware accelerator, incorporating advanced High-Level Synthesis (HLS) techniques, marks a significant advancement in the realm of image processing for deep learning systems. This innovative accelerator addresses critical inefficiencies associated with traditional host processor-based approaches by integrating image processing functions directly into the hardware. The result is a marked improvement in latency and throughput, which is crucial for applications involving big data and edge computing.

Our hardware design leverages HLS techniques to achieve a high degree of parallelism and memory efficiency, enabling faster and more effective handling of image pre-processing and post-processing tasks. By being adaptable and reconfigurable, the accelerator can be suited for various deep learning applications, making it a versatile solution for different scenarios. It is developed in such a way that allows it to be synthesized for both FPGA and ASIC implementations, broadening its applicability and integration potential.

The experimental results highlight the hardware accelerator's superior performance compared to existing solutions, particularly in terms of reducing processing time and enhancing throughput. This makes it an invaluable tool for high-performance computing environments where efficient image processing is crucial.

Looking ahead, there are opportunities to further refine the accelerator by addressing its limitations, such as making it dynamically reconfigurable. Future research will also explore additional features and optimizations to enhance the accelerator's capabilities and applicability across a wider range of applications.

In conclusion, the proposed accelerator represents a significant leap forward in hardware design for image processing in deep learning frameworks. Its advancements in efficiency, flexibility, and performance make it a promising solution for the growing demands of big data and AI applications, paving the way for more rapid and effective image processing in future computing environments.

References

1. Moosa, A.K.T., Sarma, N., Karfa, C.: ImageSpec: efficient high-level synthesis of image processing applications. In: 2022 25th Euromicro Conference on Digital System Design (DSD), Maspalomas, Spain, pp. 67–74 (2022)
2. Cortes, A., Velez, I., Irizar, A.: High level synthesis using Vivado HLS for Zynq SoC: image processing case studies. In: 2016 Conference on Design of Circuits and Integrated Systems (DCIS), Granada, Spain, pp. 1–6 (2016)
3. HLS Arbitrary Precision Types Library. https://github.com/Xilinx/HLS_arbitrary_Precision_Types
4. HLSLIBS Algorithmic C Datatypes. https://github.com/hlslibs/ac_types
5. Algorithmic C Datatypes Reference Manual. https://github.com/hlslibs/-ac_types/blob/master/pdfdocs/ac_datatypes_ref.pdf
6. Chen, Y.-H., Emer, J., Sze, V.: Eyeriss: a spatial architecture for energy-efficient dataflow for convolutional neural networks. In: International Symposium on Computer Architecture (ISCA) (2016)
7. Du, Z., et al.: Shidiannao: shifting vision processing closer to the sensor. In: International Symposium on Computer Architecture (ISCA) (2015)
8. Bhagirath, K., Pant, D., Tiwari, A., Khaneja, V.: Towards scalability and performance: framework for heterogeneous cluster integration in deep learning accelerators. In: 2024 IEEE 4th International Conference on VLSI Systems, Architecture, Technology and Applications (VLSI SATA), Bangalore, India, pp. 1–6 (2024)

Applications-1

Explainable Detection and Analysis of Cauliflower Leaf Diseases

Sharia Arfin Tanim[1], Rubaba Binte Rahman[1], Kazi Tanvir[2](✉), Md. Sayem Kabir[1], Tasnim Sultana Sintheia[1], and Md Saef Ullah Miah[1]

[1] Department of Computer Science, American International University-Bangladesh(AIUB), Dhaka, Bangladesh
shariaarfin096@gmail.com, rubabarahman209@gmail.com, {22-46985-1,22-46039-1}@student.aiub.edu, saef@aiub.edu

[2] Department of Mathematics, School of Advanced Sciences (SAS), Vellore Institute of Technology, Vellore, Tamil Nadu, India
kazi.tanvir2023@vitstudent.ac.in

Abstract. This study presents a novel approach for detecting diseases in cauliflower using advanced deep-learning techniques. Based on the scope of neural networks such as NasNet Mobile and InceptionV3, cauliflower's distinct leaf diseases were first detected and then classified in this study. To generate the dataset, pictures were taken under various weather conditions. The integration of NasNet Mobile and InceptionV3 not only improved the detection rate but also increased computational efficiency. This strategy will immensely help if employed during practical activities such as farming. The results elicited the use of a deep learning model to improve existing plant disease diagnosis techniques. It applied of LIME (Local Interpretable Model-agnostic Explanations) and Grad-CAM (Gradient-weighted Class Activation Mapping) as explainable artificial intelligence techniques to enhance the understanding of the suggested model's output and, therefore, foster decision-making transparency. In terms of the suggested model's performance metrics, we got a precision of 99.80%, a recall of 100% and F1-score of 99.90%. This innovative method, with its high accuracy and speed, holds promise for widespread adoption by farmers to maintain plant health and optimize crop yields.

Keywords: Cauliflower leaf disease · Explainable artificial intelligence · Computational efficiency · LIME · Grad-CAM

1 Introduction

Agriculture has a significant impact on the global economy. Many farmers rely on cauliflower cultivation as a source of nutrition and income. However, cauliflower production is frequently hampered by a variety of leaf diseases, which may significantly reduce both production and quality [1]. Fungal diseases like black rot and downy mildew, along with bacterial diseases such as black rot and bacterial spot

A. Dasgupta et al. (Eds.): BDA 2024, LNCS 15526, pp. 165–182, 2025.
https://doi.org/10.1007/978-3-031-81821-9_10

rot, pose serious threats to cauliflower plants. Although there are limited cures for these diseases, early detection and preventive measures can help mitigate their impact.

Black Rot [2] infected plants become stunted and may not form heads at all which leads to significant yield losses. Even if heads do form, they may be smaller, misshapen, and have internal discolouration, making them unmarketable. To prevent outbreaks, farmers may need to invest in disease-resistant seeds, copper-based bactericides, and stricter sanitation practices, all of which increase production costs. Bacterial Spot Rot [3], similar to black rot, the spots and blemishes caused by bacterial spot rot make cauliflowers less visually appealing and reduce their marketability. The open wounds created by the spots can become entry points for other diseases, further damaging the crop. Bacterial spot rot can spread easily through water splashing and contaminated tools, making it challenging to control once established. On the other hand, Downy Mildew [4] weakens the plants, hindering their ability to grow and develop healthy heads. The fungal growth on leaves disrupts photosynthesis, limiting the plant's energy production and affecting head formation. Downy mildew spores can spread quickly in cool, moist weather, leading to rapid infection throughout the field. The impact of these diseases extends beyond the field. For instance, yield losses due to diseases like Black Rot and Downy Mildew not only affect farmers' immediate income but also disrupt supply chains, leading to increased prices for consumers. Reduced quality and marketability can result in lower prices for affected crops, affecting farmers' profit margins. Annually, post-harvest losses in vegetable crops can range from 40% to 60% due to their perishable nature. Losses occur during various stages, including field conditions (15 to 20%), packaging and storage (15 to 20%), and transport (30 to 40%) [5]. Implementing effective disease management practices and ensuring proper handling and storage can help reduce these losses and improve the overall profitability of cauliflower farming. By understanding the nature of these diseases and adopting preventive measures, farmers can better protect their cauliflower crops, enhance yield and quality, and ultimately increase their profitability.

Early and precise identification of these diseases is critical. Traditional disease detection methods are time-consuming and prone to human error, which can result in significant economic losses. Conventional disease detection systems rely mainly on manual examination, which is both labour-intensive and inaccurate [1]. Due to these reasons, there is an urgent need for an automated, dependable, and effective approach to detecting cauliflower leaf diseases early and correctly. Implementing convolutional neural networks (CNNs) trained on large datasets of images of healthy and diseased leaves can enable rapid and accurate identification of symptoms like leaf spots or discolouration. Several studies have been conducted regarding the use of deep learning algorithms for recognizing cauliflower disorders from images [6]. Researchers have demonstrated that convolutional neural networks (CNNs) and exchange learning techniques can accurately diagnose common cauliflower diseases such as black rot, downy mould, and bacterial delicate rot [7]. To assist progress in this area of cauliflower disease detection, this study intends to develop a robust and effective deep-learning model for identifying and classifying disease in cauliflower leaves. Deep learning has revolutionized image classification tasks by offering remarkable accuracy and efficiency [8,9]. In this study, we leveraged the NASNet [10] and InceptionV3 [11] architectures due to their proven performance in image recognition. To enhance the trustworthiness and interpretability of our model, we integrated explainable AI (XAI) techniques, which provide insights into the model's decision-making process [12]. The primary objectives of this study are to develop an accurate and reliable model for detecting cauliflower leaf diseases and to enhance its interpretability using XAI. The key contributions are listed below:

1. We developed a customized application of NASNet and InceptionV3 architectures for cauliflower disease detection.
2. We applied the integration of XAI for model interpretability and validated our suggested model's performance by comparing it to current studies.

The remainder of this paper is organized as follows: Sect. 2 reviews related work in the field of plant disease detection. Section 3 details the methodology, including data collection, preprocessing, and model training. Section 4 presents the results and discussions with experimental results and analysis. Finally, Sect. 5 concludes the paper and outlines future work.

2 Related Works

The use of deep learning in plant and agricultural sciences is rapidly advancing due to the need for modernization and increased efficiency in these fields. Researchers are employing various deep-learning models to improve the accuracy and effectiveness of disease detection and crop management. For instance, Pradhan et al. [13] evaluated four leading deep learning models-EfficientNetB3, DenseNet121, VGG19 CNN, and ResNet50-to identify and classify cauliflower diseases. They found that EfficientNetB3 achieved the highest accuracy at 98%, followed by VGG19 CNN at 84%, DenseNet121 at 81%, and ResNet50 at 78%. This highlights the varying performance of different models and the superiority of EfficientNetB3 in this context.

Barman et al. [14] used various image processing techniques to extract ten statistical and GLCM features from cauliflower images. They employed the SelectKBest method with ANOVA F-value to identify the most important attributes for classification. Their machine learning models achieved an accuracy of 82.35%. In another study, Shakil et al. [15] introduced an agro-medical expert system that simplifies cauliflower disease diagnosis using digital images from phones or handheld devices. They utilized data augmentation to create a large dataset, segmented the disease-affected areas using k-means clustering, and extracted ten statistical and grey-level co-occurrence matrix (GLCM) features. Five machine learning algorithms were tested, with logistic regression on the augmented dataset achieving the highest accuracy of 90.77%.

Maria et al. [16] compared traditional machine learning techniques, utilizing k-means clustering for image segmentation and subsequent feature extraction, with transfer learning methods leveraging deep learning models such as InceptionV3, MobileNetV2, ResNet50, and VGG16. Their results indicate that Random Forest achieved 81.68% accuracy in disease classification, while InceptionV3 outperformed with 90.08% accuracy among CNN-based models. Abdul Malek et al. [17] investigated the effectiveness of CNNs and transfer learning using state-of-the-art architectures including VGG16, VGG19, ResNet50, and InceptionV3. The study specifically targeted the classification of four cauliflower diseases-bacterial soft rot, black rot, buttoning, and white rust-utilizing a dataset comprising approximately 2500 images. Their results demonstrated that InceptionV3 achieves the highest test accuracy of 93.93%. Li et al. [18] compared the performance of the MobileNet model with other models like VGG19, InceptionV3, and NASNetMobile. Their experimental results showed that the MobileNet model achieved a loss value of 0.033, an accuracy of 99.27%, and an F1 score of 99.24% on the test set, with a learning rate of 0.001, dropout set to 0.5, and 80 frozen layers.

Zhong et al. [19] introduced LiteMixer, a lightweight model designed to address efficiency issues in disease detection. The model includes a Lightweight Mixed-Domain Feature Extraction (LMFE) module and a Multi-Scale Lightweight Convolutional Attention Fusion (MLCAF) module. LiteMixer achieved a remarkable accuracy of 99.86% in identifying cauliflower diseases, demonstrating the effectiveness of combining LMFE and MLCAF modules. Kanna et al. [20] examined the effectiveness of ten deep transfer learning models, including EfficientNetB0, Xception, MobileNetV2, DenseNet201, InceptionResNetV2, and ResNet152V2, using the VegNet dataset. EfficientNetB1 stood out with the highest validation accuracy of 99.90%, the lowest loss of 0.16, and a root mean square error of 0.40. These studies collectively underscore the significant potential of deep learning models in enhancing agricultural practices, particularly in disease detection and management, ultimately benefiting farmers and improving crop yields. Uddin et al. [7] presented a custom YOLOv8 model that was finetuned for cauliflower disease detection. They achieved impressive results with a precision of 93.2%, recall of 82.6%, and mean average precision (mAP) of 91.1% on a test dataset.

Suryavanshi et al. [21] employed federated averaging to combine the local learnings from each client's dataset into a single global model. The study demonstrated the model's effectiveness, showing consistent improvements in performance metrics. The model's performance was evaluated using macro and weighted averages, with results ranging from 91.83% to 97.42% for macro averages and 91.88% to 97.41% for weighted averages across different client datasets. Additionally, micro averages, indicating overall accuracy, showed an improvement from 91.87% for px-1 to outstanding results nearing 97% for px-5. Similarly, Chattopadhyay et al. [22] introduced a categorization scheme for cauliflower leaf diseases based on four severity levels, utilizing data from four different clients. Federated Learning enabled the integration of local data from two clients (fq1 to fq4) into a global model using Federated Averaging, ensuring the model benefited from diverse data while maintaining privacy. Their results showed a consistent improvement in classification accuracy across all metrics. The Macro Average, treating all classes equally, improved from 96.35% for fq-1 to 99.11% for fq-4. The Weighted Average, accounting for class representation, increased from 96.45% for fq-1 to 99.10% for fq-4. Similarly, the Micro Average, which sums contributions from all classes, rose from 96.44% for fq-1 to 99.10% for fq-4. These findings highlight the potential of combining federated learning with CNNs to create a powerful tool for accurately diagnosing and classifying cauliflower diseases from diverse data sources. Moreover, recent advancements in plant disease detection have focused on deep learning models combined with effective segmentation techniques to enhance accuracy and performance. Traditional methods, including manual inspection and basic image processing, have been replaced by CNN-based approaches, with EfficientNet emerging as a state-of-the-art architecture due to its efficient scaling of network depth, width, and resolution. Segmentation models like U-Net and its modified versions have significantly improved image preprocessing by isolating diseased leaf regions from the background. Chowdhury et al. [23] modified U-Net achieved high segmentation accuracy (98.66% accuracy, 98.5% IoU), while EfficientNet-B7 and EfficientNet-B4 outperformed earlier models, reaching accuracies of 99.95% for binary classification and 99.89% for ten-class classification. These results highlight the importance of deeper networks and effective segmentation, though further research is needed to enhance model generalizability and interpretability.

The proposed method, combining NasNet and Inception V3, stands out with a 99.90% accuracy and includes Explainable AI (XAI), addressing the need for model transparency. Table 3 highlights a significant gap in previous studies, which focused primarily on achieving high accuracy without sufficient emphasis on explainability. For example, methods like Inception V3, EfficientNetB3, and MobileNet achieved accuracies ranging from 90.08% to 99.27% but did not incorporate XAI. This lack of explainability poses challenges for practical applications, as users and stakeholders need to understand how and why decisions are made by the model, particularly in critical fields like agriculture. The integration of XAI in the proposed method not only enhances performance but also ensures that the model's decision-making process is transparent and interpretable. This is crucial for gaining the trust of farmers and other users, enabling them to make informed decisions based on the model's output. Therefore, the proposed method fills this gap by providing both high accuracy and a clear understanding of the model's workings, making it a more robust and practical solution for cauliflower disease detection.

3 Methodology

In this study, we focused on improving the accuracy of our classification tasks by using sophisticated feature extraction approaches. We started our work by collecting an appropriate dataset and then we performed data pre-processing to ensure the accuracy of the data. The NasNet [10] and Inception v3 [24] architectures are well-known for their capacity to identify complex patterns in images. These two models were used for feature extraction. The forward feed layers of these models were then used to carry out decision-making and classification, using their capacity to analyze extracted features and generate precise predictions. We used XAI approaches, such as GradCAM to see the location of the models' focus within images during classification and LIME for local interpretability, which reveals the contribution of individual features to predictions, to validate and give transparency to these predictions. The workflow of this study is given below in Fig. 1.

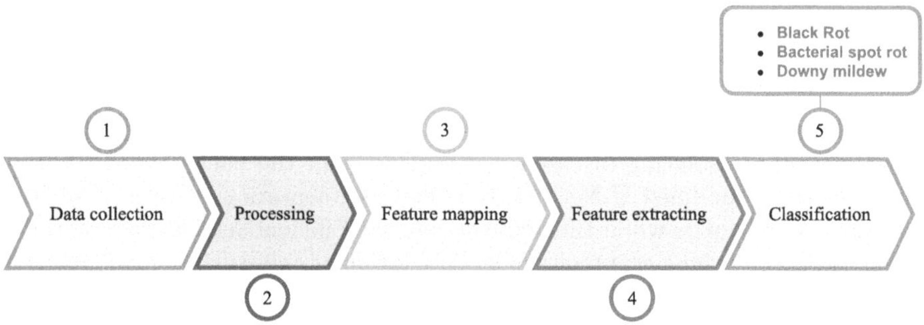

Fig. 1. Detailed diagram presenting the overview of the proposed methodology.

3.1 Dataset Overview

In this research, we used a comprehensive dataset to detect diseases in cauliflower leaves. The dataset [25] was collected from the Manikganj vegetable-producing area near Dhaka, Bangladesh. All photographs were obtained with digital camera equipment. From December 2021 to January 2022, the photographs are recorded in a natural setting with various viewpoints and backdrops. The images were manually captured between December 20th and January 15th, coinciding with the full bloom of the flowers and the clear manifestation of most diseases. The authors carefully took all the images during the peak of cauliflower maturity, following the guidance of a plant pathology expert from the Bangladesh Rice Research Institute (BARI) in Gazipur. This dataset is meticulously organized, making it ideal for the development and validation of machine learning-based automated cauliflower disease detection algorithms. The dataset is hosted by the Department of Computer Science at the National Institute of Textile Engineering and Research (NITER). It includes a total of 656 images of both healthy and diseased leaves. The images are categorized into four classes: Downy Mildew, Black Rot, Bacterial Spot, and Healthy Leaves. To increase the dataset size, we applied data augmentation techniques, resulting in an augmented dataset with 7,360 images. A batch of collected images is shown below in Fig. 2.

(a) (b) (c) (d)

Fig. 2. Sample images of the collected dataset. a) Bacterial spot rot, b) Downy Mildew, c) Black rot, d) No disease, e) No disease.

3.2 Data Preprocessing

Data processing is a crucial first step in improving the classification performance of deep learning models. Initially, we resized the original images in the dataset and applied colour inversion. To increase the volume and diversity of the data, we used various data augmentation techniques on the original images. These techniques included centring, normalization, rotation, shifting, and flipping, which expanded the dataset to 7,360 images. This ensured a diverse set of images that enhanced model performance. Additionally, we employed the Interquartile Range (IQR) method to detect outliers and the Mean Absolute Deviation (MAD) method to handle them. MAD is effective for identifying and managing extreme deviations in image characteristics. The formula for MAD is,

$$\text{MAD} = \frac{1}{N}\sum_{i=1}^{N} |x_i - \text{median}(x)| \tag{1}$$

3.3 Proposed Architecture

NasNet Mobile & Inception V3 Architecture

NASNet is designed using a method called neural architecture search (NAS), which employs reinforcement learning to automatically identify the most effective network structure for a specific task [26]. This approach allows for the discovery of architectures that may not be intuitively designed by humans, optimizing performance across various datasets. NASNet is particularly well-suited for tasks involving image classification due to its ability to adaptively find the best configurations for feature extraction.

InceptionV3, on the other hand, is a convolutional neural network architecture that excels in image classification. It utilizes a series of convolutional and pooling layers to detect fundamental features like edges and textures in images [11]. A key feature of InceptionV3 is its use of inception modules, which allow the model to analyze data at multiple scales simultaneously. This multi-scale analysis helps the model capture a wide range of features, improving its ability to understand complex images. Additionally, InceptionV3 includes "reduction blocks" that compress the size of feature maps while preserving essential information. These blocks help to maintain computational efficiency, reducing the processing load without compromising the model's performance [11]. In Fig. 3, we have illustrated the architecture through a diagram, showcasing how these components work together. In our study, we employed NASNet Mobile and InceptionV3 as the backbone models for extracting features and performing classification tasks on cauliflower leaf diseases. We initialized both models by excluding their final fully connected layers and utilizing weights pre-trained on the ImageNet dataset, a large benchmark dataset used widely in computer vision tasks. To ensure stability during training, we set both backbones to "trainable=False," preventing the updates of their pre-trained weights. To extract features, we passed input data through each backbone and then applied "GlobalAveragePooling2D," which reduces the spatial dimensions of the feature maps while retaining important information. This pooling operation helps to summarize the spatial information, making the subsequent classification task more efficient. The extracted features from NASNet Mobile and InceptionV3 were concatenated, followed by flattening and normalization processes to prepare the data for classification. In our forward feed layer, we utilized the "ReLU" (Rectified Linear Unit) activation function, which enhances non-linearity and aids in capturing complex relationships within the data. To further stabilize and accelerate the training process, we applied batch normalization.

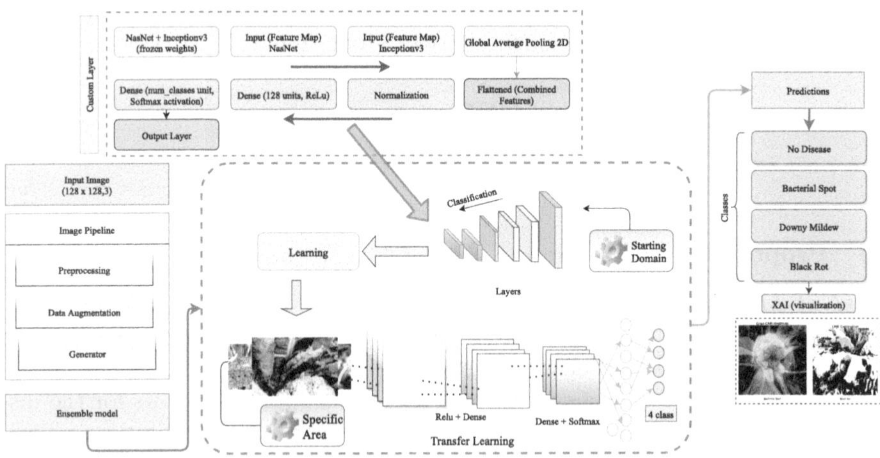

Fig. 3. Architecture of the proposed model of NasNet and Inception V3.

Our final output layer consisted of a dense layer with "softmax" activation, which is crucial for multi-class classification tasks. This layer provides probability distributions across the different classes, enabling the model to effectively classify the input images of cauliflower leaves. Overall, our custom NASNet Inception model demonstrated enhanced feature extraction capabilities and improved accuracy in classifying cauliflower leaf diseases.

3.4 Explainable AI

GradCAM [27] and LIME [28] are two significant tools in AI that help to make complicated deep learning models easier to understand. They offer different ways to show why and how these models make predictions, which is really helpful for people trying to trust and use AI in various fields. GradCAM provides insights into how a neural network makes decisions by highlighting the regions of input data that are most influential for a particular prediction. It achieves this by computing gradients of the prediction score y_c (before applying softmax) with respect to the activations A_k of a convolutional layer.

$$\frac{\partial y^c}{\partial A^k} \tag{2}$$

These gradients indicate how changes in the activation maps A_k affect the prediction score for a specific class. To highlight important regions, GradCAM then calculates weights α_c^k through global average pooling of these gradients.

$$\alpha_k^c = \frac{1}{Z}\sum_i \sum_j \frac{\partial y^c}{\partial A_{ij}^k} \tag{3}$$

These weights represent the importance of each activation map A_k in the final prediction. Using these weights, GradCAM generates a heatmap L^c_{GradCAM} that visually shows which parts of the input image contributed the most to predicting class.

$$L^c_{Grad-CAM} = \text{ReLU}\left(\sum_k \alpha_k^c A^k\right) \tag{4}$$

On the other hand, LIME focuses on generating local, interpretable approximations of a model's behaviour for individual predictions. It achieves this by creating perturbed samples $\{x'_i\}$ around a given instance x. Each perturbed sample x'_i is a modified version of x where features are either included or excluded. For these samples, LIME obtains predictions $f(x'_i)$ from the original complex model f. To provide a simpler explanation, LIME fits an interpretable model g (such as a linear model) that approximates the predictions of f around x. This fitting process minimizes a loss function that balances the proximity of x to the perturbed samples z_i, the difference between $f(z_i)$ $g(z_i)$, and complexity penalty $\Omega(g)$ on g.

$$\xi(x) = \arg\min_{g\in G} \sum_i \pi_x(z_i)\left(f(z_i) - g(z_i)\right)^2 + \Omega(g) \tag{5}$$

The resulting model g provides insights into the local behaviour of the original model f at x, with its coefficients indicating feature importance.

4 Results and Findings

The proposed deep learning and XAI-based approach for cauliflower disease detection was evaluated using a benchmark dataset. Our method demonstrated significant improvements in performance metrics compared to existing approaches. Specifically, our strategy achieved promising precision, recall, F1-score and accuracy.

4.1 Evaluation Criteria

Accuracy is a popular evaluation metric, but it's not always enough to measure a model's performance. Therefore, we also considered other metrics like precision, recall, F1-score, ROC, AUC score, weighted average, macro average, and accuracy. We generated a ROC curve and calculated AUC scores for each class.

$$\mathcal{P} = \frac{\sum_{i=1}^{N} \mathrm{TP}_i}{\sum_{i=1}^{N} \mathrm{FP}_i + \sum_{i=1}^{N} \mathrm{TP}_i} \tag{6}$$

$$\mathcal{R} = \frac{\sum_{i=1}^{N} \mathrm{TP}_i}{\sum_{i=1}^{N} \mathrm{FN}_i + \sum_{i=1}^{N} \mathrm{TP}_i} \tag{7}$$

$$\mathcal{F}_1 = \frac{2 \cdot \mathcal{R} \cdot \mathcal{P}}{\mathcal{R} + \mathcal{P}} \tag{8}$$

$$\mathcal{A} = \frac{\sum_{i=1}^{N} (\mathrm{TP}_i + \mathrm{TN}_i)}{\sum_{i=1}^{N} (\mathrm{TP}_i + \mathrm{TN}_i + \mathrm{FP}_i + \mathrm{FN}_i)} \tag{9}$$

Precision measures the correctness of positive predictions by calculating the ratio of true positive results to the total predicted positives. In disease detection, precision helps to measure how many of the predicted positive cases (disease detections) are actually correct. For example, in diagnosing critical diseases, higher precision reduces the number of false alarms. Recall assesses a model's ability to identify all actual positive instances by dividing the true positives by the sum of true positives and false negatives. The F1-score provides a balance between precision and recall by calculating their harmonic mean. Accuracy is the proportion of correctly classified instances among all instances.

The ROC-AUC score is a performance measurement for classification problems at various threshold settings. The ROC curve plots the true positive rate against the false positive rate, while the AUC score quantifies this relationship, with higher values indicating better performance. Weighted average considers the support (the number of true instances) of each class to provide a balanced performance measure, accounting for class imbalances. Macro average treats all classes equally, providing an average performance measure without considering class imbalance. By employing these metrics, we ensure a more holistic evaluation of our model, particularly in the context of disease detection, where class imbalance and false negatives can have significant consequences.

4.2 Experimental Results

All settings were configured to ensure consistency while evaluating feature extraction and classification. The experiment was conducted using Google Colaboratory [29], which is a cloud-based service provided by Google. We used Tensorflow and GlobalAveragePooling2D services in the Python 3 programming language. Moreover, we used a virtual TPU for processing, provided by Google Colab. The dataset was split into training, testing, and validation sets with an 80:10:10 ratio. Images were resized to 128 × 128 pixels and normalized by rescaling with a factor of 1./255.

Table 1. Hyperparameters and values of the proposed model

Hyperparameter	Value
Batch Size	32
Optimizer	Nadam
Learning Rate	0.0001
Weight Decay	640
Beta_1	0.9
Beta_2	0.999
Epsilon	1e–07
Epochs	25
Classes	4

Features from the backbones were combined and processed through dense layers with batch normalization and the "ReLU" activation function. In the output layer, the "Softmax" activation function has been used. During training, we used ReduceLROnPlateau to lower the learning rate when needed and EarlyStopping to stop training if validation loss didn't improve for five epochs. Table 1 illustrates all the hyperparameters we used in our experiment.

4.3 Performance Analysis of Proposed Architecture

The proposed deep learning architecture for classifying cauliflower leaf disease performed extremely well. It achieved a test accuracy of 99.15% after 25 epochs. This shows the model's ability to recognize complex patterns in the data and accurately predict new, unseen data. During training, both training and validation losses consistently decreased. The final test loss was 0.0280. The accuracy and loss curves in Fig. 4 shows the model's improving accuracy and decreasing loss over time.

Figure 5 shows the ROC curve of the model. The confusion matrix in Fig. 6 provides more details on how well the model classified the diseases, showing both correct and incorrect predictions. Overall, these results demonstrate the effectiveness of the model in classifying cauliflower leaf disease.

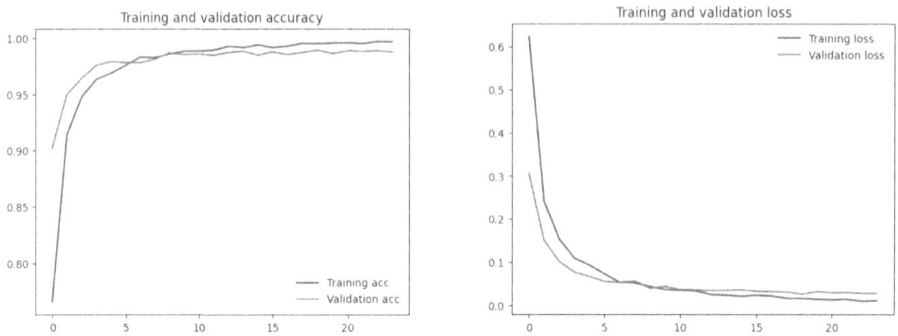

(a) (b)

Fig. 4. Graphical representation of (a) Training-Validation and, (b) Training-Loss accuracy

Both training and validation accuracy improved quickly within the first few epochs which indicates an effective initial learning. The validation accuracy stabilizes around 97%. It indicates good generalization to unseen data with minimal overfitting. On the other hand, both training and validation losses drop significantly within the first few epochs. Moreover, the validation loss follows the training loss closely and remains low. It also indicating good generalization with minimal overfitting.

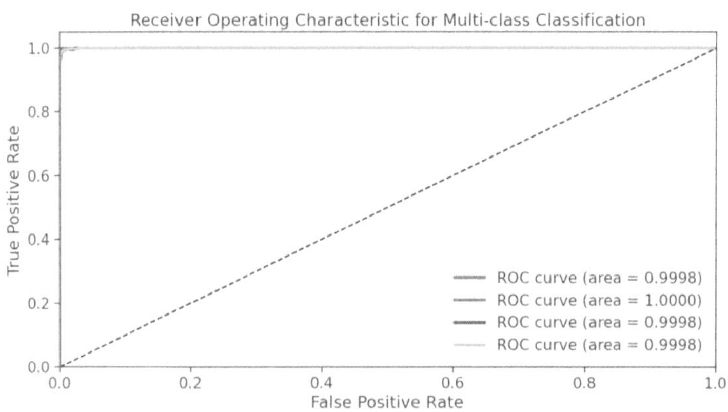

Fig. 5. ROC curve with AUC scores.

The ROC curve for multi-class classification shown in the image illustrates the performance of the classifier for different classes of cauliflower diseases. Each coloured line represents the ROC curve for one class, with the area under the curve (AUC) values indicating the classifier's effectiveness. The green curve, with an AUC of 1.0000, signifies perfect classification performance for that class. The other curves, with AUC values of 0.9998, also show near-perfect performance. This indicates that the classifier is highly effective and reliable, with an excellent ability to distinguish between the different disease classes, making it a valuable tool for accurate and efficient disease detection in cauliflower.

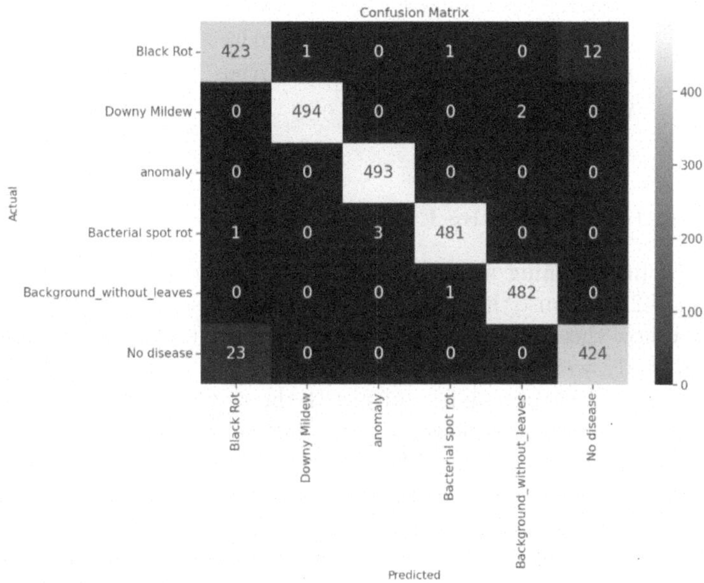

Fig. 6. Confusion Matrix of proposed methods.

In Fig. 6, the confusion matrix shows high classification accuracy across all categories, as indicated by the high values along the diagonal. Specifically, the model correctly classified 493 out of 500 samples with "No disease," 500 out of 500 samples with "Black Rot," 496 out of 500 samples with "Downy Mildew," and 494 out of 500 samples with "Bacterial spot rot." Although the model performs exceptionally well, focusing on reducing the misclassification of "No disease" and "Downy Mildew" could further enhance performance. Fine-tuning the model or incorporating more diverse training data could help address these misclassifications.

Table 2 shows the performance of a model in detecting four classes of conditions in cauliflower plants: No Disease, Black Rot, Downy Mildew, and Bacterial Spot Rot. The metrics used are Precision, Recall, and F1-Score, all of which are very high (0.99 or 1.00), indicating excellent performance. The "Support" column indicates that each class has 500 samples, totalling 2000 samples. Overall accuracy, macro average (the average performance across all classes), and weighted average (considering the number of samples per class) are all 0.99, further demonstrating the model's high performance.

Table 2. Precision, Recall, F1-Score, and Support for Each Class

	Precision	Recall	F1-Score	Support
No Disease	0.99	0.99	0.99	500
Black Rot	0.99	1.00	1.00	500
Downy Mildew	0.99	0.99	0.99	500
Bacterial Spot Rot	0.99	0.99	0.99	500
Accuracy	0.99	0.99	0.99	2000
Macro Avg.	0.99	0.99	0.99	2000
Weighted Avg.	0.99	0.99	0.99	2000

4.4 Result Justification by Explainable AI

Complex machine learning models and deep neural networks are often seen as "black boxes" [28] because their decision-making processes are hidden. XAI techniques reveal how these models work and explain why they make certain predictions. In our study, we used LIME (Local Interpretable Model-agnostic Explanations) and Grad-CAM to explain and justify the results. NasNet and Inception V3 predicted whether a cauliflower leaf is healthy or diseased. We used LIME to understand these predictions. LIME slightly changed the input image and watched how the model's predictions change. It then creates a simple model to mimic the complex model's behaviour around the input image. By doing this, LIME showed which parts of the image influenced the model's decision the most [27]. Grad-CAM, on the other hand, produced a heatmap that highlighted the important areas of the input image for the model's prediction. Grad-CAM calculated the gradients of the target class (like diseased or healthy) with respect to the final convolutional layer's feature maps. These weighted activation maps are combined to create a heatmap, which is then placed over the original image to show the areas the model focused on [27]. Overall, LIME and Grad-CAM help build trust, ensure accountability, and improve the models through better understanding and debugging. In Fig. 7, we have shown the interpretation of LIME and Grad-CAM.

From Table 2, we can see that the performance of the model is excellent. The metrics, which are Precision, Recall, and F1-Score, are all very high (either 0.99 or 1.00) for every class. This means the model is very accurate in identifying whether a cauliflower plant has no disease, Black Rot, Downy Mildew, or Bacterial Spot Rot. The "Support" column shows the number of samples for each class, which is 500 samples per class, making a total of 2000 samples. In Table 3, we see a summary of previous studies on detecting diseases in cauliflower plants using deep learning. Different models were used, such as Logistic Regression, InceptionV3, LiteNet, and EfficientNetB3. These results are very impressive, but our model surpassed their results. Also, none of these studies used Explainable AI (XAI). In our work, we applied Explainable AI to our model. This means we not only have high accuracy but also made sure our model is trustworthy and

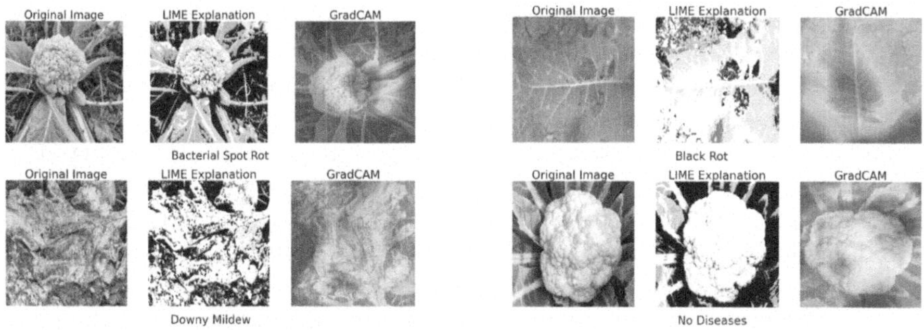

Fig. 7. Interpretation of LIME & Grad-CAM.

its decisions can be understood and explained. This makes our approach more reliable and effective.

Table 3. Summary of Related Works on Cauliflower Disease Detection Using Deep Learning

Year	Best Method	Results	Use of XAI
2021 [16]	Inception V3	90.08%	×
2021 [23]	U-Net Segmentation	98.66%	×
2021 [23]	EfficientNetB7	99.95%	×
2024 [13]	EfficientNetB3	98%	×
2024 [14]	Logistic Regression	90.77%	×
2022 [17]	InceptionV3	93.93%	×
2024 [19]	LiteMixer	99.86%	×
2024 [20]	EfficientNetB1	99.90%	×
2024 [7]	Custom YOLOv8	mAP 91.1%	×
2024 [21]	Federated Learning with CNN	97.42%	×
2024 [22]	Federated Learning with CNN	99.11%	×
2024 [18]	MobileNet	99.27%	×
Proposed	**NasNet, Inception V3**	**99.90%**	✓

4.5 Limitations and Future Work

One limitation of our study is the potential variability in the model's performance under different environmental conditions and variations in image quality, such as lighting, weather, and camera resolution, which can affect the accuracy of disease detection. To improve the model's robustness, future work will focus on advanced data augmentation techniques that simulate various environmental conditions, enhancing the model's generalization ability. We also plan to implement real-time data collection with different unknown classes and feedback mechanisms, allowing the model to continuously adapt and improve based on new data.

Furthermore, integrating multi-spectral and hyperspectral imaging data could provide additional features for more precise disease detection by capturing information beyond the visible spectrum. Expanding the model's scope to cover a broader range of crops and plant diseases will further enhance its practical applicability in agriculture. Incorporating the model into decision support systems (DSS) could provide farmers with real-time, actionable insights for managing crop health. Finally, collaborating with agricultural experts and conducting extensive field trials will be essential to validate the model's effectiveness in real-world scenarios, ensuring its reliability and utility in diverse conditions.

5 Conclusion

Our proposed model has the potential to transform cauliflower farming by providing a reliable and timely identification system for diseases such as Bacterial Spot, Black Rot, and Downy Mildew. Designed for both novice and experienced farmers, it seamlessly integrates with smartphone technology for real-time monitoring, enabling swift action against emerging threats. Its cost-effectiveness and user-friendly interface ensure broad accessibility, protecting crops from damage and minimizing financial losses. The primary advantage of our model is its ability to quickly and accurately identify diseases, allowing farmers to implement timely interventions that enhance crop health. This leads to higher yields and increased profits, contributing to the sustainability of cauliflower farming. However, deploying this model presents challenges, such as variations in image quality caused by poor lighting, camera resolution, or weather conditions. These factors can reduce the model's accuracy. Differences in smartphone camera quality among users may further exacerbate these issues. Overall, this technology represents a significant advancement in disease prevention and crop management for cauliflower cultivation. By integrating this model into their practices, farmers can achieve better outcomes and contribute to sustainable agriculture. The continued development of this technology will further solidify its role as a valuable asset in the future of cauliflower farming.

References

1. Sharma, R., Mittal, M., Gupta, V., Vasdev, D.: Detection of plant leaf disease using advanced deep learning architectures. Int. J. Inf. Technol., 1–18 (2024)
2. Gupta, M., Vikram, A., Bharat, N.: Black rot-a devastating disease of crucifers: a review. Agric. Rev. **34**(4), 269–278 (2013)
3. Sharma, B.R., Dhiman, J.S., Thakur, J.C., Singh, A., Bajaj, K.L.: Multiple disease resistance in cauliflower. Adv. Hort. Sci. **5**(1), 30–34 (1991)

4. Shaw, R.K., Shen, Y., Zhao, Z., Sheng, X., Wang, J., Yu, H., Gu, H.: Molecular breeding strategy and challenges towards improvement of downy mildew resistance in cauliflower (Brassica oleracea var. botrytis L.). Front. Plant Sci. **12**, 667757 (2021)
5. Tripathi, A.N., Tiwari, S.K., Behera, T.K.: Postharvest diseases of vegetable crops and their management. In: Postharvest Technology-Recent Advances, New Perspectives and Applications. IntechOpen (2022)
6. Pradhan, N.R., Ghosh, H., Rahat, I.S., Ramesh, J.V.N., Yesubabu, M.: Enhancing agricultural sustainability with deep learning: a case study of cauliflower disease classification. EAI Endorsed Trans. Internet Things **10** (2024)
7. Uddin, M.S., et al.: Cauli-Det: enhancing cauliflower disease detection with modified YOLOv8. Front. Plant Sci. **15**, 1373590 (2024)
8. Jahan, I., et al.: 4d: a real-time driver drowsiness detector using deep learning. Electronics **12**(1), 235 (2023)
9. Sumit, S.S., et al.: Restinet: on improving the performance of tiny-yolo-based cnn architecture for applications in human detection. Appl. Sci. **12**(18), 9331 (2022)
10. Adedoja, A.O., Owolawi, P.A., Mapayi, T., Tu, C.: Intelligent mobile plant disease diagnostic system using nasnet-mobile deep learning. IAENG Int. J. Comput. Sci. **49**(1), 216–231 (2022)
11. Joshi, K., Tripathi, V., Bose, C., Bhardwaj, C.: Robust sports image classification using inceptionv3 and neural networks. Procedia Comput. Sci. **167**, 2374–2381 (2020)
12. Gao, Y., Gu, S., Jiang, J., Hong, S.R., Yu, D., Zhao, L.: Going beyond xai: a systematic survey for explanation-guided learning. ACM Comput. Surv. **56**(7), 1–39 (2024)
13. NR Pradhan, H Ghosh, IS Rahat, JVN Ramesh, and M Yesubabu. Enhancing agricultural sustainability with deep learning: A case study of cauliflower disease classification. EAI Endorsed Trans. Internet Things **10** (2024)
14. Barman, M.R., Biswas, A.A., Sultana, M., Rajbongshi, A., Zulfiker, M.S., Tabassum, T.: Automated classification of diseased cauliflower: a feature-driven machine learning approach. TELKOMNIKA (Telecommun. Comput. Electron. Control) **22**(4), 887–896 (2024)
15. Shakil, R., Akter, B., Shamrat, F.J.M., Noori, S.R.H.: A novel automated feature selection based approach to recognize cauliflower disease. Bull. Electr. Eng. Inf. **12**(6), 3541–3551 (2023)
16. Maria, S.K., Taki, S.S., Mia, M.J., Biswas, A.A., Majumder, A., Hasan, F.: Cauliflower disease recognition using machine learning and transfer learning. In: Somani, A.K., Mundra, A., Doss, R., Bhattacharya, S. (eds.) Smart Systems: Innovations in Computing. SIST, vol. 235, pp. 359–375. Springer, Singapore (2022). https://doi.org/10.1007/978-981-16-2877-1_33
17. Malek, M.A., Reya, S.S., Zahan, N., Hasan, M.Z., Uddin, M.S.: Deep learning-based cauliflower disease classification. In: Computer Vision and Machine Learning in Agriculture, vol. 2, pp. 171–186. Springer, Heidelberg (2022)
18. Li, Y., Xue, J., Wang, K., Zhang, M., Li, Z.: Surface defect detection of fresh-cut cauliflowers based on convolutional neural network with transfer learning. Foods **11**(18), 2915 (2022)
19. Zhong, Y., Teng, Z., Tong, M.: LiteMixer: cauliflower disease diagnosis based on a novel lightweight neural network. Comput. J., bxae010 (2024)
20. Kanna, G.P., et al.: Advanced deep learning techniques for early disease prediction in cauliflower plants. Sci. Rep. **13**(1), 18475 (2023)

21. Suryavanshi, A., Kukreja, V., Malhotra, S., Mehta, S., Choudhary, A.: Transforming crop diagnosis: the role of federated learning cnn in cauliflower leaf disease detection. In: 2024 3rd International Conference for Innovation in Technology (INOCON), pp. 1–6. IEEE (2024)
22. Chattopadhyay, S., Verma, A., Srivastava, A., Kukreja, V., Mehta, S., Hariharan, S.: Cauliflower leaf disease: unraveling severity levels with federated learning cnn. In: 2024 3rd International Conference for Innovation in Technology (INOCON), pp. 1–6. IEEE (2024)
23. Chowdhury, M.E.H., et al.: Automatic and reliable leaf disease detection using deep learning techniques. AgriEngineering **3**(2), 294–312 (2021)
24. Bazgir, E., Haque, E., Maniruzzaman, M., Hoque, R.: Skin cancer classification using Inception Network. World J. Adv. Res. Rev. **21**(2), 839–849 (2024)
25. Sara, U., Rajbongshi, A., Shakil, R., Akter, B., Uddin, M.S.: VegNet: an organized dataset of cauliflower disease for a sustainable agro-based automation system. Data Brief **43**, 108422 (2022)
26. Zoph, B., Vasudevan, V., Shlens, J., Le, Q.V.: Learning transferable architectures for scalable image recognition. In: Proceedings of the IEEE Conference on Computer Vision and Pattern Recognition, pp. 8697–8710 (2018)
27. Nguyen, H.T.T., Cao, H.D., Nguyen, K.V.T., Pham, N.D.H.: Evaluation of explainable artificial intelligence: Shap, lime, and cam. In: Proceedings of the FPT AI Conference, pp. 1–6 (2021)
28. Dieber, J., Kirrane, S.: Why model why? assessing the strengths and limitations of lime. *arXiv preprint*arXiv:2012.00093 (2020)
29. Bisong, E., Bisong, E.: Google colaboratory. In: Building Machine Learning and Deep Learning Models on Google Cloud Platform: A Comprehensive Guide for Beginners, pp. 59–64 (2019)

MARINE: A Computer Vision Model for Detecting Rare Predator-Prey Interactions in Animal Videos

Zsófia Katona[1], Seyed Sahand Mohammadi Ziabari[1,2(✉)], and Fatemeh Karimi Nejadasl[1]

[1] University of Amsterdam, Amsterdam, The Netherlands
zsofia.katona@student.uva.nl, {s.s.mohammadiziabari, f.kariminejadasl}@uva.nl

[2] SUNY Empire State University, 111 West Ave, Saratoga Springs, NY 12866, USA

Abstract. Encounters between predator and prey play an essential role in ecosystems, but their rarity makes them difficult to detect in video recordings. Although advances in action recognition (AR) and temporal action detection (AD), especially transformer-based models and vision foundation models, have achieved high performance on human action datasets, animal videos remain relatively underresearched. This paper addresses this gap by proposing the model MARINE, which utilizes motion-based frame selection designed for fast animal actions and DINOv2 feature extraction with a trainable classification head for action recognition. MARINE outperforms VideoMAE in identifying predator attacks in videos of fish, both on a small and specific coral reef dataset (81.53% against 52.64% accuracy), and on a subset of the more extensive Animal Kingdom dataset (94.86% against 83.14% accuracy). In a multi-label setting on a representative sample of Animal Kingdom, MARINE achieves 23.79% mAP, positioning it mid-field among existing benchmarks. Furthermore, in an AD task on the coral reef dataset, MARINE achieves 80.78% AP (against VideoMAE's 34.89%) although at a lowered t-IoU threshold of 25%. Therefore, despite room for improvement, MARINE offers an effective starter framework to apply to AR and AD tasks on animal recordings and thus contribute to the study of natural ecosystems.

Keywords: Action recognition · Large vision models · Frame selection · Animal videos · Predator attacks

1 Introduction

Encounters between predator and prey play an essential role in ecosystem dynamics. While it has been long understood that predators play a crucial role in the regulation of prey populations, the last decades have shown a growing interest in the complexities of predator-prey interactions and their importance in terms of the ecosystem and biodiversity [14]. An understanding of predator-prey interactions is especially important in light of the ongoing climate change, as a changing climate has been shown to alter

A. Dasgupta et al. (Eds.): BDA 2024, LNCS 15526, pp. 183–199, 2025.
https://doi.org/10.1007/978-3-031-81821-9_11

predator-prey dynamics across a range of different habitats and species [4, 17, 25], and it is also implied that the understanding of predator-prey interactions is crucial in designing sustainable solutions in biological systems [13, 27].

At the same time, the observation of predator-prey interactions presents a significant challenge even in a time when video footage from wildlife camera traps is available. The reason behind this is the rarity of these interaction events (such as attacks), which means that relevant sections of the camera footage typically comprise only an extremely small fraction of the total length of videos, making manual review unfeasible and calling for the use of artificial intelligence to detect relevant events in the footage [26].

Automatic action recognition (AR) and temporal action detection (AD) may be effective approaches for the identification of relevant events. Advanced techniques in the field of computer vision, especially pre-trained transformer [7, 19, 29] and foundation models (large vision models, LVMs) [24], have achieved high performance in AR and AD tasks. However, most of the studies in this field concern only the domain of human actions (with large benchmarking datasets such as Kinetics 400 [15], Something Something [12], and UCF101 [28]). In comparison, AR and AD in animal videos appear to be a much less researched field.

Existing papers on animal AR and AD tend to focus on a specific habitat or species, which are most often terrestrial large mammals [9, 23, 26, 31], while other types of animals (fish, birds, etc.) and habitats (water environments) are relatively neglected. Animal AR and AD studies most often apply custom datasets collected for a narrow application, which is understandable considering that the video analysis often serves the purpose of some specific animal study. On the other hand, this makes the results of these studies difficult to generalize over a wider domain. At the same time, there have been attempts to tackle AR and AD tasks on more diverse animal datasets (such as in [20]), most notably on Animal Kingdom [22], which comprises videos of a variety of different species and environments. Such studies, however, are still relatively scarce compared to the abundance of human action recognition research.

Therefore, there is a scientific gap in approaches to identify actions (especially rare ones, such as predator attacks) in animal videos which can efficiently be applied to small and specific datasets (especially in scarcely researched domains, such as marine habitats), while also being able to generalize over larger and more diverse ones. This paper addresses this gap by introducing the model MARINE (Motion-based Action Recognition In Natural Environments), which comprises motion-based frame selection in videos to identify the most relevant frames for fast animal actions, a pre-trained DINOv2 LVM backbone for feature extraction, and a lightweight trainable classification head to adapt to custom datasets. To test its applicability to small and specific datasets, MARINE is applied in both an AR and AD task to a custom dataset of fish predator attacks at a coral reef, and to test its generalizability to more species and more varied environments, it is applied in an AR task to a subset of the Animal Kingdom dataset [22] comprising videos of predation-related actions of diverse fish species. MARINE's scalability is an important consideration for future deployment in larger datasets or diverse ecological contexts. While the model demonstrates strong performance on relatively small, specific datasets, applying MARINE to broader datasets may require additional computational resources. Varying video quality, environmental conditions, and species diversity could impact the model's effectiveness. These challenges suggest the need for further optimizations, such as dynamic frame selection methods or preprocessing adjustments, to ensure robust performance across different settings.

With this approach, the paper aims to answer the following research question: *To what extent can computer vision techniques be used to automatically detect rare ecological events, such as predation, in video footage of fish species?*

2 Related Work

The paper is directed at filling the research gap on approaches in the automatic recognition of rare ecological events, specifically predator attacks among fish. Due to the relatively low amount of studies on this problem and the often narrow scope of the existing ones, it is expected that the project could offer a valuable contribution to the domain.

Action recognition (AR) is the task of classifying an action seen on video. Closely related is the task of action detection or temporal action localization (AD), which is the task of selecting frames depicting a certain action in a longer video sequence (i.e., finding the timestamps marking the beginning and end of an action within the video). In both cases, essential is the ability to encode and combine spatial and temporal information. Earlier attempts at action recognition proposed 2D and 3D convolutional frameworks [3, 32], which were designed to pool information from both the spatial and temporal dimension. As a notable example, the SlowFast family is known for applying two pathways, one at a lower frame rate to capture spatial information and one at high temporal resolution to capture fine motion [8]. More recently, inspired by successes in the language processing domain, transformer-based models (vision transformers, ViT) made their way into the computer vision domain with excellent results [6]. Members of this family, such as MViT [7] and Swin [19] consistently outperform their convolutional counterparts on common benchmark datasets. For instance, MViT and Swin both achieve over 81% top-1 accuracy on Kinetics-400 and over 67% on Something-Something v2, against Slowfast's 79.8% and 63.1%, respectively. What is more, the model VideoMAE utilizes an effective pre-training method using input masking with autoencoders, which results in state-of-the-art performance in both AR (75.4% on Something-Something v2 and 87.4% on Kinetics-400) and AD (39.3% mean average precision, mAP, on AVA v2.2) tasks [29]. It is important to note, however, that transformer models are computationally heavy with a vast amount of parameters (e.g., over 600 million for VideoMAE with the largest, ViTH, backbone), which means that applying them to custom datasets usually involves fine-tuning some trainable parameters instead of the complete retraining of the model.

Furthermore, the emergence of foundation models, particularly Large Vision Models (LVM), is also an asset that can be capitalized upon in video understanding. These models are trained at scale on vast amounts of data, meaning that they may be used for diverse downstream tasks, potentially without the need of a task-specific labelled dataset (zero-shot learning) [1]. A notable example of the state-of-the-art in LVMs is DINOv2 [24], which is trained on a curated dataset created using self-supervised methods (a total of 142 million images), thus improving the quality of the extracted features, as opposed to using uncurated data. DINOv2's design builds on DINO [2] and iBOT [36] and comprises a student and teacher network for both image-level and patch-level objectives (with separate loss terms for these two tasks). An important aspect of DINOv2's training process is a short high-resolution period at the end of training, which ensures that small objects do not disappear in the embedding.

DINOv2, despite being trained solely on images, achieves comparable performance to the state-of-the-art in action recognition tasks through feature extraction on equally spaced frames (78.4% accuracy on Kinetics-400) [24]. In this approach, the embeddings produced by DINOv2 are passed through a simple linear classifier, which suggests that video understanding tasks may be solved without the need of training complex classifiers on video data, thus saving significantly on computational costs. What is more, a more recent addition to the DINOv2 architecture, register tokens, appear to enhance performance further by removing feature artifacts from non-informative background areas in an image [5], which might be particularly useful in recordings of natural environments, where the background may be dense and noisy.

Efficiency is at the heart of many studies in the domain of video understanding, as these tasks generally require large computational resources. One important implication of this is that it is rarely possible for a model to consider every frame in a video: it is much more common to select only some of them. To resolve this problem, Liu et al. [18] propose a full-video action recognition approach which efficiently utilizes information from every frame through dynamic temporal clustering. At the same time, informed frame selection may also be an effective and computationally efficient method to identify the most relevant frames for a video understanding task. Notably, Gowda et al. [11] propose an intelligent frame selection approach, SMART, which outperforms other frame selection approaches (such as AdaFrame [33] and FrameGlimpse [35]) and even full video action recognition on the UCF-101 dataset. The reasoning behind the latter is that it discards frames which are difficult to classify and thus confuse the decision. However, the implementation of these frame selection techniques is either not published ([11, 33]), or comprises more of an end-to-end model which can be difficult to integrate into a custom AR or AD pipeline such as MARINE, necessitating the construction of a lightweight custom frame selection method inspired by these techniques.

Research on action recognition in the animal domain is still sparse compared to human actions. What is more, existing studies focus mainly on a narrow range of (or even a single) species, which are most often large mammals. In [26], Schindler et al. introduce MAROON, an action recognition system relying on segmentation and deployed on a self-assembled dataset of deer camera trap footage. In [23], Nishioka et al. use temporal action localization to monitor elephant eating behavior in a zoo, while in [9], Feng et al. focus specifically on feline action recognition in the wild. These studies rely heavily on the identification of the actor, an aspect which is also highlighted in [31], a study on current and future directions in computer vision for primate behavior studies. In such approaches, first the (potential) actor is identified in the footage, then action classification follows based primarily on the actor's visual features [23, 26] or estimated pose [9].

This type of approach, however, carries disadvantages in settings when a single actor is difficult to identify, or when its appearance or pose cannot easily be discerned. Schools of fish in underwater video footage are such an example, and this might partly be the reason why video understanding studies on fish videos are especially rare. Existing research mainly concerns aquacultures, such as [21], in which the combination of a 3D convolutional network and Long Short-Term Memory is used to classify whether salmon are feeding or not. While this model achieves good performance on the (small) dataset used in the study (80.0% average accuracy), it is likely that more advanced methods such as vision transformers could achieve higher results and generalize better over other data samples.

MSQNet is a welcome example of an action recognition model which achieves high performance on a diverse animal dataset, Animal Kingdom [22], and which does not rely on the identification of the actor [20]. Due to the CLIP-based image-language embedder included in its architecture, MSQNet is capable of utilizing textual cues, thus possessing zero-shot capabilities. As such, it offers a viable solution for tackling video understanding tasks even in animal footage which depicts previously unseen species. On Animal Kingdom, MSQNet achieves 73.10% mAP [20], which is significantly higher than I3D's 16.48%, Slowfast's 20.45%, or X3D's 25.25% (as reported in [22]).

Although ecological and ethological studies could greatly benefit from computational video understanding through the automation of tasks, there is a relative scarcity in publicly available animal video datasets. Many of the studies in this area utilize only a small and specific dataset which is often compiled particularly for the study's purpose (e.g., [9, 21, 23, 26]). Besides focusing on a specific environment or species, these datasets are often unpublished, making it difficult for future studies to build on. Of the existing public datasets, a majority target image-based tasks or animal recognition in videos, less so the understanding of temporal dynamics. Good examples are iNaturalist [30] and Animals with Attributes [16], both of which encompass a wide range of species and conditions, but comprise only images. Therefore, if used in video understanding tasks, the temporal dimension would have to be neglected. In [22], Ng et al. introduce Animal Kingdom, a comprehensive animal video dataset collected from YouTube and complete with annotations for several different tasks, among them action recognition. The true advantage of this dataset is not only its relatively large size (over 30,000 clips in case of action recognition), but also its diversity. The videos depict a total of 850 species from six large animal groups in varying environments (underwater, desert, forest, etc.) and conditions (snow, night vision, etc.). The annotations include 140 different action classes. Such a varied dataset can be an important asset in building models for action classification tasks in the animal domain. At the same time, as explained in [22], broad datasets such as Animal Kingdom tend to possess a long-tailed distribution of annotated action classes, as rare animal behaviors are inherently less documented. It is yet to be explored how well such a diverse dataset can be adopted in ecological studies focusing only on a single or few animal action(s), species, or habitat(s), especially if these belong to minority classes. In summary, video understanding is a dynamically developing field within computer vision, but most of the research in the AR and AD area has been dedicated to human actions, with animal videos being neglected in comparison. Existing research in the animal domain predominantly focuses on actor-based action recognition and often relies on rather outdated methods and small custom datasets. What is more, water habitats and non-mammal species are generally underrepresented in animal video research. This paper aims to address this gap by introducing a methodology to utilize state-of-the-art computer vision techniques in AR and AD tasks to identify fish predator attacks, using both a small and specific dataset and a subset of a larger and more diverse animal benchmark dataset.

3 Methodology

The methodology concerns the construction of the MARINE model, comprising a motion-based frame selection module, DINOv2 feature extraction, and a trainable classification head, and its evaluation in a binary AR (classifying whether a predator attack occurs in a short video clip) and AD (temporally localizing predator attacks in longer recordings) task on two datasets of fish videos.

The study relies mainly on two datasets: Animal Kingdom (AK), as presented by Ng et al. [22], and a small custom dataset referred to as the coral reef dataset. Animal Kingdom (AK) is a dataset annotated for several different video understanding tasks, of which the action recognition dataset, comprising over 850 animal species and 140 fine-grained action classes [22], is utilized in this study. As the paper focuses on predator attacks observed in underwater habitats, the AK action recognition dataset is filtered for videos which depict fish. To avoid confusion, this narrower dataset will be referred to as the AK fish dataset throughout the paper. The AK fish dataset is filtered from the AK action recognition dataset by selecting those videos in which a fish species performs some (any) action. Out of these, videos containing predator attacks are specified as ones in which one of the following eight predation-related actions, taken from the fine-grained action labels of AK, is performed by a fish species:

• Attacking • Being eaten • Biting• Chasing • Fighting• Fleeing• Retaliating • Struggling.

There are 887 video clips in the AK fish dataset, out of which 81 contain an attack-related action performed by fish. Table 1 shows a summary of the dataset. For the purpose of the AR task (i.e., identifying whether a video contains an attack or not), the videos in the AK fish dataset receive binary labels: positive if they contain one of the predation-related actions performed by a fish species, and negative otherwise. The resulting dataset suffers from high class imbalance, where only 9.13% of videos are labelled as positive. This factor must be considered during model training.

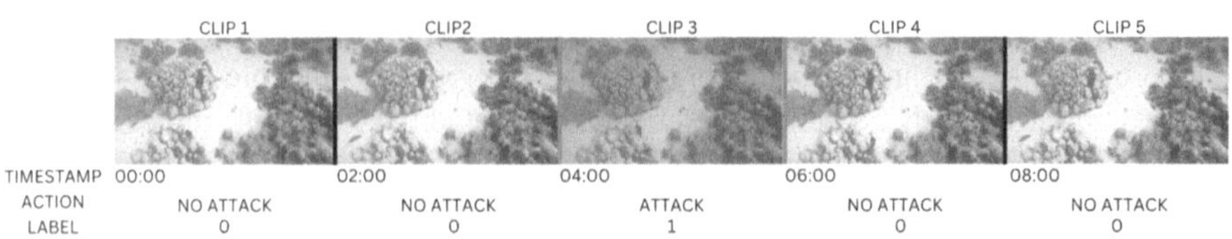

Fig. 1. Processing the original coral reef dataset. Each of the 44 10-s-long videos are separated into 5 clips, with the middle one containing a predator attack. The resulting 2-s-long clips (220 in total) will comprise the processed coral reef dataset, where the clips originally situated at the centre of the uncut videos receive the positive label and all others the negative label.

Table 1. Summary table of the AK fish and coral reef datasets. Both datasets comprise binary classes ("attack" or "no attack").

	AK fish	Coral reef
Number of videos	887	220
Positive ("attack") class (% of sample)	9.13	20.00
Number of frames (mean)	117.68	239.55
Number of frames (st. dev.)	92.71	0.61

The coral reef dataset, recorded by members of the Institute of Biodiversity and Ecosystem Dynamics at the University of Amsterdam, comprises 44 videos, each of which are 10 s long. These videos were recorded in high spatial and temporal resolution (1080 × 1920 pixels and 120 frames per second). Each of the videos show a similar coral reef environment and contain a fish predator attack in the central section of the video (within a 2-s time window around the middle point), as illustrated in Fig. 1.

For the purpose of the binary action recognition task, the coral reef dataset is processed through the following steps, which are also explained in Fig. 1:

1. The videos are separated into 2-s-long clips without overlaps, resulting in 5 clips cut from each 10-s long video.
2. As a predator attack is contained in the middle section of each video, the resulting clips are labelled so that the centre clip is labelled as positive, and the four other clips are labelled as negative.

The resulting dataset comprises 220 videos, of which 44 (the centre clips of the original 44 videos) contain an attack and receive a positive label. When mentioning the coral reef dataset from here onwards in the paper, this processed dataset is referred. The coral reef dataset, summarized in Table 1, also suffers from class imbalance, with only 20% of instances comprising the positive class. The action recognition task of identifying whether a video contains a predator attack can be understood as a binary classification problem. In general terms, the model proposed for this task, MARINE, comprises the following components (also referred to as modules), which are illustrated in Fig. 2:

1. Frame selection module. To increase computational efficiency, not every frame from the video is considered for action recognition. Instead, only a limited number of frames are selected and processed for classification.
2. Image embedding using DINOv2. The image foundation model DINOv2 [24] is utilized for feature extraction on the selected frames. The resulting frame features are concatenated to retain information from the temporal dimension of the video.
3. Shallow classification head. Using the concatenated features resulting from the previous step as input, a relatively simple classifier network is trained to distinguish between videos which do or do not contain a predator attack.

The novelty and contribution of this general approach is three-fold. Firstly, it integrates two important techniques in action recognition: intelligent frame selection methods, which enable classification models to focus on the most relevant frames of a video, and large vision models, which provide a robust solution for extracting meaningful features from images without the need for extensive training. Secondly, it tackles videos in the domain of animal actions, which are considerably less researched than videos depicting human actions. Lastly, it promotes computational efficiency in several of its aspects:

- Allowing to process only a limited number of frames from the video, which requires considerably less resources than processing the entire clip, especially if it is high resolution, as the coral reef dataset is.
- Utilizing a pre-trained image foundation model without additional training. DINOv2 has been shown to achieve.
- state-of-the-art performance in action recognition tasks without being fine-tuned on the video dataset [24].
- Using a shallow classifier network at the head of the action recognition pipeline. Using the extracted features as input, a small classifier is trained instead of a computationally heavy deep learning network.

The details of MARINE's modules are as follows.

- Frame selection module. Predator attacks are characterized by swift movements, which can be especially pronounced visually if the attack happens in a school of fish. Therefore, in the proposed frame selection method, those frames are selected for inspection in which rapid movement can be detected. This is done by assessing the dissimilarity between two consecutive frames and selecting a set number of frames (specifically 10) which are most dissimilar to the one before them. The dissimilarity score $\alpha_{p,q}$ can be calculated between two frames represented as tensors W^p and W^q, each with n elements, in the following way.

$$\alpha_{p,q} = \sum_{i}^{n} \left| W_i^p - W_i^q \right| \tag{1}$$

As changes in individual color channels may not indicate motion, frames are converted to greyscale images and represented as two-dimensional matrices during motionbased frame selection.

- **Feature extraction using DINOv2.** Image embeddings of the selected frames are obtained by retrieving the classifier tokens from the DINOv2 backbone's output, without fine-tuning the model to maximize computational efficiency. The features are then concatenated into a single vector to retain temporal information from the video. Four different DINOv2 backbones are tested as part of the MARINE model.
- Classification model. After obtaining the concatenated frame features, the resulting video-level vectors are used as input to a classifier head, a shallow neural network with a final sigmoid activation. The exact settings of the network are determined through cross-validation.

Besides being able to classify a single short clip as containing a predator attack or not, a method to localize attacks in longer untrimmed videos (AD task) is proposed. This approach builds on the AR model MARINE and consists of the following steps, also summarized in Fig. 2:

1. Separate the untrimmed video into shorter clips and store the delimiting timestamps of each clip in the original video.
2. Classify each of the short clips as containing a predator attack or not using MARINE.
3. Using the predicted label of each short clip, as well as the timestamps preserved when cutting the clips from the untrimmed footage, identify the sections in the original video which are classified as containing an attack.

4 Evaluation

For evaluating MARINE's performance on the AR task (both in the AK fish and coral reef datasets), the commonly used metrics accuracy, precision, recall, and F1-score are applied. The latter three are especially important to consider due to the high class imbalance observed in the data. ROC curves will further help clarify the models' ability to balance between recall and maintaining a low false positive rate.

We evaluate MARINE's performance in detecting predator-prey interactions using the key metrics of accuracy, precision, recall, F1-score, and mean average precision (mAP). Accuracy reflects the proportion of correctly identified predator attacks in the coral reef and AK fish datasets. Given the class imbalance in these datasets, precision is critical in measuring how well the model avoids false positives, while recall assesses its ability to correctly detect actual attacks. The F1-score is particularly relevant for balancing precision and recall, as misclassifications could skew results due to the rarity of attack events. For temporal action detection on longer videos, we utilize mAP, which averages precision over various temporal Intersection-over-Union (t-IoU) thresholds, providing a robust measure of MARINE's ability to localize predator attacks within untrimmed video footage.

To assess the performance of temporal action detection, mean average precision (mAP) is most commonly used [34]. This metric is suitable for multi-label classification in a binary case as in this study, average precision (AP, the mean precision over all videos) suffices. To calculate precision, the correctness of each predicted action time interval must be decided. Following [34], t-IoU (temporal Intersection over Union, as shown in Eq. 2) with a threshold of 50% is applied for this purpose.

$$t - IoU = \frac{predicted\ time\ interval \cap true\ time\ interval}{predicted\ time\ interval \cup true\ time\ interval} \tag{2}$$

Note that a post-hoc test with a t-IoU threshold of 25% is also conducted.

While extensive benchmarks exist for human action recognition problems, it is less so the case for animal datasets. On human action datasets in both AR and AD tasks (on the Kinetics 400 and AVA v2.2 datasets), VideoMAE achieves state-of-the-art performance, obtaining the highest results of all models. As VideoMAE also demonstrates remarkable transfer learning capabilites [29], it was chosen to serve as the baseline model to apply to the AK fish and coral reef datasets to benchmark MARINE's performance.

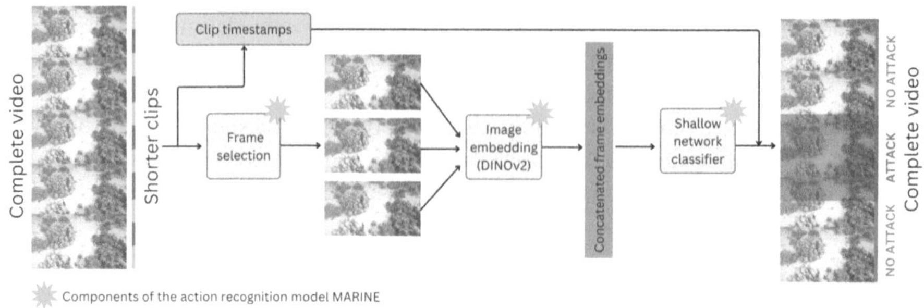

Fig. 2. Action detection pipeline. The untrimmed video is separated into shorter clips, which are passed to the action recognition model. Combining the predictions of this model and the timestamps which locate the clips in the original video, the temporal location of the predator attack can be predicted in the original footage.

At the same time, limited benchmarks do exists on animal video datasets, specifically for Animal Kingdom, including the models MSQNet, I3D, Slowfast and X3D. As an additional test of its applicability across animal domains, MARINE is adapted to a multi-label setting and compared to these benchmarks on a representative subsample of Animal Kingdom.

The performance on the action recognition task is measured on two datasets: the AK fish dataset and the coral reef dataset. As both of these datasets can be considered quite small in the domain of video understanding, a bootstrapped test sample is used, so that results can be reported with confidence intervals. This is done by defining training and test samples within both datasets: as Animal Kingdom is published with predefined train and test sets, the same split is kept within the AK fish sample, while in the coral reef dataset, a random split with 20% test size is performed (ensuring that clips separated from the same video, as shown in Fig. 1, are assigned to the same split). After training the different MARINE setups (and fine-tuning VideoMAE), all models are evaluated on bootstrapped test sets of the two datasets: the test set is resampled with replacement 100 times (with the number of samples remaining identical to the original test set), and the mean accuracy, precision, recall, and F1-score are reported with standard deviation in the bootstrapped test set.

The training process of MARINE starts with using the motion-based frame selection method to identify the 10 most relevant frames, then applying DINOv2 feature extraction to the selected frames (as described in 3.2.2). The experiments use four different DINOv2 backbones: ViT-S/14 (MARINE S14), comprising 21 million parameters, and ViT-G/14 (MARINE G14), comprising 1,100 million parameters, both architectures with and without register tokens. In MARINE, feature extraction refers to obtaining the classifier token of the DINOv2 output when the frame is passed to the model. After concatenating the extracted frame features, the classifier head is trained using 3-fold cross-validation. This classifier is a shallow neural network with a 10-node dense input layer with ReLu activation, between 0 and 3 dense hidden layers with 128 ReLu-activated nodes each, and a final sigmoid output node. Importantly, class weights are utilized during training to address the imbalance in both datasets. The threshold to consider the sigmoid output of the network as predicted positive is determined posttraining by optimizing for F1-score. Regarding the benchmark model, the classification head of a VideoMAE model with ViT-B backbone pretrained on Kinetics 400 is fine-tuned on the training sets of the AK fish and coral reef datasets, using video frames selected through the same motion-based method as in MARINE. The threshold to consider the continuous output of the model as predicted positive is also determined identically to MARINE, by optimizing for F1-score.

The contribution of the motion-based frame selection module in MARINE is assessed in an ablation study setting. In this experiment, the performance of the MARINE G14 model is compared to a MARINE G14 model where the motion-based frame selection module is swapped for one which selects 10 evenly spaced frames, starting from the first one in the video. Performance in the action detection task is measured through AP relying on t-IoU, as formulated in Eq. 2. As localized action annotation is available only for the coral reef dataset, the models (MARINE setups and VideoMAE) will be evaluated on this dataset for this task. Importantly, the test set for this task comprises only 9 videos (20% of the 44 untrimmed videos in the coral reef dataset), and to report results, a bootstrapped test set is once more used. This means that a sample of 9 videos is selected from the test set, with replacement, 100 times, and the mean of the average precisions on these 100 samples is reported, as well as its standard deviation. The size of the individual test samples is kept identical to the original test set, as small sample sizes are characteristic to real-world biological datasets. At the same time, resampling 100 times helps introduce variation, so that the results may be reported with statistical confidence.

The performance is tested on a representative sample of Animal Kingdom in a multi-label setting. This random sample comprises 1000 videos and 93 classes of the Animal Kingdom action recognition dataset, and its distribution of labels is confirmed to be representative of the original dataset at 95% confidence with a chi-squared test of homogeneity (p-value = 0.68). MARINE G14 (without registers) is adapted to this multi-label setting by replacing the single sigmoid output of the classification head with a layer of 93 sigmoid nodes. Using the predefined train and test split of the Animal Kingdom sample, motion-based frame selection and DINOv2 feature extraction is performed on both splits, while the multi-label classification head is trained using cross validation on the training set. Finally, predictions are obtained for the test set and mAP is calculated as the main performance metric to compare with existing benchmarks published in [20] and [22].

5 Results

To address research questions (1) and (3), specifically whether MARINE can outperform VideoMAE in the AR task and how different DINOv2 backbones influence its performance, Table 2 summarizes the outcomes of the experiments. For clarity's sake, and as the presence of register tokens in the DINOv2 backbones does not result in any obvious performance increase, Table 2 only reports the performance of the MARINE S14 and G14 models without registers. Appendix D shows the outcomes of experiments run using MARINE models containing DINOv2 backbones with registers. Table 2 shows the performance of the proposed models in terms of accuracy, recall, precision, and F1-score (mean ± 1.96 standard deviations to indicate the 95% confidence interval) on the binary classification task of identifying videos which contain a predator attack. The results indicate that MARINE G14 is capable of better performance than VideoMAE in terms of accuracy, precision and F1-score on both datasets, and in terms of all four metrics on the AK fish dataset. On this dataset, MARINE G14 achieves over 10% points increase in accuracy compared to the other two models, and over 30% points in F1-score.

Notably, the performance of all three models is considerably higher on the AK fish dataset. Achieving a good balance between precision and recall appears especially problematic on the coral reef dataset, where MARINE S14 achieves 100.0% recall with 0.00% points variation, but low precision (19.51%). The fact that the accuracy in this case (19.51%) is very close to the proportion of positive samples in the dataset (20.0%) indicates that the model essentially classifies all instances as positive. At the same time, MARINE G14 appears more capable of balancing precision and recall although at rather low values (52.52% and 53.01%, respectively).

While these results suggest that MARINE G14 could be capable of state-of-the-art performance on relatively small animal video datasets of specific animal types and environments, it must be noted that the limited test sample size results in large variations in the outcomes, as indicated by the wide confidence intervals. The challenging nature of the coral reef dataset is highlighted by Fig. 3, which shows the ROC curves of MARINE G14 on both test sets. The area-under-the-curve (AUC) is a relevant measure in the AR task as the target action, a predator attack, is extremely rare, which results in highly imbalanced datasets. Therefore, it is essential that the model is able to find an appropriate balance between identifying all relevant instances (true positive rate) without extensively compromising on specificity (or, conversely, on the false positive rate). Figure 3 clearly shows that MARINE G14 was more capable of this on the AK fish dataset than on the coral reef dataset (AUC of 0.82 and 0.68, respectively). The fact that this outcome is consistent across all configurations of the MARINE model indicates that achieving an appropriate recall-false positive rate balance is generally more challenging in case of the coral reef dataset.

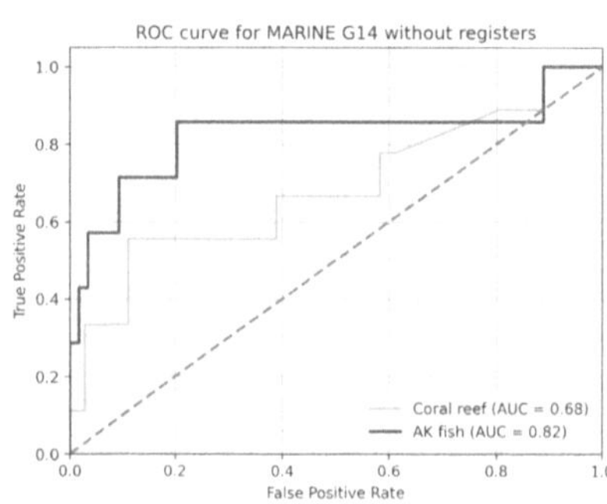

Fig. 3. ROC curves of MARINE G14 on the coral reef and AK fish datasets, without registers in the DINOv2 backbone. The higher AUC value for the AK fish dataset is consistent across all configurations of the MARINE model.

In this experiment, MARINE S14 and G14 were tested against their counterparts using evenly selected frames instead of the motion-based frame selection method. The results are presented in Table 3 (without the use of register tokens). As the table shows, applying motion-based frame selection does not clearly contribute to higher performance. Specifically, accuracy decreases on the AK fish dataset but increases significantly (from 23.71% to 81.53%) on the coral reef datase when motion-based frame selection is introduced in MARINE G14. At the same time, while motion-based frame selection does not consistently improve F1-scores, it appears to help avoid an extreme imbalance between precision and recall, where 100% performance is achieved in one of the metrics and (extremely) low performance on the other, indicating that the model either classifies virtually all instances as positive (100% recall, low precision) or as negative (100% precision, low recall). Notably, however, this behaviour occurs even using motion-based frame selection with MARINE S14 on the coral reef dataset, which indicates that such a challenging dataset requires not only more sophisticated frame selection but also more complex feature extraction using larger LVM backbones. To address how the proposed MARINE model performs in the AD task in comparison to VideoMAE, Table 4 presents the results of this task on the coral reef dataset in terms of average precision (mean over bootstrapped test samples and standard deviation). The presented results were obtained applying motion-based frame selection in all models, and applying DINOv2 backbones without registers in the MARINE models.

The most important outcome of the AD task experiments is that applying a threshold of 50% t-IoU for considering a predicted time interval as correct yielded no correct predictions for any of the models (MARINE or VideoMAE), highlights the challenging nature of the coral reef dataset. Therefore, it is crucial to understand that the results reported in Table 4 were obtained using a lower t-IoU threshold of 25%, meaning that a predicted time interval is considered correct if it coincides with a true attack in at least 25% of its duration.

Using the lowered threshold, MARINE G14 outperforms both other models by a substantial margin (80.78% against VideoMAE's 34.89% and MARINE S14's 0.00%). At the same time, the non-varyingly low performance of MARINE S14 (0.00% with a standard deviation of 0.00%) shows not only this model's poor ability to temporally localize predator attacks in the coral reef dataset, but also the limitations introduced by the small size of the test sample. The fact that MARINE S14's performance is non-varyingly 0.00% indicates that it cannot correctly predict a single attack timeframe at a threshold of 25% t-IoU in any of the 9 test videos from which the bootstrapped samples are taken. To investigate the extent to which MARINE may generalize across varied animal domains, its performance was tested in a multi-label setting on a representative sample of Animal Kingdom comprising 93 classes. In this experiment, MARINE G14 (without registers) achieves 23.79% mAP, which is significantly lower than MSQNet's 73.10% [20], comparable to X3D's 25.25%, and higher than Slowfast's 20.45% and I3D's 16.48% [22]. This shows that although MARINE could be a viable option in a binary classification setting, it cannot currently compete with the best of the state-of-the-art on a diverse multi-label animal dataset.

Table 2. Model performances on the action recognition task, using motion-based frame selection with each model and DINOv2 backbones without registers in the MARINE models. The table shows the performance on the AR task in terms of accuracy, recall and precision (± 1.96 standard deviations to indicate the 95% confidence intervals).

Dataset	Model	Accuracy (%)	Recall (%)	Precision (%)	F1-score (%)
Coral reef	MARINE S14	19.51 +/ 12.25	**100.00 +/ 0.00**	19.51 +/ 12.25	32.19 +/ 17.30
Coral reef	MARINE G14	**81.53 +/ 10.36**	52.52 +/ 33.77	**53.01 +/ 37.24**	**51.06 +/ 29.60**
Coral reef	VideoMAE	52.64 +/ 13.92	69.72 +/ 31.33	24.51 +/ 16.77	35.49 +/ 20.33
AK fish	MARINE S14	81.74 +/ 7.05	57.40 +/ 41.50	16.07 +/ 14.68	24.62 +/ 20.54
AK fish	MARINE G14	**94.86 +/ 3.97**	**63.31 +/ 41.60**	**51.64 +/ 36.86**	**55.14 +/ 33.18**
AK fish	VideoMAE	83.14 +/ 7.70	42.73 +/ 43.08	14.33 +/ 16.76	20.92 +/ 22.76

Table 3. Results of the ablation experiment to test the contribution of the motion-based frame selection method to the AR task using the MARINE models without registers in the DINOv2 backbones. The highest performances in terms of each metric on both datasets are highlighted in bold.

Dataset	Frame selection	Model	Accuracy (%)	Recall (%)	Precision (%)	F1-score (%)
Coral reef	Evenly spaced	MARINE S14	19.51 +/ 12.25	100.00 +/ 0.00	19.51 +/ 12.25	32.19 +/ 17.30
Coral reef	Motion-based	MARINE S14	19.51 +/ 12.25	100.00 +/ 0.00	19.51 +/ 12.25	32.19 +/ 17.30
Coral reef	Evenly spaced	MARINE G14	23.71 +/ 12.53	**100.00 +/ 0.00**	20.35 +/ 12.70	33.34 +/ 17.66
Coral reef	Motion-based	MARINE G14	**81.53 +/ 10.36**	52.52 +/ 33.77	**53.01 +/ 37.24**	**51.06 +/ 29.60**
AK fish	Evenly spaced	MARINE S14	**96.65 +/ 3.18**	45.99 +/ 32.53	**100.00 +/ 0.00**	**61.27 +/ 30.12**
AK fish	Motion-based	MARINE S14	81.74 +/ 7.05	**57.40 +/ 41.50**	16.07 +/ 14.68	24.62 +/ 20.54
AK fish	Evenly spaced	MARINE G14	**97.12 +/ 3.01**	52.63 +/ 36.55	**100.00 +/ 0.00**	**67.07 +/ 30.66**
AK fish	Motion-based	MARINE G14	94.86 +/ 3.97	**63.31 +/ 41.60**	51.64 +/ 36.86	55.14 +/ 33.18

Table 4. Model performances on the action detection task on the coral reef dataset, in terms of average precision (AP) at a lowered t-IoU threshold of 25%. The values relate to the mean and standard deviation of average precisions in the bootstrapped samples of test videos. Motion-based frame selection is applied with all models, and DINOv2 backbones do not apply registers. The highest performance is highlighted in bold.

Model	Average precision	Standard deviation
MARINE S14	0.00%	0.00%
MARINE G14	**80.78%**	12.05%
VideoMAE	34.89%	14.32%

6 Conclusion

The study of encounters between predators and prey is of importance in ecosystem dynamics. This paper investigates the extent to which computer vision techniques could be used to detect such rare ecological events, specifically predator attacks, in videos of fish species. The model MARINE is proposed, which comprises motion-based frame selection, feature extraction of selected frames using DINOv2 backbones, and a lightweight trainable classifier head to label video clips as containing a predator attack or not. In the action recognition task on the coral reef and AK fish datasets, MARINE

G14 is found to outperform state-of-the-art benchmark VideoMAE in terms of accuracy and F1-score on both the small and specific coral reef dataset (by close to 30% points and 15% points, respectively), and the larger and more diverse AK fish dataset (by close to 12% points and 35% points, respectively). Additionally, it is confirmed that larger LVM backbones (DINOv2 ViTG/14 against ViT-S/14) increase performance in the AR task (by over 60% points in accuracy for the coral reef, and 13% points for the AK fish dataset), while register tokens do not prove to be of particular advantage. Similarly, the contribution of the motion-based frame selection method remains unclear, as it does not consistently improve MARINE's performance in an ablation study. Additionally, a multi-label AR experiment on a representative sample of Animal Kingdom reveals that MARINE cannot exceed the current state-of-the-art on a more diverse animal dataset, but it does perform comparably to other benchmarked methods (23.79% mAP compared to existing benchmarks between 16.48% and 73.10% [20, 22]). In terms of the AD task on the coral reef dataset, MARINE achieves rather low performance (0.00% AP at 50% t-IoU and 80.78% AP at 25% t-IoU), but still higher than benchmark model VideoMAE (34.89% AP at 25% t-IoU).

These results must be taken in light of the limitations introduced by the small size and narrow scope of the coral reef and AK fish datasets, as well as some technical aspects of MARINE with room for improvement. Consequently, future research could be directed towards two general areas. Firstly, MARINE's performance, especially in the AD task or in a multi-label setting, could be enhanced by applying more sophisticated frame selection methods, frame handling with focus on potential actors (e.g., feature extraction on cutouts of target animals), and foundation models utilizing non-visual modularities, such as textual cues. Secondly, MARINE could be tested on datasets specific to new animal domains, possibly a range of similarly small and narrow ones, to assess its suitability to be used in varied animal studies focusing on specific behaviours. These directions could each bring the state of research closer to a widely applicable model for identifying rare ecological events.

In conclusion, MARINE is a convincing starter model for modular computer vision frameworks applied to animal video understanding tasks. By expanding it, future research could bring much-needed further insight into the field of animal computer vision problems. This paper hopes to offer a valuable contribution to this field and an expansion to the range of tools available for a better understanding of our natural ecosystems. MARINE's scalability is an important consideration for future deployment in larger datasets or diverse ecological contexts. While the model demonstrates strong performance on relatively small, specific datasets, applying MARINE to broader datasets may require additional computational resources. Varying video quality, environmental conditions, and species diversity could impact the model's effectiveness. These challenges suggest the need for further optimizations, such as dynamic frame selection methods or preprocessing adjustments, to ensure robust performance across different settings.

References

1. Bommasani, R., et al.: On the Opportunities and Risks of Foundation Models. ArXiv (2021)
2. Caron, M., et al.: emerging properties in self-supervised vision transformers (2021). arXiv: 2104.14294 [cs]

3. Carreira, J., Zisserman, A.: Quo vadis, action recognition? a new model and the kinetics dataset, pp. 6299–6308 (2017)
4. Damien, M., Tougeron, K.: Prey–predator phenological mismatch under climate change. Curr. Opin. Insect Sci. **35**, 60–68 (2019)
5. Darcet, T.L., Oquab, M., Mairal, J., Bojanowski, P.: Vision Transformers Need Registers (2024). arXiv:2309.16588 [cs]
6. Dosovitskiy, A., et al.: An image is worth 16x16 words: transformers for image recognition at scale (2021). arXiv:2010.11929 [cs]
7. Fan, H., et al.: multiscale vision transformers. In: 2021 IEEE/CVF International Conference on Computer Vision (ICCV), pp. 6804– 6815 (2021). ISSN: 2380–7504
8. Feichtenhofer, C., Fan, H., Malik, J., He, K.: SlowFast networks for video recognition (2019). arXiv:1812.03982 [cs]
9. Feng, L., Zhao, Y., Sun, Y., Zhao, W., Tang, J.: Action recognition using a spatial-temporal network for wild felines. Animals **11**(2), 485 (2021)
10. Gao, R., Oh, T.H., Grauman, K., Torresani, L.: Listen to look: action recognition by previewing audio. In: 2020 IEEE/CVF Conference on Computer Vision and Pattern Recognition (CVPR), pp. 10454–10464 (2020). ISSN: 2575–7075
11. Shreyank N. Gowda, Marcus Rohrbach, and Laura Sevilla-Lara. SMART Frame Selection for Action Recognition. Proceedings of the AAAI Conference on Artificial Intelligence, 35(2):1451–1459, 2021. Number: 2
12. Goyal, R., et al.: The "something something" video database for learning and evaluating visual common sense (2017). arXiv:1706.04261 [cs]
13. Hunsicker, M.E., et al.: Functional responses and scaling in predator-prey interactions of marine fishes: contemporary issues and emerging concepts. Ecol. Lett. **14**(12), 1288–1299 (2011)
14. Ives, A.R., Cardinale, B.J., Snyder, W.E.: A synthesis of subdisciplines: predator–prey interactions, and biodiversity and ecosystem functioning. Ecol. Lett. **8**(1), 102–116 (2005).https://doi.org/10.1111/j.1461-0248.2004.00698.x
15. Kay, W., et al.: The kinetics human action video dataset (2017). arXiv:1705.06950 [cs]
16. Lampert, C.H., Nickisch, H., Harmeling, S.: Attribute-based classification for zero-shot visual object categorization. IEEE Trans. Pattern Anal. Mach. Intell. **36**(3), 453–465 (2014)
17. Laws, A.N.: Climate change effects on predator–prey interactions. Curr. Opin. Insect Sci. **23**, 28–34 (2017)
18. Liu, X., et al.: No frame left behind: full video action recognition, pp. 14892–14901 (2021)
19. Liu, Z., Ning, J., Cao, Y., Wei, Y., Zhang, Z., Lin, S., Hu, H.: Video Swin Transformer, pp. 3202–3211 (2022)
20. Mondal, A., Nag, S., Prada, J.M., Zhu, X., Dutta, A.: Actor-agnostic multi-label action recognition with multi-modal query. In: 2023 IEEE/CVF International Conference on Computer Vision Workshops (ICCVW), pp. 784–794 (2023). arXiv:2307.10763
21. Måløy, H., Aamodt, A., Misimi, E.: A spatio-temporal recurrent network for salmon feeding action recognition from underwater videos in aquaculture. Comput. Electron. Agric. **167**, 105087 (2019). https://doi.org/10.1016/j.compag.2019.105087
22. Ng, X.L., Ong, K.E., Zheng, Q., Ni, Y., Yeo, S.Y., Liu, J.: Animal kingdom: a large and diverse dataset for animal behavior understanding. In: 2022 IEEE/CVF Conference on Computer Vision and Pattern Recognition (CVPR), pp. 19001–19012. IEEE, New Orleans (2022)
23. Nishioka, K., Noguchi, W., Izuka, H., Yamamoto, M.: Detecting eating behavior of elephants in a zoo using temporal action localization. Sens. Mater. **35**(11), 3927 (2023)
24. Oquab, M., et al.: DINOv2: learning robust visual features without supervision (2023). arXiv e-prints ADS Bibcode: 2023arXiv230407193O

25. Sadykova, D., Scott, B.E., De Dominicis, M., Wakelin, S.L., Wolf, J., Sadykov, A.: Ecological costs of climate change on marine predator–prey population distributions by 2050. Ecol. Evol. **10**(2), 1069–1086 (2020)
26. Schindler, F., Steinhage, V., van Beeck Calkoen, S.T., Heurich, M.: Action detection for wildlife monitoring with camera traps based on segmentation with filtering of tracklets (SWIFT) and mask-guided action recognition (MAROON). Appl. Sci. **14**(2), 514 (2024)
27. Schmitz, O.J., Barton, B.T.: Climate change effects on behavioral and physiological ecology of predator– prey interactions: Implications for conservation biological control. Biol. Control **75**, 87–96 (2014)
28. Soomro, K., Zamir, A.R., Shah, M.: UCF101: a dataset of 101 human actions classes from videos in the wild (2012). arXiv:1212.0402
29. Tong, Z., Song, Y., Wang, J., Wang, L.: VideoMAE: masked autoencoders are data-efficient learners for self-supervised video pre-training (2022). arXiv:2203.12602
30. Van Horn, G., et al.: The INaturalist species classification and detection dataset, pp. 8769–8778 (2018)
31. Vogg, R., et al.: Computer vision for primate behavior analysis in the wild (2024). arXiv: 2401.16424 [cs, q-bio]
32. Wang, Y., Long, M., Wang, J., Yu, P.S.: Spatiotemporal pyramid network for video action recognition, pp. 1529–1538 (2017)
33. Wu, Z., Xiong, C., Ma, C.Y., Socher, R., Davis, L.S.: AdaFrame: adaptive frame selection for fast video recognition. In: 2019 IEEE/CVF Conference on Computer Vision and Pattern Recognition (CVPR), pp. 1278–1287 (2019). ISSN: 2575–7075
34. Xia, H., Zhan, Y.: A survey on temporal action localization. IEEE Access **8**, 70477–70487 (2020)
35. Yeung, S., Russakovsky, O., Mori, G., Fei-Fei, L.: End-to-end learning of action detection from frame glimpses in videos, pp. 2678–2687. IEEE Computer Society (2016). ISSN: 1063–6919
36. Zhou, J., et al.: iBOT: image BERT pre-training with online tokenizer (2022). arXiv:2111.07832 [cs]

A Novel Multi-task Learning Framework for Predicting Traffic Congestion

Charan Teja Marrimanu(✉) and Arjun Chakravarthi Pogaku

Madanapalle Institute of Technology and Science, Madanapalle, Andhra Pradesh, India
charantejacherry2003@gmail.com

Abstract. Traffic congestion is a persistent problem in many cities worldwide, creating a strong need for efficient models to predict traffic conditions across large transportation networks. Most previous studies have relied on single-task learning-based machine learning models, which often struggle with accuracy and involve high computational costs. In this paper, we propose a novel multi-task learning-based neural network framework to forecast traffic congestion large-scale transportation networks. The proposed model utilizes "hard parameter tuning with ReLU" to achieve improved performance across various road segments. Experimental results demonstrate that the proposed framework has outperformed state-of-the-art algorithms in terms of both accuracy and efficiency, making it a promising solution for traffic prediction in urban environments.

Keywords: Neural Networks · Multi-Task Learning · Traffic Congestion · Intelligent Transportation Systems

1 Introduction

Adapting to the modern world's needs and requirements is mandatory for survival. A few years ago, there were hardly a million automobiles worldwide. Today, each major city has more than a million vehicles. Road transportation has become a daily routine for humans, leading to several problems, from traffic congestion to pollution to health issues. Governments worldwide have been investing significant amounts to conduct quality research in this field to manage the traffic uncertainties caused by the enormous number of vehicles on the road daily. To improve the quality of life, we must take preventive and precautionary measures to minimize traffic congestion.

Previous articles have described several statistical, machine learning (ML) [1,2], and deep learning (DL) [3] models to predict traffic congestion in transportation networks. The basic framework used in most of these models is "single task learning (STL)." It involves building a model and making predictions for every road segment in isolation [4,5]. A key advantage of STL is its simplicity in implementation. However, STL suffers from reduced accuracy and increased

A. Dasgupta et al. (Eds.): BDA 2024, LNCS 15526, pp. 200–216, 2025.
https://doi.org/10.1007/978-3-031-81821-9_12

computation costs, limiting its implementation in large-scale transportation networks [6,7]. This paper aims to improve traffic congestion prediction by addressing the limitations of STL.

Recently, the "multi-task learning (MTL) framework" has received considerable attention from both industry and academia [8,9]. Unlike STL, where the model is focused on a particular task during training, MTL identifies the underlying relations within the entire dataset to enhance prediction accuracy [10]. Few machine learning models have been designed based on MTL in the literature. With this motivation, this paper takes the supportive step to develop an efficient traffic congestion prediction system based on the MTL framework.

This paper proposes a novel MTL-based Neural Network (MTL-NN) framework to predict traffic congestion in large-scale transportation networks effectively. This proposed framework involves the following steps: (*i*) Normalizing the traffic data (*ii*) Feeding the data into the MTL-NN model, and (*iii*) Predicting the traffic congestion for multiple roads. The experimental dataset consists of 5-minute traffic congestion data collected over 12 months. The performance of the developed model has been compared with 10 other ML and DL models, showcasing its efficiency by outperforming all the algorithms tested. In particular, the results demonstrate that the proposed framework is efficient, with 10% time consumption and 40% memory consumption. The 10% time consumption may not be a significant difference with normal datasets, but in industrial scenarios, this makes a huge difference in terms of response time.

The remainder of the paper is organized as follows: Sect. 2 discusses previous work related to our current research. Section 3 introduces the proposed framework for traffic congestion forecasting. Section 4 demonstrates the operational design of the framework. Section 5 presents experimental results showcasing performance efficiency. Section 6 concludes the paper with all the findings of the research.

2 Related Works

Traffic prediction has been the subject of extensive research since the late 1970s. Several prediction methods are utilized for both short-term and long-term traffic flow forecasting. In this focus, we primarily examine short-term traffic congestion prediction models. Manibardo et al. [11] performed a comparative study of various shallow and deep learning algorithms and claimed that deep learning algorithms perform similarly to shallow learning algorithms. Furthermore, the authors stated that the performance of the selected algorithm varies depending on the dataset used. Fan et al. [12] suggested CNN, recurrent neural networks (RNN), and generative adversarial networks (GANs) for predicting traffic congestion. They tested various datasets, emphasizing the challenges and needs that are to be addressed. A detailed comparison of RNN and LSTM was performed for a similar real-world traffic dataset in [13]. While their RNN model achieved the desired results, the computation costs were not optimal.

Kothai et al. [14] suggested a hybrid DL model consisting of CNN and boosted LSTM ensemble (BLSTME), which can predict traffic congestion with 10%

higher efficiency compared to existing models. However, the limitation of the model was that it was trained entirely on a simulated dataset. Meanwhile, Alkarim [15] demonstrated other ensemble methods such as LSTM, random forests, and linear regression, achieving a 93.5% improvement in accuracy compared to previous models. Cheng et al. [16] demonstrated a Vector Auto-Regression (VAR) model to understand the relationship between the variables or factors causing traffic congestion. Utilizing the identified factors, the historical data on speed was given as input to train a hybrid CNN-LSTM model.

Several other studies have focused on improving traffic prediction by leveraging real-time spatiotemporal data. For instance, Barros et al. [17] discusses various model-driven and data-driven methods, highlighting the strengths of both approaches. In another study, He et al. [18] proposed a Spatio-Temporal Neural Network (STNN) that accounts for spatial dependencies like road types and external factors. The use of rush-hour behaviors and event data in pan et al. [19] showed a significant improvement in both short-term and long-term prediction accuracy by up to 78% and 67%, respectively.

The role of external factors like weather has also been explored in improving prediction models. Dunne et al. [20] developed a weather-adaptive traffic forecasting algorithm using stationary wavelet transform (SWT) to integrate rainfall data into the model. The neurowavelet model outperformed standard artificial neural networks, especially during inclement weather conditions. Similarly, RNNs and their variants, like LSTM and GRU, have proven highly effective for traffic prediction. As highlighted in [21], these models outperform traditional statistical methods in multiple traffic prediction tasks.

In Liu et al. [22], deep learning techniques were used to predict various traffic indicators, such as speed and flow, with a focus on uncovering hidden patterns from mobility data. This deep learning approach is further extended in [23], where a Spatio-Temporal Convolutional Neural Network (STCNN) was used to capture both spatial and temporal dependencies, achieving higher accuracy in long-term traffic prediction. On the other hand, mobile network traffic prediction was studied in Trinh et al. [24], where LSTM networks were applied to cellular data to optimize resource allocation in real time.

Intelligent Transportation Systems (ITS) Boukerche et al. [25] has gained significant attention due to advancements in vehicular technology. These systems aim to create safer and more efficient transportation environments, relying heavily on accurate traffic flow prediction models. Various methodologies, including mathematical, parametric, and non-parametric methods, have been employed to enhance traffic prediction accuracy. A comprehensive evaluation of deep neural networks under common datasets has highlighted the need for specialized models that effectively predict congestion formation rather than just overall traffic conditions. Recent studies have shown that exploiting spatiotemporal data enhances prediction accuracy, especially under congested conditions. Among the various models, error recurrent models, such as the error recurrent CNN (eRCNN), have outperformed traditional deep learning approaches, achieving high accuracy with lower computational costs Men-Oreja et al. [26].

Table 1. Summary of traffic prediction studies and techniques

Ref.	Study	Technique
[11]	Comparative study of shallow and deep learning algorithms for traffic congestion prediction	Deep learning and shallow learning algorithms
[12]	Suggested CNN, RNN, and GANs for traffic congestion prediction	CNN, RNN, GANs
[13]	Comparison of RNN and LSTM for real-world traffic dataset; RNN performed well but high computation cost	RNN, LSTM
[14] et al.	Proposed hybrid DL model (CNN + BLSTME) for traffic congestion prediction with higher efficiency	CNN, Boosted LSTM Ensemble (BLSTME)
[15]	Demonstrated ensemble methods like LSTM, random forests, and linear regression for traffic prediction	LSTM, Random Forests, Linear Regression
[16]	Used VAR model to find factors causing congestion and trained a CNN-LSTM hybrid model	VAR, CNN-LSTM Hybrid
[17]	Discussed model-driven and data-driven approaches to traffic prediction	Model-driven and data-driven methods
[18]	Proposed Spatio-Temporal Neural Network (STNN) accounting for spatial dependencies in traffic	Spatio-Temporal Neural Network (STNN)
[19]	Integrated rush-hour and event data for improved traffic prediction accuracy	Real-time event and rush-hour data
[20]	Developed a weather-adaptive traffic forecasting algorithm using rainfall data	Stationary Wavelet Transform (SWT), Neurowavelet Model
[21]	Used RNNs and variants for effective traffic prediction outperforming traditional methods	RNN, LSTM, GRU
[22]	Predicted traffic indicators using deep learning, uncovering patterns from mobility data	Deep Learning for traffic speed and flow
[23]	Used Spatio-Temporal CNN for capturing spatial and temporal dependencies in long-term traffic prediction	Spatio-Temporal CNN (STCNN)
[24]	Applied LSTM networks to mobile network traffic for real-time resource allocation optimization	LSTM
[25]	Discussed ITS and traffic prediction methodologies, emphasizing accuracy improvements	Mathematical, parametric, and non-parametric methods
[26]	Introduced error recurrent CNN for lower computational cost and higher accuracy	Error Recurrent CNN (eRCNN)
[27] et al.	Proposed T-GCN model combining GCN and GRU to capture spatial and temporal dependencies	Temporal Graph Convolutional Network (T-GCN)
[28]	Used GRU for traffic matrix prediction to optimize resource allocation in optical backbone networks	Gated Recurrent Units (GRU)
[29]	Developed a dynamic traffic simulator for real-time adjustments based on sensor data	Dynamic Traffic Simulator
[30]	Provided a comprehensive survey on traffic prediction, comparing models and emphasizing preprocessing	Comparative analysis of various predictors, including time series
[31]	Highlighted double exponential smoothing as a computationally efficient predictor for traffic flow	Double Exponential Smoothing
Our paper	Predicting traffic congestion for large-scale transportation networks applying neural networks	Multi-task Learning Neural Networks

Furthermore, in Zhao et al. [27], the temporal graph convolutional network (T-GCN) model has been proposed to capture spatial and temporal dependencies in traffic data simultaneously. By integrating graph convolutional networks (GCN) with gated recurrent units (GRU), the T-GCN has demonstrated superior performance on real-world datasets compared to existing baselines, emphasizing the importance of modeling complex urban road networks. Additionally, research has extended beyond traditional road traffic forecasting to include applications in mobile and metro-core networks. Troia et al. [28], utilizing deep learning methods, such as Gated Recurrent Units (GRU), for traffic matrix predictions allows for proactive optimization of resource allocation in optical backbone networks. By implementing dynamic resource allocation based on accurate traffic forecasts, significant savings in network capacity can be achieved during unexpected traffic peaks.

Predicting traffic flow during unpredictable events, such as accidents or road closures, remains a challenging task. Abadi et al. [29] proposed a dynamic traffic simulator that adjusts predictions based on real-time sensor data. This model maintained a low error rate, even in the presence of unforeseen disruptions. Additionally, Yuan et al. [30] provided a comprehensive survey of traffic prediction, focusing on spatiotemporal data and preprocessing techniques. A comparative analysis of various predictors, including classic time series models and artificial neural networks, has underscored the importance of balancing accuracy with computational efficiency and power consumption. For instance, double exponential smoothing has emerged as a reliable predictor that offers a favorable trade-off between performance and computational cost Iqbal et al. [31]. All the related works are summarized in Table 1, along with our paper, mentioning the type of work and technique used to obtain the results.

However, existing models often struggle with adapting to new datasets and emerging traffic scenarios. Most research has concentrated on applying single algorithms or hybrid models based on single-task learning (STL). In this context, we aim to demonstrate the superiority of the Multi-Task Learning (MTL) approach compared to STL models, which may offer better adaptability and accuracy across various traffic scenarios.

3 Proposed Framework

Predicting traffic congestion is a critical and challenging task in the field of intelligent traffic management. The objective is to forecast changes in congestion lengths over time across multiple road segments, enabling more efficient traffic control and planning. To address this, we propose a novel framework utilizing a Multi-Task Learning-based Neural Network (MTL-NN). This framework is designed to simultaneously predict multiple traffic-related outputs, leveraging shared information across tasks to enhance accuracy. The architecture of the proposed model is illustrated in Fig. 1 and described in detail below.

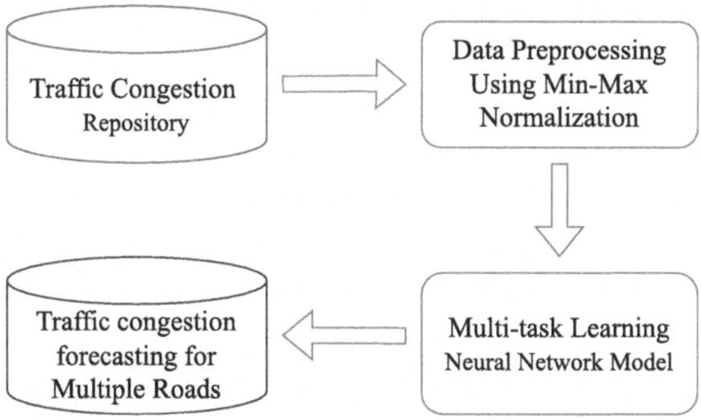

Fig. 1. Steps in traffic congestion forecasting

Definition 1. ***(Traffic congestion dataset:)*** *Let $N = \{R_1, R_2, \cdots, R_m\}$ be a transportation road network containing a set of road segments R_1 to R_m, where $m \geq 1$. Let $VR_j = \{v^1_{R_j}, v^2_{R_j}, \cdots, v^n_{R_j}\}$, $n \geq 1, v^i_{R_j} \in R^+$, denote the values (or congestion lengths) recorded on a road segment R_j, $R_j \in N$ over a time duration n. A traffic congestion dataset, TCD, represents the collection of congestion lengths observed on all road segments in N over the time duration of n. That is, $TCD = \{VR_1, VR_2, \cdots, VR_m\}$.*

Example 1. Let $N = \{R_1, R_2, R_3\}$ be the three road segments in a road network. If the traffic congestion lengths observed on the road segment R_1 at three different (uniform) time intervals are 270 m, 410 m, and 230 m, then $VR_1 = \{270, 410, 230\}$. The hypothetical traffic congestion dataset generated on these three road segments over a duration of three units of time is shown in Fig. 2a.

Definition 2. ***(Normalized traffic congestion dataset:)*** *Let $\widehat{VR_j} = \{\widehat{v^1_{R_j}}, \widehat{v^2_{R_j}}, \cdots, \widehat{v^n_{R_j}}\}$ be the normalized timeseries data generated for a road-segment R_j using a function M (say, min-max normalization), where $\widehat{v^i_{R_j}} \in (0, 1)$ and $1 \leq i \leq n$. Once the data has been normalized, it shall be transformed into tensors since the proposed MTL model is based on NN. Therefore, let $T(.)$ be a function that converts the normalized values of a road segment R_j into a tensor, $\widehat{TR_j}$ i.e., $T(\widehat{VR_j}) = \widehat{TR_j}$.*

Example 2. If *min-max* is the normalization function, then $\widehat{VR_1} = \{0.22\ (= \frac{270-230}{410-230}), 1\ (= \frac{410-230}{410-230}), 0\ (= \frac{230-230}{410-230})\}$. The normalized traffic congestion lengths for all the roads in R are shown in Fig. 2(b). The following is the Python code representation to transform the normalized values into tensors.

$$\widehat{TR_j} = torch.tensor(\widehat{VR_j}, dtype = torch.float)$$

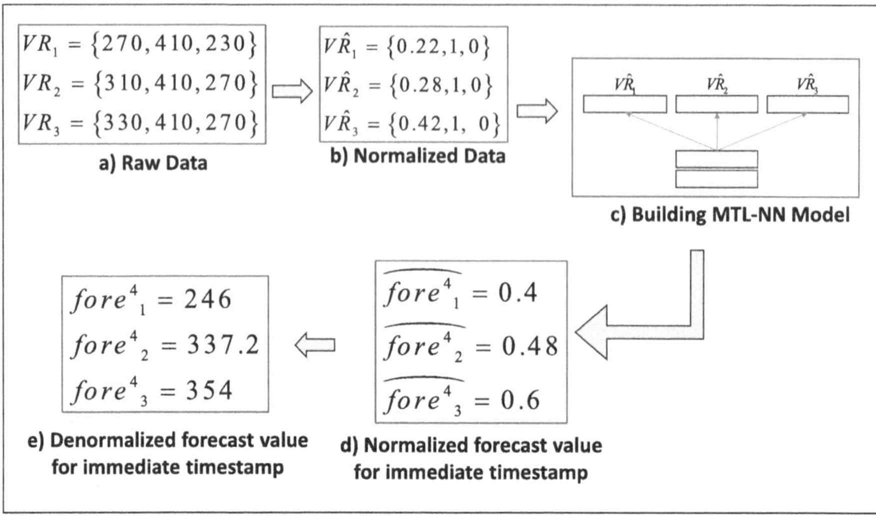

Fig. 2. A step by step illustration of the proposed framework

Definition 3. ***(Multi-task learning model.)*** *Let* $MTL(\widehat{TR_1}, \widehat{TR_2}, \cdots, \widehat{TR_m}) \rightarrow \{\widehat{fore_1^{n+1}}, \widehat{fore_2^{n+1}}, \cdots, \widehat{fore_m^{n+1}}\}$ *be the proposed single-step multi-task learning model (discussed in the next section) developed by hard-parameter tuning. The proposed* MTL *model accepts the normalized traffic congestion lengths of all road segments in the dataset and outputs the normalized forecast congestion lengths of all the road segments at the subsequent time unit* $(n+1)$.

Example 3. Figure 2(c) shows the proposed single-step MTL model. The proposed model accepts all road segments' normalized traffic congestion lengths as an input, performs hard parameter tuning, and outputs the traffic congestion lengths of all road segments for the next time unit. That is, $MTL([0.22\ 0.28\ 0.42], [1\ 1\ 1], [0\ 0\ 0]) \rightarrow [0.40.480.6]$.

Definition 4. ***(Denormalized predicted traffic congestion lengths)*** *Outputs obtained from the normalized values stay in the range of (0,1). To obtain the predicted traffic congestion length values, denormalization should be performed. Let* $DN(\widehat{fore_1^{n+1}}, \widehat{fore_2^{n+1}}, \cdots, \widehat{fore_m^{n+1}}) \rightarrow fore_1^{n+1}, fore_2^{n+1}, \cdots, fore_m^{n+1}$ *be a denormalization function that converts the normalized predicted traffic congestion length values into denormalized.*

Example 4. The predicted normalized traffic congestion length for R_1 by MTL is 0.4. That is, $\widehat{fore_{R1}^4} = 0.4$. The denormalized value of $\widehat{fore_{R_1}^4}$, i.e., $fore_{R_1}^4 = (0.4 \times (270 - 230)) + 230 = 246$. The denormalized traffic congestion lengths predicted for remaining road segments at $timestamp = 4$ are shown in Fig. 2(e).

Algorithm 1. Multi-task learning applied Neural Networks
Input: trafficCongestionDataset (N)
Output: predictedTrafficCongestionLengthDataset ($\widehat{fore_{R_j}^{n+1}}$)

1: Let $\widehat{N}$ be the Min-Max normalized dataset generated from N.
2: Split the normalized dataset $\widehat{N}$ into training and testing datasets using 80-20 rule. Let $\widehat{TRAIN_j}$ and $\widehat{TEST_j}$ denote the training and test datasets, respectively.
3: Simultaneously build an MTL-NN model for all the road segments in $\widehat{TR_j}$. The proposed MTL-NN model contains two hidden shared layers and a task-specific layer as shown in Fig 2. The hidden layers utilize linear transformation and *ReLU* activation functions. In the task-specific layer, based on the outputs from the hidden layers, the predictions are generated for each specific task.
4: Using the test dataset, $\widehat{TEST_J}$ simultaneously predict the normalized traffic congestion lengths for all road segments in $\widehat{TRAIN_j}$.
5: Evalaute the performance of the MTL-NN model using the evaluation metrics, such as *root mean square error* ($RMSE$), *mean absolute error* (MAE) and *R Squarred* (R^2).
6: Perform denormalization to obtain the actual predicted traffic congestion length values.
7: Measure the resource computation metrics, such as *time* and *memory* required for computation.

4 Model Design

Consider the normalized traffic data samples of m road segments $\{\widehat{TR_1}, \widehat{TR_2}, \cdots, \widehat{TR_m}\}$, each traffic congestion length data sample is divided into two segments including *training samples* and *testing samples*. The proposed MTL-NN model consists of three main components: shared layers, task-specific layers, and the output layer.

The *training samples* are input into the MTL-NN model, which consists of two hidden shared layers. These layers perform computations based on the multi-task learning objective, following Eq. 1 and Eq. 2. Specifically, the shared layers capture common patterns across multiple tasks, allowing the model to learn general representations applicable to all tasks. After passing through the shared layers, the outputs are directed to task-specific layers designed to handle the unique aspects of each task. These layers further refine the representations to produce the final predictions for each specific task. The task-specific layers ensure that while the model benefits from shared knowledge, it can still specialize for each output, leading to more accurate predictions across multiple traffic parameters.

- The shared layers are composed of two hidden layers with linear transformations followed by ReLU activation functions which are responsible for learning the shared representation of the input data.

- Each road segment has its task-specific layer that takes the output from the shared layers and processes it further to generate predictions. Task-specific layers consist of a linear transformation and are unique for each task.
- The output layer generates the predicted traffic congestion lengths for all road segments. The predictions are in normalized form and are later denormalized to obtain the actual values.

$$H_i = W_i(\widehat{TR_j}) + b_i \tag{1}$$

$$A_i = \max(0, H_i) \tag{2}$$

- W_i: Weight matrix for the ith linear layer
- b_i: Bias vector for the ith linear layer
- A_i: Output after applying ReLU Activation in i^{th} linear layer
- H_i: Linear Transformation output

Finally, we predict the traffic congestion for all the roads present in the dataset using Eq. 3.

$$O_{T_k} = W_{T_k} A_F + b_{T_k} \tag{3}$$

- O_{T_k}: Output of the kth task-specific layer
- W_{T_k}: Weight matrix for the kth task-specific layer
- b_{T_k}: Bias vector for the kth task-specific layer
- A_F: Final layer ReLU activation output

For example, Let $(\widehat{TR_j})$ be the input tensor containing normalized traffic congestion lengths for all road segments. The output O for road segment R_j can be expressed as:

$$O_{R_j} = W_{R_j} \cdot ReLU(W_2 \cdot ReLU(W_1 \cdot \widehat{TR_j} + b_1) + b_2) + b_{R_j}$$

where,

- W_1, W_2 are the weight matrices for the shared layers
- W_{R_j} is the weight matrix for the task-specific layer of road segment R_j
- b_1, b_2, b_{R_j} are the bias terms.

The loss function used is the Mean Squared Error (MSE), which measures the difference between the predicted and actual traffic congestion lengths. The loss function can be determined using Eq. 4.

$$Loss = \sum_{i=1}^{N} MSE(O_i, Y_i) \tag{4}$$

- N: Number of tasks
- O_i: Predicted output of ith task
- Y_i: Actual value of ith task
- MSE: Mean Squared Error function for comparing predicted and actual values

Finally, we use the gradient descent algorithm using Eq. 5, to optimize the parameters and reduce the loss function.

$$\theta = \theta - \eta \nabla_\theta \text{Loss} \tag{5}$$

- θ: Model Parameters (weights and biases)
- η: Learning Rate
- ∇_θLoss: Gradient of the loss function with respect to model parameters

Following are the sequence of the steps the training loop will take place.

- **Forward Pass**: In the forward pass, the input data is passed through the shared layers and then through the task-specific layers to obtain the predictions O_i for each task.
- **Loss Computation**: The loss for each task is computed using the Mean Squared Error (MSE) between the predicted and true values. The total loss is the sum of individual task losses.
- **Backward Pass**: The gradients of the total loss concerning the model parameters are computed in the backward pass.
- **Parameter Update**: The model parameters are optimized using the *Adam optimizer* with a learning rate of 0.001. The training process involves iterating over the training dataset multiple times (epochs) until the model converges to a minimum loss value.

Once the predictions are obtained, they will be denormalized into the original scale using the inverse of the Min-Max Scaling method shown in Eq. 6 (Fig. 3).

$$VR_P = \left(\widehat{VR_P} \times (\max(VR_j) - \min(VR_j))\right) + \min(VR_j) \tag{6}$$

- VR_P: Predictions in the original data format
- $\widehat{VR_P}$: Normalized predictions from the model.

The various predicted outputs of different traffic samples are analyzed and evaluated using the root mean squared error ($RMSE$) & mean absolute error (MAE) i.e., Eq. 7 & 8 for m data samples respective to each data sample. Additionally, the coefficient of determination (R^2) is used to assess the goodness-of-fit of the model i.e., Eq. 9

$$\text{RMSE} = \sqrt{\frac{1}{N}\sum_{i=1}^{N}(Y_i - \hat{Y}_i)^2} \tag{7}$$

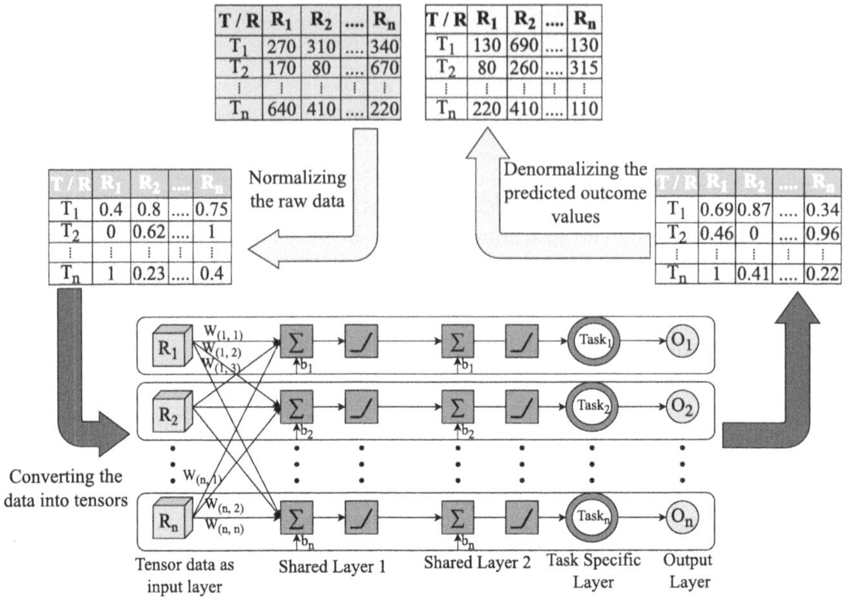

Fig. 3. MTL-NN Forecast Model

$$\mathrm{MAE} = \frac{1}{N}\sum_{i=1}^{N}|Y_i - \hat{Y}_i| \tag{8}$$

$$R^2 = 1 - \frac{\sum_{i=1}^{N}(Y_i - \hat{Y}_i)^2}{\sum_{i=1}^{N}(Y_i - \bar{Y})^2} \tag{9}$$

- N: Number of Samples
- Y_i: Actual Value
- $\hat{Y}_i$: Predicted Value
- $\bar{Y}$: Mean of Actual Values

5 Performance Evaluation

5.1 Experimental Setup

The experiments were conducted on a Gigabyte r282-z93 rack server with an AMD Exynos Processor, an A5000 GPU, and 300GB of RAM. This machine's operating system is Ubuntu server OS 24.04. The extensive dataset used for the experiments was provided by an anonymous company. The dataset contained 5-minute traffic congestion data gathered for one year (1-4-2019 to 31-3-2020) on ten road segments.

The proposed model, *MTL-NN*, has been evaluated through extensive experimentation. To demonstrate the efficiency of our model, we compared it with the following ten different machine learning and deep learning models: *Decision Tree Regression* (DTR), *Gradient Boosting Regression* (GBR), *K-Neighbors Regression* (KNR), *Extreme Gradient Boosting* (XGBoost), *Autoregressive Integrated Moving Average* (ARIMA), *Long Short-Term Memory* (LSTM), *Bidirectional LSTM* (BiLSTM), *Gated Recurrent Unit* (GRU), *Convolutional Neural Network* (CNN), and *Neural Networks* (NN).

5.2 Results and Comparison

Figures 4a and 4b illustrate that the proposed MTL-NN consistently outperforms other models regarding RMSE and MAE values. RMSE and MAE are standard measures to evaluate the accuracy of predictive models, with lower values indicating higher accuracy. The MTL-NN model achieves shallow RMSE values between 0.0002 and 0.0005 across various roads, while other models exhibit much higher RMSE values, ranging from 0.2 to 0.45. Similarly, the MTL-NN model shows significantly lower errors for MAE than the alternatives. It indicates that MTL-NN significantly minimizes prediction errors for traffic congestion length estimation across multiple road segments. The reason behind this strong performance is that MTL-NN can capture complex patterns in traffic data over time and space. It does this by using shared layers, which help the model learn common features across different road segments. This shared learning reduces the errors across tasks and makes the predictions more reliable.

The R^2 metric in Fig. 4c further emphasizes the strength of the MTL-NN model. With R^2 values close to 1 across the road segments, the MTL-NN provides near-perfect fits to the data, demonstrating that it captures almost all the variability in the traffic data. In contrast, other models show much lower R^2 values, with some even negative, indicating a poor fit. The negative values suggest that the predictions of these models are worse than a simple mean prediction, further highlighting the under-performance of traditional models when dealing with complex traffic scenarios. The near-perfect R^2 values of the MTL-NN model suggest that it is not only more accurate but also more reliable when applied to real-world traffic datasets. The MTL-NN's ability to balance between task-specific learning and shared learning among different tasks (i.e., predicting traffic congestion lengths for different road segments) contributes to this improved performance.

Table 2 compares the training time and memory consumption of different models, revealing some significant insights into the efficiency of the proposed MTL-NN model. One of the key aspects of the MTL-NN is its balance between computational efficiency and memory usage. While the training time of the MTL-NN (6.06 s) is slightly higher than lightweight models such as KNR (0.047 s) and DTR (0.436 s), it is still considerably faster than deep learning models like LSTM, Bi-LSTM, and GRU, which take significantly longer, with training times of 566.49 s, 736.03 s, and 582.08 s, respectively. The relatively faster training time of the MTL-NN can be attributed to its efficient architecture, which leverages

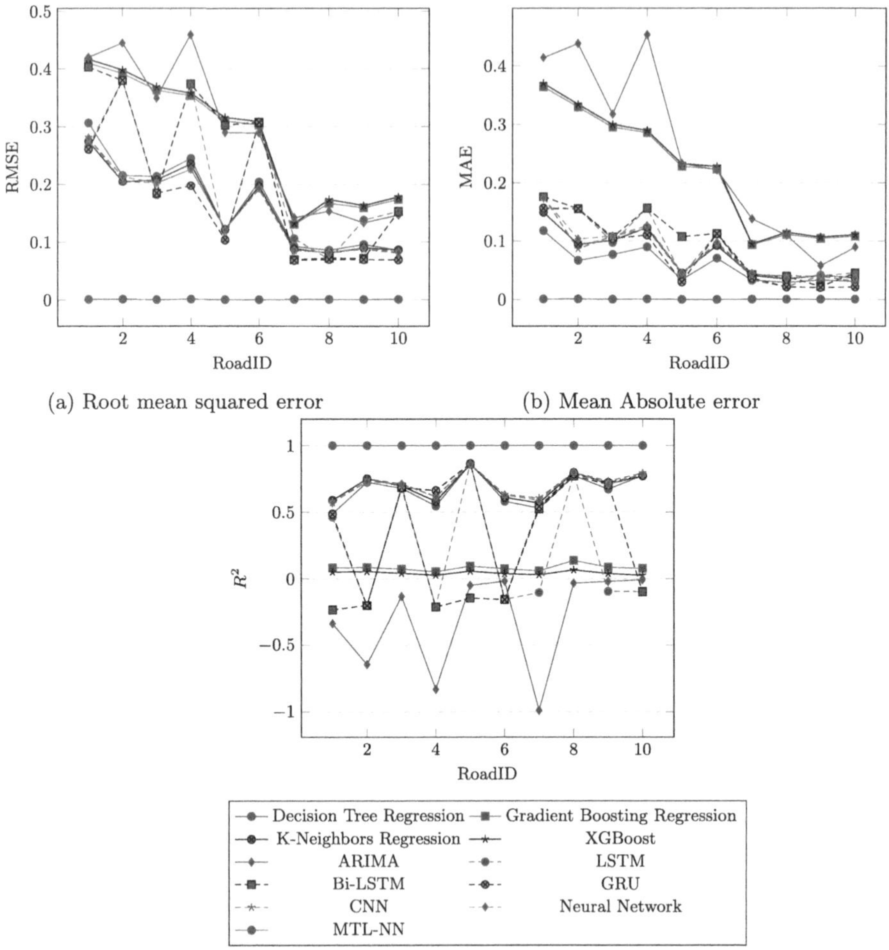

(a) Root mean squared error (b) Mean Absolute error

(c) R^2

Fig. 4. Comparison of Performance Paremetrics

shared layers. This design reduces the computational overhead that typically comes with training separate models for each task, allowing the MTL-NN to maintain competitive training speeds while handling multiple tasks simultaneously.

In terms of memory consumption, the MTL-NN model demonstrates an impressive efficiency, utilizing only 0.0001 MB, far lower than any other model, including KNR (0.023 MB) and DTR (0.039 MB). This substantial reduction in memory usage underscores the resource efficiency of the MTL-NN, making it a suitable option for real-time traffic congestion prediction in systems with limited computational resources. High-memory models like NN (24.33 MB), CNN (17.66 MB), and Bi-LSTM (22.17 MB) significantly exceed the resource requirements of the MTL-NN, which further emphasizes the superiority of the MTL-NN in handling large-scale traffic prediction tasks efficiently. This combination of relatively fast training time and minimal memory usage makes the MTL-NN model an ideal choice for deployment in real-world traffic management systems, where both accuracy and efficiency are critical. For better understanding, Fig. 5 illustrates the scatter plot of Memory Utilization (MBs) and Training Time (Sec) for all models.

The comprehensive comparison across RMSE, MAE, R^2, training time, and memory consumption clearly demonstrates that the MTL-NN model is not only the most accurate model but also one of the most efficient. Its ability to balance prediction accuracy with computational efficiency sets it apart from other deep learning models such as LSTM, Bi-LSTM, and GRU, which are known for their high accuracy but also for their significant computational demands. Moreover, the results suggest that MTL-NN is highly adaptable and scalable across different traffic scenarios. While traditional models may perform well under certain conditions or on specific road segments, the MTL-NN's multi-task approach ensures that it generalizes better to different road conditions and traffic patterns. This makes it a strong candidate for practical applications in intelligent traffic management systems, where adaptability and scalability are critical for real-world performance.

Table 2. Comaprision of Training Time and Memory Consumed

Model	Training Time (Sec)	Memory Utilized (MBs)
DTR	0.43616	0.039063
GBR	4.72629	0.038672
KNR	0.04739	0.023047
XGBoost	0.15246	0.671484
ARIMA	11.62217	2.253125
LSTM	566.4938	7.403906
Bi-LSTM	736.0347	22.17305
GRU	582.0781	14.79258
CNN	186.2783	17.66328
NN	0.921483	24.33125
MTL-NN	**6.0632**	**0.0001**

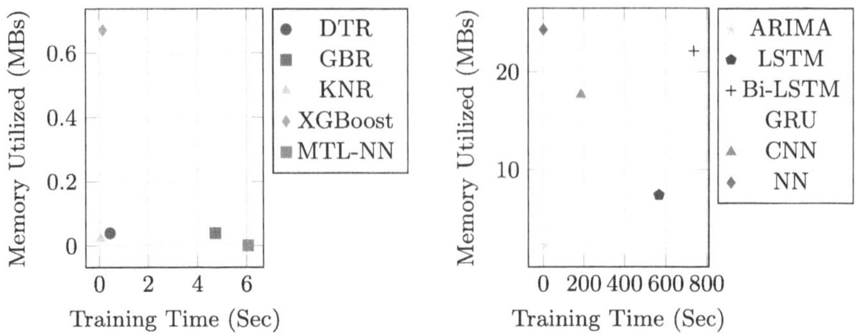

Fig. 5. Training Time vs Memory Utilized for Different Models

6 Conclusion

Traffic congestion continues to be a significant challenge in many urban areas, underscoring the need for effective predictive models. In this paper, we introduced a novel multi-task learning-based neural network (MTL-NN) framework to forecast road traffic congestion using a large-scale dataset, addressing the limitations of single-task learning models. Our MTL-NN model was rigorously evaluated and compared with various state-of-the-art ML and DL models, including DTR, GBR, KNR, XGBoost, ARIMA, LSTM, BiLSTM, GRU, CNN, and NN. The MTL-NN model consistently achieved superior accuracy with the lowest RMSE values (ranging from 0.0002 to 0.0005), outperforming other models (with RMSE values ranging from 0.2 to 0.45). Additionally, the MTL-NN model demonstrated efficiency in terms of both memory and time consumption. In conclusion, the MTL-NN framework presents a powerful and efficient solution for predicting traffic congestion, surpassing existing methods in both accuracy and resource optimization. This makes it a precious tool for improving urban traffic management and planning efforts.

References

1. Gilani, S.A.U., Al-Rajab, M., Bakka, M.: Challenges and opportunities in traffic flow prediction: review of machine learning and deep learning perspectives. Data Metadata **3**, 378–378 (2024)
2. Hazarika, A., Choudhury, N., Nasralla, M.M., Khattak, S.B.A., Rehman, I.U.: Edge ML technique for smart traffic management in intelligent transportation systems. IEEE Access (2024)
3. Guarino, I., Aceto, G., Ciuonzo, D., Montieri, A., Persico, V., Pescapè, A.: Explainable deep-learning approaches for packet-level traffic prediction of collaboration and communication mobile apps. IEEE Open J. Commun. Soc. (2024)
4. Buroni, G., Lebichot, B., Bontempi, G.: AST-MTL: an attention-based multi-task learning strategy for traffic forecasting. IEEE Access **9**, 77359–77370 (2021)

5. Wang, A., Ye, Y., Song, X., Zhang, S., James, J.Q.: Traffic prediction with missing data: a multi-task learning approach. IEEE Trans. Intell. Transp. Syst. **24**(4), 4189–4202 (2023)
6. Abdillah, A.F., Hamidi, M.Z., Anggraeni, R.N.E., Sarno, R.: Comparative study of single-task and multi-task learning on research protocol document classification. In: 2021 13th International Conference on Information & Communication Technology and System (ICTS), pp. 213–217. IEEE (2021)
7. Huang, H., Mao, J., Liu, R., Lu, W., Tang, T., Liu, L.: MTLMetro: a deep multi-task learning model for metro passenger demands prediction. IEEE Trans. Intell. Transp. Syst. (2024)
8. Manas, K., Zwicklbauer, S., Paschke, A.: TR2MTL: LLM based framework for Metric Temporal Logic Formalization of Traffic Rules. arXiv preprint arXiv:2406.05709 (2024)
9. Tejima, K., Saxena, D., Rage, U.K.: A novel multi-task single-step traffic congestion forecasting framework for large-scale road networks. In: International Conference on Industrial, Engineering and Other Applications of Applied Intelligent Systems, pp. 277–288. Springer, Singapore (2024)
10. Zhang, Y., Yang, Q.: An overview of multi-task learning. Natl. Sci. Rev. **5**(1), 30–43 (2018)
11. Manibardo, E.L., Laña, I., Del Ser, J.: Deep learning for road traffic forecasting: does it make a difference? IEEE Trans. Intell. Transp. Syst. **23**(7), 6164–6188 (2021)
12. Fan, X., Xiang, C., Gong, L., et al.: Deep learning for intelligent traffic sensing and prediction: recent advances and future challenges. CCF Trans. Pervasive Comp. Interact. **2**, 240–260 (2020). https://doi.org/10.1007/s42486-020-00039-x
13. Vinayakumar, R., Soman, K.P., Poornachandran, P.: Applying deep learning approaches for network traffic prediction. In: 2017 International Conference on Advances in Computing, Communications and Informatics (ICACCI), pp. 2353–2358. IEEE (2017)
14. Kothai, G., et al.: A new hybrid deep learning algorithm for prediction of wide traffic congestion in smart cities. Wirel. Commun. Mob. Comput. **2021**(1), 5583874 (2021)
15. Alkarim, A.S., Al-Ghamdi, A.S.A.M., Ragab, M.: Ensemble learning-based algorithms for traffic flow prediction in smart traffic systems. Eng. Technol. Appl. Sci. Res. **14**(2), 13090–13094 (2024)
16. Cheng, Z., Lu, J., Zhou, H., Zhang, Y., Zhang, L.: Short-term traffic flow prediction: an integrated method of econometrics and hybrid deep learning. IEEE Trans. Intell. Transp. Syst. **23**(6), 5231–5244 (2021)
17. Barros, J., Araujo, M., Rossetti, R.J.: Short-term real-time traffic prediction methods: a survey. In: 2015 International Conference on Models and Technologies for Intelligent Transportation Systems (MT-ITS), pp. 132–139. IEEE (2015)
18. He, Z., Chow, C.Y., Zhang, J.D.: STNN: a spatio-temporal neural network for traffic predictions. IEEE Trans. Intell. Transp. Syst. **22**(12), 7642–7651 (2020)
19. Pan, B., Demiryurek, U., Shahabi, C.: Utilizing real-world transportation data for accurate traffic prediction. In: 2012 IEEE 12th International Conference on Data Mining, pp. 595–604. IEEE (2012)
20. Dunne, S., Ghosh, B.: Weather adaptive traffic prediction using neurowavelet models. IEEE Trans. Intell. Transp. Syst. **14**(1), 370–379 (2013)
21. Ramakrishnan, N., Soni, T.: Network traffic prediction using recurrent neural networks. In: 2018 17th IEEE International Conference on Machine Learning and Applications (ICMLA), pp. 187–193. IEEE (2018)

22. Liu, Z., Li, Z., Wu, K., Li, M.: Urban traffic prediction from mobility data using deep learning. IEEE Network **32**(4), 40–46 (2018)
23. He, Z., Chow, C.Y., Zhang, J.D.: STCNN: a spatio-temporal convolutional neural network for long-term traffic prediction. In: 2019 20th IEEE International Conference on Mobile Data Management (MDM), pp. 226–233. IEEE (2019)
24. Trinh, H.D., Giupponi, L., Dini, P.: Mobile traffic prediction from raw data using LSTM networks. In: 2018 IEEE 29th Annual International Symposium on Personal, Indoor and Mobile Radio Communications (PIMRC), pp. 1827–1832. IEEE (2018)
25. Boukerche, A., Wang, J.: Machine learning-based traffic prediction models for intelligent transportation systems. Comput. Netw. **181**, 107530 (2020)
26. Mena-Oreja, J., Gozalvez, J.: A comprehensive evaluation of deep learning-based techniques for traffic prediction. IEEE Access **8**, 91188–91212 (2020)
27. Zhao, L., et al.: T-GCN: a temporal graph convolutional network for traffic prediction. IEEE Trans. Intell. Transp. Syst. **21**(9), 3848–3858 (2019)
28. Troia, S., Alvizu, R., Zhou, Y., Maier, G., Pattavina, A.: Deep learning-based traffic prediction for network optimization. In: 2018 20th International Conference on Transparent Optical Networks (ICTON), pp. 1–4. IEEE (2018)
29. Abadi, A., Rajabioun, T., Ioannou, P.A.: Traffic flow prediction for road transportation networks with limited traffic data. IEEE Trans. Intell. Transp. Syst. **16**(2), 653–662 (2014)
30. Yuan, H., Li, G.: A survey of traffic prediction: from spatio-temporal data to intelligent transportation. Data Sci. Eng. **6**(1), 63–85 (2021)
31. Iqbal, M.F., Zahid, M., Habib, D., John, L.K.: Efficient prediction of network traffic for real-time applications. J. Comput. Netw. Commun. **2019**(1), 4067135 (2019)

Data Science

MLN-geeWhiz: Supporting Complex Data Analysis Including Visualization

Amey Shinde[1], Viraj Sabhaya[1], Kevin Farokhrouz[1], Fariba Irany[2], Ali Khan[2], Sanjukta Bhowmick[2], Abhishek Santra[1(✉)], and Sharma Chakravarthy[1]

[1] IT Laboratory and CSE Department, UT Arlington, Arlington, USA
{amey.shinde,virajvipinbhai.sabhaya,kaf2886, abhishek.santra}@mavs.uta.edu, sharmac@cse.uta.edu

[2] University of North Texas, Denton, USA
{FaribaAfrinIrany,AliKhan}@my.unt.edu, sanjukta.bhowmick@unt.edu

Abstract. The availability of numerous algorithms and techniques developed for complex data analysis makes it difficult for domain users to navigate the life cycle workflow for their analysis. One way to overcome this is by providing interactive tools that are intuitive to use, graphical, and can be used for all stages of the life cycle by non-experts – from modeling to complex analysis to drill down & visualization. If these tools are also extensible and modular, they allow additions of new models and algorithms as they become available for the benefit of the larger community.

Graphs have been extensively used to study the complex systems of interacting entities from diverse disciplines. However, when studying complex data with multiple types of entities and relationships, simple or even attributed graphs are not always ideal for modeling them. Currently, to derive knowledge from complex data with multiple entity types and relationships, multilayer networks (or MLNs) are being increasingly used.

MLN-geeWhiz, is conceived to be such an end-to-end interactive tool with its initial version available for public use. Its purpose is to fill a void and allow non-experts from diverse disciplines to analyze their complex data using state-of-the-art analysis techniques and visualize the results. This dashboard has been developed to use complex analysis techniques with ease. It is modular, so new algorithms and approaches can be added transparently. The user can upload, generate needed MLNs (or graphs), analyze them and visualize results in many ways. In this paper, we discuss the challenges of a modular architecture needed for supporting complex analysis underneath and visualization to understand the results.

Keywords: Multilayer Networks · Modular/Extensible Architecture · Complex Data Analysis · Visualization

1 Motivation

Unlike modeling a relational or object-oriented database where a schema is generated first and then populated for analysis, in the graph models, the graph itself contains data. The schema is implicit. Hence, all the necessary (and hopefully minimal) information

A. Dasgupta et al. (Eds.): BDA 2024, LNCS 15526, pp. 219–239, 2025.
https://doi.org/10.1007/978-3-031-81821-9_13

for analysis need to be included in the graphs generated. Typically, nodes of the graph represent entities and edges represent relationships. Although labels can be used both for nodes and edges, it is important to keep the amount of information minimal to satisfy the objectives specified. For example, if you are clustering authors, an author can be represented by an ID and not include other attributes which are not needed for analysis. Similarly, for edges. However, the rest of the information of entities and relationships is needed for drill-down in order to visualize results for understand. We focus on the MLN graph model, its analysis, and visualization in this paper.

1.1 Multilayer Networks and Their Utility

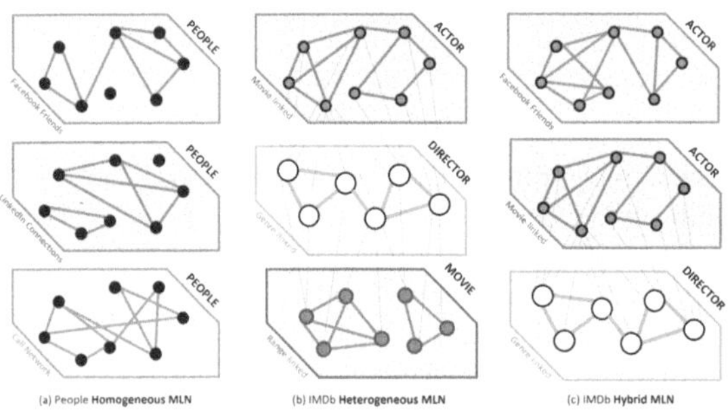

Fig. 1. MLN Types

Multilayer networks (MLNs) provide an alternate way to model complex data sets. It can capture each relationship as a layer which is a separate simple graph increasing understandability. It also affords itself for flexible analysis of layers. Nodes from different layers can also be connected by an edge if relationships exist between two entity types. For example, actor and director layers can be connected by the 'direct-actor' relationship where a director entity has an edge with each actor entity s/he has directed in a movie. Design of MLN model for a data set has been addressed separately in the literature [37]. MLNs, by their structure, also offer flexibility to analyze each layer individually (and in parallel) as well as arbitrary subsets of layer combinations. Based on entity types in each layer and connectivity within and across layers, MLNs can be classified as homogeneous MLNs (HoMLNs), heterogeneous MLNs (HeMLNs), and hybrid MLNs (HyMLNs), illustrated in Fig. 1(a), (b), and (c), respectively.

Definition 1. *A multilayer network $MLN(G, X)$, is defined by two sets of graphs. The set $G = \{G_1, G_2, \ldots, G_n\}$ contains* ***simple*** *graphs (one for each layer), where $G_i(V_i, E_i)$ is defined by a set of vertices V_i and a set of edges E_i. An edge $e(v, u) \in E_i$, connects vertices v and u, where $v, u \in V_i$. The set $X = \{X_{1,2}, X_{1,3}, \ldots, X_{m-1,m}\}$ consists of* ***bipartite*** *graphs*[1]*. Each graph $X_{i,j}(V_i, V_j, L_{i,j})$ is defined by two sets of vertices V_i and V_j and a set of edges (or links) $L_{i,j}$, such that for every link $l(a, b) \in L_{i,j}$, $a \in V_i$ and $b \in V_j$, where V_i (V_j) is the vertex set of graph G_i (G_j). Some of $X_{i,j}$ can be empty.*

MLNs find use in diverse disciplines that require modeling complex interactions, including drug design [21], understanding collaboration patterns [17], understanding

[1] The number of bipartite graphs can be less or even more than the number of graphs/layers. That is, m can be $<=$ or even $> n$.

the interaction between spoken and written language [24], discovering long-term coexistence of viruses [33], comparing the evolution of species [43], finding vulnerabilities in power grids [29], and identifying illegal activities via social interactions [14]. More recently, MLNs have been used to model the spread of COVID-19, where each layer represents a different social interaction [39] and how international supply chains have been affected by the restrictions imposed due to COVID-19 [20].

The use of MLNs for modeling is a new frontier for data analysis. Analysis algorithms and even the definitions of metrics, such as communities, centrality, substructures, and k-core are still evolving. Thus, a platform where researchers can access the latest algorithms and apply them to their data sets is essential for advancing this field. Although the analysis of multilayer networks is a fast-growing area, there is yet no tool for the generation, analysis, and visualization of MLNs, let alone the sharing of data and results or community interaction.

Problem Addressed: Given the complexity of complex data analysis workflow as shown in Fig. 2 and the availability of numerous algorithms (as shown in Table 1) for analysis of MLNs (as elaborated in Sect. 2), develop a *modular and extensible* interactive tool that will support the life cycle. In addition to modeling and analysis, the tool should also support *drill-down of results & visualization.*

The contributions of this paper are:

- An **interactive, web-based dashboard**[2] supporting life cycle of complex data analysis using graphs and MLN data models.
- Generation of **graphs and MLNs**, their analysis, drill-down, and visualization.
- **Extensibility** of modules to extend layer generation, analysis, and visualization.
- **Modular** approach for concurrent development by groups with different skill sets.
- **Multiple (interactive) Visualization Alternatives** for base and analyzed results.
- Support for **multiple concurrent users** and **Optimization** of response time

This paper is organized as follows. Section 2 discusses analysis life cycle, MLN analysis alternatives, and a *sample* of algorithms to demonstrate the vastness of the analysis space. Section 3 discusses related work. Section 4 discusses the challenges to be addressed during the design of the interactive tool (**MLN-geeWhiz** dashboard.) Sect. 5 details the architecture of the dashboard and the front-end design. Section 6 presents critical back-end modules of this dashboard: layer generator, graph and MLN analysis, and drill-down & visualization. Dashboard's usage on some real-world data sets are highlighted in Sect. 7. Conclusions are in Sect. 8.

2 Life Cycle of Complex Data Analysis

In contrast to data mining, big data analysis is more holistic and far more complex. Multiple types of analysis need to be performed based on the data domain and analysis objectives which may vary significantly from one data set to another. Also, data needs to be modeled using an appropriate alternative to make the required analysis possible. The life cycle of big data analysis using graph models (other models are possible, but

[2] Dashboard: https://itlab.uta.edu/mlndash-live/.

not addressed in this paper) is shown in Fig. 2. The dashboard detailed in this paper includes all the components of the life cycle shown in Fig. 2. Configuration files are used to generate graphs or MLN layers. Configuration files are also used for MLN analysis. Drill-down information is maintained and used along with visualization for easy understanding of results.

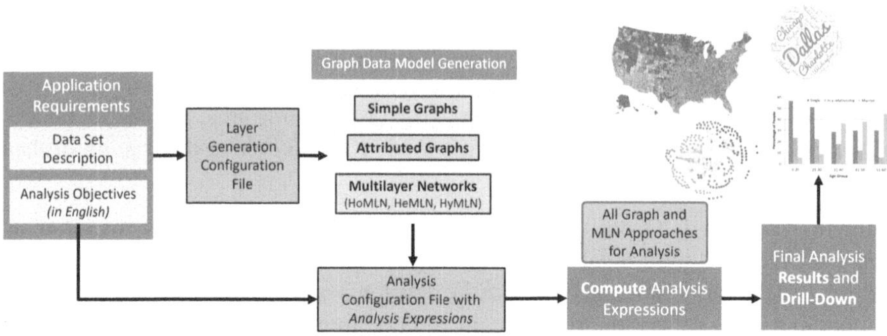

Fig. 2. Complex Data Analysis Life Cycle that need to be Supported

2.1 Space and Variety of MLN Algorithms

Several types of algorithms can be developed for MLN analysis ranging from reducing an MLN to a simple graph to use existing algorithms to developing **new** algorithms using the entire MLN, Between these two, there is another option that can leverage the first approach without developing **totally new** algorithms. They are briefly described below to show the space of algorithms that need to be supported by the interactive tool.

1. Traditional Aggregation Approach: In this approach (shown in Fig. 3(a)), layers of an MLN are conflated into a simple graph using type-independent or projection-based approaches. The aggregation process can be costly (depending on the number of layers) and the resulting graph can be large (for OR composition). The conversion, in either case, loses information [19,

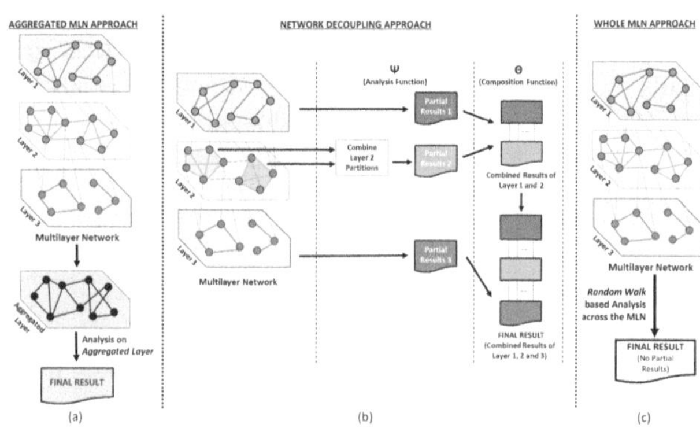

Fig. 3. Traditional vs. Decoupling vs. Holistic Approaches

22] for nodes, edges, labels, and further their relationships are not differentiated[3]. The advantage of this alternative is that extant algorithms (e.g., Louvain, Infomap, Degree Centrality, Substructure Discovery, etc.) can be used for analysis. This approach is not amenable to parallel processing as each layer cannot be processed independently. The loss in MLN structure, due to aggregation, also makes it difficult to drill-down results without maintaining extensive mapping. Finally, for processing a subset of layers, separate aggregation is required – making the approach less flexible and inefficient.

2. Decoupling-Based Approach: It is a "divide and conquer" approach (shown in Fig. 3(b)) for analyzing multilayer networks. The goal is to find the desired metric in each layer independently and then use a *separate* **composition function** to combine the results to generate what is missing for that analysis (e.g., community, centrality, substructure, motif, etc.) The *network decoupling* approach has been established as one of the better ways for analyzing MLNs in terms of efficiency and flexibility [34–36].

3. Holistic Approach: In this approach (shown in Fig. 3(c)), neither the MLN is aggregated nor the decoupling approach is used. An algorithm is developed from scratch (as available graph analysis algorithms cannot be used as in the previous two approaches), keeping the structure and semantics intact. However, a new algorithm needs to be developed for each analysis making this approach complex as the algorithm needs to traverse back and forth across layers. This approach has been used for coherent clustering in [11].

Given the number of approaches and the many metrics used for analysis, making these available in one place (along with descriptions, advantages, disadvantages, and even examples) is critical for data analyzers from non-computer science domains. This is a clear justification for not only the tool, but its modularity and extensibility.

2.2 Sample MLN Algorithms to be Supported

Table 1 shows a subset of analysis algorithms (for community, centrality, ...) available for graphs and MLNs that need to be part of any tool that supports the complex data analysis life cycle.

Currently, the dashboard described in this paper includes a large subset of the available sample algorithms shown in Table 1. This and the workflow shown in Fig. 2 have allowed us to test the modularity and extensibility of the approach described in this paper.

3 Related Work

Graphical interfaces and visualizations are not new and there exists a wide variety of tools for visualizing both base data, results, and drilled-down information in multiple ways [1,4,6,12]. Our focus, in this paper, is to make use of available visualization tools in the best way possible and not propose new ones. For example, we have experimented

[3] If one does not want to lose information, extensive mappings are needed for converting the MLN and its analysis results back to MLN.

Table 1. Sample set of Graph and MLN Analysis algorithms. Only a sample set is included due to space constraints

Analysis Type	Graph Type	Algorithm Description
Community Detection	Simple Graphs/ MLN Layers	`Louvain` [10], `Infomap` [12], `Walktrap` [32], `Fast-Greedy` [15], `Multilevel` [10], `Leading-Eigenvector` [28]: Simple graph community detection algorithms
	HoMLN	`CE-AND`: Intersection of Community Edges [34,36]
		`CE-OR`: Community detection on an OR-meta graph with AND-communities as meta nodes [36]
		`MiMAG`: Coherent subgraph (Community detection) for the whole MLN [11]
	HeMLN	`Maximum Weight Matching (MWM), Maximum Weight Matching with Ties (MWMT), Maximum Weight Perfect Matching (MWPM), Maximum Weight Relaxed Matching (MWRM)`: Matching algorithm for bipartite meta graph with communities as nodes and edge weights as the number of inter-community edges (ω_e), hub participation (ω_h) or density (ω_d) [38]
Degree Centrality	HoMLN	`DC-AND`: Estimated high degree nodes across 2 layers [35]
		`DC-OR`: Estimated nodes with high degree in at least 1 layer [30]
	HeMLN	`HeMLN-DC-AG`: Accuracy Guarantee Heuristic [26]
		`HeMLN-DC-PG`: Precision Guarantee Heuristic [26]
Betweenness Centrality	HeMLN	`HeMLN-BC`: Estimated high betweenness MLN nodes [25]
Closeness Centrality	HoMLN	`CC-AND`: Estimated high closeness nodes across 2 layers [31]
Substructure Discovery	Simple Graphs/ MLN Layers	`SUBDUE`, `MLN-SUBDUE`: Map/Reduce Algorithm for substructure discovery in large graphs [13,16,40]

with a wide variety of tools including, maps, individual graph and community visualization, animation of features in different ways, hovering to highlight data, and real-time data fetching and display, based on user input from a menu. The main contribution is the life cycle approach we have taken for analyzing complex data using graph and MLN models. To the best of our knowledge, we have not come across a dashboard that serves all aspects of MLN usage including modeling, generation of layers, analysis of generated layers as well as drill down & visualization. In addition, our architecture uses a common back-end to drive different user interaction and visualization front-ends. We have also paid attention to efficiency at the back-end by caching previously generated results with an efficient data structure for lookup.

Existing dashboards (Py3Plex [41], MultiNetx [27], Muxviz [18], and Pymnet [23]) are limited in their ability to analyze MLNs and mainly focus on visualizing the results as graphs. Some dashboards focus on reporting and visualizing raw data using maps [1,7,8] or time series plots and statistical modeling [4,5]. In contrast, this dashboard is about complex data analysis in the way the end user wants. That is, the end user should be able to control graphs or layers generated, the kind of analysis to be performed once the graphs/layers are generated, and further visualize the results (or even base data) in multiple ways. Drill-down is integrated with visualization to make the model representation efficient for analysis. For example, minimal information is used for the model and mapping of label information is **generated** to be used for drill-down as part of visualization. A number of existing simple graph and MLN algorithms have been incorporated. The decoupling-based ones make it easier to perform drill-down using only layer information for recreating the structure [42]. The schema generation also separates information needed for drill-down (Relations) and information needed for analysis (MLNs) from the same Enhanced Entity Relationship (EER) diagram [37].

As this dashboard focuses on facilitating each user to perform the analysis appropriate for application data in their domain and hide the complexity of algorithms and their understanding, a flexible configuration file strategy is used for specifications (along with examples.) Layer generation can be done using multiple metrics and are easy to specify. Similarly, MLN analysis, whether simple or complex, can be specified using known operators (e.g., community, substructure, centrality, etc.) and precedence for evaluation. Input and output formats are transparently handled by the dashboard. Multiple visualizations can be generated and compared side-by-side and the ability to create files and directories for proper organization and management of data and results is provided.

4 Challenges to be Addressed

A graphical tool that caters to end users (who are likely to be domain experts, but may not be experts in the details of data analysis algorithms) by providing support for all stages of complex data analysis (i.e., life cycle) requires an easy-to-use interface that aids in the clear understanding of results of each step that can be stored for future lookup or sharing. The overall requirement can be succinctly expressed as: i) interface should be intuitively navigable, ii) minimize the number of clicks used to get a job done, iii) pro-actively prompt for available actions at any point (to prevent mistakes), and iv) provide minimal but useful meta information. These principles were used not only for the front-end, but also for individual module usage. The choice of tech stack and packages used for each module are discussed in Sect. 6.

The broader steps of data analysis involve: i) loading the data in the *specified* format (for the time being which will be extended to other formats), ii) generating appropriate layers of MLN required for analysis, iii) expressing and performing the analysis, and iv) visualizing base data and/or results to understand them. Data is assumed to be formatted as a csv (comma separated values) file which is widely used in mining and data analysis applications. Loading of data is done by choosing a local file and uploading it into a specific destination folder within the user space of the dashboard. Dashboard provides a navigable user directory structure (similar to what users are familiar with) as created by the user. Below, we discuss the main challenges for each of the back-end modules.

Modeling or Layer/Graph Generation: The primary challenge for the creation of layers is to choose a metric for generating layer graphs. It should support widely used metrics as well as allow users to customize their metrics as needed. The metric needs to be specified on the input data including the generation of inter-layer edges for HeMLNs.

In addition to the columns used for a node (it can be more than one column as in lat and long), a computation metric (such as distance) and a threshold (such as distance being less than the threshold) need to be specified. In addition to supporting widely used similarity metrics (such as equality, Haversine, Jaccard etc.), we want this to be customizable as well. Different feature types, such as nominal, categorical, numeric, date also need to be supported. Generation of inter-layer edges, on the other hand, requires two nodes that are likely to be from two different files along with a join condition for determining whether the nodes are connected. Apart from this, input file locations, the output locations need to be specified unambiguously. Details of how this is accomplished are described in Sect. 6.1.

Graph and MLN Analysis: Once the appropriate layers or graphs are generated and automatically stored in the directory specified, many kinds of analysis can be performed. Again, different analyses can be performed on the same generated layers. For example, if we generate 3 separate layers for the social networking applications – Facebook, LinkedIn, and X (formerly, Twitter), these layers can be analyzed individually or in any combination. These need to be specified in some way to compute and store the results. Multiple algorithms may exist for analysis. For example, there are several community detection algorithms (Louvain and Infomap being more popular) and it is the user's choice as to what is best for their analysis. Note that each algorithm may generate different communities (based on the approach and heuristics used). Multiple analyses as well as different combined analyses can be performed which needs to be expressed clearly. We use the infix form for expressing analysis expressions with precedence and each analysis expression is computed and its result stored as specified. An infix to post-fix conversion is done prior to evaluating the given expression. Optimization is incorporated by storing analysis sub-expression results to be re-used in later expressions by the same user. Details are in Sect. 6.2.

Drill-Down and Visualization: Analyzing results of complex analysis resulting in large output is extremely difficult and is further exacerbated by the fact that the result is in some encoded form without the semantics (e.g., labels) to understand it. Hence, a visualization that includes the original labels (drill-down) is more appropriate. Even when it is visualized, some visualizations may be more useful for understanding compared to others. For example, to identify an accident-prone region, visually, a word cloud representation may be better than looking at the dense graph (unless it is plotted on a map). Hence, it is imperative that in addition to drill down, alternate visualizations be provided. Another important issue is the size of results and scalability of visualization. This is a hard problem and there is ongoing research to address this. Interactive pan and zoom features can help deal with size. Rendering response time is also a consideration.

In this dashboard, we support drill-down transparently keeping mapping information during layer generation. This is superimposed on the visualization to show relevant attributes as part of visualization. We have also chosen different visualization alternatives, such as word cloud, Bokeh, interactive graphs, etc. to provide meaningful alternatives based on what the user is looking at. They can also be visualized side-by-side in different windows for better understanding. More details are provided in Sect. 6.3.

Finally, designing each module to work with other modules and sharing minimal information is a challenge in itself. Information is shared by persisting relevant data using efficient data structures. This will allow independent and parallel development of modules rather than waiting on each other. As each module needs slightly different skill sets, it will also make it easier for different groups to work on different modules by robustly sharing data. This will allow the dashboard to automatically detect and enforce certain dependencies. For example, if an analysis is specified on a layer that has not been generated, it will be detected and notified to the user including what needs to be done. This is critical as actions can be taken independently, and the user may not be aware of the inherent dependencies.

5 Modular and Extensible MLN-GeeWhiz Architecture

As shown in Fig. 4, front-end only interacts with the user and displays results. The back-end has several modules including registration and login. A lightweight and efficient relational database (SQLite) is used to store user meta information (email, password, etc.)

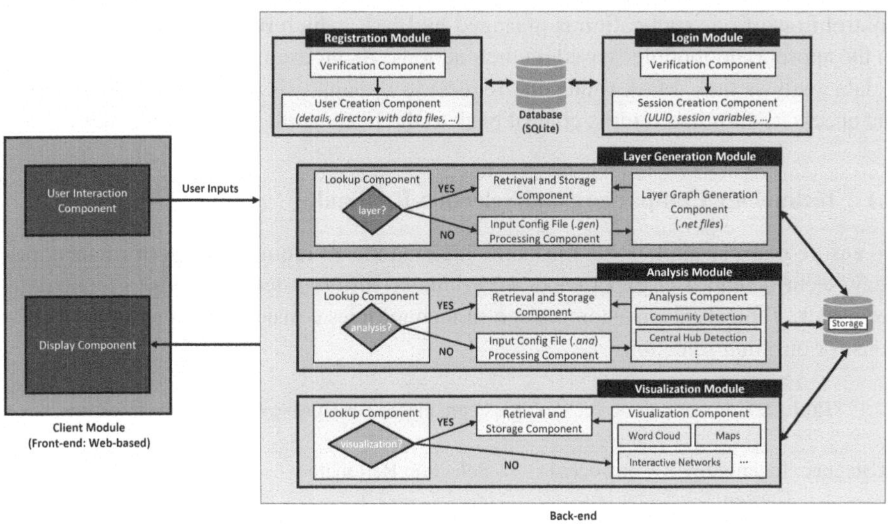

Fig. 4. MLN-geeWhiz Dashboard Architecture

Upon user registration, each user is allocated his/her own directory space. This ensures the isolation of user data for privacy and selective sharing as desired by the user. Each user starts with a pre-specified directory structure that can be extended. All files generated by the dashboard are also stored in the specified user directory.

The dashboard prompts actions based on the file selected and its extension. For example, the dashboard allows the generation of layer files by utilizing the "Generate Layers" button only when a configuration file with .gen extension is selected. The dashboard pro-actively shows the possible actions based on the file chosen. This is further extended to choose the visualization option on files that can be visualized. The log data created by any module is shown in the display panel where information regarding the files generated, such as file name, destination folder, generation time (in seconds), number of nodes, number of edges, and number of connected components is shown.

The back-end includes major modules for analysis: layer generation, graph and MLN analysis, and drill-down & visualization. Each is separate and is implemented using different skill sets and in parallel. The flow in each module is kept as similar as possible. Communication between modules is through persistent data created and used by all modules. For example, Python dictionary is used to store layer names generated along with additional information (e.g., timestamp) and is accessed by other modules through the import command.

We use Flask [2] as the core framework for our application. Flask handles user authentication, registration, and routing requests to the appropriate modules based on user interactions. This seamless integration enables smooth data flow and interaction between different components of our application. HTML and CSS are used for front-end development, providing an intuitive and user-friendly interface for interacting with our application. Our front-end components are designed to facilitate easy navigation and interaction with the data analysis functionalities offered by our application. The control flow of our application is managed by Flask, which routes requests from users to the appropriate modules based on their actions. As an example, when a user initiates a data analysis task, Flask routes the request to the analysis module, which generates the necessary files and returns control back to the Flask application upon completion.

5.1 Technology Comparison and Selection Rationale

To ensure that our application meets the high standards required for performance, privacy, security, and user experience, we evaluated different technologies to finalize the tech stack. Table 2 shows a comparison of technologies considered, along with the reasons for our final selections.

Table 2. Comparison of Technologies considered for `MLN-geeWhiz` Architecture

Category	Technology Chosen	Considered Options	Selection Rationale
Front-end	**HTML/CSS**	Angular, React, Vue.js	HTML/CSS has been chosen for its simplicity and broad support. It allows for rapid prototyping and is sufficient for our current UI needs without the overhead of a more complex framework
Back-end	**Flask**	Django, Express.js, Ruby on Rails	Flask's lightweight and modular nature makes it ideal for our application, offering flexibility and ease of development for our RESTful API design
Database	**SQLite**	PostgreSQL, MySQL, MongoDB	SQLite DB provides a lightweight, file-based solution, eliminating the need for a separate database server, which suits the scale of our current user base and data

5.2 Session Management

Session management is important when multiple users are using the dashboard. It needs to be thread-safe to make sure session data is isolated and routed properly. User-specific data, including file paths and user identifiers need to be *isolated* and not shared with other users. For this purpose, we use the session management capabilities provided by Flask. Flask offers a secure and scalable way to manage user sessions by assigning a unique session ID to each user upon login. This session ID is then used to track and store user-specific data in a way that is segregated from other users' data.

The implementation of dedicated Flask sessions results in a secure and reliable user experience. Each user's session is properly isolated, ensuring that users can only access their own files, even when multiple users are logged in simultaneously. This not only takes care of the security issues but also enhances the overall integrity and trustworthiness of our application.

6 Back-End Modules

In this section, we will summarize the functionality of each module, module architecture, and interaction with other modules and the front-end.

6.1 MLN Layer Generation

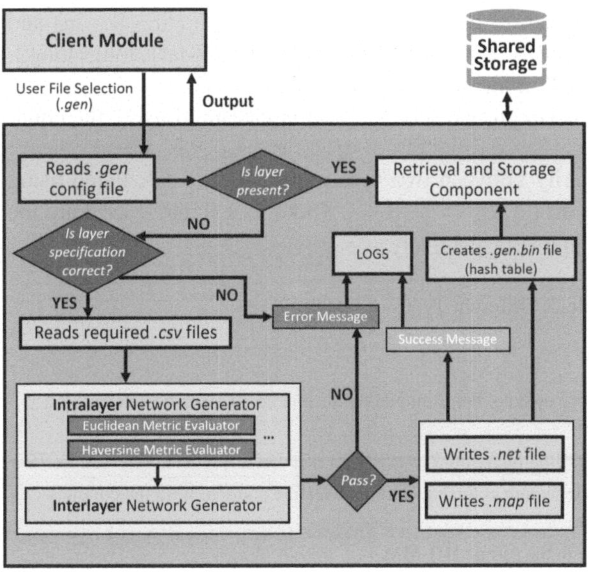

Fig. 5. Layer Generation Module Flow

Developing the layer generation tool for the multilayer network entailed users to specify precisely how nodes and edges of layers/graphs are generated using input data files. In addition to widely used metrics (given later in this section), we want to support custom metrics that can be provided as Python code. Currently, every line in the input data file (csv) becomes a node in the layer. For edges, one issue involves the use of multiple attributes for nodes where each combination needs to have non-null values for a specific set of node attributes. The node attribute combinations for layer generation are designed to allow users to choose the appropriate one based on their needs. We devise a method to enable users to generate multiple layers utilizing a single configuration file. The configuration file (.gen extension) contains information regarding the input directory, output directory, username, and the property specifications used for each layer generation. Both intra-layer and inter-layer can be generated using the same input configuration file. Each layer generation is specified separately. Specification for the intra-layer generation differs from that of inter-layer, allowing for two distinct types of layer generations with a single configuration file. Figure 5 shows the workflow of the layer generator module. Appropriate error codes are sent to the client module which displays messages for recovering from the error. Successful generation of layers for a configuration file is indicated.

Intra-Layer Edge Generation: Several widely-used metrics are supported: equality, Jaccard, Euclidean, Haversine, cosine, range, multi-range, fixed range for determining whether an edge exists between two nodes. An edge between two node IDs is established using a similarity metric which compares the similarity between two node IDs using a comparison of column attribute(s). The value returned by the comparator is checked against the threshold value specified by the user. If the value returned is below the threshold value given, an edge is established between two node IDs. The following similarity metrics are supported:

- **EQUALITY**: This metric checks whether two values are equal or not.
- **EUCLIDEAN**: The Euclidean distance between the two values, a and b, is calculated by taking the absolute value of their difference, $|a - b|$.
- **HAVERSINE**: The haversine distance is calculated by formula $d = R * c$, Here R is earth's radius (mean radius = 6,371km) and $c = 2 * a * tan2(\sqrt{a}, \sqrt{1-a})$ and $a = sin^2(\frac{lat}{2}) + cos(lat1) * cos(lat2) * sin^2(\frac{long}{2})$ where lat=lat2-lat1, long=long2-long1 (lat and long represents latitude and longitude respectively).
- **JACCARD**: The similarity between two sets A and B is measured using the equation: $J(A, B) = \frac{A \cap B}{A \cup B}$.
- **COSINE**: The cosine similarity between two vectors A and B is calculated using the following equation: $Cosine(A, B) = \frac{A*B}{||A||||B||}$ where $|| A ||$ and $|| B ||$ are the euclidean norm of vector A and B respectively.

Table 3. Different Types of Feature

Feature type	Description
NOMINAL	It represents a category or state. For example: political belief of a person can be **democratic**
NUMERIC	It represents integer or floating point. For example: speed of a vehicle in mph can be **75**
GEOGRAPHIC	It represents longitude feature column and latitude feature column. Both the columns contain numeric value
TIME	It represents the time in the format **HH-MM**
DATE	It represents the date in the format **DD-MM-YYYY** or **MM-DD-YYYY**
SET	A set with m members has the format: {value1, value2, ... valuem}. For example: genre of a movie is represented by {**ACTION, COMEDY, DRAMA**}
TEXT	It represents string of any length. For example: **Tweet**: **Congratulations, France FIFA 2018 Winners, IMDB Review**: **The movie was crap**

For similarity metrics - range, multi-range and fixed range, the generator verifies if, for a given pair of node IDs, the associated comparison column value falls within the specified interval for establishing the connection between two node IDs. For the multi-range, the user inputs numerous ranges. Two node IDs are connected if their comparison column value lies inside the same range. The user-provided parameter fixed-range is separated into equal-sized parts using the number of segments attribute. Two node IDs are connected if their comparison column values fall within the same segment. The user has the option to select either an open or closed interval for the range, and multi-range depending on their specific needs. Another challenge was to accommodate the tool to support using single or multiple columns as nodes IDs. For instance, a single column $Accident\_id$ can be used as node column. Again, two location columns ($Location\_Easting\_OSGR$, $Location\_Northing\_OSGR$) can be used as nodes IDs for layer generation. To solve this issue, the tool maps both single and multiple-column values to a single node ID. Mapping is necessary for analysis purposes like community detection. The tool also supports a variety of data types: nominal, numeric, geographic, date, time, set, text. Table 3 provides the description of different data types allowed by the layer generation tool for comparison column. Below is an example of a layer generation configuration, that connects two accident nodes which occurred within 50 km of each other based on haversine distance.

```
BEGIN_LAYER
INPUT_FILE_NAME= large-5KUKAccident-2014.csv
LAYER_NAME=distance_between_accidents
PRIMARY_KEY_COLUMN=Location_Easting_OSGR,Location_Northing_OSGR
LONGITUDE_FEATURE_COLUMN=Longitude
LATITUDE_FEATURE_COLUMN=Latitude
FEATURE_TYPE=NUMERIC
SIMILARITY_METRIC=HAVERSINE
THRESHOLD=50
END_LAYER
```

Inter-Layer Generation: Requires merging two node IDs based on a join column to depict a specific relation between the merged node pairs. It is done as follows. The layer generation tool needs to generate two layers before generating an inter-layer. Then, the tool creates an inter-layer file from the input csv files required to generate layers. The inter-layer is generated by linking the node IDs from both layers. The two node IDs from two csv files used for layer generation are connected by matching them based on a join attribute specified in the configuration file. The relation between two node IDs from two layers is depicted in the generated inter-layer using the relationship attribute from the configuration file. Below is an example of a configuration file with property specifications for the inter-layer generation between the Actor and Director layers. This will connect nodes from the actor and director layers who have worked in the same movie.

```
BEGIN_INTERLAYER
LAYER_1_NAME=AvgRating_Actors
LAYER_1_INPUT_FILE_NAME=AvgRating_Actors.csv
LAYER_2_NAME=AvgRating_Directors
LAYER_2_INPUT_FILE_NAME=AvgRating_Directors.csv
JOIN_COLUMN_NAME=Movie_ID
RELATIONSHIP_NAME=works_with
INTER_LAYER_NAME=actor_director_works_with
INTER_LAYER_GENERATION_TYPE=System_Generated
END_INTERLAYER
```

Although part of the MLN dashboard, the layer generation tool is an independent module. The generating tool accepts the user information and configuration file path as input parameters. The client module invokes the generation module by providing these two parameters to execute layer generation.

6.2 MLN Analysis Module

Although an independent module, the analysis module seamlessly integrates with the layer generation, directly utilizing the generated layers for analysis. This is done by persistent dictionaries written by each module and used by the other modules. Upon completion of the analysis, the results are readily available for the visualization module, where users can view and interpret the findings through various visualizations.

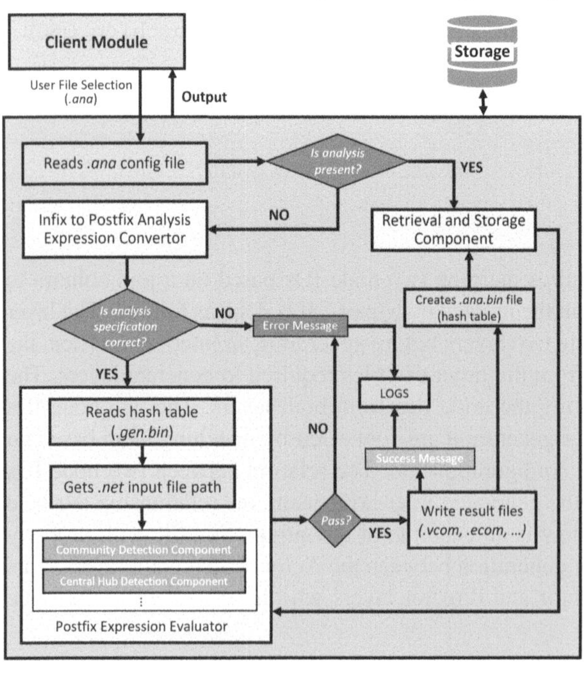

Fig. 6. Graph/MLN Analysis Module Flow

This ensures a smooth and efficient workflow from data input to result visualization. The analysis process is carried out through the utilization of an analysis configuration file (with extension.ana). This file serves as a blueprint, containing expressions that specify the analysis computations to be performed. An infix expression (with explicit precedence) is used for specifying the analysis and a name for storing the results. Workflow within the analysis module is shown in Fig. 6.

Analysis configuration file (.ana extension) indicates the layers to be used by importing the layer generation configuration file name that was used for generating layers. This can be specified unambiguously and multiple configuration files can be imported to analyze layers generated at different times by different configuration files. The analysis expression is specified in the widely used infix form and is easy to understand. The operators are specific to graph/MLN analysis and the domain user knows the analysis being performed. A sample infix expression along with its translated post-fix expression is shown below.

Infix Analysis Expression: `Louvain(NOT layer1) CV-AND (Louvain (layer2 OR layer1) CE-AND Infomap(layer3 AND layer1))`

Postfix of the Above Expression: `layer1 NOT Louvain layer2 layer1 OR Louvain layer3 layer1 AND Infomap CE-AND CV-AND`

Observe the use of analysis operators (Louvain [10] and Infomap [12]), Boolean operators (AND, OR, NOT) [34], and multilayer composition operators CE-AND [36] and CV-AND [34]. In order to honor the precedence, an infix expression needs to be converted into its postfix form for evaluation. The operands here are files containing graphs or results of previous computations.

Analysis Optimization: Some of the algorithms used for graph analysis take a significant amount of time to compute depending on the number of nodes, edges, and other graph properties (e.g., degree). Also, observe that the same analysis is likely to be done on the same layer for two different (sub) expressions. Redoing the same analysis on the same layer using the same algorithm increases analysis time and hence response time.

To overcome the above, primitive layer computations are time-stamped and checked for their presence using the dictionary. If they have been generated earlier, they are not regenerated during the analysis. This has reduced the time for analysis significantly, especially when the graph sizes are large.

Use of Multiple Languages for Analysis: Many available simple graph analysis algorithms are in C++. Integrating C++ methods with the main analysis Python driver presented some challenges in communication and data handling between the two languages. We utilized C++ for computationally intensive analysis methods while leveraging Python for overall control and orchestration. A robust system was developed to facilitate seamless data exchange and function calls between the two languages.

6.3 Drill-Down and Visualization

Developing an effective drill-down and visualization is a challenge due to the complex nature of multilayer networks (MLNs) and the various requirements of users. Key challenges include:

- **Availability of alternatives**: We aim to provide users with multiple visualization options to cater to different analysis needs. Implementing various visualization techniques while ensuring user-friendly interaction is the key.

- **Handling Large and Complex Data Sets**: MLNs often involve large graphs from various domains, making it difficult to visualize them in the screen space available.
- **Integrating Interactivity**: Enabling interactive features such as pan, zoom, hover, and drill-down capabilities within the visualization interface required careful design and implementation to ensure a smooth user experience.

To handle large graphs meaningfully, we adopt a design where the options provided are based on the number of nodes and edges. To further enhance user engagement and exploration, the dashboard incorporates interactive visualization features. Users can dynamically interact with visualizations, such as zooming in on specific network regions, hovering over nodes for additional information, and drilling down into analysis results.

The Visualizer module relies on layer generation and analysis modules for input file types such as `.net` files (Network Files), `.ecom` files (Edge Community Files), and `.vcom` files (Vertex Community Files). Additionally, the module utilizes a primary mapping input file (`.map`) generated by the layer generation module to map labels or node attributes with respective node IDs in the visualization when hovered. The following types of visualizations are offered to facilitate comprehensive analysis:

1. **Network Visualization Type**: Utilizes the Plotly Python module to generate a static graph where node colors are determined based on the number of connections.
2. **Degree-Centrality Visualization Type**: Focuses on visualizing degree centrality.
3. **Community Visualization Type**: Provides insights into community structures within the network, designed for `.ecom` and `.vcom` input files.
4. **Word Cloud Visualization Type**: Specifically tailored for `.ecom` input files, generates word clouds representing node attributes and ledger information such as degree centrality and top 10 communities of the layer.
5. **Interactive Visualization Type**: Utilizing the PyVis Python module to offer an interactive experience where users can hover over nodes, toggle node visibility, and explore network structures in detail.
6. **Map Visualization Type**: Available for input mapper files containing respective longitude and latitude data, this visualization allows users to visualize node locations on the map, enhancing the understanding of the analyzed data.

Table 4 shows the packages used for different visualization needs and the rationale for choosing them.

Table 4. Comparison and rationale of Visualizations Used in MLN-geeWhiz

Packages Used	**Reason for selection and Use Cases**	**Visualization Snippet**
Plotly	Generates a static network graph with node colors to reflect the number of connections, enabling easy identification of network hubs and isolated nodes.	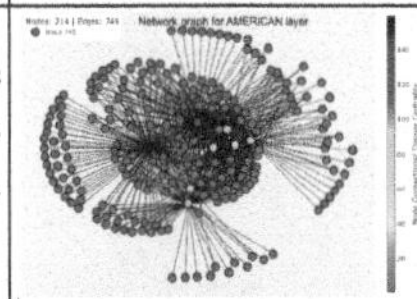
Plotly (Map)	Employs Plotly's robust mapping tools to visually represent geographic data, making it excellent for highlighting relationships and patterns, such as accident-prone areas.	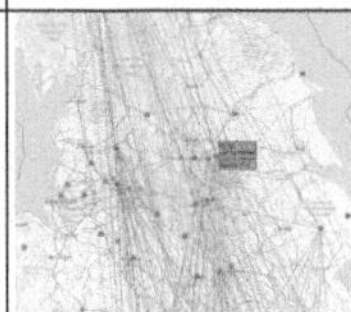
Bokeh (Community Detection)	Provides interactive capabilities, making it suitable for visualizing community structures within networks.	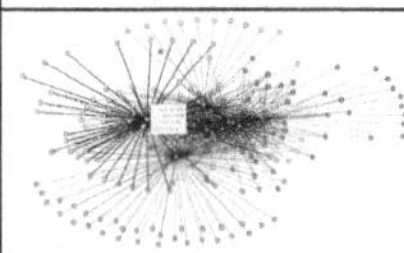
Bokeh (Degree Centrality)	Renders networks with nodes sized according to their degree centrality, providing a clear visual hierarchy that identifies and ranks key influencers or hubs.	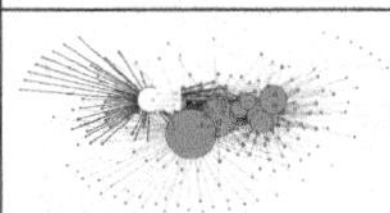
Wordcloud	Creates an engaging word cloud that summarizes key attributes of network communities, such as the density and average degree.	
PyVis	Seamlessly integrates with NetworkX to provide a fully interactive network visualization experience, where users can manipulate nodes and edges in real time to gain deeper insights into the network structure.	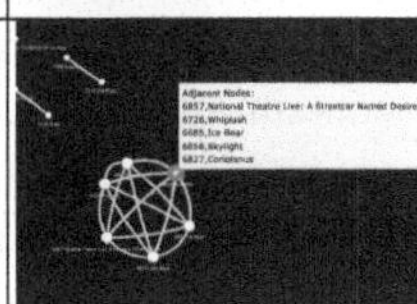

7 Real-Word Data Analysis Using the MLN-GeeWhiz Dashboard

In this section, we will demonstrate the use of the dashboard for some applications to show its utility, functionality, and visualization options supported.

UK Accident Use Case: We have used the UK accident data set [9] for this. Some sample requirements for an accident data set are finding out the (i) accident-prone regions, (ii) days of the week and time periods that witness an increase in the number of accidents, (iii) the most unsafe lighting, weather, and road conditions, (iv) type of locations that become the epicenter of accidents, and so on. Such insights can help city planners to take safety precautions.

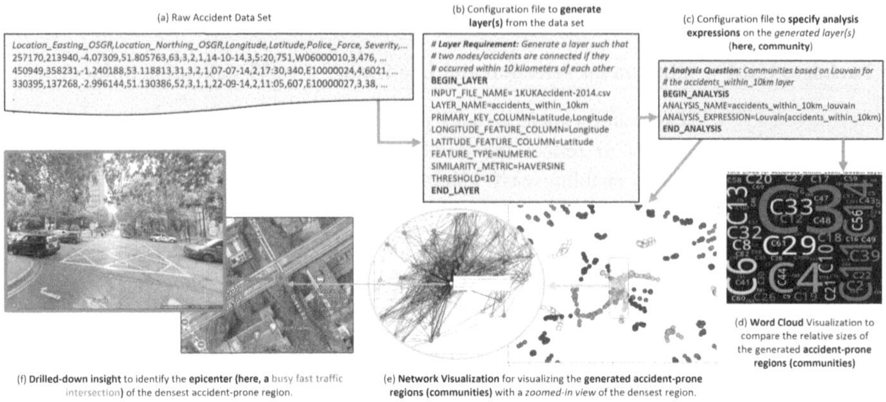

Fig. 7. Showcasing the use of the MLN-Dashboard to model, analyze, drill-down and visualize a raw accident data set.

Figure 7 shows the entire workflow from a raw accident data set to the final drill-down visualized analysis insight. Figure 7 (a) shows the snapshot of the UK Accident data set which for every accident occurrence captures the location (latitude, longitude), light conditions, weather conditions, date of occurrence, and so on.

The user-specified configuration file in Fig. 7(b) connects two accident nodes if they have occurred within 10 km of each other based on Haversine distance to generate a layer. Then, the configuration file in Fig. 7(c)[4] detects communities (i.e., in this case accident-prone regions) using the Louvain algorithm [10] with the output stored as a text file containing *(edge, communityID)* pairs. These accident-prone regions (communities) have been visualized using a word cloud to understand the relative severity of the regions in Fig. 7(d). Alternatively, network visualizations of the accident-prone regions with zoom-in, hover, and pan facility help us understand the connectivity within the *densest* communities (Fig. 7(e)). Finally, Fig. 7(f) gives us drilled-down insight to identify that the epicenter of the densest accident-prone region (i.e., node with the highest degree in the densest community) is a fast traffic intersection on the Google Maps using the latitude, longitude node label.

In addition to what is shown in this paper, the MLN-Dashboard has been used to analyze other raw data sets, such as IMDb [3] to find the groups of movies that are most similar to each other based on the genres or the highly rated actors who have never worked together; US Airline data sets have been analyzed to identify airlines hubs or cities from which you can reach other cities with least number of hops or flight changes using closeness centrality analysis, and so on. The DBLP data set has been used to discover author groups that have published significantly in some conferences over 20 years, authors who publish in top-tier conferences, but not in other conferences, etc.

[4] A complete example of the layer generation configuration file is shown in Sect. 6.1.

8 Conclusions

In this paper, we have presented an interactive tool architecture for supporting complex data analysis using graph models including multilayer networks. The dashboard is conceived for the whole life cycle and designed to cover all stages of life cycle in an easy-to-use, pro-active manner. Special attention has been paid to prevent user mistakes by prompting what is allowed. For example, only permitted actions are highlighted at each step. The architecture and modularity, based on functionality, provide *flexibility* (of development and optimization), *extensibility* (of visualizations, analysis, layer and inter-layer graph generation, and data set formats), *consistency for multiple concurrent users who may share data and results*, and *efficiency* (for response time by using previously computed results using timestamp checks).

Future work includes extending the dashboard into a full-fledged portal providing selective uploads and sharing (of data, algorithm code, as well as results), community interaction, and comparison of algorithms and creation of benchmarks while taking care of challenges like security, load balancing, performance issues and so on.

Acknowledgments. This work was partially supported by NSF awards #1955798 and #1916084.

References

1. The centre for disease control covid dashboard. https://covid.cdc.gov/covid-data-tracker/
2. Flask web framework. https://palletsprojects.com/p/flask/
3. The internet movie database. ftp://ftp.fu-berlin.de/pub/misc/movies/database/
4. Johns hopkins university covid dashboard. https://www.arcgis.com/apps/opsdashboard/index.html#/bda7594740fd40299423467b48e9ecf6
5. The new york times covid dashboard. https://www.nytimes.com/interactive/2020/us/coronavirus-us-cases.html
6. The university of washington covid dashboard. https://hgis.uw.edu/virus/
7. The world health organization covid dashboard. https://covid19.who.int/
8. Worldometer covid statistics. https://www.worldometers.info/coronavirus/country/us/
9. Road safety - accidents 2014 (2014). https://data.gov.uk/dataset/road-accidents-safety-data/resource/1ae84544-6b06-425d-ad62-c85716a80022
10. Blondel, V.D., Guillaume, J., Lambiotte, R., Lefebvre, E.: Fast unfolding of community hierarchies in large networks. CoRR **abs/0803.0476** (2008). http://arxiv.org/abs/0803.0476
11. Boden, B., Günnemann, S., Hoffmann, H., Seidl, T.: Mining coherent subgraphs in multi-layer graphs with edge labels. In: Proceedings of the 18th ACM SIGKDD International Conference on Knowledge Discovery and Data Mining, pp. 1258–1266. KDD 2012, ACM (2012). https://doi.org/10.1145/2339530.2339726
12. Bohlin, L., Edler, D., Lancichinei, A., Rosvall, M.: Community detection and visualization of networks with the map equation framework (2014). http://www.mapequation.org/assets/publications/mapequationtutorial.pdf
13. Bolaj, K., Santra, A., Chakravarthy, S.: Substructure discovery in heterogeneous multilayer networks. In: IEEE International Conference on Knowledge Graph, ICKG 2024, Abu Dhabi, UAE, December 11-12, 2024. p. accepted. IEEE (2024)
14. Calvó-Armengol, A., Zenou, Y.: Social networks and crime decisions: the role of social structure in facilitating delinquent behavior. Int. Econ. Rev. **45**(3), 939–958 (2004)

15. Clauset, A., Newman, M.E., Moore, C.: Finding community structure in very large networks. Phys. Rev. E **70**(6), 066111 (2004)
16. Das, S., Chakravarthy, S.: Duplicate reduction in graph mining: approaches, analysis, and evaluation. IEEE Trans. Knowl. Data Eng. **30**(8), 1454–1466 (2018). https://doi.org/10.1109/TKDE.2018.2795003
17. De Domenico, M., Lancichinetti, A., Arenas, A., Rosvall, M.: Identifying modular flows on multilayer networks reveals highly overlapping organization in interconnected systems. Phys. Rev. X **5**(1), 011027 (2015)
18. De Domenico, M., Porter, M.A., Arenas, A.: Muxviz: a tool for multilayer analysis and visualization of networks. J. Complex Netw. cnu038 (2014)
19. De Domenico, M., Solé-Ribalta, A., Gómez, S., Arenas, A.: Navigability of interconnected networks under random failures. Proc. National Acad. Sciences (2014). https://doi.org/10.1073/pnas.1318469111, https://www.pnas.org/content/early/2014/05/21/1318469111
20. Gomez, M., Garcia, S., Rajtmajer, S., Grady, C., Mejia, A.: Fragility of a multilayer network of intranational supply chains. Appl. Netw. Sci. **5**(1) (2020). https://doi.org/10.1007/s41109-020-00310-1
21. Hopkins, A.L.: Network pharmacology: the next paradigm in drug discovery. Nat. Chem. Biol. **4**(11), 682 (2008)
22. Kivelä, M., Arenas, A., Barthelemy, M., Gleeson, J.P., Moreno, Y., Porter, M.A.: Multilayer networks. CoRR **abs/1309.7233** (2013). http://arxiv.org/abs/1309.7233
23. Kivelä, M.: Pymnet: free library for analysing multilayer networks. http://people.maths.ox.ac.uk/kivela/mln_library/
24. Lara-Martínez, P., Obregón-Quintana, B., Reyes-Manzano, C., López-Rodríguez, I., Guzmán-Vargas, L.: A multiplex analysis of phonological and orthographic networks. PLoS ONE **17**(9), e0274617 (2022)
25. Mukunda, K., Roy, A., Santra, A., Chakravarthy, S.: Stress centrality in heterogeneous multilayer networks: heuristics-based detection. In: Ninth IEEE International Conference on Big Data Computing Service and Applications, BigDataService 2023, Athens, Greece, July 17-20. IEEE (2023)
26. Mukunda, K., Santra, A., Chakravarthy, S.: The challenge of finding degree centrality nodes in heterogeneous multilayer networks. In: Proceedings of the 31st Italian Symposium on Advanced Database Systems, SEBD 2023, Galzignano Terme, Italy, July 2-5, 2023. CEUR Workshop Proceedings (2023)
27. multiNetx github Page (2019). https://github.com/nkoub/multinetx/blob/master/README.md
28. Newman, M.E.: Finding community structure in networks using the eigenvectors of matrices. Phys. Rev. E–Stat. Nonlinear Soft Matter Phys. **74**(3), 036104 (2006)
29. Pagani, G.A., Aiello, M.: The power grid as a complex network: a survey. Phys. A **392**(11), 2688–2700 (2013)
30. Pavel, H., Roy, A., Santra, A., Chakravarthy, S.: Degree centrality definition, and its computation for homogeneous multilayer networks using heuristics-based algorithms. In: Knowledge Discovery, Knowledge Engineering and Knowledge Management - 14th International Joint Conference, IC3K 2022, Valletta, Malta, October 24-26, 2022, Revised Selected Papers. p. Accepted. Communications in Computer and Information Science, Springer (2023)
31. Pavel, H.R., Roy, A., Santra, A., Chakravarthy, S.: Closeness centrality detection in homogeneous multilayer networks. In: Proceedings of the 15th International Joint Conference on Knowledge Discovery, Knowledge Engineering and Knowledge Management, IC3K 2023, KDIR, Rome, Italy, November 13-15, 2023 (2023)
32. Pons, P., Latapy, M.: Computing communities in large networks using random walks. J. Graph Algorithms Appl. **10**(2), 191–218 (2006)

33. Sahneh, F.D., Scoglio, C.: Competitive epidemic spreading over arbitrary multilayer networks. Phys. Rev. E **89**(6), 062817 (2014)
34. Santra, A., Bhowmick, S., Chakravarthy, S.: Efficient community re-creation in multilayer networks using boolean operations. In: International Conference on Computational Science, pp. 58–67. Zurich, Switzerland (2017)
35. Santra, A., Bhowmick, S., Chakravarthy, S.: Hubify: efficient estimation of central entities across multiplex layer compositions. In: IEEE ICDM Workshops (2017)
36. Santra, A., Bhowmick, S., Irany, F.A., Madduri, K., Chakravarthy, S.: Efficient community detection in multilayer networks using boolean compositions. Front. Big Data (2023)
37. Santra, A., Komar, K., Bhowmick, S., Chakravarthy, S.: From base data to knowledge discovery – a life cycle approach – using multilayer networks. Data Knowl. Eng. 102058 (2022). https://doi.org/10.1016/j.datak.2022.102058, https://www.sciencedirect.com/science/article/pii/S0169023X22000556
38. Santra, A., Komar, K.S., Bhowmick, S., Chakravarthy, S.: A new community definition for multilayer networks and a novel approach for its efficient computation. arXiv preprint arXiv:2004.09625 (2020)
39. Scabini, L.F., Ribas, L.C., Neiva, M.B., Junior, A.G., Farfán, A.J., Bruno, O.M.: Social interaction layers in complex networks for the dynamical epidemic modeling of COVID-19 in Brazil. Physica A: Stat. Mech. Appl. **564**, 125498 (2021)
40. Singh, A., Santra, A., Chakravarthy, S.: Scalable substructure discovery algorithm for homogeneous multilayer networks. In: 2024 IEEE International Conference on Big Data (BigData). IEEE (2024)
41. Skrlj, B., Kralj, J., Lavrac, N.: Py3plex toolkit for visualization and analysis of multilayer networks. Appl. Netw. Sci. **4**(1), 94:1–94:24 (2019)
42. Vu, X.S., Santra, A., Chakravarthy, S., Jiang, L.: Generic multilayer network data analysis with the fusion of content and structure. In: CICLing 2019, La Rochelle, France (2019)
43. Wang, Z., Wang, L., Szolnoki, A., Perc, M.: Evolutionary games on multilayer networks: a colloquium. Eur. Phys. J. B **88**(5), 124 (2015)

gCDR: A Group Aided Cross-Domain Recommendation Framework

Adamya Shyam[1(✉)], Kavita Kanwar[2], Vikas Kumar[1], and Venkateswara Rao Kagita[2]

[1] University of Delhi, Delhi, India
{ashyam,vikas}@cs.du.ac.in
[2] National Institute of Technology, Warangal, Warangal, India
kkcs21124@student.nitw.ac.in, venkat.kagita@nitw.ac.in

Abstract. Cross-domain recommendation (CDR) tasks aim to improve the recommendation performance in domains with limited data by leveraging information from other domains. The goal is to capture and then transfer the knowledge available in the information-rich domains to the sparse domain. However, traditional CDR techniques struggle when there is little overlap among users, making it hard to map preferences accurately. This paper introduces the Group-aided Cross-Domain Recommender System (gCDR), which integrates group information from overlapping users across domains. By identifying and utilizing shared preference patterns among user groups, gCDR enhances recommendation stability even with minimal user overlap across domains. To assess the efficacy of our gCDR framework, we have conducted experiments across two key settings, one with overlapping users and the other with overlapping items. For both scenarios, the gCDR model outperforms the best benchmark model over diverse evaluation metrics by a significant margin. The results highlight the importance of group dynamics in optimizing cross-domain recommendations and provide a robust solution to enhance recommender system performance in different data environments.

Keywords: Cross-domain recommendation · Clustering · Semi-supervised Learning

1 Introduction

In the current context of rapidly growing technologies, the volume of data continues to grow exponentially, making recommender systems increasingly essential for delivering filtered and relevant information to users. Collaborative filtering (CF) [1,16,31] is a widely adopted approach in this scenario, leveraging collective user data with the premise that users with similar interests will exhibit comparable preferences. One of the prominent approaches utilized in CF is matrix factorization (MF) [3,15,17,34], which maps users and items into a shared latent space, enabling recommendations based on their proximity within that space. In real-world scenarios, recommender systems often encounter sparse rating data,

A. Dasgupta et al. (Eds.): BDA 2024, LNCS 15526, pp. 240–256, 2025.
https://doi.org/10.1007/978-3-031-81821-9_14

where some users or items lack sufficient ratings to make accurate recommendations. This data sparsity issue leads to the problem of cold-start [13,18,22]. For example, consider a new user joining a streaming platform without any prior movie ratings, posing a classic cold-start problem for the recommender system. Similarly, newly released films on a movie streaming platform often have sparse viewer ratings compared to popular classics, illustrating another aspect of the data sparsity challenges in recommendation systems.

To address data sparsity, specialized techniques such as Cross-domain recommendations (CDR) [7,14] are utilized. These techniques are particularly effective in situations where one domain (e.g., movies) has abundant data (source domain) while another domain (e.g., books) suffers from sparse data (target domain). CDR methods enhance the recommendation performance in the target domain by exploiting the knowledge extracted from the source domain. For example, a CDR approach might use the viewing habits of users who rate movies frequently to predict preferences for books that share similar themes or genres. However, capturing and transferring such knowledge about users and items across different domains is a non-trivial task. Recent developments in this field have introduced various approaches to address this challenge by establishing relationships between users and items in both domains [4,37,39,47]. These methods can be classified as follows: graph-based techniques [5,33,44], matrix factorization-based models [28,30,36], and semantic-based methodologies [21,38]. Most of these techniques rely on overlapping user or item information to learn effective mappings. However, they may struggle to provide relevant recommendations in scenarios with limited users or items overlap. Recent studies suggest that incorporating group information into the learning process can guide and enhance recommendation outcomes [27].

This paper introduces a new framework, Group-aided Cross-Domain Recommender System (gCDR), which is designed to enhance cross-domain recommendation (CDR) outcomes by leveraging group information and overlapping entities while establishing a bridge for knowledge transfer. While overlapping entities provide a direct bridge for knowledge transfer, focusing solely on them restricts the amount of transferable knowledge as they are often limited in size. Including group information helps generalize the mapping from the source to the target domain and prevents overfitting due to limited entity overlap. This is because groups, such as user groups, often exhibit similar preference patterns across different domains. For instance, a user group like *travel enthusiasts* exhibits consistent preferences across various domains. On travel booking platforms, they may search for destinations rich in cultural experiences and on *photography apps*, they curate collections centered around travel themes. Similarly, in the entertainment domain, a user group like *cinephiles* may frequently watch and highly rate classic films from renowned directors and show a strong inclination towards film scores and soundtracks, creating playlists themed around specific movie genres or directors on *music streaming services*. Building on this concept, we formed a group corresponding to each overlapping entity, considering them as representative of that group. The rationale behind constructing groups around overlapping entities is the inherent availability of one-to-one mapping between the source and target domains. We are not claiming that this approach is the best approach, instead, there can be several choices to form the groups. However, the groups formed using the traditional clustering approach may further require a sophisticated approach to resolve the one-to-one mapping.

In each domain, the embedding of groups is learned by considering the embedding of overlapping entities as centroids and then assigning the remaining users to a group based on their proximity to the centroid. Finally, a group embedding is obtained by taking the mean of all users assigned to that group. The group embedding is then used to learn a more generalized mapping from the source to the target domain. This enables our gCDR framework to obtain individual user preferences as well as the collective influence of user groups, leading to improved accuracy and relevance of recommendations across domains.

Our main contributions are as follows:

- We propose a novel framework called the Group-aided Cross-Domain Recommender System (gCDR). This framework can be integrated with any existing cross-domain recommendation methods that rely on overlapping user information, thus enhancing recommendation accuracy and relevance.
- The gCDR framework underscores the significance of incorporating group information to address the issue of sparse overlap data, thereby making the learning process more effective.
- We validate our framework through comprehensive experiments in two scenarios, using four evaluation metrics. The outcomes signify that our framework remarkably enhances the CDR performance.

The structure of the rest of the paper is as follows. We briefly outline the existing CDR techniques, in Sect. 2. In Sect. 3, we describe our proposed gCDR framework. We present a comprehensive experimental analysis to assess the efficacy of the gCDR framework and compare it with the benchmark techniques in Sect. 4. In Sect. 5, we conclude our work and discuss the future possibilities.

2 Related Works

The existing CDR methods can be broadly classified as (i) Content-based, and (ii) Transfer-based. The content-based techniques typically link various domains by identifying common features like user-generated texts, social tags, etc. [6,26,29]. Such techniques fail in the absence of any common feature associating both domains. Meanwhile, the transfer-based techniques mainly focus on transferring the embeddings or rating patterns across domains [38,45]. The former first learns the patterns observed in the source domain's interactions and utilizes it to provide recommendations in the target domain [10,19,23], whereas the latter employs machine learning models to obtain embeddings of entities in both domains and uses them for mapping purposes. In this work, we mainly emphasize on transfer-based methods involving embeddings. In [24], an embedding and mapping framework (EMCDR) is proposed, where the embeddings in both domains are learned independently using latent factor modeling. Further, the extracted source embeddings are utilized to model a mapping function to recommend items in the target domain. To ensure the privacy of users, [9] uses only item features for mapping and provides recommendations accordingly. The authors of [36], perform latent factor alignment before the mapping process for accurate knowledge transfer. Fusing the review texts and additional side information of items, [8] propose a deep fusion model (RC-DFM) to preserve more semantic information in the learned latent factors. A multi-layer perceptron (MLP) is further utilized for the transfer process between domains. In addition to CDR, [43] provides Cross-system recommendations (CSR) by combining the latent features of both domains. The factors are combined by considering the sparsity degrees of each entity across both domains. The DCDCSR framework then utilizes these combined features to align the target domain embeddings through a deep network. In [42], the authors propose a dual-target approach (DTCDR) with an adaptable embedding-sharing strategy by combining the embeddings of overlapping entities across the domains. In [35], the authors propose a framework that exploits both inter-site and intra-site information to learn complex interactions and provide movie recommendations across multiple sites. To deal with the problem of overlap sparsity, a latent orthogonal mapping approach is proposed in [20] to obtain user preferences while sustaining user relationships across different latent spaces. In an extension to EMCDR, [13] proposed a semi-supervised approach, by learning a mapping function over the overlapping entity's embeddings acting as labeled data, and further mapping the embeddings of non-overlapping entity between domains posing as unlabeled data. Several other works have utilized the concept of embedding and mapping approach for CDR tasks [2,32,39,46]. However, these existing techniques rely on a certain percentage of overlap between the two domains, restricting the models from utilizing the feedback data completely. In our work, we study the incorporation of group information to achieve better recommendation accuracy.

3 Proposed Work

This section describes the proposed group-aided cross-domain recommendation (gCDR) framework. This framework can be combined with existing CDR methods that rely on overlapping entity information and enhance the recommendation quality. In this paper, we have showcased the benefits of the proposed approach using a recent method presented in [13]. The proposed method encompasses three main steps. First, the latent representations of both entities are obtained independently for the source as well as the target domain. Second, the groups are formed considering overlapping users' representation as centroid, and their latent representation is computed using the user embeddings. Finally, these group representations along with overlap information are utilized to effectively learn the mapping from the source domain latent space to that of the target domain. In the subsequent section, we first present the notations used in the paper and then discuss each step in detail.

3.1 Notations and Problem Formulation

We define the notions for two domains: the source domain, from which knowledge is transferred, and the target domain, where recommendations are made based on this knowledge transfer. In our notations, we utilize superscripts s and t to represent the source and target domains respectively. Notation without any superscript refers to a general notation applicable to any domain. The user sets in the source and target domains are represented by U^s and U^t, respectively. Similarly, I^s and I^t represent the item sets. In most of the scenarios, any one or both of the entities (users and items) partially overlap. For example, consider a scenario where "Movies" serve as the source domain and "Books" as the target domain. Users engage with both movies and books, resulting in an overlap between these domains. Similarly, music enthusiasts may also find interest in podcasts, thereby establishing an overlap between music and podcast domains. We use $O = U^s \cap U^t$ (or, $I^s \cap I^t$) to denote the overlapping entities. Given a user-item implicit interaction matrix $Y_{n \times m} \in \{0, 1\}$, where n and m represent the cardinalities of U and I respectively. An entry $y_{ij} \in Y$ is 1 if user i has interacted with item j, otherwise 0. The interaction matrices are represented by Y^s and Y^t. Additionally, for a source domain (resp., target domain) user u, its set of interacted items is represented by I_i^s (resp., I_i^t). Similarly, U_j^s (resp., U_j^t) denotes the user set for item j in the source domain (resp., target domain). Given the partially observed rating matrices of both domains, our goal is to recommend top-k items to users in the target domain.

3.2 Learning of Embeddings

First, we learn embeddings for each domain separately. In this process, we represent users and items in a low-dimensional latent space such that a user's preference for an item is inversely related to the distance between them in this new space. In other words, if an item is located close to a user in this space, the user prefers it more than items that are located farther away. Given the embeddings of the ith user, denoted as u_i and the jth item, denoted as v_j, we compute squared Euclidean distance between the embeddings using Eq. 1.

$$d(u_i, v_j) = \|u_i - v_j\|^2. \tag{1}$$

In a latent space, users with similar preferences are closer to each other, and items with similar features are also closer. Furthermore, items closer to a user in the space are more likely to be favored by that user. In a CDR scenario, the basic assumption is that the target domain will likely exhibit patterns observed in the latent space of a source domain. For example, if a group of users enjoy the same movies, they might also have similar preferences in books. To achieve effective cross-domain recommendations, the latent space representation of both domains should accurately capture user-user correlations and user-item associations. In other words, we map users and items from both domains into a shared latent space while preserving user-user and user-item relationships. To model the users

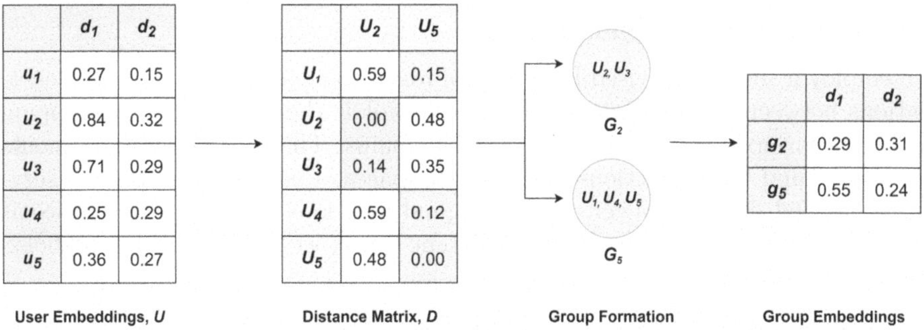

Fig. 1. An example of group formation for proposed gCDR approach.

and items in such space, we optimize the triplet margin loss [13] to ensure that the user and her interacted items are closer to each other than those she has not interacted with. The loss function is defined as follows,

$$L = \sum_{i \in U} \sum_{j \in I_i} \sum_{z \notin I_i} [m + d(u_i, v_j) - d(u_i, v_z)]_+,$$
$$\text{s.t. } \|u_*\|^2 \leq 1 \text{ and } \|v_*\|^2 \leq 1, \tag{2}$$

where m denotes the margin size and $[x]_+ = max(x, 0)$ is the standard hinge loss. To avoid the wide spread of the embeddings, we enforce unit sphere constraint by normalizing the embeddings ($x_* = x_*/max(1, \|x_*\|^2)$, as done in [12].

3.3 Formation of Groups

The extracted embeddings of users and items for both domains are utilized to create clusters based on overlapping entities. These entities could be either users or items. In the subsequent discussion, for readability, we explained the methodology by taking overlapping entities as users. However, the framework is flexible and can alternatively consider overlapping items. These overlapping users' embeddings represent shared attributes between the two domains. By using them as cluster centroids, we can identify users with similar preferences across the domains, thereby improving the accuracy of recommendations. We create a total of $|O|$ clusters, with each overlapping user serving as a cluster centroid. Then, we calculate the distance (Eq. 1) between each overlapping user and all non-overlapping users. The non-overlapping users are assigned to clusters based on these distances. The group embedding is the mean of the embeddings of all the members of a cluster.

For instance, given the learned embeddings for a user set of 5 users with users U_2 and U_5 serving as overlapping users. As shown in Fig. 1, the Euclidean distance of each non-overlapping user is calculated from both overlapping users. The groups so formed are $G_2 = \{U_1, U_2, U_4\}$ and $G_5 = \{U_3, U_5\}$. Finally, the group embedding for both groups is computed as the average of the individual user embeddings of their corresponding members.

3.4 Cross-Domain Mapping

We employ a semi-supervised learning approach to effectively capture the associations between the two latent spaces. We exploit the presence of overlapping users and groups that are active in both domains. These overlapping elements provide ground truth positions in the latent spaces of both domains, hence serving as labeled data. However, as the original locations of source domain items are unavailable in the target domain's latent space, we treat them as unlabeled data.

The proposed approach hinges on the assumption of local consistency in semi-supervised learning, which states that highly similar data instances will likely belong to the same class [41]. This aligns with the concept of CDR, which suggests that users with matching choices in one domain (Movies) will have similar tastes in another domain (Books). Extending on the formulation given in [13], we assume that the embeddings of groups in the target domain will also follow the same behavior as shown in the source domain. Hence, we optimize the supervised loss L_s such that the mapped positions of the source domain user and group vectors are close to their actual locations in the target domain latent space. We formulate the supervised loss function as given below.

$$L_s = \sum_{i \in O} d(f_\theta(u_i^s), u_i^t) + d(f_\theta(g_i^s), g_i^t), \tag{3}$$

where g_i^s and g_i^t are the group's embeddings in source and target domain, respectively, considering user i as the cluster centroid and $f_\theta(\cdot)$ denoting the mapping function parameterized by θ. For the mapping, we have employed a multilayer perceptron (MLP) with its architecture discussed in the hyper-parameter tuning section; please refer to Sect. 4.4 for details.

Now as the mapping function, $f_\theta(\cdot)$, is learned using a supervised loss, the transferred source user and group vectors will be confined close to the overlapping users and groups, respectively. This can lead to an over-reliance on these overlapping elements for the mapping function, $f_\theta(\cdot)$. To mitigate this problem, we utilize the unsupervised loss proposed in the underlying base algorithm. The unsupervised loss, L_u, transfers the source domain items' embeddings to the target domain latent space and adjusts the locations according to their interaction with the overlapping users. By incorporating the source domain interactions, the mapping function eliminates the over-reliance on overlapping elements and further fine-tunes the location of transferred users toward their true preferences in the target domain. As outlined in Sect. 3.2, a user should be less distant from its interacted items than from the items it has not interacted with. The unsupervised loss is computed as follows,

$$L_u = \sum_{i \in O} \sum_{j \in I_i^s} \sum_{z \notin I_i^s} [m + d(f_\theta(v_j^s), u_i^t) - d(f_\theta(v_z^s), u_i^t)]_+, \tag{4}$$

where $f_\theta(v_j^s)$ is the mapped location of source domain item j. The complete loss function, first maps the overlap and group data to the target domain and further adjusts the mapped locations using unsupervised data, ensuring that the unlabeled data is distributed across the target domain latent space based on their proximate similarities to the labeled data. The complete loss function is defined as,

$$L = L_s + \lambda.L_u,$$
$$\text{s.t. } \|f_\theta(u_*)\|^2 \leq 1 \text{ and } \|f_\theta(v_*)\|^2 \leq 1, \tag{5}$$

where λ controls the importance of the unsupervised loss.

3.5 Cross-Domain Recommendations

We utilize the affine latent factors obtained from the mapping process to provide recommendations to a specific user. For a target user i, its distance from target item j is calculated as

$$d_{ij} = -2(u_i^t \times v_j^t) + \|u_i^t\|^2 + \|v_j^t\|^2. \tag{6}$$

Once the distances to all items from user i are calculated, we recommend the top-k items to the user.

4 Experiments

This section discusses the experimental setup used to evaluate our work compared to the baselines and presents a comparative analysis. First, we describe the datasets used for our experiments and the required preprocessing steps in Sect. 4.1. Next, we discuss the evaluation metrics used for comparative analysis in Sect. 4.2. We compared our work to several benchmark CDR methods described in Sect. 4.3 to ensure a fair evaluation. The hyperparameter tuning and the criteria for selecting optimal values are explained in Sect. 4.4. Finally, we reported our results in Sect. 4.5.

4.1 Datasets

We have utilized two pairs of datasets to conduct the experiments. The first one is the Douban-Book and Douban-Music datasets[1], which contain overlapping users, and the other one is the MovieLens 1M and MovieLens 100K datasets[2], which have overlapping items. We treat each observed interaction as an implicit record as done in [11,25].

We have established two CDR scenarios: 1) Douban Book → Douban Music and 2) MovieLens 1M → MovieLens 100K[3]. For each scenario, we are focusing on

[1] https://github.com/RUCAIBox/RecBole-CDR.
[2] https://grouplens.org/datasets/movielens.
[3] $A \rightarrow B$ indicates that A is the source domain and B is the target domain.

Table 1. Statistics of the CDR datasets.

		#Users	#Items	#Interactions	Density	Overlap
Scenario 1	Douban Book	6,139	11,073	7,53,245	1.11%	3631
	Douban Music	6,507	8,328	4,74,606	0.87%	
Scenario 2	MovieLens 1M	6,039	3,308	8,35,789	4.18%	1154
	MovieLens 100K	944	1,197	81,705	7.23%	

the users who have at least 20 item interactions [8,11]. We summarize the statistics of our scenarios in Table 1. The dataset is split into train, test, and validation sets in the proportion of 70 : 20 : 10. Furthermore, to enhance the learning process for models, we employ uniform negative sampling by selecting one negative instance per positive instance from all possible negative examples [12].

4.2 Evaluation Metrics

To evaluate the accuracy of the top-k recommendations provided by different models, we have used four standard evaluation metrics: Mean Reciprocal Rank (MRR), Normalized Discounted Cumulative Gain (NDCG), Precision, and Hit Ratio (HR). Given the test set of users U_{test}, MRR quantifies a recommender's efficacy by averaging the inverse of the ranks of the first relevant item for each user in the retrieved top-k items.

$$MRR@k = \frac{1}{|U_{test}|} \sum_{i \in U_{test}} \frac{1}{r_i}, \tag{7}$$

where r_i represents the first relevant item's rank for user i. NDCG evaluates the ranking quality by comparing the positions of relevant items in the retrieved list to an ideal order, where all relevant items are positioned at the top. It gives higher scores to relevant items that appear in higher ranks.

$$NDCG@k = \frac{\sum_{i=1}^{k} \frac{rel_i}{log_2(i+1)}}{\sum_{i=1}^{|R|} \frac{1}{log_2(i+1)}}, \tag{8}$$

where, rel_i denotes the relevance score of the ith ranked item, and R is the set containing all relevant items. To measure whether at least one relevant item is present in the top-k retrieved items, we compute $HR@k$ as follows.

$$HR@k = \frac{1}{|U_{test}|} \sum_{i \in U_{test}} 1(hit_i), \tag{9}$$

where the indicator function, $1(hit_i)$ equals 1 if at least one of the retrieved top-k items is relevant for user i, otherwise 0. $Precision@k$ evaluates the ratio of relevant items to the top-k items recommended, indicating the predictive quality of the models.

$$Precision@k = \frac{1}{|U_{test}|} \sum_{i \in U_{test}} \frac{d_i(k)}{k}, \tag{10}$$

where $d_i(k)$ is the total relevant items for the ith user out of the top-k suggestions. The value of all four metrics lie in $[0, 1]$, where 1 represents best recommendation accuracy. These metrics provide a comprehensive assessment of the recommendation models, highlighting their effectiveness in ranking relevant items and ensuring the inclusion of relevant items within the top-k positions.

4.3 Baselines

To validate the efficiency of our model, we have compared it with the following five benchmark CDR models.

- **EMCDR**$_{MF}$ and **EMCDR**$_{MLP}$ [24]: First, it employs matrix factorization as a latent factor model over both domains to extract the respective users' and items' embeddings. Next, to learn the association between two domains, it utilizes a linear matrix translation (EMCDR$_{MF}$) and multi-layer perceptron (EMCDR$_{MLP}$).
- **CoNet** [12]: Collaborative cross networks (CoNet) is a deep transfer learning method that learns complex user-item associations by allowing the transfer of knowledge across domains with the help of cross-connections between base networks.
- **DTCDR** [42]: This approach focuses on achieving a dual target recommendation task. The DTCDR framework creates rating and document embeddings from the rating and multi-source content data using neural matrix factorization (NeuMF). It combines and transfers the embeddings of overlapping users across domains.
- **SSCDR** [13]: This is a semi-supervised CDR approach that leverages both labeled (overlap data) and unlabeled data to enhance the CDR task. The approach investigates the distribution of overlapping users to deliver recommendations to cold-start users.

4.4 Hyperparameter Tuning

For a fair comparison of the baselines with our proposed model, we have implemented all the models using the RecBole[4] library [40]. We performed hyperparameter tuning for each method using the hyperopt[5] library over the validation set. We fixed the embedding size, d as 100. Adam optimizer with batch size 256 and a maximum of 100 epochs is utilized to train all models. We employed early stopping with $patience = 10$ over $N@10$ metric on the validation set to

[4] https://recbole.io/subpackage.html.

[5] https://hyperopt.github.io/hyperopt/.

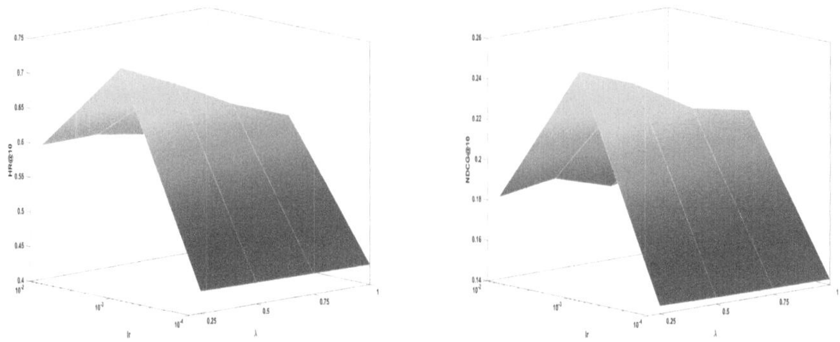

(a) HR@10 (b) NDCG@10

Fig. 2. Effect of λ and lr on gCDR over Scenario 2.

avoid overfitting. The learning rate, lr is searched in $[0.01, 0.001, 0.0001]$. For CoNet, DTCDR, and EMCDR models, the regularization weight γ is tuned in $[0.01, 0.001, 0.005, 0.0001, 0.0005]$. As reported in [13], we search the control parameter, λ for DTCDR, SSCDR and gCDR models in $[0.25, 0.5, 0.75, 1.0]$. The margin size, m, for SSCDR and gCDR models, is tuned in $[0.5, 1.0]$. The architecture of the underlying mapping functions is taken from the respective base papers, as $[64 \rightarrow 32 \rightarrow 16 \rightarrow 8]$ for CoNet, $[d \rightarrow 32 \rightarrow 16 \rightarrow d]$ for DTCDR. For EMCDR, SSCDR, and our approach, we use $[d \rightarrow 2 * d \rightarrow d]$ architecture, as mentioned in [13]. For each algorithm, we have reported the optimal set of hyperparameters by computing the average of $M@k$, $N@k$, $HR@k$, and $Precision@k$ for $k = [5, 10, 20]$ over the three runs. The combination of hyperparameters corresponding to the highest score is chosen as the optimal set as reported in Table 2. We have shown the effect of varying lr and λ for our proposed model over Scenario 2 in Fig. 2. The proposed approach performs better for lower values of the control parameter λ, giving more influence to the supervised part. This signifies that the group data plays a vital role in the mapping process from the source domain to the target domain, effectively enhancing the CDR performance. It also reduces the reliance on unsupervised data for capturing cross-domain relationships.

Table 2. Optimal values of hyperparameters for each model.

	Model	lr	γ	λ	m
Scenario 1	EMCDR_{MF}	0.0001	0.005	–	–
	EMCDR_{MLP}	0.0001	0.0005	–	–
	CoNet	0.01	0.0001	–	–
	DTCDR	0.0001	0.0001	0.75	–
	SSCDR	0.0001	–	1	0.5
	gCDR	0.001	–	0.25	0.5
Scenario 2	EMCDR_{MF}	0.001	0.0001	–	–
	EMCDR_{MLP}	0.001	0.005	–	–
	CoNet	0.01	0.0001	–	–
	DTCDR	0.001	0.0001	0.5	–
	SSCDR	0.001	–	0.25	0.5
	gCDR	0.001	–	0.25	0.5

4.5 Results

The experimental results reported in this section show that the proposed framework significantly outperforms the underlying SSCDR model and other baselines across both scenarios. Figure 3 illustrates the Hit Ratio ($HR@k$) for different k values. It demonstrates the gCDR approach's superiority over the baseline algorithms, indicating a more robust understanding of user preferences. It also delivers a better user experience by presenting more relevant items early in the recommendation list, enhancing user engagement.

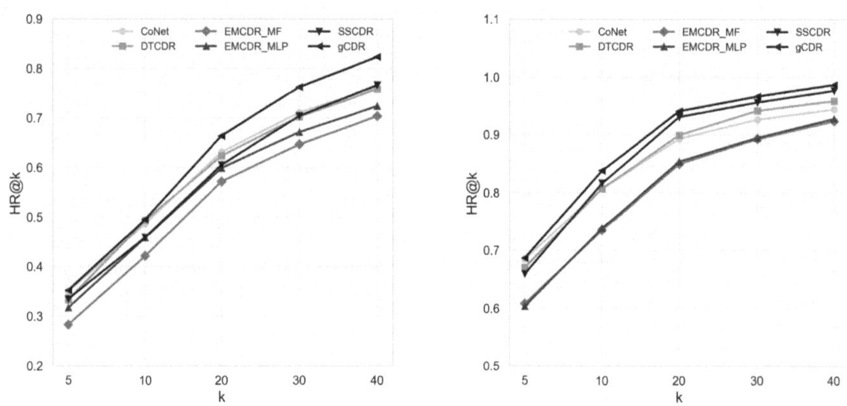

(a) Scenario 1 (b) Scenario 2

Fig. 3. Comparative analysis of algorithms on the basis of $HR@k$.

Table 3. Comparative analysis of algorithms in terms of $M@k$ and $N@k$. (*Imp.* denotes the relative improvement over the best baseline).

	Metrics	$EMCDR_{MF}$	$EMCDR_{MLP}$	CoNet	DTCDR	SSCDR	gCDR	Imp.
Scenario 1	M@5	0.1649	0.1821	0.2037	0.1764	0.2175	**0.2260**	3.91%
	N@5	0.0827	0.0919	0.0997	0.0979	0.1236	**0.1284**	3.88%
	M@10	0.1832	0.2009	0.2220	0.1909	0.2342	**0.2448**	4.53%
	N@10	0.0866	0.0947	0.0996	0.0953	0.1290	**0.1334**	3.41%
	M@20	0.1937	0.2108	0.2319	0.1998	0.2442	**0.2564**	5.00%
	N@20	0.0989	0.1068	0.1106	0.1011	0.1415	**0.1478**	4.45%
	M@30	0.1968	0.2137	0.2352	0.2030	0.2482	**0.2604**	4.92%
	N@30	0.1098	0.1175	0.1222	0.1088	0.1530	**0.1602**	4.71%
	M@40	0.1984	0.2152	0.2366	0.2046	0.2500	**0.2621**	4.84%
	N@40	0.1193	0.1274	0.1320	0.1165	0.1625	**0.1708**	5.11%
Scenario 2	M@5	0.3913	0.4090	0.4802	0.4580	0.4483	**0.4841**	0.81%
	N@5	0.2531	0.2626	0.2977	0.2848	0.2963	**0.3258**	9.44%
	M@10	0.4082	0.4269	0.4967	0.4767	0.4695	**0.5045**	1.57%
	N@10	0.2482	0.2576	0.2896	0.2806	0.2997	**0.3281**	9.48%
	M@20	0.4161	0.4348	0.5029	0.4833	0.4777	**0.5120**	1.81%
	N@20	0.2559	0.2633	0.3003	0.2919	0.3207	**0.3444**	7.39%
	M@30	0.4179	0.4365	0.5043	0.4851	0.4796	**0.5137**	1.86%
	N@30	0.2687	0.2773	0.3186	0.3098	0.3415	**0.3645**	6.73%
	M@40	0.4187	0.4374	0.5048	0.4856	0.4802	**0.5143**	1.88%
	N@40	0.2849	0.2946	0.3371	0.3264	0.3589	**0.3838**	6.94%

The experimental results relevant to comparative analysis in terms of MRR ($M@k$) and NDCG ($N@k$) are presented in Table 3. These results signify the effective mapping of group and overlap information from source to target domain, leading to more relevant and accurate recommendations. In Fig. 4, the predictive accuracy of the comparing algorithms is depicted based on $Precision@k$ scores. The experimental results exhibit the efficacy of the gCDR framework across diverse evaluation metrics and showcase its outstanding predictive accuracy. Additionally, it signifies that incorporating group information can significantly enhance the learning process. Furthermore, the framework addresses the issue of sparsity to a certain extent, laying the groundwork for developing more efficient algorithms.

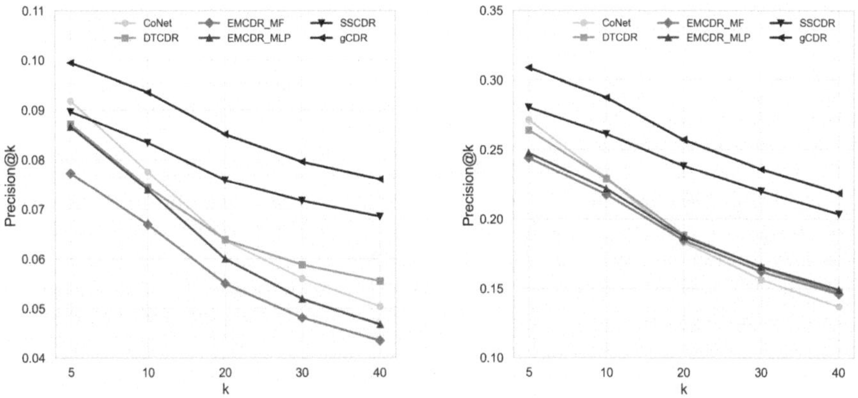

(a) Scenario 1 (b) Scenario 2

Fig. 4. Comparative analysis of algorithms on the basis of *Precision@k*.

5 Conclusion and Future Works

The proposed work introduces gCDR, a novel framework that can be integrated with most existing cross-domain recommendation systems. By incorporating group and overlap information, gCDR enhances recommendation accuracy. We extend the SSCDR [13] method by using group information in the mapping process, where overlapping users serve as cluster centroids to identify users with similar preferences across domains. These clusters and overlapping users provide labeled data with known ground-truth locations in the target domain. We hypothesize that like overlapping users, a group in the target domain is likely to exhibit similar patterns to those observed in the source domain, aiding in handling sparse overlap data. Through extensive experiments across two scenarios and four prominent evaluation metrics, we have demonstrated significant performance improvements. This validates that gCDR effectively enhances the mapping process by jointly exploiting the overlapping entities and group information.

Despite the framework's promising results, there are areas for improvement. Extracting group embeddings adds computational overhead, particularly challenging for large-scale datasets. To address this, we plan to explore hierarchical clustering and sampling-based approaches, which could significantly reduce computational complexity and improve scalability. Also, for group construction, the framework will need to work on the cases with no overlap data. To overcome this, we aim to investigate domain adaptation techniques for group construction in such scenarios. This approach could enable effective cross-domain recommendations even when direct user or item connections are absent. In future work, we will focus on extending the framework to multi-target recommendations, broadening its applicability and potential impact in diverse recommendation scenarios.

References

1. Adomavicius, G., Tuzhilin, A.: Toward the next generation of recommender systems: a survey of the state-of-the-art and possible extensions. IEEE Trans. Knowl. Data Eng. **17**(6), 734–749 (2005)
2. Bi, Y., Song, L., Yao, M., Wu, Z., Wang, J., Xiao, J.: DCDIR: a deep cross-domain recommendation system for cold start users in insurance domain. In: Proceedings of the 43rd International ACM SIGIR Conference on Research and Development in Information Retrieval, pp. 1661–1664 (2020)
3. Bokde, D., Girase, S., Mukhopadhyay, D.: Matrix factorization model in collaborative filtering algorithms: a survey. Procedia Comput. Sci. **49**, 136–146 (2015)
4. Chen, X., Zhang, Y., Tsang, I.W., Pan, Y., Su, J.: Toward equivalent transformation of user preferences in cross domain recommendation. ACM Trans. Inf. Syst. **41**(1), 1–31 (2023)
5. Cui, Q., Wei, T., Zhang, Y., Zhang, Q.: HeroGraph: a heterogeneous graph framework for multi-target cross-domain recommendation. In: ORSUM@ RecSys (2020)
6. Fernández-Tobías, I., Cantador, I.: Exploiting social tags in matrix factorization models for cross-domain collaborative filtering. In: CBRecSys@ RecSys, pp. 34–41. Citeseer (2014)
7. Fernández-Tobías, I., Cantador, I., Kaminskas, M., Ricci, F.: Cross-domain recommender systems: a survey of the state of the art. In: Spanish conference on information retrieval, vol. 24 (2012)
8. Fu, W., Peng, Z., Wang, S., Xu, Y., Li, J.: Deeply fusing reviews and contents for cold start users in cross-domain recommendation systems. In: Proceedings of the AAAI Conference on Artificial Intelligence, vol. 33, pp. 94–101 (2019)
9. Gao, C., et al.: Cross-domain recommendation without sharing user-relevant data. In: The World Wide Web Conference, pp. 491–502 (2019)
10. Gao, S., Luo, H., Chen, D., Li, S., Gallinari, P., Guo, J.: Cross-domain recommendation via cluster-level latent factor model. In: Machine Learning and Knowledge Discovery in Databases: European Conference, ECML PKDD 2013, Prague, Czech Republic, 23–27 September 2013, Part II, pp. 161–176. Springer (2013)
11. He, X., Liao, L., Zhang, H., Nie, L., Hu, X., Chua, T.S.: Neural collaborative filtering. In: Proceedings of the 26th International Conference on World Wide Web, pp. 173–182 (2017)
12. Hu, G., Zhang, Y., Yang, Q.: CoNet: collaborative cross networks for cross-domain recommendation. In: Proceedings of the 27th ACM International Conference on Information and Knowledge Management, pp. 667–676 (2018)
13. Kang, S., Hwang, J., Lee, D., Yu, H.: Semi-supervised learning for cross-domain recommendation to cold-start users. In: Proceedings of the 28th ACM International Conference on Information and Knowledge Management, pp. 1563–1572 (2019)
14. Khan, M.M., Ibrahim, R., Ghani, I.: Cross domain recommender systems: a systematic literature review. ACM Comput. Surv. (CSUR) **50**(3), 1–34 (2017)
15. Koren, Y., Bell, R., Volinsky, C.: Matrix factorization techniques for recommender systems. Computer **42**(8), 30–37 (2009)
16. Kumar, V., Pujari, A.K., Sahu, S.K., Kagita, V.R., Padmanabhan, V.: Collaborative filtering using multiple binary maximum margin matrix factorizations. Inf. Sci. **380**, 1–11 (2017)
17. Kumar, V., Pujari, A.K., Sahu, S.K., Kagita, V.R., Padmanabhan, V.: Proximal maximum margin matrix factorization for collaborative filtering. Pattern Recogn. Lett. **86**, 62–67 (2017)

18. Lam, X.N., Vu, T., Le, T.D., Duong, A.D.: Addressing cold-start problem in recommendation systems. In: Proceedings of the 2nd International Conference on Ubiquitous Information Management and Communication, pp. 208–211 (2008)
19. Li, B., Yang, Q., Xue, X.: Can movies and books collaborate? Cross-domain collaborative filtering for sparsity reduction. In: Twenty-First International Joint Conference on Artificial Intelligence (2009)
20. Li, P., Tuzhilin, A.: Dual metric learning for effective and efficient cross-domain recommendations. IEEE Trans. Knowl. Data Eng. **35**(1), 321–334 (2021)
21. Li, Z., et al.: Semantic relation transfer for non-overlapped cross-domain recommendations. In: Pacific-Asia Conference on Knowledge Discovery and Data Mining, pp. 271–283. Springer (2023)
22. Lika, B., Kolomvatsos, K., Hadjiefthymiades, S.: Facing the cold start problem in recommender systems. Expert Syst. Appl. **41**(4), 2065–2073 (2014)
23. Loni, B., Shi, Y., Larson, M., Hanjalic, A.: Cross-domain collaborative filtering with factorization machines. In: Advances in Information Retrieval: 36th European Conference on IR Research, ECIR 2014, Amsterdam, The Netherlands, 13–16 April 2014, pp. 656–661. Springer (2014)
24. Man, T., Shen, H., Jin, X., Cheng, X.: Cross-domain recommendation: an embedding and mapping approach. In: IJCAI, vol. 17, pp. 2464–2470 (2017)
25. Park, C., Kim, D., Xie, X., Yu, H.: Collaborative translational metric learning. In: 2018 IEEE International Conference on Data Mining (ICDM), pp. 367–376. IEEE (2018)
26. Sahebi, S., Walker, T.: Content-based cross-domain recommendations using segmented models. In: CBRecSys@ RecSys, pp. 57–64 (2014)
27. Shyam, A., Kumar, V., Kagita, V.R., Pujari, A.K.: UniRecSys: a unified framework for personalized, group, package, and package-to-group recommendations. Knowl.-Based Syst. 111552 (2024)
28. Song, T., Peng, Z., Wang, S., Fu, W., Hong, X., Yu, P.S.: Based cross-domain recommendation through joint tensor factorization. In: Database Systems for Advanced Applications: 22nd International Conference, DASFAA 2017, Suzhou, China, 27–30 March 2017, Part I 22. pp. 525–540. Springer (2017)
29. Tan, S., Bu, J., Qin, X., Chen, C., Cai, D.: Cross domain recommendation based on multi-type media fusion. Neurocomputing **127**, 124–134 (2014)
30. Taneja, A., Arora, A.: Cross domain recommendation using multidimensional tensor factorization. Expert Syst. Appl. **92**, 304–316 (2018)
31. Veeramachaneni, S.D., Pujari, A.K., Padmanabhan, V., Kumar, V.: A maximum margin matrix factorization based transfer learning approach for cross-domain recommendation. Appl. Soft Comput. **85**, 105751 (2019)
32. Wang, X., Peng, Z., Wang, S., Yu, P.S., Fu, W., Hong, X.: Cross-domain recommendation for cold-start users via neighborhood based feature mapping. In: Database Systems for Advanced Applications: 23rd International Conference, DASFAA 2018, Gold Coast, QLD, Australia, 21–24 May 2018, Part I, pp. 158–165. Springer (2018)
33. Xie, R., et al.: Improving accuracy and diversity in matching of recommendation with diversified preference network. IEEE Trans. Big Data **8**(4), 955–967 (2021)
34. Xu, M., Zhu, J., Zhang, B.: Fast max-margin matrix factorization with data augmentation. In: International Conference on Machine Learning, pp. 978–986. PMLR (2013)
35. Yan, H., Chen, X., Gao, C., Li, Y., Jin, D.: DeepAPF: deep attentive probabilistic factorization for multi-site video recommendation. TC **2**(130), 17–883 (2019)
36. Yu, X., Hu, Q., Li, H., Du, J., Gao, J., Sun, L.: Cross-domain recommendation based on latent factor alignment. Neural Comput. Appl. 1–12 (2022)

37. Zang, T., Zhu, Y., Liu, H., Zhang, R., Yu, J.: A survey on cross-domain recommendation: taxonomies, methods, and future directions. ACM Trans. Inf. Syst. **41**(2), 1–39 (2022)
38. Zhang, Q., Hao, P., Lu, J., Zhang, G.: Cross-domain recommendation with semantic correlation in tagging systems. In: 2019 International Joint Conference on Neural Networks (IJCNN), pp. 1–8. IEEE (2019)
39. Zhao, C., Zhao, H., He, M., Zhang, J., Fan, J.: Cross-domain recommendation via user interest alignment. In: Proceedings of the ACM Web Conference 2023, pp. 887–896 (2023)
40. Zhao, W.X., et al.: Recbole: towards a unified, comprehensive and efficient framework for recommendation algorithms. In: CIKM (2021)
41. Zhou, D., Bousquet, O., Lal, T., Weston, J., Schölkopf, B.: Learning with local and global consistency. In: Advances in Neural Information Processing Systems, vol. 16 (2003)
42. Zhu, F., Chen, C., Wang, Y., Liu, G., Zheng, X.: DTCDR: a framework for dual-target cross-domain recommendation. In: Proceedings of the 28th ACM International Conference on Information and Knowledge Management, pp. 1533–1542 (2019)
43. Zhu, F., Wang, Y., Chen, C., Liu, G., Orgun, M., Wu, J.: A deep framework for cross-domain and cross-system recommendations. arXiv preprint arXiv:2009.06215 (2020)
44. Zhu, F., Wang, Y., Chen, C., Liu, G., Zheng, X.: A graphical and attentional framework for dual-target cross-domain recommendation. In: IJCAI, pp. 3001–3008 (2020)
45. Zhu, F., Wang, Y., Chen, C., Zhou, J., Li, L., Liu, G.: Cross-domain recommendation: challenges, progress, and prospects. arXiv preprint arXiv:2103.01696 (2021)
46. Zhu, Y., et al.: Transfer-meta framework for cross-domain recommendation to cold-start users. In: Proceedings of the 44th International ACM SIGIR Conference on Research and Development in Information Retrieval, pp. 1813–1817 (2021)
47. Zhu, Y., et al.: Personalized transfer of user preferences for cross-domain recommendation. In: Proceedings of the Fifteenth ACM International Conference on Web Search and Data Mining, pp. 1507–1515 (2022)

CKELM: Unified Approach to Parking Space Classification for All Weather Types

Navpreet(✉), Rajendra Kumar Roul, and Rinkle Rani

Department of Computer Science and Engineering, Thapar Institute of Engineering and Technology, Patiala, Punjab, India
{nnavpreet_phd21,raj.roul,raggarwal}@thapar.edu

Abstract. Recently, surging urbanization and the subsequent increase in vehicular movement have amplified the popularity of parking management systems. A crucial hurdle in effective parking management is the optimal utilization of available parking spaces. Although deep learning techniques have showcased their prowess in accurately identifying parking spaces, they sometimes falter when partial obstructions and diverse lighting conditions are present. This challenge is further exacerbated during foggy or hazy weather, resulting in a decline in their performance. This research article introduces a novel approach that leverages the benefits of CNN and Kernel Extreme Learning Machine (KELM). The approach not only enhances accuracy but also significantly reduces training time. The key innovation lies in replacing the fully connected layer of CNN with KELM, which circumvents the backpropagation process and reduces the training time. A purpose-built data set tailored to classify parking spaces in hazy conditions and PKLot is employed for the validation. Within the proposed approach, images from the hazy dataset undergo preprocessing using Light-DehazeNet (LD-Net) to counteract the effects of haze. Subsequently, CNN is trained on this preprocessed dataset to extract features and generate a feature vector. This feature vector is input into the KELM to classify parking spaces as occupied or vacant. Compared with standalone CNN, KELM, machine learning classifiers, and similar existing models, the proposed model showcases superior performance and better computational efficiency. This study demonstrates that the integration of CNN with KELM offers a promising solution for parking space classification in challenging weather conditions.

Keywords: CNN · Kernel ELM · Hazy Parking Dataset · Parking Space · Classification · PKLot

1 Introduction

Approximately 30% of vehicular traffic in densely populated cities continuously seeks parking slots, as indicated by previous research [2]. This continuous hunt for parking places has put immense stress on parking management systems, resulting in increased commute times as drivers struggle to find parking spots [20].

A. Dasgupta et al. (Eds.): BDA 2024, LNCS 15526, pp. 257–278, 2025.
https://doi.org/10.1007/978-3-031-81821-9_15

Efficient parking classification techniques play a pivotal role in addressing this challenge. These techniques provide critical information about parking space availability, enabling effective parking lot management. By informing drivers about the number of vacant parking spaces, these systems can substantially reduce long waiting queues, enhance scalability, and reduce the duration needed to find a parking space. Finding and maneuvering a vehicle into a suitable parking spot can be daunting for many drivers, especially new ones. This challenge is exacerbated during hazy weather conditions, as reduced visibility makes it difficult for drivers to assess the size and configuration of parking spaces and obstacles around their vehicles.

Many research studies have already been done for parking space classification under non-hazy conditions. Parking space classification systems were either manually keeping track of vehicles leaving/entering or sensor-based [8]. Manual-based tracking of vehicles is a time-consuming process. The problem of using sensors is not suitable for large parking spaces, requires time-consuming tasks for installing sensor updates, has high installation and maintenance costs, is susceptible to physical tampering, and raises cyber security concerns [21].

Image processing techniques are better options to deal with the shortcomings mentioned earlier about sensors [7]. It has achieved notable advancements in computer vision in recent times [21, 22]. Deep learning techniques have already been employed to classify parking spaces under clear or non-hazy conditions [6]. Vision-based techniques can experience a decline in performance due to obstacles in parking spots like trees and adverse weather conditions (i.e., haze), where the fog is an obstacle for sensors or cameras. Also, Real-time datasets for parking space classification under hazy conditions are unavailable. Hence, researchers have not considered parking space classification under foggy conditions.

Fully connected layers in a CNN typically have many parameters and involve backpropagation to adjust parameters. The backpropagation computes gradients and updates the layer weights based on these gradients, which leads to longer training times and increased computational complexity. As a result, a very high computational time is required to train CNN.

Huang GB et al. [18] proposed a Kernel Extreme Learning Machine (KELM), a simple and efficient solution for the backpropagation process. KELM offers significant advantages over traditional Extreme Learning Machine (ELM), particularly in handling non-linear data. KELM maps inputs to a higher-dimensional space by utilizing kernel functions, effectively managing non-linearly separable data that traditional ELMs struggle with due to their linear nature. This approach improves generalization by creating more flexible and robust decision boundaries and eliminates the need for random weight initialization, resulting in more consistent and reproducible performance. KELM is also more efficient in high-dimensional spaces, avoiding the excessive computational and memory demands that ELMs may face. Furthermore, its natural incorporation of regularization techniques helps reduce overfitting, making it a powerful alternative for complex and high-dimensional data tasks. It has been demonstrated that

ELM's training speed is considerably faster than that of most popular machine learning (ML) techniques like RF, SVM, DT, etc. [1,4,17].

Researchers have either improved their model performances or reduced the computational time but have failed to optimize both of these parameters for parking slot classification under non-hazy conditions. This is the first study in which parking space classification has been done for hazy weather. The proposed model has taken care of both these parameters (i.e., model performance and computational time) to a better extent. Due to the advantages mentioned above and the excellent performance of CNN and KELM, the proposed work has combined CNN with KELM to classify the parking space under hazy conditions.

1.1 Our Contribution

The paper's primary contributions are summarized as follows:

- The proposed approach utilizes CNN and KELM for feature extraction and classification, reducing the computational time and increasing the model's performance.
- The primary dataset is generated and pre-processed for parking space classification under hazy conditions.
- The proposed approach undergoes validation on our primary hazy parking dataset and PKLot, with performance assessment conducted in terms of various parameters such as precision, sensitivity, specificity, accuracy, and F1-Score.
- A comparative analysis of the proposed CKELM is performed with ML classifiers. Further, the CKELM model's efficacy is compared with similar existing models.

2 Proposed Model: CKELM

The proposed model comprises CNN, a well-known feature extractor, and a KELM parking slot classifier. Images are dehazed through LD-Net to remove the visual effects of haze. The output of the dehazing network is fed to CNN for feature extraction and then classified through KELM. LD-Net and CKELM work together as a single unit and can be employed to get better results in hazy weather.

2.1 CKELM Architecture

The proposed approach integrated CNN with KELM to reduce computational time and enhance model performance. The detailed proposed architecture of CKELM is shown in Fig. 1. The specific details of each layer in the CKELM architecture are further discussed below and are summarized in Table 1.

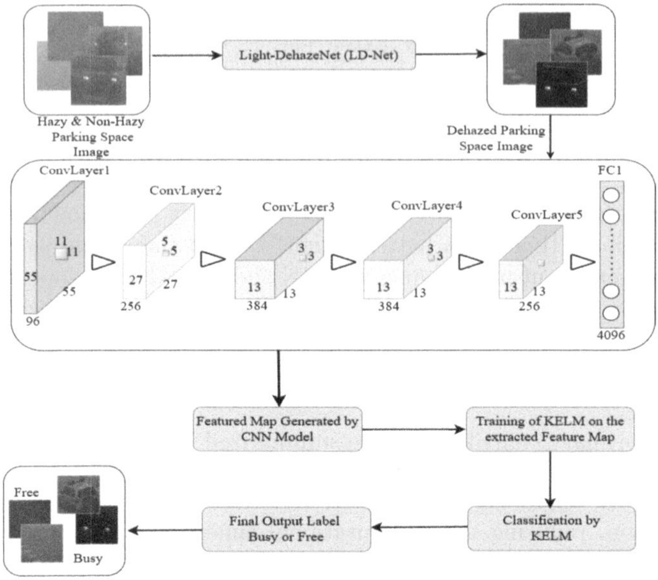

Fig. 1. Architecture of CKELM Model

- **Input Layer:** Dehazed images of dimensions 227 × 227 × 3 are fed to the input layer where 227 × 227 signifies the number of pixels along the horizontal and vertical axes, and 3 represents color channels, to store color information for each pixel of the image.
- **CL:** Images from the input layer are passed to the CL for extracting features such as edges, curves, and shapes by applying filters (kernels) of different sizes. The number of kernels in the first CL is 96, with a filter size of 11 × 11. For the second and fifth layers, the kernel size is 256 with a filter size of 3 × 3 and 5 × 5, respectively. For the third and fourth CLs, the kernel size is 384 with a filter size of 3 × 3. The padding is added at each CL for a more accurate analysis of images.
- **ReLU:** Each CL is followed by a ReLU layer. ReLU aims to get the activation values corresponding to neurons by increasing the non-linearity in the CNN. It solves the vanishing gradient problem as the network grows. It is represented by the Eq. 1.

$$f(y) = \begin{cases} 0, if\ y < 0 \\ y, if\ y \geq 0 \end{cases} \tag{1}$$

  where y is an input value. The output of ReLU is either zero when y is negative or the same input value when y is positive.
- **BatchNormalization Layer:** Batch normalization is a normalization technique applied to the intermediate outputs of a neural network, helping in stabilize and accelerate the training process by normalizing inputs and introducing learnable parameters. It has become a standard component in many deep-learning architectures. The Eq. 2 gives the batch normalization formula.

$$\alpha^* = \frac{\alpha - E[\alpha]}{\sqrt{var(\alpha)}} \quad (2)$$

where α^* is the new value of α, $E[\alpha]$ is the mean for a batch, and $var(\alpha)$ is the variance for a batch.

- **Flattened Layer:** The flattening layer is introduced to prepare the data for input into the FCL. It takes the output from the convolutional layer (in the form of a 3D array) and transforms it into a one-dimensional vector. This transformation involves "flattening" the multi-dimensional array into a single line of values. The flattening layer acts as a bridge between the convolutional layers, which capture hierarchical features, and the fully connected layers, which make use of these features for decision-making. It reshapes the data to facilitate the transition from spatial hierarchies in the convolutional layers to a format suitable for traditional neural network layers.
- **FCL:** Fully connected implies a layer in which each neuron is linked to every neuron in the preceding and subsequent layers. The proposed model consists of one FCL with 4096 features and the softmax as an activation function.
- **Dropout Layer:** A dropout layer follows the FCL to avoid over-fitting. It is a mask that nullifies the participation of some neurons toward the next layer and leaves all other neurons unmodified. The dropout rate is decided to be 0.5 based on the experiment.

KELM is plugged at the FCL of CNN, consisting of the featured map as input, a hidden layer, and a classifier layer. The featured vector generated by CNN is fed to the KELM's input layer. KELM's classifier layer applies the different kernels with varying gamma values.

Table 1. Architecture details of proposed CKELM

Layer	Kernel size	Filters	Stride	Padding	Feature Map	Activation Function	Trainable Params
Input	-	-	-	-	$227 \times 227 \times 3$	-	-
Conv-layer1	11×11	96	4	same	$55 \times 55 \times 96$	ReLU	34944
MaxPool-layer1	3×3	2	-	-	$27 \times 27 \times 96$	-	0
Conv-layer2	5×5	256	1	same	$27 \times 27 \times 256$	ReLU	614656
MaxPool-layer2	3×3	2	-	-	$13 \times 13 \times 256$	-	0
Conv-layer3	3×3	384	1	same	$13 \times 13 \times 384$	ReLU	885120
Conv-layer4	3×3	384	1	same	$13 \times 13 \times 384$	ReLU	1327488
Conv-layer5	3×3	384	1	same	$13 \times 13 \times 256$	ReLU	884992
MaxPool-layer3	3×3	-	2	-	$6 \times 6 \times 256$	-	0
Flatten-layer1	-	-	-	-	9216	ReLU	0
Fully-connected layer1	-	-	-	-	4096	-	37752832
Dropout-layer1	-	-	-	-	4096	-	0
Softmax layer	-	-	-	-	2	Softmax	8194
Hidden layer	-	-	-	-	256	-	-
KELM Classifier	-	-	-	-	2	RBF Kernel	-

2.2 Working of CKELM

An in-depth discussion of the proposed model and its workings is provided in the subsequent subsections. The working of proposed model comprised six steps, as follows:

1. **Data Collection:** The proposed model is validated on the Hazy parking dataset (HPD) and the PKLot Dataset (PKD) and is discussed as follows:
   a. HPD is constructed for the proposed model. Cameras are used to capture hazy parking images. There are 752 captured image patches which have not undergone any form of data augmentation. The dataset is partitioned into HPD_training and HPD_Testing sets. The HPD_training set includes 70% of the dataset and the remaining in the HPD_Testing set. Among these images, 502 are hazy busy images, while the remaining 250 are hazy free images. Table 2 represents the details of the Hazy Parking Dataset. To demonstrate the proposed dataset, we have kept around 10% of the Hazy Parking Dataset in GitHub for free download and is available at the link[1].

Table 2. Summary of Hazy Parking Dataset

Images Type	Free Slot Images	Busy Slot Images	Total Images
Unaugmented Training Images	150	380	532
Unaugmented Testing Images	100	122	222
Total Unaugmented Images	250	502	752
Augmented Training Images	1900	1500	3400
Augmented Testing Images	1000	610	1610
Total Augmented Images	2500	2510	5010

   b. PKD [10] is an expanded version of the parking lot dataset first released in Almeida 2013 [5]. The dataset comprises images captured within the parking premises of PUCPR (Pontifical Catholic University of Parana) and UFPR (Federal University of Parana) located in Curitiba, Brazil. To reduce the likelihood of occlusion of nearby vehicles, cameras were strategically positioned on the rooftops of buildings. Table 3 represents the occupied and empty parking slots summary.

[1] https://github.com/GauravS9776/Hazy-parking-system.

Table 3. Overview of the PKD

Parking Lot	Weather Condition	Parking Capacity		
		Occupied Parking Slots	Empty Parking Slots	Total Parking Slots
UFPR04	Sunny	32,166	26,334	58,400
	Rainy	2,351	5,607	7,958
	Cloudy	11,608	27,779	39,387
UFPR05	Sunny	57,584	42,306	99,890
	Rainy	6,078	2,851	8,929
	Cloudy	33,764	23,202	56,966
PUCPR	Sunny	96,762	1,11,672	2,08,433
	Rainy	55,104	27,951	83,056
	Cloudy	42,363	90,417	132,780
Total Parking Capacity		3,37,780	3,58,119	6,95,899

2. **Data Augmentation:** Keras, a deep learning Framework, was applied to augment and preprocess the dataset. The following are steps applied for data augmentation.
   a. **Cropping and Flipping:** Cropping, horizontal, and vertical flipping are randomly applied to create image variations. During cropping, only 10% of the image width and height is cut. These data augmentation techniques helped the model's ability to generalize to unseen data and increased the variability of training images. The same methods are applied to the testing images.
   b. **Rotation:** As part of the data augmentation process, images are rotated by angles range from 0 to 45, thereby introducing valuable diversity in parking images and effectively simulating many distinct viewing angles.

   After applying augmentation, there are 5010 augmented labeled images as busy and free as indicated in Table 2, captured in extreme weather conditions like fog and snow. The dataset is specifically curated to include hazy occupied and unoccupied parking space images. This process involves manual extraction, ensuring a comprehensive representation of foggy conditions within the parking system.

3. **Data Pre-processing:** Captured hazy images are fragmented into individual parking spaces, and each hazy image is pre-processed. A meticulous and thoughtfully devised series of pre-processing steps are applied before the model training and testing and are discussed as follows:
   a. **Image Resizing:** Each hazy image is resized to a standardized resolution of 224 × 224 pixels and a consistent input size for the CKELM model, ensuring the smooth operation of its convolutional layers.

b. **Monochrome Conversion:** The CKELM model is designed for grayscale images, a crucial preprocessing step. The hazy, colorful images are converted into grayscale representations before passing into the CNN for feature extraction.
c. **Data Normalization:** The normalization of grayscale images is achieved by scaling each pixel's value to fit within the [0, 1] range through division by 255.

4. **Dehazing of Images:** For the dehazing of images, a CNN-based model called LD-Net is used. The ASM is transformed into a new representation, where the parameters $t(y)$ and A are estimated into a new variable called $K(y)$. The reformulated ASM is combined with the proposed LD-Net, reducing the error in dehazed image reconstruction. LD-Net comprises eight CLs, followed by the ReLU and concatenation layers. RGB images of size ($640 \times 480 \times 3$) are passed as input to the CL through filters (kernels) of different sizes to extract relevant features. The concatenation layers preserve earlier learned semantics, which are lost as they pass through various CLs.

5. **Feature Extractor and Classification:** The features from dehazed images are extracted using CNN. Figure 2 represents the visuals of feature maps from each convolutional layer. The key observation from visualization is that the feature map of $ConvLayer1$, which is closer to the model's input, captures intricate details from the image. As the model progresses, the feature maps from $ConvLayer2$ and $ConvLayer3$ exhibit fewer fine-grained details and transform abstract features from the image into progressively more complex and abstract representations. The feature maps become more abstract and complex upon reaching $ConvLayer4$ and $ConvLayer5$. At the same time, it is not straightforward to interpret these deeper feature maps visually.
The extracted features are fed as input to KELM and classified into busy or free classes. The integrated proposed model benefits from the fast training property of ELM and offers efficient training with less computational overhead. Kernel ELM eliminates the need for random weight initialization as it directly computes the kernel matrix, leading to more stable and repro-

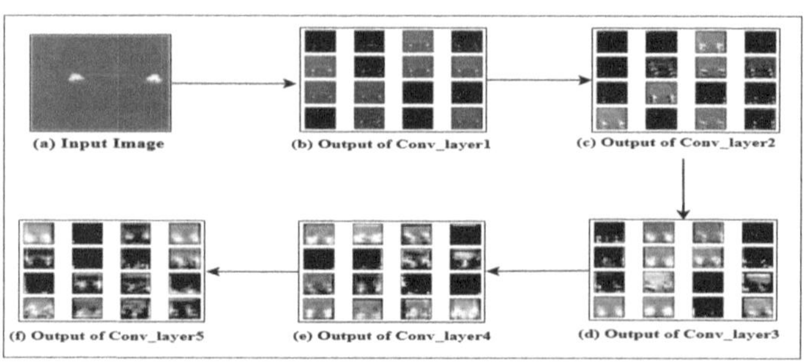

Fig. 2. Sample of extracted feature map (Visuals) from convolution layer of CNN

ducible results. The kernel trick allows KELM to operate efficiently in high-dimensional spaces without explicitly computing the coordinates of the data in that space, thus reducing computational and memory overhead. CKELM is good at generalizing, which means it can make accurate predictions on data that hasn't been seen before with less computational overhead and accelerate the learning speed by creating more flexible and robust decision boundaries.

6. **Performance Evaluation:** Finally, results are aggregated, and CKELM is compared with ML classifiers integrated with CNN and with similar existing models in terms of performance metrics and is discussed in detail in the Sect. 3.

2.3 Methodology of the Proposed Model

The proposed model is implemented on the Visual Studio Code IDE and the Python 3.9.9 platform with 12GB RAM and 24GB GPU with an Intel Core i11 processor. The model is trained and validated on the HPD and PKD. Further, Algorithm 1 represents the algorithm for implementing the CKELM model. Both datasets are partitioned into ratios of 70 and 30. Table 4 lists the parameters used for feature extraction by the CNN.

Table 4. Parameters setting during feature extraction by CNN

Parameter	Value and description
Learning Rate	0.001
Optimization	SGD
Batch Size	32
Epochs	5–20
Iteration Per Epoch	106

Firstly, images are dehazed by CNN based model called LD-Net. In the next step, dehazed images are fed to CNN for feature extraction. During feature extraction, the loss is calculated by binary cross-entropy, and network weights are updated by stochastic gradient descent (SGD) optimizer. The dropout function is applied to avoid overfitting, ensuring that it is only applied during training, not classification. While training, the callback function is used as ModelCheckpoint to save the model's best validation accuracy. The extracted feature map is flattened and fed to KELM for classification. KELM is trained on this extracted feature map with a poly kernel and classifies images into output class labels as busy or free.

Algorithm 1. CKELM Classification Algorithim

Input: Hazy Parking Dataset HPD, HPD_train training set of images, HPD_test testing set of images, and PKLot Dataset PKD, n number of iterations required for training of the CKELM model.

Output: Target class for each input image - busy or free and accuracy.

begin
Partition HPD into HPD_train, HPD_test, and HPD_val and PKD into PKD_train, PKD_test, and PKD_val.
Build CNN model with five convolutional layers and one FCL. ReLU, max-pooling, batch-normalization, and the dropout layer follow each convolutional layer.
Prepare featured vector using CNN and do the hyper-tunning of training parameters.
// Train CNN for n epochs.
for $i \leftarrow 1$ to n **do**
    Training of CNN model and validate it using HPD_val.
end for
Build KELM Model using the featured vector and set different kernels
Initialize a bias vector b of size l, matrix w and vector b.
Use the generated values of w and b as the input weights and bias in KELM.
Compute $H(x)$ and β. *//End of ELM algorithm.*
Classify the dehazed image to the target class as busy or free using the CKELM model.
Analysis of the classification accuracy through a Confusion Matrix.
return image class label, accuracy.
end

3 Results and Discussion

CKELM is trained for 5–20 epochs with different kernels in ELM and is validated on the hazy and non-hazy datasets. The performances are analyzed by plotting the confusion matrix (CM), comparing precision, accuracy, sensitivity, specificity, and F1-Score at different kernels. Finally, a performance comparison is made between CKELM and similar existing models.

3.1 Analysis by Confusion Matrix

Figure 3 illustrates the confusion matrices of the CKELM model for both HPD and PKD. The matrices show the model's performance across different epochs, highlighting each class's accuracy and misclassification rates.

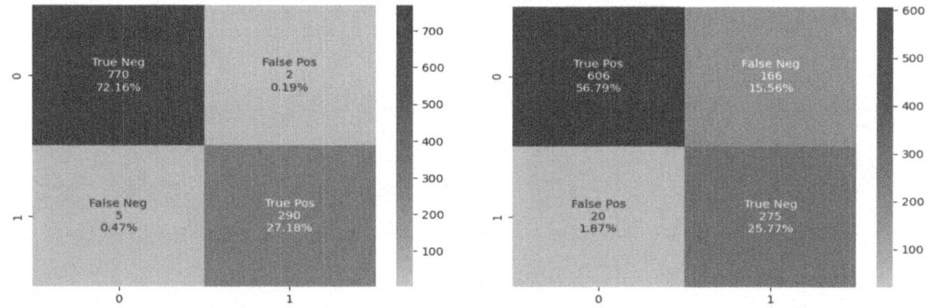

(a) (b)

Fig. 3. Analysis of Confusion Matrix of CKELM for HPD and PKD

3.2 Comparison of Performance at Different Kernels

The performance of CKELM is analyzed using different kernel functions. Table 5 represents the accuracy and specificity comparison of CKELM for HPD and PKD with different kernel functions. Figures 4a, 4b, and 4c, respectively, represent Precision, Recall, and F1-Score metrics comparison with different kernel functions for HPD. Figures 5a, 5b, and 5c, respectively, represent Precision, Recall, and F1-Score metrics comparison with different kernel functions for PKD.

Table 5. Comparison of performance across different Kernels

	Hazy Parking Dataset		PKLot Dataset	
Kernel used	Accuracy (%)	Specificity (%)	Accuracy (%)	Specificity (%)
Linear	94.38	**99.87**	92.36	95.87
RBF	73.01	90.00	69.23	85.65
Poly	**99.34**	97.74	**95.36**	**96.65**

3.3 Performance Comparison of CKELM with ML Classifiers

A performance comparison is made between CKELM and various ML classifiers, including ET, MLP, SVM, LDA, QDA, XgBoost, and DT, among others, combined with CNN. Each model has been trained for ten epochs on a hazy training subset (HPD_Training) and PKD. Table 6 lists hyper-parameter settings for the training of CNN-ML. Table 7 represents the accuracy comparison of CKELM with ml classifiers trained and tested on the hazy and non-hazy datasets. Figures 6a, 6b, and 6c represent the performance metric comparison between CKELM and CNN-ML Classifiers for HPD. Figures 6a, 6b, and 6c represent the performance metric comparison between CKELM and CNN-ML Classifiers for HPD. Figures 7a, 7b , and 7c represents the performance comparison of CKELM with ml classifiers across subsets of PKD.

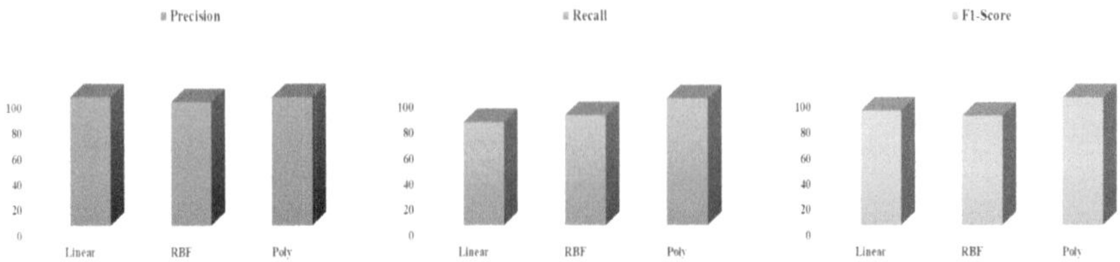

(a) Precision (b) Recall (c) F1-Score

Fig. 4. Performance comparison of CKELM for HPD

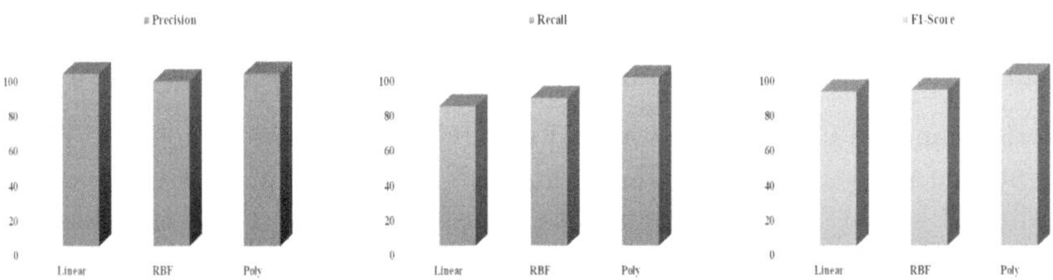

(a) Precision (b) Recall (c) F1-Score

Fig. 5. Performance comparison of CKELM for PKD

Table 6. Hyper-parameter Setting for ML Classifier training

Classifier	Hyper-parameters
ET	max_depth = 40, n_estimators = 100, random_state = 42, criterion = 'gini', min_samples_leaf = 4, in_samples_split = 10, n_jobs = 4, max_features = 50
MLP	max_iter = 300, hidden_layer_sizes = (100,50), learning_rate_init = 0.1, activation='relu', random_state = 42, solver='adam'
LDA	priors = None, covariance_estimator = None, solver = 'svd'
QDA	reg_param = 0.0, priors = None
SVM	C = 1.0, verbose = False, probability = False, random_state = 0, kernel = 'linear', max_iter = −1,shrinking = True, tol = 0.001
Decision Trees	criterion = 'gini', max_depth = 5, class_weight = None, max_features = None, max_leaf_nodes = None, min_samples_leaf = 1, min_samples_split = 10, presort = False, random_state = 0, splitter = 'best'

- The empirical findings demonstrate that CKELM outperforms CNN-ML classifiers in terms of accuracy for the UFPR05 and PUCPR subsets, as illustrated in Table 7. Specifically, CNN-SVM achieved the highest accuracy for the hazy dataset, while CNN-LR achieved the highest accuracy for the UFPR04 dataset.

Table 7. Comparison of Accuracy of CKELM vs CNN-ML classifiers

Model→ Datasets ↓	CNN	KELM	CNN-LR	CNN-RF	CNN-KNN	CNN-GNB	CNN-XgBoost	CNN-SVM	CKELM
HPD	94.65	82.56	99.08	98.93	99.33	93.6	99.06	**99.78**	99.34
UFPR04	92.51	80.23	**99.96**	97.85	87.85	82.56	95.36	94.36	97.68
UFPR05	91.32	85.32	92.56	92.65	82.56	95.36	94.36	97.68	**97.96**
PUCPR	95.26	84.63	99.53	95.65	82.56	95.36	92.36	96.68	**98.54**

- CNN-RF has the highest precision compared to CNN-ML classifiers. The reason may be that ET arbitrarily chooses a subset of features for each node during the construction of decision trees. The randomization process introduces diversity and reduces the correlation between the decision trees. Due to Randomization, RF captures specific patterns in the data, which exhibits higher specificity.
- In comparison to other ML classifiers, CKELM achieves the highest specificity and F1-Score.
- CNN-LR shows a higher recall than CKELM. The expanded ability of LR to effectively capture the relationships between features and classes makes it particularly sensitive to identifying positive instances. Although some of the classifiers show better results in terms of recall and specificity, their precision, accuracy, and F1-score are less than the proposed CKELM model.

3.4 Comparison with Similar Existing Models

Table 8 compares the performance of the proposed model with CNN, KELM, AOD-Net, and AOD-Net+modified mCNN for the hazy parking dataset. The evaluation was conducted on the Hazy Parking Dataset, and the performance metrics were compared. Table 9 compares the precision and recall values of CKELM with existing deep learning models. Table 10 demonstrates that CKELM outperforms existing models in accuracy, showcasing its superiority over conventional methods.

- The CKELM performance is compared with existing deep learning models in terms of accuracy. CKELM attained the highest accuracy values as compared with existing models.
- In Table 9, we comprehensively compare CKELM with established deep learning models. The assessment is based on critical performance metrics such as

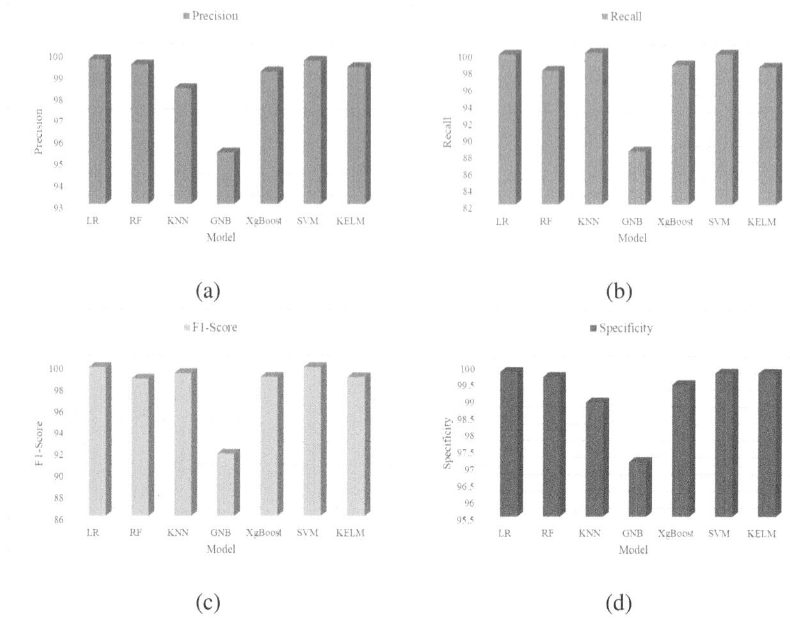

Fig. 6. Comparison of Precision, Recall, F1-Score, and Specificity for CKELM vs CNN-ML Classifiers for HPD

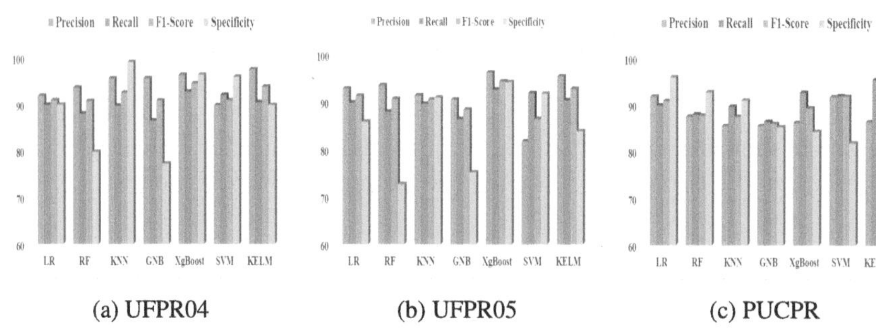

Fig. 7. Comparison of performance for CKELM vs. CNN-ML Classifiers for PKD

precision and recall. Remarkably, when considering precision and recall, CNN integrated with KELM emerges as the standout performer, boasting the highest scores for recall. Its precision and recall values signify its exceptional ability to identify positive cases precisely while effectively capturing all relevant instances. In essence, Table 9 (in the Recall column, '-' indicates 'not computed') encapsulates the superior performance of CKELM, showcasing their prowess when pitted against both established benchmarks and contemporary methods.

Table 8. Accuracy Comparison of the CKELM with existing models for HPD

Model	Accuracy (training and testing on hazy)
CNN	94.65%
KELM	82.56%
AOD-Net [26]	86.71%
AOD-Net + modified mAlexnet [26]	88.39%
CKELM (Proposed)	**97.56%**

Table 9. Precision and Recall comparison of CKELM with existing Models for PKD

Author(s)	Model	Precision	Recall
Polprasert et al. [24]	mAlexNet	79.00%	98.00%
Dhuri et al. [12]	VGG16	87.20%	-
Mettupally et al. [21]	FMR-CNN	91.90%	-
Sairam et al. [25]	Mask R-CNN	92.33%	-
Ding et al. [13]	YOLO v3	93.30%	90.90%
Agrawal et al. [3]	Mask R-CNN	95.00%	-
Kolhar et al. [19]	mAlexNet	**97.90%**	-
Proposed Model	CKELM	97.32%	**99.32%**

- CKELM is benchmarked against existing parking classification models [6, 11, 23, 27]. Table 10 demonstrates that CKELM outperforms existing models in terms of accuracy, showcasing its superiority over conventional methods. CKELM exhibits the highest accuracy across subsets such as UFPR04\UFPR04 (99.47%), UFPR04\PUCPR (97.77%), PUCPR\PUCPR (99.09%), UFPR05\UFPR05 (99.26%), and remains competitive for other subset combinations. Mean Ensemble from De et al. [11] exhibits superior performance for the subset UFPR04\UFPR05 compared to existing methods. Amato et al.'s mAlexNet [7] excels for the UFPR05\PUCPR and PUCPR\UFPR04. CKELM demonstrates remarkable performance across all subsets of PKLot compared to ResNet50 proposed by He et al. [15]. Referring to Table 10, CKELM outperforms other methods for UFPR05\UFPR05, UFPR04\UFPR04, UFPR04\PUCPR, and PUCPR\PUCPR with accuracies of 99.26%, 99.47%, 97.77%, and 99.09% respectively and is comparable for other subsets.

3.5 Runtime Analysis

During the training phase, FCL of CNN consumes ten percent of the computation time for adjusting layer parameters [14] and the training time is computed as follows, represented in the Eq. 3.

Table 10. Accuracy comparison of CKELM with existing state-of-the-art models

Author(s)	Model	Training	Testing		
			UFPR04	UFPR05	PUCPR
			Accuracy (%)		
Simonyan et al. [27]	VGG19	UFPR04	80.40	92.30	91.90
		UFPR05	88.80	95.10	95.90
		PUCPR	81.50	93.80	94.60
He et al. [15]	ResNet50	UFPR04	93.70	94.80	93.30
		UFPR05	92.20	94.80	95.50
		PUCPR	90.50	93.90	94.10
De et al. [11]	LPQ, LBP and its variants	UFPR04	84.25	99.55	84.92
		UFPR05	87.74	85.76	98.90
		PUCPR	99.58	87.15	82.78
De et al. [11]	Mean Ensemble	UFPR04	88.40	**99.64**	88.33
		UFPR05	89.83	85.53	99.30
		PUCPR	99.61	88.88	84.20
Nurullayev et al. [23]	CarNet	UFPR04	98.30	95.60	97.60
		UFPR05	**98.40**	95.20	97.50
		PUCPR	98.80	94.40	97.70
Amato et al. [7]	mAlexnet	UFPR04	98.27	99.54	93.29
		UFPR05	92.72	93.69	**99.49**
		PUCPR	**99.90**	98.03	96.00
Chollet et al. [9]	Xception	UFPR04	94.00	94.60	93.40
		UFPR05	95.70	90.90	91.20
		PUCPR	96.30	92.50	93.30
Szegedy et al. [28]	Inception V3	UFPR04	91.70	95.20	92.40
		UFPR05	94.30	92.90	93.70
		PUCPR	90.80	91.10	94.20
Proposed Model	CKELM	UFPR04	**99.47**	98.97	**97.77**
		UFPR05	94.90	**99.26**	95.90
		PUCPR	96.83	96.08	**99.09**

$$\sum_{l=1}^{k} a_{l-1}.b_l^2.a_l.c_l^2 \tag{3}$$

where, l is the index of the convets, a_l is the number of kernels at the lth layer, a_{l-1} is the number of input channels of the lth layer, k is the number of convets, b_l is size of filter, and c_l is size of the output feature vector.

Previous work has demonstrated that ELM classifiers require less training time than traditional ML classifiers because ELM is trained in a single iteration, eliminating the need for costly backpropagation [16]. Hence, the FCL of CNN is replaced with KELM, and due to this, CNN exhibits the minimum training and testing time compared with all other models. Table 11 depicts the training time comparison of CKELM with other ML classifiers combined with CNN (bold indicates minimum). As shown in Table 11, CNN obtained the highest training and testing time, which may be due to the following reasons:

- CNN has a deeper architecture of eight layers, including five convolutional and three fully connecting layers. Due to more parameters and complex calculations, each layer exhibits the maximum training time.
- Backpropagation is applied to adjust the output layers' weights in the fully connecting layer. This process entails numerous iterations, which in turn extends the duration of the training process.

Table 11. Training Time Comparison

Model	HPD	Training time (in sec)		
		PKD		
		UFPR04	UFPR05	PUCPR
CNN	4728	9785	8526	9654
KELM	1.669	1.289	1.456	1.401
CNN-LR	1.700	2.451	1.589	1.653
CNN-RF	6.945	1.356	1.326	1.421
CNN-KNN	1.989	1.547	1.897	**1.123**
CNN-GNB	1.451	1.145	1.987	1.235
CNN-XgBoost	1.897	1.564	1.789	1.954
CNN-SVM	0945	4895	9785	8754
CKELM (Proposed)	**0.131**	**1.111**	**1.045**	1.651

3.6 Discussion

CNN attained an accuracy of 94.09% after ten epochs of training. When trained for fifty or more epochs with a very high computational time, CNN achieved an even greater degree of accuracy. In the same terms, KELM attained an accuracy of 82.56% with a 'poly' kernel. Further enhancing the accuracy of KELM is achieved by training for more epochs. The proposed model has leveraged the benefits of CNN and KELM, resulting in a maximum accuracy of 99.34% for HPD and 98.54% for PKD with few training epochs. The model performed better with relatively less training time (discussed in Sect. 3.5).

- Figure 8 and 9 represents the confusion matrix for CNN, KELM, and CKELM for HPD and PKD datasets.

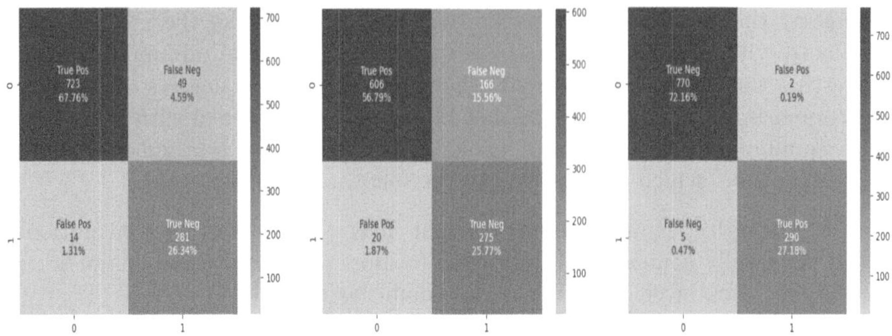

(a) (b) (c)

Fig. 8. Confusion Matrix for (a) CNN (b) KELM (c) CKELM for HPD

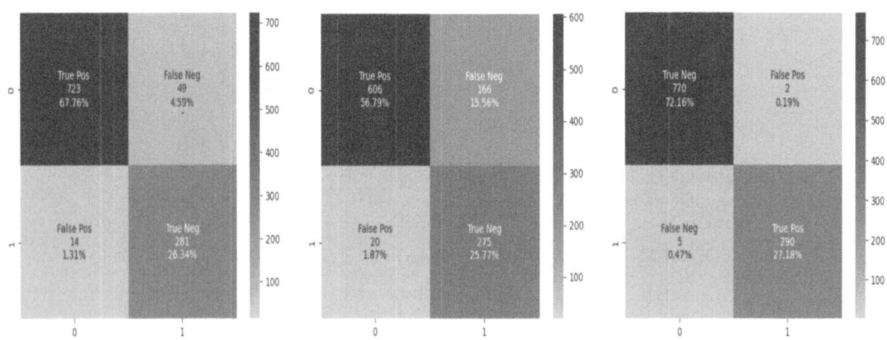

(a) (b) (c)

Fig. 9. Confusion Matrix for (a) CNN (b) KELM (c) CKELM for PKD

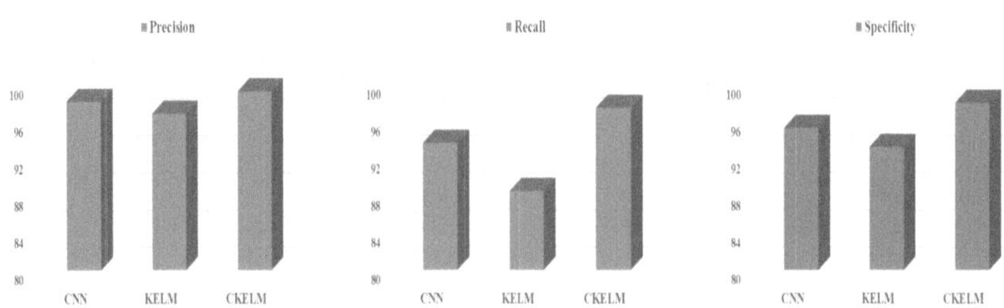

(a) (b) (c)

Fig. 10. Performance comparison for CNN, KELM, and CKELM for HPD

- Figure 10a, 10b, and 10c represents precision, recall, and specificity comparison for CNN, KELM, and CKELM for 20 epochs for HPD. Figure 11a, 11b, and 11c represents precision, recall, and specificity comparison for CNN, KELM, and CKELM for 20 epochs for PKD dataset.

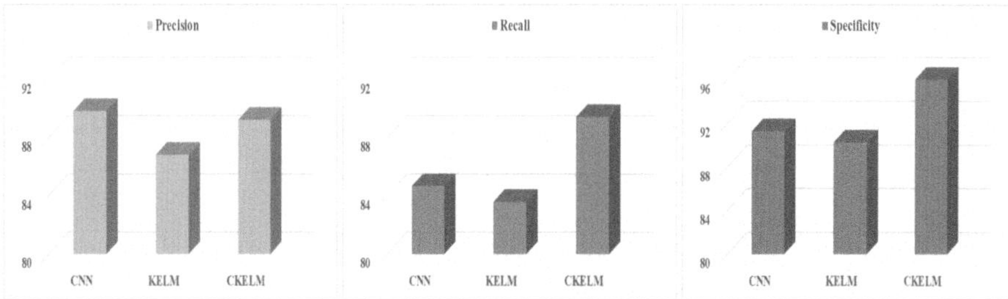

(a) (b) (c)

Fig. 11. Performance comparison for CNN, KELM, and CKELM for PKD

4 Conclusion and Future Directions

This study introduces an innovative ensemble approach called CKELM, designed to classify parking spaces under hazy and non-hazy conditions. The hazy parking classification process consists of three key stages. Firstly, images are dehazed using LD-Net, a CNN-based model. LD-Net effectively removes haze and corrects image distortions by combining convolutional and recurrent layers. It extracts essential features for estimating parameters like A and $t(x)$, reconstructing the image using a modified version of the Atmospheric Scattering Model (ASM), which describes how light interacts with atmospheric particles. In the second stage, the dehazed image is input into CNN for feature extraction. Finally, the extracted feature vector is passed to the Kernel ELM (KELM) for classification, with KELM replacing CNN's FCL. To validate the effectiveness of this approach, a Hazy Parking Dataset was curated, comprising images captured in various hazy conditions, including haze, fog, and snow. The CKELM is tested on the PKLot dataset for non-hazy weather conditions. Comparative performance analysis was made with similar existing models.

The results demonstrate that CKELM achieved an impressive accuracy rate of 97.56% for HPD and 99.34% for PKD. Additionally, the computational time required for the entire classification process either outperformed or was comparable with similar existing models. These findings underscore the potential of CKELM as a promising approach for accurately classifying parking spaces in hazy and non-hazy weather conditions. Future extensions of this work may encompass optimizing parameters essential for the classification operation within the KELM. Furthermore, utilizing CNN variants for feature extraction may reduce computational time further.

Abbreviations

Table 12 lists all the abbreviations used in the paper.

Table 12. Abbreviations used in the paper

CNN	Convolution Neural Network	ELM	Extreme Learning Machine
LD-Net	Lightweight Dehazing Network	ET	Extra Trees
MLP	Multilayer Perceptron	LDA	Linear Discriminant Analysis
QDA	Quadratic Discriminant Analysis	SVM	Support Vector Machine
DT	Decision Tree	VGGNet	Visual Geometry Group Network
ResNet	Residual Network	ASM	Atmospheric Scattering Model
ReLU	Rectified Linear Unit	tanh	hyperbolic tangent
FCL	Fully Connected Layer	CL	Convolution Layer

References

1. Huang, G.B.: What are extreme learning machines? filling the gap between Frank Rosenblatt's dream and John von Neumann's puzzle. Cogn. Comput. **7**(3), 263–278 (2015)
2. Aftab, K., et al.: Reducing parking space search time and environmental impacts: a technology driven smart parking case study. IEEE Technol. Soc. Mag. **39**(3), 62–75 (2020)
3. Agrawal, T., Urolagin, S.: Multi-angle parking detection system using mask r-CNN. In: Proceedings of the 2020 2nd International Conference on Big Data Engineering and Technology, pp. 76–80 (2020)
4. Ahmad, I., Basheri, M., Iqbal, M.J., Rahim, A.: Performance comparison of support vector machine, random forest, and extreme learning machine for intrusion detection. IEEE Access **6**, 33789–33795 (2018)
5. Almeida, P., Oliveira, L.S., Silva, E., Britto, A., Koerich, A.: Parking space detection using textural descriptors. In: 2013 IEEE International Conference on Systems, Man, and Cybernetics, pp. 3603–3608. IEEE (2013)
6. Amato, G., Carrara, F., Falchi, F., Gennaro, C., Meghini, C., Vairo, C.: Deep learning for decentralized parking lot occupancy detection. Expert Syst. Appl. **72**, 327–334 (2017)
7. Amato, G., Carrara, F., Falchi, F., Gennaro, C., Vairo, C.: Car parking occupancy detection using smart camera networks and deep learning. In: 2016 IEEE Symposium on Computers and Communication (ISCC), pp. 1212–1217. IEEE (2016)
8. Chiu, M.Y., Depommier, R., Spindler, T.: An embedded real-time vision system for 24-hour indoor/outdoor car-counting applications. In: Proceedings of the 17th International Conference on Pattern Recognition, 2004. ICPR 2004, vol. 3, pp. 338–341. IEEE (2004)
9. Chollet, F.: Xception: deep learning with depthwise separable convolutions. In: Proceedings of the IEEE Conference on Computer Vision and Pattern Recognition, pp. 1251–1258 (2017)

10. de Almeida, P.R., Oliveira, L.S., Britto, A.S., Silva, E.J., Koerich, A.L.: Pklot - a robust dataset for parking lot classification. Expert Syst. Appl. **42**(11), 4937–4949 (2015)
11. De Almeida, P.R., Oliveira, L.S., Britto, A.S., Jr., Silva, E.J., Jr., Koerich, A.L.: Pklot-a robust dataset for parking lot classification. Expert Syst. Appl. **42**(11), 4937–4949 (2015)
12. Dhuri, V., Khan, A., Kamtekar, Y., Patel, D., Jaiswal, I.: Real-time parking lot occupancy detection system with vgg16 deep neural network using decentralized processing for public, private parking facilities. In: 2021 International Conference on Advances in Electrical, Computing, Communication and Sustainable Technologies (ICAECT), pp. 1–8. IEEE (2021)
13. Ding, X., Yang, R.: Vehicle and parking space detection based on improved yolo network model. J. Phys. Conf. Ser. **1325**, 012084 (2019)
14. He, K., Sun, J.: Convolutional neural networks at constrained time cost. In: Proceedings of the IEEE Conference on Computer Vision and Pattern Recognition, pp. 5353–5360 (2015)
15. He, K., Zhang, X., Ren, S., Sun, J.: Deep residual learning for image recognition. In: Proceedings of the IEEE Conference on Computer Vision and Pattern Recognition, pp. 770–778 (2016)
16. Huang, G.B., Ding, X., Zhou, H.: Optimization method based extreme learning machine for classification. Neurocomputing **74**(1–3), 155–163 (2010)
17. Huang, G.B., Zhou, H., Ding, X., Zhang, R.: Extreme learning machine for regression and multiclass classification. IEEE Trans. Syst. Man Cybern. Part B (Cybern.) **42**(2), 513–529 (2011)
18. Huang, G.B., Zhou, H., Ding, X., Zhang, R.: Extreme learning machine for regression and multiclass classification. IEEE Trans. Syst. Man Cybern. Part B (Cybern.) **42**(2), 513–529 (2012). https://doi.org/10.1109/TSMCB.2011.2168604
19. Kolhar, M.S., Alameen, A.: Multi criteria decision making system for parking system. Comput. Syst. Sci. Eng. **36**(1), 101–116 (2021)
20. Mangiaracina, R., Tumino, A., Miragliotta, G., Salvadori, G., Perego, A.: Smart parking management in a smart city: costs and benefits. In: 2017 IEEE International Conference on Service Operations and Logistics, and Informatics (SOLI), pp. 27–32. IEEE (2017)
21. Mettupally, S.N.R., Menon, V.: A smart eco-system for parking detection using deep learning and big data analytics. In: 2019 SoutheastCon, pp. 1–4. IEEE (2019)
22. Navpreet, R.K.R., Rani, R.: Cultivating road safety: a comprehensive examination of intelligent ensemble-based road crack detection. Multimed. Tools Appl. 1–34 (2024)
23. Nurullayev, S., Lee, S.W.: Generalized parking occupancy analysis based on dilated convolutional neural network. Sensors **19**(2) (2019)
24. Polprasert, C., Sruayiam, C., Pisawongprakan, P., Teravetchakarn, S.: A camera-based smart parking system employing low-complexity deep learning for outdoor environments. In: 2019 17th International Conference on ICT and Knowledge Engineering (ICT&KE), pp. 1–5. IEEE (2019)
25. Sairam, B., Agrawal, A., Krishna, G., Sahu, S.P.: Automated vehicle parking slot detection system using deep learning. In: 2020 Fourth International Conference on Computing Methodologies and Communication (ICCMC), pp. 750–755. IEEE (2020)
26. Satyanath, G., Sahoo, J.K., Roul, R.K.: Smart parking space detection under hazy conditions using convolutional neural networks: a novel approach. Multimed. Tools Appl. **82**(10), 15415–15438 (2023)

27. Simonyan, K., Zisserman, A.: Very deep convolutional networks for large-scale image recognition. arXiv preprint arXiv:1409.1556 (2014)
28. Szegedy, C., Vanhoucke, V., Ioffe, S., Shlens, J., Wojna, Z.: Rethinking the inception architecture for computer vision. In: Proceedings of the IEEE Conference on Computer Vision and Pattern Recognition, pp. 2818–2826 (2016)

Health-Care Analytics

GCN-MLP Framework for EEG-Based Vigilance Estimation for Variable Data Length

Ravi Shekhar Tiwari, Shabnam Samima(✉), and Tauheed Ahmed

Mahindra University, Hyderabad, India
{shabnam.samima,tauheed.ahmed}@mahindrauniversity.edu.in

Abstract. Estimating vigilance has always been a non-trivial challenge in safety-critical domains such as surveillance, driving, security monitoring, and air traffic control. Traditional methods have been enhanced by machine learning and deep learning approaches to seek effective solutions. However, despite several notable approaches, the problem persists due to the binary nature of classification. To gain a comprehensive understanding of an operator's mental state, it is crucial to categorize vigilance into multiple levels for deeper insights. Another challenge is handling variable-sized data, as human performance leads to fluctuating data sizes. While many researchers address this issue by data padding or truncation, which often results in introducing biases. In this research, we address both issues by proposing a Graph Convolution Network (GCN)-based approach to estimate vigilance from varying lengths of EEG signals. The use of GCNs is motivated by their flexibility and scalability, allowing them to manage datasets of varying lengths effectively and adapt to diverse data patterns by focusing on localized node relationships. Features extracted by the GCN are classified into six vigilance levels using a Multilayer Perceptron (MLP) and a Liquid Neural Network (LNN). Our results show that the MLP outperforms the LNN, achieving an accuracy of 93.23% in training and 91.45% in testing when combined with the GCN, while the GCN paired with the LNN achieves 81.94% in training and 82.16% in testing. These findings suggest that the GCN-MLP framework is a promising tool for real-time vigilance monitoring in high-stakes environments.

Keywords: Brain Computer Interface · Vigilance · Graph Neural Network · Multi Layer Perceptron · Liquid Neural Network

1 Introduction

Safety critical events such as driving, air traffic control, automated navigation, military and border surveillance, life-guarding, cyber operations, space exploration, and much more demand a high degree of prolonged attention or vigilance. According to the reports of the Ministry of Road Transport and Highways during 2022, a total of 4,61,312 road accidents have been reported by States and

A. Dasgupta et al. (Eds.): BDA 2024, LNCS 15526, pp. 281–294, 2025.
https://doi.org/10.1007/978-3-031-81821-9_16

Union Territories, which claimed 1,68,491 lives and caused injuries to 4,43,366 persons [10,11]. The situation raises the alarm for proactively seeking mitigation strategies and highlights vigilance's importance in this crucial domain. Over the years, many techniques have evolved to measure vigilance [1,8,9,15,16]. Amongst them, remarkable performance has been obtained from physiological-based measures such as eye-blink rate, electroencephalography (EEG), and electrocardiography (ECG) [13,14]. Understanding mental state directly from brain data gives a better insight [12]. In this regard, EEG data serves as a valuable reservoir of information for suitable analysis. Additionally, EEG technology is safe, non-invasive, cost-effective, easy to use, and offers high temporal resolution.

Many researchers have explored the potential of brain dynamics to develop vigilance estimation approaches [5,12]. In these methods, the key role in distinguishing the mental states is played by the hand-crafted features. While these features are carefully chosen based on the deep-layer functionality of the brain region, they often fail to capture subtle patterns, which can lead to lower prediction accuracy. With the advancement of technology, neural networks and gradually deep neural networks have eased the tasks of feature engineering and also better prediction accuracy. For example, the authors in [7] investigated an inter-subject transfer learning approach for recognizing mental fatigue without requiring calibration. They compared single-channel and multi-channel EEG data using the random forest algorithm to identify the most distinctive channels. For classifying semantic vigilance levels, the authors in [1] have proposed two novel classification mechanisms using EEG-directed connectivity patterns, relative wavelet transform entropy and partial directed coherence quantifying information flow, and graph theory analysis for summarizing network topology. Further, the study in [19] addressed noisy EEG data using neural processes using the SEED-VIG dataset, demonstrating robustness in real-world conditions. In [3], a compact convolution neural network (CNN) for driver drowsiness detection has been proposed that utilized a global average pooling layer and class activation map methods for feature localization. A hybrid ensemble CNN framework in [6] enhanced EEG signal decoding by decomposing signals and using component-specific batch normalization to reduce subject variability. A study conducted by the authors in [17] compared EEG patterns for vigilance estimation across BCI tasks using visual evoked potential-based and rapid serial visual presentation-based systems. In another study [4], a novel CNN with separable convolutions is proposed which allows for spatial-temporal EEG signal processing and sample-wise feature analysis. A recent study conducted in [18] developed a multimodal vigilance estimating network for cursor control, integrating EEG and EOG signals with spatial-temporal convolution, attention mechanisms, and long short-term memory modules for effective vigilance state estimation. These studies collectively advance EEG-based applications by improving accuracy, robustness, and interpretability.

While most researchers have estimated vigilance by exploring the benefits of deep networks and achieving out-performance accuracy, they mainly focus on distinguishing between low and high classes. They often do not concentrate on grading the different levels of vigilance. In life-critical domains like driving, air traffic control, and power plant monitoring, evaluating a person's suitability becomes easier if their vigilance level is known. Moreover, they often fail to address the issue of variable data sizes. To address this, the present study proposes a novel technique for estimating vigilance across six levels which is independent of the data length. The publicly available driving simulator dataset [2] which consists of variable length data size has been utilized in this research for classifying the levels of vigilance. The proposed method entails transforming EEG signals into a graph network, which is subsequently analyzed using a blend of a *graph convolutional network (GCN)*, a *multi-layer perceptron (MLP)*, and a *liquid neural network (LNN)* to classify vigilance into different categories. Also, it helps to work in an environment where the data size is variable in nature. While working with human data, the dynamicity of human nature causes the data to be of variable length. The motivation for using GCN in the proposed work stems from this variable length of the dataset, as GCNs are particularly adept at handling datasets of varying lengths due to their inherent flexibility and scalability with different graph sizes and structures. By focusing on localized node neighborhoods, GCNs adapt well to diverse data patterns and capture rich contextual information through node relationships. This way, the inherent data information remains intact without introducing any bias into the data quality. After feature extraction by the GCN, the extracted features are processed separately through MLP and LNN layers. The MLP learns higher-level representations from the GCN-extracted features, while the LNN applies linear transformations to further refine them. This combination enables the model to capture both nonlinear and linear patterns in the data, improving its ability to accurately classify vigilance states. The proposed GCN model achieved a promising performance by attaining accuracy of 93.23% in training and 91.45% in testing when combined with MLP. Furthermore, when paired with LNN, the model obtained an accuracy of 81.94% in training and 82.16% in testing.

Rest of the paper is organized as follows: Details of the dataset and its preprocessing steps are discussed in Sect. 2. The details of the proposed methodology along with model architecture is given in Sect. 3. This is followed by the details of the experiments and results obtained in Sect. 4. Next, the proposed work is compared with the existing state of the art approaches in Sect. 5. Finally Sect. 6 concludes the paper.

2 Data Preparation

2.1 Dataset Description

For the proposed work, a publicly available dataset by Cao et al. [2] has been utilized. The dataset suitably adheres to the requirements of the problem at hand. The inherent data collection procedure has beautifully captured the variable data length nature of each participant across various sessions. In this dataset, the data is collected from 27 participants across multiple sessions, totaling 62 samples. For data collection, a dynamic driving simulator has been used, allowing the authors to quantitatively measure brain EEG dynamics and fluctuations in behavioral performance. During the experiment, participants drove on a four-lane highway and were instructed to keep the car cruising in the center of the lane. Lane-departure events were randomly induced, causing the car to drift from the original lane towards either the left or right lane. Participants were instructed to quickly compensate for this perturbation by steering the wheel (response onset) to return the car to the original lane (response offset). A complete trial included events with deviation onset, response onset, and response offset. The event types in the dataset are denoted by 251, 252, 253, and 254, corresponding to deviation onset (left), deviation onset (right), response onset, and response offset, respectively.

During the experiment, the stimulus computer generating the virtual reality scene recorded the car's trajectories and events with timestamps in a "log" file and sent synchronized triggers to the Neuroscan EEG acquisition system, which concurrently recorded EEG data with the trigger timestamps. In the collected EEG dataset, the first 32 signals were from the $Fp1$, $Fp2$, $F7$, $F3$, Fz, $F4$, $F8$, $FT7$, $FC3$, FCZ, $FC4$, $FT8$, $T3$, $C3$, Cz, $C4$, $T4$, $TP7$, $CP3$, CPz, $CP4$, $TP8$, $A1$, $T5$, $P3$, PZ, $P4$, $T6$, $A2$, $O1$, Oz, and $O2$ electrodes, with two reference electrodes ($A1$ and $A2$) placed on the mastoid bones. The 33rd signal described the position of the simulated vehicle.

2.2 Data Preprocessing

In the preprocessing step, raw EEG signals are prepared for further analysis by applying several techniques. Initially, noise and artifacts are removed using filtering methods such as band-pass filters to retain relevant frequency ranges. The signals are then segmented into epochs based on annotated information in the dataset, categorizing the EEG data into three types: *left turn*, *right turn*, and *no deviation*. Finally, normalization techniques are applied to standardize the signal amplitudes, ensuring consistency across different recordings.

2.3 Transformation of Tabular EEG Data Into Graphs

The considered multi-channel EEG signals are of shape $32 \times D$ shape where 32 represents the channels of the EEG signal and D represents the duration of the experiment (variable data length). As the duration of experiment depends on the promptness and time taken by an individual to complete a task so it varies throughout the experiment. To preserve the sequence of action performed during the experiment, the directed acyclic graph (DAG) has been created based on the multi-channel EEG signals where each node represents the specific point of time in the multi-Channel EEG signal and there features as the 32 EEG channel values (Refer Fig. 1). The nodes are connected with a covariance value with 32 EEG channel value as node features and the response event of the vehicle driver (refer Eq. 1) as nodes. The Algorithm 1 shows the process for transforming the multi-channel EEG tabular data into EEG graphs.

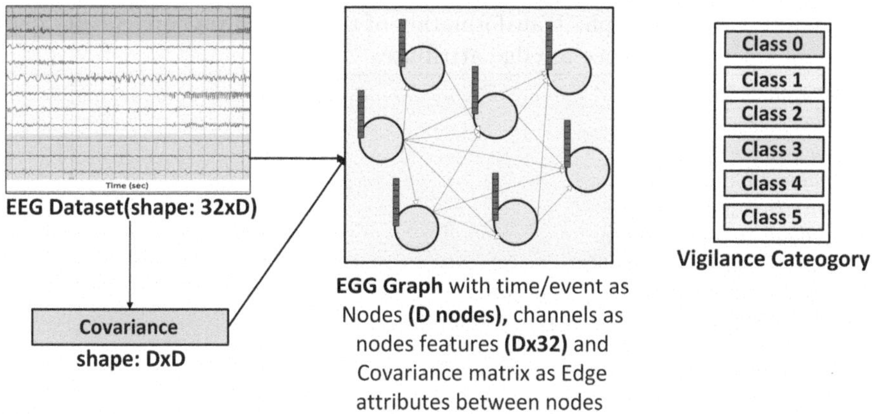

Fig. 1. Transformation of EEG tabular data into EEG graph with edges as covariance matrices and node features as 32 EEG channel values

$$\text{Response Event} = \frac{\text{Duration of the experiment(left, right, straight)}}{(2 * \text{sampling rate})} \tag{1}$$

In Algorithm 1 the graph labels are extracted as the response event 1, and an integer(Z) in the range from 0.009 to 315.455. Next, using the obtained values, the *response event* is categorized into six different classes based on the vehicle driver's vigilance (refer Eq. 2).

$$Vigilance \frac{1}{\propto} \text{response event} \tag{2}$$

The *response event* is categorized into six levels of vigilance classes from class 0 to class 5 as shown in Eq. 2. Here, class 5 is the lowest level of vigilance and class 0 is the highest level of vigilance.

$$\text{Class (C)} = \begin{pmatrix} \text{response event} < 0.5, C \leftarrow 0 \\ 0.5 \leq \text{response event} < 1.5, C \leftarrow 1 \\ 1 \leq \text{response event} < 2.5, C \leftarrow 2 \\ 2.5 \leq \text{response event} < 3.5, C \leftarrow 3 \\ 3.5 \leq \text{response event} < 4.5, C \leftarrow 4 \\ \text{Response Event} \geq 4.5, C \leftarrow 5 \end{pmatrix} \tag{3}$$

3 Methodology

The proposed methodology involves converting EEG signals into a graph network, which is then analyzed using a combination of a GCN, along with a MLP, and an LNN for six-class vigilance classification. The detailed explanation about the proposed model architecture is described next.

Algorithm 1. EEG Graph: Transformation of tabular multi-channel EEG data into graphs with covariance as edge attributes

$file_{data} \leftarrow []$
$Edge_{attr} \leftarrow []$
$Edge_{index} \leftarrow []$
$Nodes_{features} \leftarrow []$
$Sampling_{rate} \leftarrow 500$
$Path_{EEG} \leftarrow$ path of file
$Read_{Path_{EEG}} \leftarrow$ reads file as pandas dataframe
$No_{columns} \leftarrow Read_{Path_{EEG}}.columns$
$label \leftarrow Read_{Path_{EEG}}.shape[1]/(2 * Sampling_{rate})$
for i in range(0,$Read_{Path_{EEG}}$.shape[0]-1) **do**
  $x \leftarrow []$
  $row_{dataframe} \leftarrow Read_{Path_{EEG}}.iloc[l]$
  for m in range(0,$Read_{Path_{EEG}}$.shape[1]-1) **do**
    $x.append(row_{dataframe}[m])$
    $Correlation_{matrix} \leftarrow dataframe(x).corr()$
    for j1 in range($No_{columns}$) **do**
      for j2 in range(j1+1,$N0_{columns}$) **do**
        $Edge_{index}.append([j1, j2])$
        $Edge_{attr}.append(Correlation_{matrix}[j1][j2])$
      end for
    end for
  end for
  $EEG_{values} \leftarrow dataframe(x)$
  for i in $EEG_{values}.columns$ **do**
    $normalized_{features} \leftarrow \frac{(EEG_{values}[i] - EEG_{values}.mean()}{(EEG_{values}[i].std()}$
    $Nodes_{features}.append(normalized_{features}.values)$
  end for
end for

3.1 GCN Model Architecture

The motivation behind considering GCN lies in the fact that the considered dataset is of variable length and GCNs excel with variable-length datasets due to their inherent flexibility and scalability in handling graphs of different sizes and structures. They focus on localized node neighborhoods, making them adaptable to varying data patterns while capturing rich contextual information through node relationships.

The proposed model shown in Fig. 2 uses two GCN layer where each node in a EEG graph aggregates information from its neighboring EEG nodes to update it's own features. ReLu as non-linear activation function, introduces non-linearity into the model and enables the network to learn more complex patterns along with dropout of 0.5 to reduce the overfitting. The last layer in the GCN is global mean-pooling layer (GMP) to calculate the mean of the features across all nodes in the EEG graph. The GMP layer reduces the entire set of node features into a single feature vector by reducing the dimensionality of the EEG

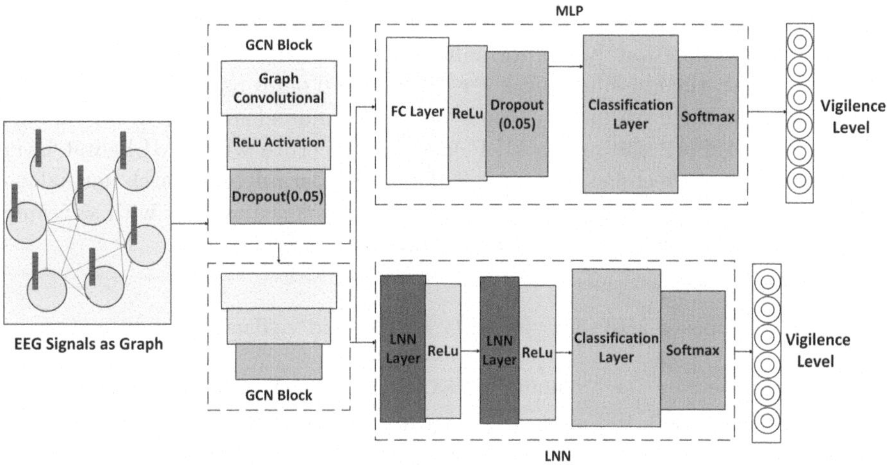

Fig. 2. Proposed model architecture to classify vigilance of four-wheeler driver based on the EEG graphs with GCN, MLP, and LNN

graphs. The below Table 1 shows the model architecture for the proposed GCN layer. The first layer of GCN is composed of 32 input nodes and 400 output nodes with ReLu non-linear activation function and dropout of 0.5 with total trainable parameter of 12, 800 whereas the second layer of GCN is stacked just after the first layer of GCN with 400 input nodes and 600 output nodes with ReLu non-linear activation function and dropout of 0.5 with a total trainable parameter of 2, 40, 600.

Table 1. Proposed GCN layer for EEG graph-based vigilance level estimation

Layer Name	Input Nodes	Output Nodes	Trainable Parameters
GCN Layer 1	32	400	12,800
GCN Layer 2	400	600	2,40,600

Following the feature extraction by the GCN, the extracted features are passed separately through MLP and LNN layers. The MLP learns higher-level representations from the GCN-extracted features, while the LNN introduces linear transformations to further refine them. This combination allows the model to capture both nonlinear and linear patterns in the data, enhancing its ability to accurately classify vigilance states using GCN with MLP and GCN with LNN. The details of the MLP and LNN are discussed in next subsections.

Multilayer Perceptron. For classifying the $1D$-vector representation of the EEG graph we have added MLP as a fully connected layer. The MLP consists of two fully connected layer where the first layer consists of 900 neurons with ReLu activation function, batch normalization function, and dropout of 0.5. The second layer i.e., the classification layer consists of 6 neurons with *softmax* activation function since we have 6 stages of vigilance as classes for multi-channel EEG graphs. Table 2 shows the MLP layer architecture where MLP first layer with ReLu activation consists of 8, 10, 900 training parameters, batch normalization layer consists of 1, 800 parameters and the classification layer with *softmax* activation consists of 5, 406 trainable parameters.

Table 2. Proposed MLP layer for EEG graph-based vigilance level estimation

Layer Name	Input Nodes	Output Nodes	Trainable Parameters
MLP Layer 1 with ReLu	900	900	8,10,900
Batch Normalization Layer	900	900	1,800
Classification Layer with *Softmax*	900	6	5406

Liquid Neural Network. In another approach, the LNN model architecture is considered for classification. The model consists of three primary linear transformation layers have been designed to process input data and produce output predictions. The input layer transforms the input features from a size of 600 to 300 units, creating a more manageable representation of the data. The intermediate reservoir layer further processes this representation, reducing it from 300 to 100 units. This layer introduces non-linearity into the network through the leaky ReLU activation function, enhancing the model's capacity to capture complex patterns. The final output layer, maps the 100 unit representation to the desired output size of 6 units as shown in Table 3. This architectural design leverages linear transformations combined with non-linear activation to effectively transform and predict output from high-dimensional input data. In total, the model contains 210,006 trainable parameters, distributed across the three different layers.

Hyperparameter. The proposed model was trained using a backpropagation algorithm on categorical cross entropy (CCE) loss with batch size of 32, for 50 epochs. Adam optimizer was used to optimize the weight of the model with learning rate of $1e-3$ and test size of 20%. The tabulated data given in Table 4 shows the hyperparameters employed to train the model.

Table 3. Proposed LNN layer for EEG graph-based vigilance level estimation

Layer Name	Input Nodes	Output Nodes	Trainable Parameters
LNN Layer 1 with ReLu	600	300	1,80,300
LNN Layer 2 with ReLu	300	100	30,100
Classification Layer with *Softmax*	100	6	606

Table 4. Hyperparameter for EEG Graph-based vigilance level estimation

Hyperparameter	Values
Train Split	80%
Test Split	20%
Batch Size	32
Optimizer	Adam
Learning Rate	1e–3
Loss Function	CCE

4 Experimental Results and Discussion

Experiments have been conducted by clubbing GCN with MLP and GCN with LNN. The models are initially trained by using 80% of the data and the remaining data is used for testing the model. The models are evaluated on different parameters such as, accuracy, precision, recall and F1-score.

The proposed two models, namely GCN with MLP and GCN with LNN, have obtained promising results as can be observed from the Fig. 3 which shows the *CCE* over 50 epochs for training and testing dataset. On analyzing the *Loss Vs Epoch* graph of GCN with MLP it can be said that the GCN with MLP train loss (0.31) and the test loss (0.49) is much lower as compared to the GCN with LNN train loss (0.16) and test loss (0.21). The graph indicates that their is minimum gap between the train and test plots of *GCN+LNN* of the data indicating that the model is properly trained over the given data.

Further, the Fig. 4 shows the accuracy over 50 epochs for training and testing dataset. From the *Accuracy Vs Epoch* graph one can analyze that the GCN with MLP train accuracy (0.93) and test accuracy (0.91) is higher as compared to the GCN with LNN train accuracy (0.82) and test accuracy (0.82). The accuracy of the proposed model improves with the increase in number of epochs. Additionally, the accuracy of the MLP is always better than LNN model.

The Fig. 5 shows the recall over 50 epochs for training and testing dataset. The *Recall Vs Epoch* graph of GCN with MLP highlights that the GCN with MLP train recall 0.91 and test recall 0.89 is high as compared to GCN with LNN train recall 0.80 and test recall 0.79.

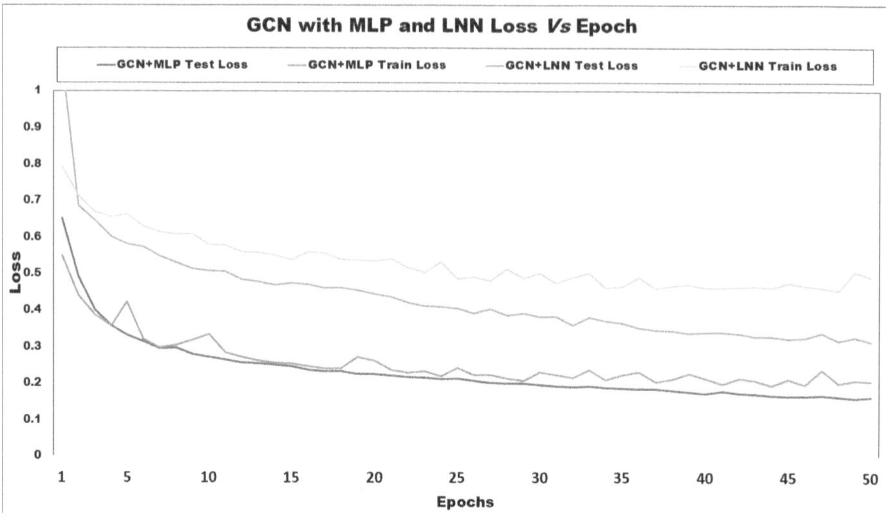

Fig. 3. Loss Vs Epoch for the proposed model - GCN with MLP and LNN on train and test set

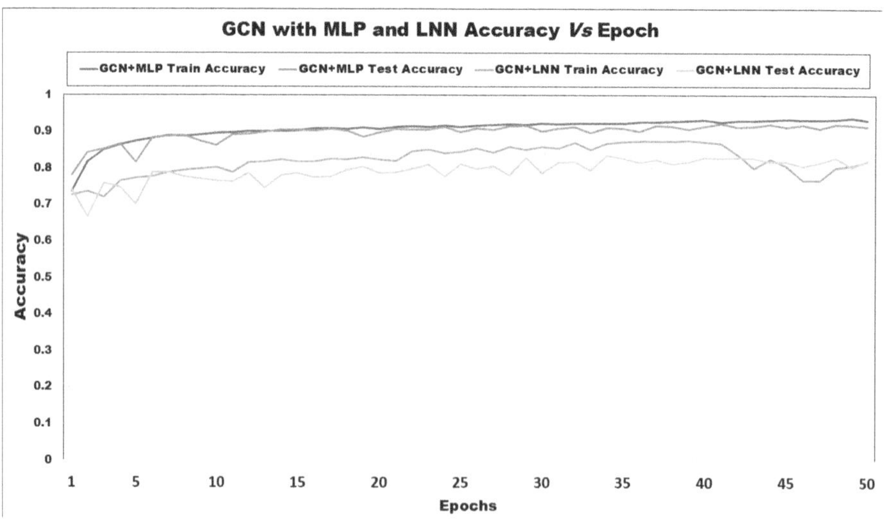

Fig. 4. Accuracy Vs Epoch for proposed model - GCN with MLP and LNN on train and test set

Additionally, the *precision* metric has been calculated to suffice the proposed model's efficacy which is depicted in Fig. 6. In the figure, the precision has been calculated over 50 epochs for the training and testing dataset. The graph clearly portrays that the GCN with MLP train precision (0.91) and test precision (0.89) is high when compared to the GCN with LNN train precision (0.80) and test

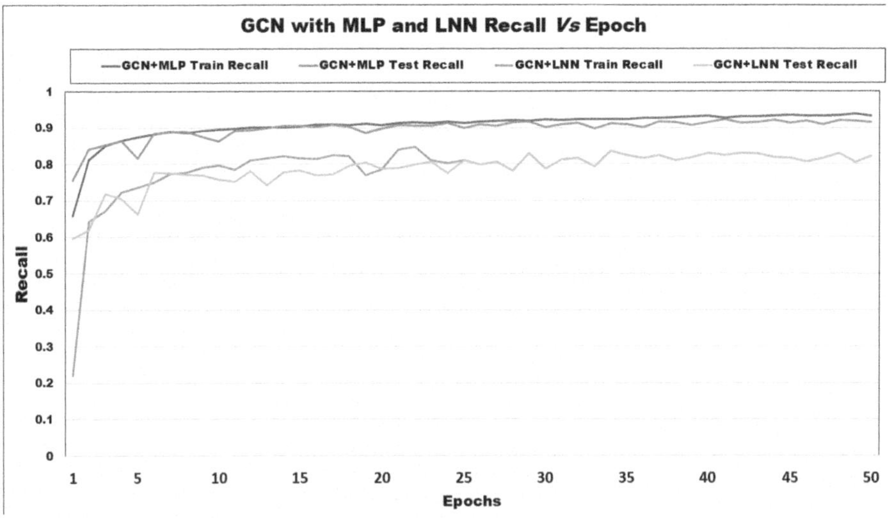

Fig. 5. Recall Vs Epoch for proposed model - GCN with MLP and LNN on train and test set

precision (0.78). From the combined observation of Fig. 5 and 6 it is indicated that the performance of both the approach improves with the increase in number of epochs and becomes almost stable by the end of epoch number '50'. However, the performance of GCN with MLP is better when compared to GCN with LNN model.

For both models, the F1-score has been calculated to show the model's prediction accuracy. Figure 7 shows the F1-score over 50 epochs for training and testing dataset. On analysis of the obtained *F1-score Vs Epoch* graph of the GCN with MLP it is observed that the GCN with MLP train F1-score (0.91) and test F1-score (0.88) is high as compared to the GCN with LNN train F1-score (0.80) and test F1-score (0.78). This indicates that the proposed GCN with MLP model obtains higher correct prediction across the entire dataset.

5 Comparison

To highlight the potential of the proposed work, the model has been compared with recent literature, as shown in Table 5. Although comparing the proposed work using the same dataset was challenging, we have attempted to compare the obtained accuracy and methodology to demonstrate its efficacy. As seen in Table 5, two experiments have been conducted on the same dataset [7,9]. In these studies, the authors performed binary classification for drowsiness and fatigue estimation, respectively. It is important to note that our proposed work focuses on multi-class classification which is independent of variability in data length, making direct comparisons with existing literature difficult. The comparison table indicates potential improvements over existing research. However, it is

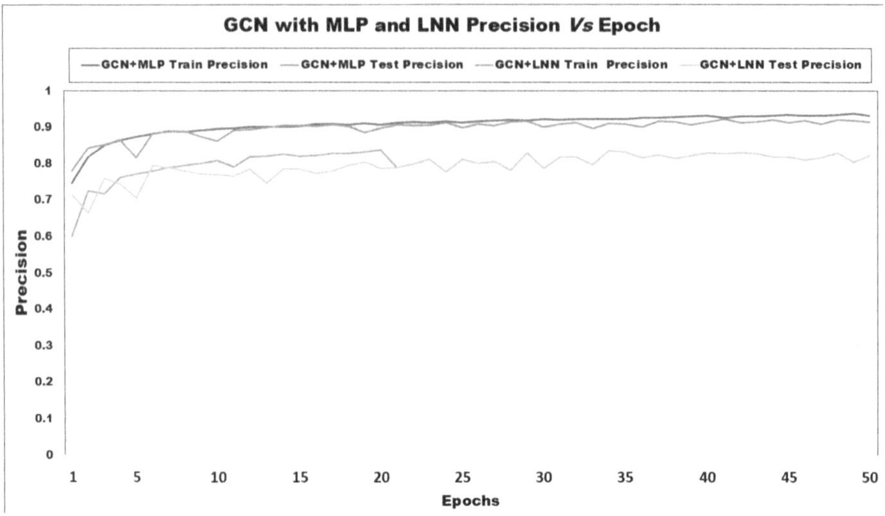

Fig. 6. Precision Vs Epoch for proposed model - GCN with MLP and LNN on train and test set

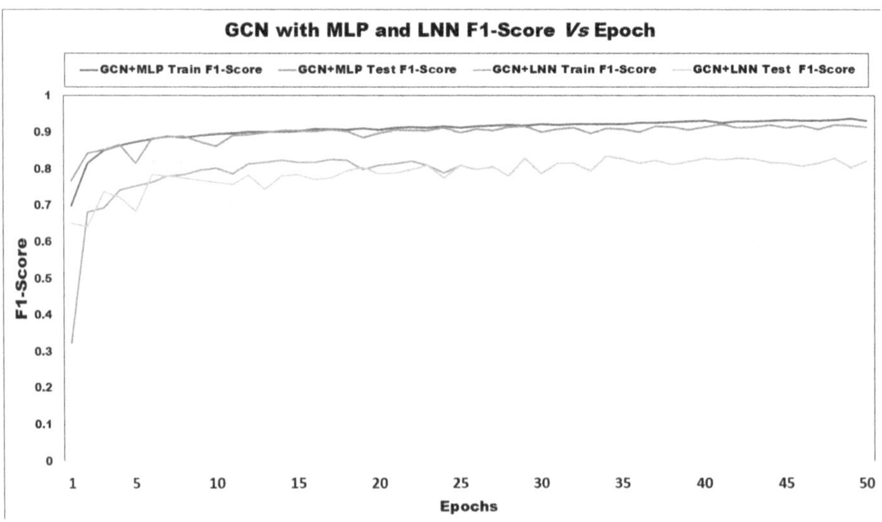

Fig. 7. F1-score Vs Epoch for proposed model - GCN with MLP and LNN on train and test set

crucial to recognize that direct comparisons may not fully reflect actual improvements due to variations in data and evaluation criteria. Therefore, while our results are promising, further experiments are necessary to conclusively demonstrate the superiority of our approach.

Table 5. Comparison with existing state-of-the-art methods

Author	Summary	Dataset	Key Method Used	Result
Lie et al. [7]	Utilizes Maximum Independence Domain Adaptation (MIDA) and Transfer Component Analysis (TCA) for calibration-free fatigue recognition using a single EEG channel from the occipital lobe. A Random Forest (RF) model is used to determine channel importance.	Cao et al. [2]	MIDA TCA Random Forest	MIDA Accuracy: 73.01% TCA Accuracy: 68.00%
Mohammed et al. [1]	Uses a Convolutional Neural Network (CNN) for calibration-free drowsiness detection, incorporating Global Average Pooling (GAP) and Class Activation Map (CAM) layers from single-channel EEG signals.	Dataset collected using 64 Ag/AgCl scalp electrodes	CNN	Average Accuracy: 73.22%
Yao et al. [19]	Proposes a novel CNN with interpretation technique for sample-wise feature analysis, using separable convolutions for spatial-temporal EEG processing.	SEED- VIG Dataset [20]	CNN	Average Accuracy: 78.35%
Jian et al. [3]	Presents a hybrid ensemble CNN framework using four decomposition methods to analyze EEG signals. Incorporates component-specific batch normalization and two ensemble modes to combine CNN outputs.	Cao et al. [2]	Discrete wavelet transform-based ensemble CNN	Average Classification Accuracy: 83.48%
Proposed Approach	Introduces a novel approach by converting EEG features into a graph-based representation for alertness classification via response time, using a Graph Neural Network (GNN).	Cao et al. [2]	GNN+MLP GNN+LNN	Accuracy of 93%

6 Conclusion

Maintaining a certain level of vigilance is a necessity for crucial time-critical work domains. The inherent variable nature of human operators mandates this evaluation process for identifying a person's suitability for work. Another major challenge in handling human datasets arises due to their variable data length. Approaches such as data truncation or data padding are commonly used to tackle such inevitable situations. However, this results in considerable information loss and leads to the possibility of data biases. To solve the aforementioned problems, the proposed work has developed a GCN-based approach to estimate and classify the vigilance state of an individual in one of the six levels that is independent of data length. By classifying the features extracted by the GCN into six vigilance levels using both MLP and LNN, the approach demonstrated that the MLP significantly outperforms the LNN model. The superior performance of the MLP highlights its potential as a reliable tool for real-time vigilance monitoring in variable data length environments. Also, it paves the way for solving more problem high are related to data length variability.

References

1. Al-Shargie, F.M., Hassanin, O., Tariq, U., Al-Nashash, H.: EEG-based semantic vigilance level classification using directed connectivity patterns and graph theory analysis. IEEE Access **8**, 115941–115956 (2020)
2. Cao, Z., Chuang, C.H., King, J.K., Lin, C.T.: Multi-channel EEG recordings during a sustained-attention driving task. Sci. Data **6**(1), 19 (2019)
3. Cui, J., Lan, Z., Liu, Y., Li, R., Li, F., Sourina, O., Müller-Wittig, W.: A compact and interpretable convolutional neural network for cross-subject driver drowsiness detection from single-channel EEG. Methods **202**, 173–184 (2022)

4. Cui, J., Lan, Z., Sourina, O., Müller-Wittig, W.: EEG-based cross-subject driver drowsiness recognition with an interpretable convolutional neural network. IEEE Trans. Neural Netw. Learn. Syst. **34**(10), 7921–7933 (2022)
5. Li, G., et al.: Driver vigilance detection based on limited EEG signals. IEEE Sens. J. **23**(12), 13387–13398 (2023)
6. Li, R., Gao, R., Suganthan, P.N.: A decomposition-based hybrid ensemble CNN framework for driver fatigue recognition. Inf. Sci. **624**, 833–848 (2023)
7. Liu, Y., Lan, Z., Cui, J., Sourina, O., Müller-Wittig, W.: Inter-subject transfer learning for EEG-based mental fatigue recognition. Adv. Eng. Inform. **46**, 101157 (2020)
8. Mathis, J., Andres, D., Schmitt, W.J., Bassetti, C.L., Hess, C.W., Schreier, D.R.: The diagnostic value of sleep and vigilance tests in central disorders of hypersomnolence. Sleep **45**(3), zsac017 (2022)
9. McWilliams, T., Ward, N.: Underload on the road: measuring vigilance decrements during partially automated driving. Front. Psychol. **12**, 631364 (2021)
10. Ministry of Road Transport and Highways: Road Accidents in India in 2022 (2023). https://morth.nic.in/sites/default/files/RA_2022_30_Oct.pdf. Accessed 31 July 2024
11. Ministry of Road Transport and Highways: Road Accidents in India (2024). https://morth.nic.in/road-accident-in-india. Accessed 31 July 2024
12. Pan, J., Cai, X., Mo, D., Yu, Y., Li, Y.: Residual attention capsule network for multimodal EEG-and EOG-based driver vigilance estimation. IEEE Trans. Instrum. Meas. (2023)
13. Saleem, A.A., Siddiqui, H.U.R., Raza, M.A., Rustam, F., Dudley, S., Ashraf, I.: A systematic review of physiological signals based driver drowsiness detection systems. Cogn. Neurodyn. **17**(5), 1229–1259 (2023)
14. Samima, S., Sarma, M., Samanta, D.: Correlation of P300 ERPs with visual stimuli and its application to vigilance detection. In: 2017 39th Annual International Conference of the IEEE Engineering in Medicine and Biology Society (EMBC), pp. 2590–2593. IEEE (2017)
15. Samima, S., Sarma, M., Samanta, D.: Detecting vigilance in people performing continual monitoring task. In: Horain, P., Achard, C., Mallem, M. (eds.) IHCI 2017. LNCS, vol. 10688, pp. 202–214. Springer, Cham (2017). https://doi.org/10.1007/978-3-319-72038-8_16
16. Samima, S., Sarma, M., Samanta, D., Prasad, G.: Estimation and quantification of vigilance using ERPs and eye blink rate with a fuzzy model-based approach. Cogn. Technol. Work **21**, 517–533 (2019)
17. Wang, K., et al.: Investigating EEG-based cross-session and cross-task vigilance estimation in BCI systems. J. Neural Eng. **20**(5), 056001 (2023)
18. Wang, K., et al.: A multimodal approach to estimating vigilance in SSVEP-based BCI. Expert Syst. Appl. **225**, 120177 (2023)
19. Yao, C.L., Lu, B.L.: A robust approach to estimating vigilance from EEG with neural processes. In: 2020 IEEE International Conference on Bioinformatics and Biomedicine (BIBM), pp. 1202–1205. IEEE (2020)
20. Zheng, W.L., Lu, B.L.: A multimodal approach to estimating vigilance using EEG and forehead EOG. J. Neural Eng. **14**(2), 026017 (2017)

Deep Learning Framework for Early Diagnosis of COPD and Respiratory Diseases Using Lung Sound Analysis

Prakash Sahu[1], Purnanand Prasad[2], Vishnu Prasad Verma[1](✉), and Santosh Kumar[1]

[1] Department of CSE, IIIT Naya Raipur, Raipur, CG, India
{prakash,vishnu,santosh}@iiitnr.edu.in
[2] Department of CSE, Silicon University, Bhubaneswar, Odisha, India

Abstract. In an era of advanced technology development, the early diagnosis of multivariate respiratory diseases is highly popular due to the unavailability of efficient solutions in the biomedical field. To solve lung respiratory diseases, an innovative structure is put forth to enable precise prediction and early diagnosis of multivariate respiratory Chronic Obstructive Pulmonary Diseases (COPD) using deep learning techniques. The proposed structure performs lung sound data preprocessing to extract the features using noise filtering techniques. After preprocessing, the framework extracts the characteristics that discriminate from the lung sound to train the model based on extracted features. The proposed framework extracts various features such as constant-Q transform, short-time Fourier transform, and Mel frequency central coefficients, the model is constructed for accurate prediction of respiratory diseases using a convolutional neural network (CNN). The effectiveness of the suggested fra has been evaluated on the open lung sound database in different benchmark settings to measure the different statistical correlations with existing methods for a prompt diagnosis and precise forecasting of COPD diseases achieving an accuracy of 95.48% with a sensitivity of 89.0% and specificity of 98.0% in predicting conditions like COPD, healthy, URTI, Bronchiolitis and pneumonia.

Keywords: Chronic Obstructive Pulmonary Diseases · Deep Learning · CNN · Lung sound · CQT

1 Introduction

Chronic obstructive pulmonary disease (COPD) is causing health-related problems globally, and now it is considered a major disease. Over time, it makes it harder for the patient to breathe. It is estimated that the disease caused over 3 million deaths worldwide in 2016 (or 5% of all fatalities worldwide in that year). Over 90% of COPD-related fatalities take place in low- and middle-income nations [1].

A. Dasgupta et al. (Eds.): BDA 2024, LNCS 15526, pp. 295–304, 2025.
https://doi.org/10.1007/978-3-031-81821-9_17

Asthma and COPD are both lung diseases but differ significantly. Asthma is triggered by allergens, irritants, or exercise, causing wheezing, chest tightness, and shortness of breath that can improve, especially in children. Treatment focuses on opening airways and reducing inflammation. COPD, mainly caused by smoking, leads to persistent shortness of breath, a mucus-producing cough, and fatigue. It develops in adults after exposure to irritants and is progressive without a cure. Treatments aim to manage symptoms and slow disease progression using bronchodilators. Emphysema and Chronic Bronchitis are two forms of COPD. In chronic bronchitis, the bronchial tubes that carry air to and from the lungs become inflamed, causing mucus buildup that narrows the airways and makes breathing difficult. Normally, tiny hair-like structures called cilia clear mucus from the airways, but in chronic bronchitis, damaged cilia can no longer perform this function effectively.

Emphysema occurs when the walls of the tiny air sacs (alveoli) in the lungs are damaged, making it difficult to breathe deeply. The alveoli, essential for oxygen intake and carbon dioxide removal, collapse, hindering proper gas exchange. This condition typically affects the lower part of the lungs, resembling the damage to leaves in an inverted tree structure of the respiratory system.

1.1 Motivation

COPD, a leading cause of illness and death worldwide, poses a significant burden on healthcare systems. Characterized by chronic respiratory symptoms and airflow restriction, early diagnosis and accurate prognosis are crucial for effective treatment and improved patient outcomes. However, current diagnostic methods are often invasive, expensive, and inaccessible, particularly in resource-limited settings.

Advances in data science and machine learning offer promising solutions for affordable, non-invasive COPD diagnostics. However, challenges remain in improving the reliability and accuracy of COPD diagnosis using multivariate data. Given COPD's complexity, a multifaceted approach is needed to analyze diverse respiratory factors. This research seeks to develop a novel machine learning framework for accurate prediction and early diagnosis of COPD, leveraging multivariate respiratory data to enhance diagnostic precision, particularly in resource-limited healthcare settings.

1.2 Contributions

The following are the research's main contributions.

1. Creation of a Multivariate Diagnostic: We represent a deep learning-based diagnostic model that improves the accuracy of COPD diagnosis by combining various respiratory features, such as lung function tests, respiratory sounds, and patient demographics. Our model makes use of multivariate data to capture the complex relationships underlying COPD, offering a more thorough diagnostic tool than traditional methods.

2. Novel Approaches to Feature Extraction and Selection: Our methodology combines statistical and machine learning-based feature selection techniques with refined feature methods of extraction such as Mel-frequency cepstral coefficients (MFCCs), Mel-spectogram, Constant Q-transform(CQT) for respiratory sounds. These methods greatly enhance the model's capacity to recognize pertinent patterns and COPD-related biomarkers.
3. Precise Forecasting of Illness Development: Our algorithm not only diagnoses but also predicts the course of the disease, allowing for early intervention and customized treatment regimens. Using longitudinal patient data, we illustrate the predictive power of the model and its ability to track the course of the disease and predict exacerbation.

2 Related Works

Numerous studies for the prediction of COPD are done using image data sets. Some of these are discussed below.

The investigation was conducted to formulate algorithms that focus on analyzing cough sounds to detect COPD, using advanced deep learning methodologies such as Convolutional Neural Networks (CNN) and Convolutional Recurrent Neural Networks (CRNN). The algorithms devised in the study leveraged both augmented and non-augmented datasets, in addition to employing various techniques for feature extraction such as Harmonic change detection function, zero crossing rate, and Mel frequency cepstral coefficient, resulting in impressive accuracy rates of 96.6% and 96.73 % for CNN and CRNN, respectively. The research findings demonstrated that the newly proposed algorithms outperformed the classification performance levels documented in previous studies, indicating the potential for the development of automated diagnostic tools that reduce the dependence on healthcare professionals [1].

CNN and LSTM architectures have been employed to predict COPD through a combination of supervised and unsupervised learning. While previous studies have focused on image datasets like CT scans, this investigation emphasizes predicting COPD using respiratory sounds. The dataset includes sounds from COPD patients, other lung conditions, and healthy individuals. Feature extraction on these audio recordings helps identify key attributes for classification. The deep learning model achieved high accuracy, offering a cost-effective solution for early COPD detection and remote monitoring [2].

The article focuses on predicting COPD severity using machine learning, with random forest showing the best performance, achieving an AUC of 0.886. The study analyzed 368 samples from mild and severe COPD categories, identifying key factors like age, the modified Medical Research Council score, and lung CO diffusion ability. The research highlights the potential of machine learning models to accurately predict COPD severity [3].

The paper reviews current medical and technological literature on respiratory sound analysis, drawing from PubMed and the researchers' professional experience. It explores various methods for collecting auscultation sounds and automated detection systems, focusing on fuzzy systems, neural networks, AI, and genetic algorithms. The study emphasizes the need for new markers to improve decision-support tools in respiratory sound analysis, highlighting AI's growing role in enhancing diagnostic accuracy [4].

3 Methods and Materials

This section outlines the steps in the proposed framework for early diagnosis, as illustrated in Fig. 1. The diagnostic model involves several phases: first, data augmentation is performed based on the time domain, followed by the arrangement of the dataset. Next, time domain features of the audio signals are extracted, and finally, a deep neural network architecture is trained and tested using the generated spectrogram data.

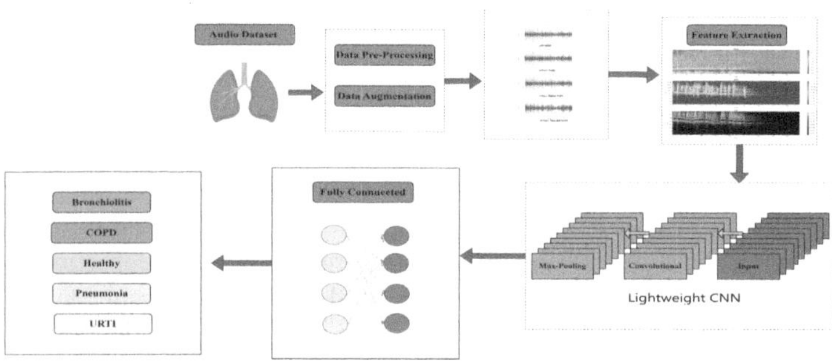

Fig. 1. Working of the proposed framework

3.1 Dataset Description

Respiratory sounds provide key insights into respiratory health, reflecting air movement, lung tissue changes, and the presence of secretions. For instance, wheezing often indicates conditions like asthma or chronic obstructive pulmonary disease (COPD). Advances in digital stethoscope technology have enabled the recording of these sounds, allowing for the development of machine learning algorithms to automatically diagnose respiratory disorders such as asthma, pneumonia, and bronchiolitis. The Respiratory Sound Database, compiled by research teams from Portugal and Greece, contains 920 annotated recordings from 126 patients, ranging from 10 to 90 s in length. These 5.5 h of recordings include 6898 respiratory cycles, with 1864 containing crackles, 886 featuring wheezes, and 506 exhibiting both. The data encompasses both clean and noisy recordings, simulating real-world conditions across all age groups, from children to the elderly [5].

3.2 Data Pre-processing

Data preprocessing is an essential stage within the machine learning process, in which raw data are converted into a structure conducive to analysis and model development. The effectiveness of a machine learning model can be greatly improved through meticulous data preprocessing. This initial phase is critical in guaranteeing the integrity and appropriateness of data for various machine-learning applications.

Data Augmentation. Data preprocessing is crucial in machine learning, transforming raw data into a suitable format for analysis and model building. Effective preprocessing enhances model performance. Figures 4 and 2 show waveforms and spectrograms of diseases after data augmentation, while Fig. 3 displays disease counts. Table 1 lists the symbols and formulas used.

Table 1. Symbols and their meanings

Symbol	Meaning
$x(t)$	Original Signal
$y(t)$	Noisy Signal
$e(t)$	Gaussian Noise
$y(s)$	Time-Shifted Signal
τ	Amount of time by which the signal is shifted
$y(u)$	Time-Stretched Signal
α	Stretching Factor
$y(v)$	Pitch-Shifted Signal
β	Pitch-Shifting Factor

1. Add Noise: Adding noise to an audio signal improves our proposed model robustness by helping it differentiate key features from background noise, enhancing its performance in varied environments. The equation for adding noise to a signal is represented as:

$$y(t) = x(t) + \epsilon(t) \tag{1}$$

2. Time Shift: Time shifting alters the temporal placement of an audio signal without changing its content, enhancing the model's ability to recognize patterns by introducing different time offsets in the signal. The equation can be represented as:

$$y(s) = x(t - \tau) \tag{2}$$

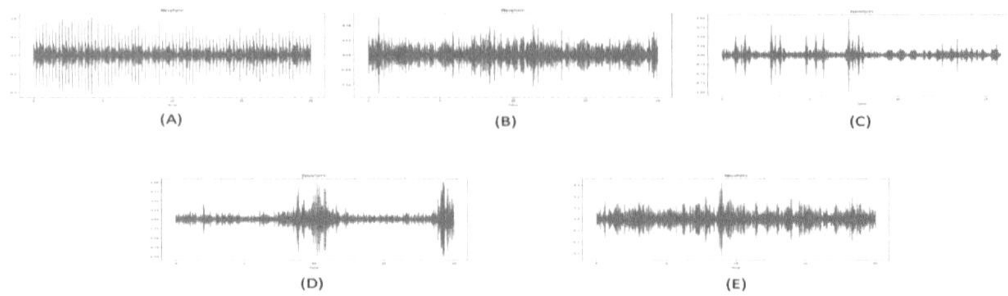

Fig. 2. Waveform of different diseases A: URTI B: Healthy C: COPD D: Bronchiolitis E: Pneumonia

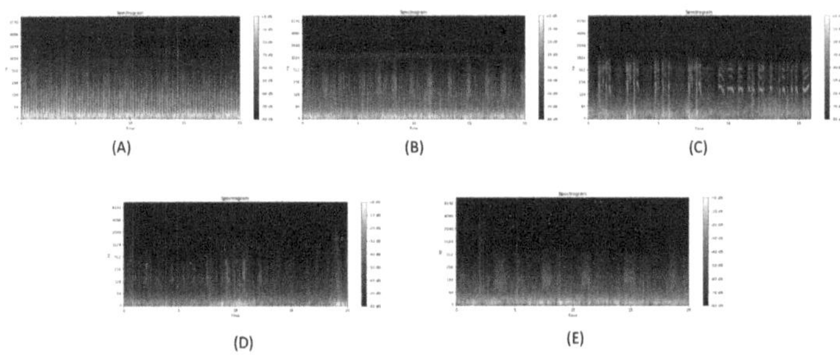

Fig. 3. Spectogram of different diseases (A) URTI (B) Healthy (C) COPD (D) Bronchiolitis (E) Pneumonia

3. Time Stretching: Time stretching alters the audio duration without affecting the pitch, making it sound faster or slower. It helps models adapt to varying speech speeds. The equation can be represented as:

$$y(u) = x(\alpha t) \tag{3}$$

4. Pitch Shifting: Pitch shifting alters the frequency of an audio signal while keeping the duration constant, making it sound higher or lower. The equation for pitch shifting a signal can be represented as:

$$y(v) = x\left(\frac{t}{\beta}\right) \tag{4}$$

3.3 Feature Extraction

MFCC: MFCC (Mel-frequency cepstral coefficients) is a vital feature extraction technique in audio processing, used for tasks like speech recognition. It represents the power spectrum of an audio signal on the mel-scale, which models human sound perception. MFCCs are computed by converting the signal to the frequency domain using the Discrete Fourier Transform and calculating cepstral coefficients from the mel-scaled spectrum. The above table represents the respective symbols and their meanings Table 2.

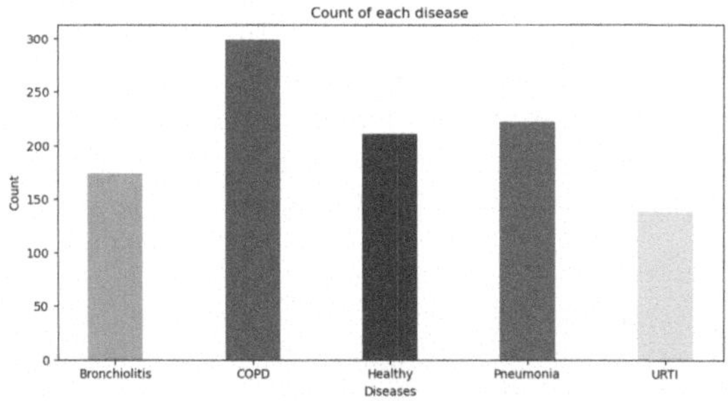

Fig. 4. Count of each disease

Table 2. Symbols and their meanings

Symbol	Meaning
$x(n)$	Input Signal
$y(n)$	Output Signal
α	Pre-emphasis Coefficient typically around 0.97
$w(n)$	Window Function
m	Time Index
$y(u)$	Time-Stretched Signal
ω	Frequency
$X(m, w)$	STFT of Signal
β	Pitch-Shifting Factor
$w[n - m_k]$	window function centered at m_k
k	Frequency Bin
N_k	Window Length for the k-th Frequency Bin
$X(k)$	CQT of the signal at Frequency Bin k

The pre-emphasis step:

$$y[n] = x[n] - \alpha x[n-1] \tag{5}$$

STFT: STFT (Short-Time Fourier Transform) analyzes signals in small segments by applying the Fourier transform to each, revealing time-varying spectral properties. It is often visualized through spectrograms and is widely used in spectrum analysis. A signal's Short-Time Fourier Transform (STFT) $x(t)$ is given by:

$$\text{STFT}\{x(t)\}(m, \omega) = X(m, \omega) = \sum_{n=-\infty}^{\infty} x[n] \cdot w[n-m] \cdot e^{-j\omega n} \tag{6}$$

CQT: The Constant-Q Transform (CQT) is a time-frequency analysis tool with logarithmic frequency resolution, ideal for musical and audio signals. Unlike STFT, it uses varying window sizes to maintain consistent Q resolution, offering a more natural representation for signals requiring logarithmic scaling.

The Constant-Q Transform (CQT) of a signal $x(t)$ is given by:

$$\mathrm{CQT}\{x(t)\}(k) = X(k) = \sum_{n=0}^{N-1} x[n] \cdot w[n - m_k] \cdot e^{-j2\pi \frac{nk}{N_k}} \tag{7}$$

3.4 Lightweight CNN Architecture

A lighweigh Convolutional Neural Network (CNN) architecture designed for lung disease classification typically consists of several layers, starting with an input layer that takes images, such as a $64 \times 64 \times 3$ RGB image of lung sound features. The first convolutional layer uses 32 filters of size 3×3, resulting in an output size of $64 \times 64 \times 32$ and approximately 896 parameters. This is followed by a max pooling layer that reduces the size to $32 \times 32 \times 32$, introducing no additional parameters. The second convolutional layer employs 64 filters of size 3×3, producing an output size of $32 \times 32 \times 64$ and around 18,496 parameters. Another max pooling layer further reduces the size to $16 \times 16 \times 64$. The architecture then includes a fully connected layer with 128 units, which contains about 2,097,280 parameters. Finally, the output layer, representing the number of classes associated with lung diseases (healthy, URTI, Bronchiolitis, COPD, pneumonia), has around 1,290 parameters. In total, this basic CNN architecture contains approximately 2.12 million parameters, making it suitable for tasks like lung disease classification while balancing depth and computational efficiency.

4 Result and Discussion

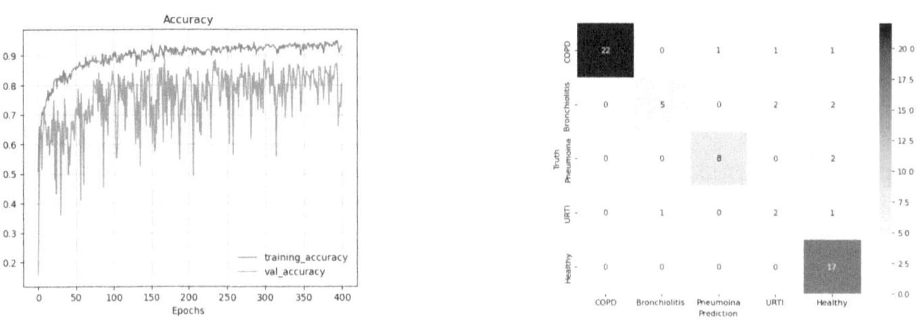

(a) Accuracy (b) Confussion Matrix

Fig. 5. Accuracy graph and confusion matrix of proposed method fig (a) Accuracy fig (b) confusion metrics

The algorithms that have been developed were obtained through the utilization of a dataset that encompasses lung sound signals, comprised of an audio sample capturing lung sounds. After the extraction of data from the database, it is planned to employ several data augmentation techniques to artificially enhance the diversity of the training dataset. This will be achieved by implementing a variety of transformations. Following the data augmentation process, the next step involves the undertaking of feature extraction. This phase will involve the utilization of three distinct methods for feature extraction to extract the features from the dataset. Once the features have been extracted from the dataset, the subsequent step will entail the initiation of the implementation of the CNN model. Upon the completion of implementing the CNN model, an accuracy of 95.48% was achieved, with a corresponding loss of 4.58%. Furthermore, the sensitivity and specificity of the model were determined to be 89.0% and 98.0% respectively. The model demonstrates proficiency in predicting conditions such as COPD, healthy status, URTI, Bronchiolitis, and pneumonia (Fig. 5).

The following table displays a selection of outcomes from earlier research endeavors, showcasing findings related to features, classifiers, accuracy rates, sensitivity, and specificity measurements (Table 3).

Table 3. Comparative study of the proposed method with existing state-of-the-art

Ref.	Features	Classifier	Accuracy	Sensitivity	Specificity
[6]	A	DBN	93.67	91.0	96.33
[7]	B	SVM	87.8	89.65	83.33
[8]	C	PCA-CA	88.0	—	—
[9]	D	FOT and ANN	90.00	91.14	98.05
Proposed Work	E	**Lightweight CNN**	95.48	89.0	98.0

Abbreviations: A: Standard deviation, Variance, and Kurtosis
B: Variance, number of peaks
C: Physiological Data cross-spectral analysis
D: forced oscillation
E: MFCC, STFT, CQT

5 Conclusion

In conclusion, the use of deep learning models in the realm of medical research has showcased a promising capability in accurately detecting COPD through the analysis of breathing sounds. The employment of the CNN model alongside the spectrogram analysis has facilitated the identification of complex patterns and unique features within the breathing sounds, thereby enabling the differentiation between individuals sickened with COPD and those who are not. To ensure the reliability and precision of these previously mentioned models, it is essential to conduct further in-depth investigations and validation studies. This approach is crucial in guaranteeing a careful evaluation of COPD patients. Furthermore, conventional diagnostic techniques should be added to the integration of deep learning models as a diagnostic tool to ensure a comprehensive assessment. Ultimately, the development and application of these advanced learning models in the realm of COPD diagnosis and treatment have the potential to remarkably enhance patient outcomes and elevate the overall efficiency of healthcare practices.

References

1. Melese, E.A., Nabaasa, E., Wondemagegn, M.T., Yonasi, S., Negasa, G.M.: Deep learning based algorithms for detecting chronic obstructive pulmonary disease. In: 2022 IST-Africa Conference (IST-Africa), pp. 1–12. IEEE (2022)
2. Khandhan, C.R., Kaviya, G.: Deep learning model for chronic obstructive pulmonary disease through breathing sound. In: 2023 4th International Conference on Electronics and Sustainable Communication Systems (ICESC), pp. 1514–1521. IEEE (2023)
3. Choi, E.-A., et al.: Prediction of COPD severity based on clinical data using machine learning. In: 2021 IEEE International Conference on Bioinformatics and Biomedicine (BIBM), pp. 1646–1648. IEEE (2021)
4. Reichert, S., Gass, R., Brandt, C., Andrès, E.: Analysis of respiratory sounds: state of the art. Clin. Med. Circulatory Respir. Pulm. Med. **2**, CCRPM-S530 (2008)
5. Rocha, B., et al.: A respiratory sound database for the development of automated classification. In: Precision Medicine Powered by pHealth and Connected Health: ICBHI 2017, Thessaloniki, Greece, 18–21 November 2017, pp. 33–37. Springer (2018)
6. Altan, G., Kutlu, Y., Allahverdi, N.: Deep learning on computerized analysis of chronic obstructive pulmonary disease. IEEE J. Biomed. Health Inform. **24**(5), 1344–1350 (2019)
7. Kanwade, A., Bairagi, V.K.: Classification of COPD and normal lung airways using feature extraction of electromyographic signals. J. King Saud Univ.-Comput. Inf. Sci. **31**(4), 506–513 (2019)
8. Newandee, D., Reisman, S., Bartels, A., De Meersman, R.: COPD severity classification using principal component and cluster analysis on HRV parameters. In: 2003 IEEE 29th Annual Proceedings of Bioengineering Conference, pp. 134–135. IEEE (2003)
9. Amaral, J.L., Faria, A.C., Lopes, A.J., Jansen, J.M., Melo, P.L.: Automatic identification of chronic obstructive pulmonary disease based on forced oscillation measurements and artificial neural networks. In: 2010 Annual International Conference of the IEEE Engineering in Medicine and Biology, pp. 1394–1397. IEEE (2010)

Evaluating the Cultural Sensitivity of Large Language Models in Mental Health Support: A Framework Inspired by Ubuntu Values

Sontaga G. Forane[1], Absalom E. Ezugwu[1](✉), and Kevin Igwe[2]

[1] Unit for Data Science and Computing, North-West University, 11 Hoffman Street, Potchefstroom 2520, North-West, South Africa
Absalom.ezugwu@nwu.ac.za

[2] Department of Psychology, Faculty of Humanities, University of Johannesburg, Bunting Road, Auckland Park, Johannesburg 2092, South Africa

Abstract. Africa's growing mental health crisis underscores a significant lack of accessible and culturally relevant mental health services, with conditions such as depression and anxiety increasingly prevalent. While Cognitive Behavioral Therapy (CBT) has proven effective, it often lacks cultural resonance in African contexts, highlighting the need for innovative approaches. This study proposes a framework for enhancing AI-driven mental health applications by integrating CBT principles with Ubuntu and fine-tuning large language models (LLMs) to embed cultural context. The goal is to improve user engagement and the effectiveness of mental health interventions in South Africa. This framework emphasizes diversity, equity, and inclusion (DEI), with a particular focus on cultural adaptation. Through theoretical exploration, the study demonstrates the potential of Africa-centric LLM applications to offer culturally sensitive support, helping to bridge the mental health service gap across the continent.

Keywords: Cultural context · mental health · emotional intelligence · foundation models · large language models · ubuntu · cognitive behavioral therapy

1 Introduction

Africa faces an intensifying mental health crisis, with an 85% access gap [1]. This crisis is compounded by the high prevalence of depression and anxiety, two of the most pressing mental health issues in the continent. Traditional therapies, such as CBT, often lack cultural relevance in African contexts. Although training health workers in African perspectives is possible, a shortage of mental healthcare workers persists across Africa [2, 3]. This exacerbates the problem, requiring innovative solutions that are not only culturally adaptive but also effective and scalable.

Recent advancements in digital mental health applications that leverage large language models (LLMs) such as Gemini and GPT-4, have sparked considerable interest in various applications [4]. The LLMs have the capabilities to produce human-like text at a large scale, perform analysis, and interpret results, to mention a few. Through fine-tuning

A. Dasgupta et al. (Eds.): BDA 2024, LNCS 15526, pp. 305–314, 2025.
https://doi.org/10.1007/978-3-031-81821-9_18

and prompt engineering, they can be adapted for specific domains including medicine, finance, and education [4]. However, despite their capabilities, the potential of LLMs that integrate African perspectives and cultural contexts to enhance mental healthcare applications in Africa remains largely untapped [5].

This position paper reviews the literature on methods for incorporating cultural context into therapeutic practices and proposes a novel framework that leverages LLM to integrate African context into digital mental health applications. By embedding Africa-centric values, such as the Ubuntu philosophy, and incorporating CBT principles, this framework provides a culturally adaptive, effective, and scalable solution to address Africa's mental health crisis.

The framework is grounded in diversity, equity, and inclusion (DEI) principles, with a particular focus on inclusion, achieved through cultural adaptation. It operationalizes Ubuntu, a philosophy rooted in community and interconnectedness, as the guiding cultural context, while also integrating emotional intelligence for empathetic, culturally sensitive interactions. Together, these elements promote a sense of belonging and support for African users, particularly as smartphone usage rises in rural areas, making mental health support more accessible and relevant [6–8].

2 Literature Review

In this section, we provide a concise yet critical review of existing studies on the adaption and integration of personalized content in mental health therapy. The discussion also extends to considerations of diversity, equity, and inclusion within mental health contexts. Additionally, we explore the role of CBT in mobile applications, the challenges posed by the dominance of Westernized therapeutic approaches in Africa, and the integration of emotional intelligence into LLMs.

2.1 Incorporating Personalized Content in Mental Health Therapy

Personalizing mental health therapy to individuals' unique needs, preferences, and cultural backgrounds has been shown to enhance engagement and improve outcomes. This approach is essential in overcoming barriers to access and providing meaningful support, especially for those in culturally diverse regions like Africa.

Diversity, Equity, and Inclusion (DEI): Digital mental health applications must be accessible, relevant, and effective for all, particularly marginalized groups facing disparities in access to culturally appropriate care. Diversity in these applications ensures cultural and linguistic sensitivity, equity addresses access disparities by aligning with resource availability, and inclusion fosters a respectful, welcoming environment. Despite the promise of culturally responsive interventions, many applications lack evidence of effectiveness due to limited DEI integration [8], having been developed mainly in Western contexts [9]. To effectively tackle mental health crises in African countries like South Africa, DEI principles centered on Africa-specific needs are essential for inclusivity and cultural competence.

Cultural Adaption vs. Culturally Grounded Interventions: Culturally grounded interventions involve adapting mental health interventions to the specific needs that are designed to be sensitive and responsive to the unique contexts in which they are implemented, often involving collaboration with community leaders and members [10]. Cultural adaptation, on the other hand, involves modifying interventions to enhance their relevance, acceptability, and effectiveness for individuals from diverse cultural backgrounds. This process includes adding superficial changes surrounding the administration of the intervention, such as language translation, and adding cultural content, such as metaphors to ensure accessibility and appropriateness for the target population [10]. Each modification can be categorized as either surface or deep structure changes. The former is a basic modification to fit interventions into specific cultures, while the latter accounts for the unique cultural patterns [11, 12].

The EVM framework [13], which includes language, persons, metaphors, and context, guides the cultural adaptation of psychological interventions, enhancing their relevance and reducing stigma. Studies have shown EVM's effectiveness in making mental health interventions culturally responsive. For related discussions, refer to the works of Sit et al. [11] and Shroff et al. [12].

Cognitive Behavioural Therapy in Mobile Applications: Ecological momentary intervention (EMI) is used to deliver treatments, addressing behaviors thoughts, or emotions in daily life. The EMI often employs CBT, which is considered the "gold standard" for many mental health conditions. The CBT techniques include cognitive restricting, behavioral activation, problem-solving, and exposure therapy, focusing on challenging negative thought patterns, engaging in positive thought patterns, and developing effective coping strategies [2].

Mobile health applications adapt these techniques into digital formats, offering interactive tools for self-help and guided therapy. For instance, cognitive restricting exercises might include journaling, while behavioral activation could involve activity scheduling with some reward mechanisms [2].

We argue that the effectiveness of CBT techniques in EMIs relies on adapting these interventions to cultural context. For example. Behavioral activation in an African context might involve communal activities such as spending time with elders, reflecting the cultural emphasis on respect for traditional knowledge. Adapting these techniques to match the target population's context is crucial for enhancing engagement and therapeutic outcomes.

2.2 Challenges of Westernized Therapy in Africa

Most psychological frameworks used in Africa are historically Western, emphasizing individualism and often overlooking the collective, communal nature of African societies. Rooted in Eurocentric theories, these Western approaches may not fully account for the socio-cultural factors influencing mental health in African contexts, focusing on individual well-being, which can misalign with the interconnected nature of African cultures [14].

In contrast, African psychologies prioritize communalism, interconnectedness, and the collective well-being of the community [14]. Grounded in African cosmology, philosophy, and cultural practices, they advocate for holistic and contextual approaches to diagnosis and treatment [10]. African psychologists view mental health as intricately linked to the health of the community, emphasizing the need for culturally grounded interventions [14]. However, colonization and Westernization have disrupted traditional healing methods, creating a disconnect between traditional practices and modern interventions. This has hindered access to culturally appropriate care, underscoring the need to integrate Africa-centric theories into mental health interventions to build trust and foster acceptance [10].

2.3 Incorporating Emotional Intelligence in LLMs

Emotional intelligence denotes the capacity to adeptly interpret and manage emotion-infused information, subsequently harnessing it to steer cognitive tasks ranging from problem-solving to behavior regulation [14]. On the one hand, prompt engineering enhances LLMs' performance by designing specific instructions without extensive retraining, utilizing techniques like zero-shot, few-shot, and chain-of-thought prompting [15].

Li et al. [16] introduced methods such as the emotional Chain of Thought (ECoT) and EmotionPrompt, integrating emotional intelligence into LLMs to enable more human-like interactions and improved performance. Integrating emotional dimensions further refines LLMs' ability to generate nuanced responses. Models like the six-dimensional model (6DE) by Ratican and Hutson [17], including dimensions like arousal, valence, and dominance, and Wortman and Wang's [18] high-coverage emotion model (HICEM), which uses techniques like UMAP and hierarchical clustering, help LLMs understand and react to emotionally complex situations across diverse cultures.

While incorporating emotional intelligence into LLMs enhances their capabilities, challenges include prioritizing emotions and ensuring cultural sensitivity due to varying cultural interpretations of emotional expressions.

It is noteworthy that despite being a recent innovation, prompt engineering shows great promise in improving LLM performance. Techniques that align responses with contextual nuances and integrate emotional intelligence can enhance AI's effectiveness in mental health support, offering exciting possibilities at the intersection of AI and psychology.

3 Analysis and Synthesis

Despite advances in mental health applications, cultural misalignment limits their effectiveness in diverse settings like South Africa. Most existing tools are based on Western psychological theories, which may not fully resonate with African cultural values [10]. Studies emphasize the importance of culturally adapted treatments, showing that incorporating cultural narratives can enhance engagement and outcomes [11, 12].

Digital tools for mental health often overlook cultural nuances, especially in emotional intelligence algorithms [19] impacting their ability to address specific needs in

African communities. For instance, spirituality, a central component of African psychology, has proven beneficial for mental health outcomes in other cultural settings and could be similarly impactful in South Africa [10].

The LLMs hold promise due to their adaptability, yet their primary training on Western data can limit their cultural sensitivity in African contexts [20]. Fine-tuning these models to integrate African values like Ubuntu, along with culturally specific prompt engineering, is essential to offer empathetic, effective support. Inclusive research that considers cultural, spiritual, and contextual differences may lead to more effective mental health solutions tailored to non-Western populations [20].

4 Proposed Framework

4.1 Deep-Level Structure Changes

Ubuntu, a South African philosophy emphasizing interconnectedness and community responsibility, aligns closely with values in African societies that prioritize social harmony. Incorporating Ubuntu into mental health frameworks offers a culturally sensitive approach centered on social bonds, shared humanity, and community support, contrasting with the individual focus of many Western therapeutic models. The Wilson and Williams' [21] core Ubuntu themes, which include connectedness, competency, and consciousness, complement CBT well. Similarly, connectedness promotes a sense of belonging and collective emotional support, competency aligns with skill-building while emphasizing community contribution, and consciousness encourages self-awareness within a cultural and communal context.

4.2 Surface-Level Structure Changes

This section outlines the high-level application of cultural adaptation in communication, focusing on both communication aspects and user engagement. It discusses the use of the Depression, Anxiety, and Stress Scale (DASS-21) and its cultural adaptation using Ubuntu theory. Finally, it explains how CBT should be applied in this assignment. This holistic approach that combines communication adaptation, culturally sensitive tools, and evidence-based psychological theory, ensures that the proposed framework delivers meaningful and impactful results.

Adaptation of Communication and User Engagement: The adaptation of communication and user engagement within the proposed framework emphasizes cultural resonance and relatability for South African users. To foster trust and familiarity, the LLM should be given a name that is widely recognized and positively regarded within local communities. It should employ culturally relevant metaphors, idioms, and proverbs to simplify complex concepts, as well as use respectful titles to honor formality and hierarchy, aligning with local social norms. Spiritual elements, such as religious encouragement, provide additional comfort and align with users' beliefs, while straightforward language ensures accessibility for those with varying levels of English literacy, considering that the majority of South Africans do not speak English as their first language. Furthermore, by using somatic symptoms (for example, feeling 'tired' or 'discouraged') instead of clinical terms like 'depressed,' the framework reduces stigma, creating a more approachable user experience. Each interaction should be designed to contribute to symptom alleviation, ensuring that every exchange is purposeful and culturally attuned.

Conceptualization of Depression and Anxiety: The depression, anxiety, and stress scale (DASS-21) is a widely used self-reported tool to assess the severity of these symptoms. For the validation of South African populations [22], the DASS-21 provides a solid foundation for understanding depression and anxiety in this framework. To enhance cultural relevance, the DASS-21 items were adapted to align with Ubuntu principles, reflecting communal values and relational aspects. For example, "I couldn't seem to experience any positive feeling at all" was rephrased to "I couldn't feel any happiness, even with the people around me," emphasizing Ubuntu's communal focus.

Using these adapted questions, a dataset of 15 test scenarios was curated, representing a range of symptom levels, including moderate anxiety, low depression, mixed symptoms, and severe depression. These scenarios, reviewed by professional African psychologists, authentically depict how someone grounded in Ubuntu might express such symptoms, incorporating cultural nuances and varying in intensity and complexity. This dataset offers a comprehensive foundation for evaluating the LLM's culturally sensitive response capabilities such as providing support aligned with Ubuntu and recognizing cultural nuances.

Cognitive Behavioral Therapy: The CBT techniques, like behavioral activation (BA), should differ across cultures. While Western BA often involves individualistic activities like meditation and self-reflection, Africa-centric BA, rooted in Ubuntu, emphasizes community and spirituality. This adapted approach includes activities such as consulting elders, prayer, and community engagement, aligning with Ubuntu's principles and enhancing cultural relevance.

4.3 Research Design

The study employs two assessment methods, namely *Qualitative Assessment*, which entails evaluating the LLM's ability to detect, understand, and respond to mental health symptoms within a cultural context, and *Quantitative Assessment*, which deals with the testing of the LLM's ability to identify and classify symptoms from culturally relevant case studies exhibiting depression and anxiety.

Conceptualized System Design: The proposed system includes two primary components: a user input module, which captures and preprocesses text, and a fine-tuned model based on a foundational language model initially trained on diverse datasets. This setup enables key functionalities, such as classification and sentiment analysis.

To promote culturally sensitive interactions, the foundational model is fine-tuned with Ubuntu-aligned datasets. This customization through fine-tuning allows the model to address specific domain tasks, with a primary focus on question answering (Task C). During the fine-tuning stage, Ubuntu-oriented data and principles are embedded to support culturally attuned responses. Prompt engineering further refines interactions to align with the framework's guidelines on culturally sensitive communication and

CBT techniques (as detailed in Sect. 4.2 on 'Adaptation of communication and user engagement'). By integrating Ubuntu values and CBT, the framework aims to build trust, reduce stigma, and provide relevant mental health support tailored to South African users.

Conceptual Evaluation Framework: The evaluation of the proposed framework employs two perspectives: clinical and research. Clinically, an African professional psychologist uses a Likert scale (0 to 5) to assess criteria like accuracy (correct symptom identification), relevance (pertinence of responses), and empathy (cultural sensitivity and compassion), with cultural appropriateness qualitatively evaluated based on alignment with Ubuntu values. The research evaluation examines the integration of Ubuntu philosophy and CBT principles by qualitatively assessing how well responses embody Ubuntu tenets and incorporate CBT strategies. Additionally, symptom detection evaluates the model's ability to identify symptoms of depression, anxiety, and stress using the adapted DASS-21 scale. Comparative analysis also investigates cultural sensitivity differences between the culturally enhanced LLM and a standard LLM based on qualitative insights (Fig. 1).

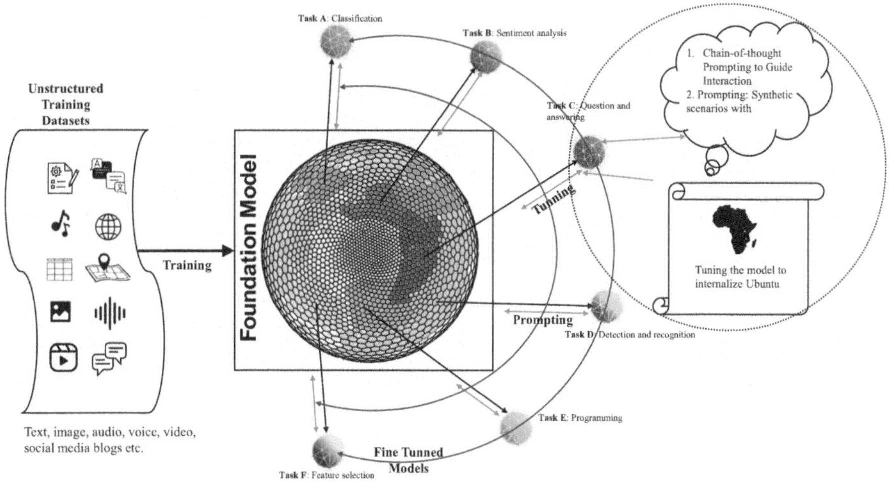

Fig. 1. Framework System Design

5 Sample Implementation and Discussions

This section demonstrates the framework's application, showing how fine-tuning and prompt engineering align LLM responses with Ubuntu values for culturally sensitive mental health support in African contexts. The presented sample interactions or conversations between a typical LLM and a human user illustrate the integration of Ubuntu and CBT principles personalized for South African users.

5.1 Integration of Ubuntu Philosophy

We fine-tuned the model using a curated dataset aligned with Ubuntu principles to address the limitations of Western-centric LLM training. This allowed the model to adapt its responses to reflect communal values like empathy and collective responsibility. Chain-of-thought prompting guided the model's tone and approach, ensuring interactions respect cultural norms and resonate with South African users.

Ubuntu-inspired behavioral activation techniques were incorporated to promote well-being through community-oriented activities. Instead of suggesting individualistic practices like meditation, the model recommends culturally relevant actions such as.

prayer, consulting elders, and engaging in community events, fostering a sense of support and belonging.

To specialize the model further, we adjusted its weights using the curated dataset, starting with a well-trained LLM to leverage existing knowledge while minimally adapting to the new domain. Data augmentation expanded the dataset by synthesizing or paraphrasing text samples. By combining fine-tuning with lightweight methods like LoRA and intelligent prompt engineering, we achieved high-quality results despite the small datasets.

5.2 Sample Conversation

Below is a condensed interaction between a human user and the LLM agent, Karabo.

User: "I feel uneasy about my role as a youth leader at church. I struggle to sleep before presentations and feel stressed about my contributions."

Karabo: "You're shouldering a lot, and it's impacting your well-being. What do you think is at the root of these feelings?".

User: "It's the group's lack of participation. When I have little to report, I feel useless."

Karabo: "Feeling unappreciated can affect self-worth. Have you tried strategies to encourage more involvement?".

User: "I've tried bringing snacks."

Karabo: "That's thoughtful. Building a community can also involve personal connections. How about one-on-one conversations to make members feel valued?".

User: "One-on-ones could help."

Karabo "Understanding each member might strengthen your role and boost engagement. How might this ease your stress as a leader?".

User: "It could help me support them better and improve attendance."

Karabo: "Individual connections can increase participation and reduce your stress. What's your plan for starting these one-on-ones?".

User: "I'll start with members who attend less often."

Karabo: "Great approach. As you build these connections, I'm here whenever you need to talk. You're making a difference."

Brief Analysis: The sample conversation reflects Ubuntu's tenets and incorporates CBT principles. Karabo fosters community by promoting one-on-one connections to strengthen empathy and participation (Connectedness). The dialogue also emphasizes effective leadership and personal growth (Competency) and helps the user gain self-awareness about stress and responsibility within the community (Consciousness). Through these strategies, which include CBT techniques like guided self-reflection and problem-solving, Karabo supports the user in creating a more inclusive and supportive environment.

6 Conclusion and Future Work

This study presents a culturally sensitive mental health framework that integrates Ubuntu philosophy with CBT principles, personalized for African contexts through fine-tuned large language models. By fostering community, personal growth, and cultural awareness, the framework enhances inclusivity and support for African users. Incorporating Ubuntu's values strengthens engagement and relevance, addressing the mental health service gap with a culturally adapted, scalable solution for South Africa and beyond.

Future research can focus on (1) validating the framework's efficacy with real-world datasets grounded in Ubuntu principles, (2) developing African psychology frameworks that incorporate cultural nuances, and (3) expanding the framework for practical applications in culturally sensitive, AI-driven mental health support. Key areas include exploring the benefits and challenges of emotionally intelligent AI within diverse African contexts, addressing trust issues in deploying such systems, and assessing large language models' readiness for mental health assessments and interventions.

References

1. Sodi, T., et al.: A research agenda for mental health in sub-Saharan Africa. Nat. Med. 1–2 (2024)
2. Denecke, K., Schmid, N., Nüssli, S.: Implementation of cognitive behavioral therapy in e–mental health apps: Literature review. J. Med. Internet Res. **24**, e27791 (2022)
3. Everitt-Penhale, B., et al.: 'I went back to being myself': acceptability of a culturally adapted task-shifted cognitive-behavioural therapy (CBT) treatment for depression (Ziphamandla) for South African HIV care settings. Psychol. Health Med. **24**, 680–690 (2019)
4. Hadi, M.U., et al.: A survey on large language models: applications, challenges, limitations, and practical usage. Authorea Preprints (2023)
5. Mindu, T., Mutero, I.T., Ngcobo, W.B., Musesengwa, R., Chimbari, M.J.: Digital mental health interventions for young people in rural South Africa: prospects and challenges for implementation. Int. J. Environ. Res. Public Health **20**, 1453 (2023)
6. Vernon, D.: Culturally competent social robotics for Africa: a case for diversity, equity, and inclusion in HRI. In: Proceedings of 2nd Workshop on Equity and Diversity in Design, Application, Methods, and Community (HRI DEI at at the Human-Robot Interaction Conference, Stockholm, Sweden (2023)
7. Naslund, J.A., et al.: Digital innovations for global mental health: opportunities for data science, task sharing, and early intervention. Current Treat. Options Psychiatry **6**, 337–351 (2019)

8. Ramos, G., Ponting, C., Labao, J.P., Sobowale, K.: Considerations of diversity, equity, and inclusion in mental health apps: a scoping review of evaluation frameworks. Behav. Res. Ther. **147**, 103990 (2021)
9. Pakomera, R., van Greunen, D., Veldsman, A.: Usability and usefulness in designing patient-centric systems in afro-centric setting. In: 2017 6th International Conference on Reliability, Infocom Technologies and Optimization (Trends and Future Directions)(ICRITO), pp. 576–579. IEEE
10. Anakwenze, O.: The cultural sensitivity continuum of mental health interventions in Sub-Saharan Africa: a systematic review. Soc Sci Med **306**, 115124 (2022)
11. Sit, H.F., Ling, R., Lam, A.I.F., Chen, W., Latkin, C.A., Hall, B.J.: The cultural adaptation of step-by-step: an intervention to address depression among Chinese young adults. Front. Psych. **11**, 650 (2020)
12. Shroff, A., et al.: A digital single-session intervention platform for youth mental health: cultural adaptation, evaluation, and dissemination. JMIR Mental Health **10**, e43062 (2023)
13. Bernal, G., Bonilla, J., Bellido, C.: Ecological validity and cultural sensitivity for outcome research: issues for the cultural adaptation and development of psychosocial treatments with Hispanics. J. Abnorm. Child Psychol. **23**, 67–82 (1995)
14. Salovey, P., Mayer, J.D., Caruso, D., Yoo, S.H.: The positive psychology of emotional intelligence. (2009)
15. Sahoo, P., Singh, A.K., Saha, S., Jain, V., Mondal, S., Chadha, A.: A systematic survey of prompt engineering in large language models: techniques and applications. arXiv preprint arXiv:2402.07927 (2024)
16. Li, C., et al.: Large language models understand and can be enhanced by emotional stimuli. arXiv preprint arXiv:2307.11760 (2023)
17. Ratican, J., Hutson, J.: The six emotional dimension (6DE) model: a multidimensional approach to analyzing human emotions and unlocking the potential of emotionally intelligent artificial intelligence (AI) via large language models (LLM). J. Artif. Intell. Robot **1** (2023)
18. Wortman, B., Wang, J.Z.: HICEM: a high-coverage emotion model for artificial emotional intelligence. IEEE Trans. Affect. Comput. (2023)
19. Ahmadi, F.B., Dehghan, M.J., Baharvand, V., Azizi, A.: Designing and evaluating a smartphone-based cognitive behavioral therapy to control anxiety and depression. Front. Health Inform. **12**, 136 (2023)
20. Bender, E.M., Gebru, T., McMillan-Major, A., Shmitchell, S.: On the dangers of stochastic parrots: can language models be too big?? In: Proceedings of the 2021 ACM Conference on Fairness, Accountability, and Transparency, pp. 610–623
21. Wilson, D., Williams, V.: Ubuntu: development and framework of a specific model of positive mental health. Psychol. J. **10** (2013)
22. Dreyer, Z., Henn, C., Hill, C.: Validation of the depression anxiety stress scale-21 (DASS-21) in a non-clinical sample of South African working adults. J. Psychol. Afr. **29**, 346–353 (2019)

Explainable Graph Neural Networks to Identify Potential Atoms and Chemical Bonds of Drug Candidates from Drug Target Interactions

Satansu Mohanty and Chandra Mohan Dasari(✉)

Computer Science and Engineering Group, Indian Institute of Information Technology, Sri City, Sri City, Andhra Pradesh, India
{satansu.m21,chandramohan.d}@iiits.in

Abstract. Drug-Target Interaction (DTI) involves the observation and recognition of interactions that occur between chemical molecules and target proteins in the human body. However, lab experimentation for DTI can be time-consuming and repetitive. Approaching these experiments from a computational perspective can save time and increase the scope of research. Graph Neural Networks (GNNs) have emerged as a leading approach in the field of DTI predictions, demonstrating state-of-the-art performance, but there is still scope for improvement in these models with new features. Thus, we propose a novel GNN-based model, GNN-DTI, for DTI predictions. The model takes the chemical molecules (as molecular graphs) and target proteins (as embedding vectors) and predicts a binding affinity score. Due to the complexity of the model, the inner workings of the models are unexplained, and it is challenging to understand the reasoning behind their outputs. Therefore, we also propose a GNN-DTI Explainer model to generate suitable explanations for drug candidates and identify the most important nodes and edges. The proposed model effectively visualizes these explanations for each drug candidate. This allows us to identify the key atoms and their bonds involved during the binding process. It provides further context for the behavior of certain drugs and their interactions with the human body, which can be used to advance drug development and research.

Keywords: Graph Neural Networks · Drug Target Interactions · Drug Candidates · Atom and Bond Explanations · Binding Affinity

1 Introduction

Drug-Target Interaction (DTI) is a pivotal process in the development of new pharmaceuticals, involving the examination and identification of interactions between chemical compounds and target proteins within the human body. Traditional laboratory methods for investigating DTI, though essential, require large

A. Dasgupta et al. (Eds.): BDA 2024, LNCS 15526, pp. 315–325, 2025.
https://doi.org/10.1007/978-3-031-81821-9_19

investments of time and money. As a result, there is a growing interest in computational approaches that can accelerate the pace of research and expand the scope of potential findings. Computational methods offer the promise of more efficient screening processes and can provide insights that might be impractical to achieve through experimental techniques alone [16].

Graph Neural Networks (GNNs) have emerged as a powerful tool for representing the complex relationships between atoms within a molecule. These networks can effectively capture the intricate structures of drug molecules and proteins, making them suitable for DTI prediction tasks. The last few years have marked the development of several GNN models built specifically for the purpose of predicting DTI binding affinities. The domain of GNN model architectures has been constantly evolving, with innovative architectures providing different avenues for graphical representation and processing of molecules. In the domain of GNNs trained specifically for DTI affinity predictions, GraphDTA [9] and DeepDTA [17] have provided suitable benchmarks to test the effectiveness of newly built DTI-GNN models.

Our contribution can be summarized as follows:

- The construction of a GNN-DTI model, along with the integration of the GNN-DTI Explainer algorithm, provides comprehensible and accessible explanations for each drug's nodes and edges, accompanied by a visual representation.
- Selection of drug candidates from DTI pairs that cross a particular binding affinity threshold.
- Presenting several observations to highlight patterns and evidence about the behavior of particular atoms and atom bonds in the drug molecules, and their effect on the drug-target binding process.

2 Related Work

We have conducted extensive literature reviews on relevant concepts that will be the main focus of this paper. A brief survey of these concepts has been described in this section.

Predicting the nature of interactions between several drugs and their interactions with target proteins, including enzymes, receptors and ion channels is a vital part of drug discovery and drug repositioning. With the advent of deep learning, neural networks are the current standard for DTI prediction. These models demonstrated the potential of neural networks to learn complex interactions from raw sequence data without extensive feature engineering. DeepDTA [17] employed Convolutional Neural Networks (CNN) to capture local patterns in drug and target representations, significantly improving predictive performance. In recent years, the swift advancement of GNNs has broadened the scope of deep learning to include graph-based data, and these methods have also found applications in drug discovery and drug repositioning. GraphDTA [9] and our model GNN-DTI employ a framework that converts the drug target into a graph before inputting it into the model.

A comparative study [8] provides an extensive taxonomy of present GNN architectures. DTI GNNs predominantly use propagation module architectures, so we will focus on that section of the taxonomy. Propagation-based architectures are based on convolution, which can be defined as generalizations of the concept of convolution in other domains, to apply it in graph domains. They can be largely divided into two main categories- spectral and spatial. Spectral graph convolution is theoretically based on graph signal processing and it involves transformation of graph signals to the spectral domain and back. GCN is a notable example of a spectral graph convolution architecture [5]. Spatial graph convolution is centred around graph topology. GraphDTA explores variants that notably include GCNs, while our model uses a modified GCN framework.

Explainability in Artificial Intelligence (XAI) is an area of work that has received increasing attention in recent years and coincides with the growth of AI. XAI allows models to be more transparent and provide explanations for their decisions to a certain extent. Explainability is defined by two key factors- interpretability and completeness. The extent to which a human can understand the internal calculations of a model is measured by its interpretability, whereas the goal of completeness is to successfully and accurately describe the operating of a network model. Post-hoc Neural Network Explainers analyze the model after the prediction stage and derive the decisions that the model took to reach that prediction [10]. Most GNN Explainers fall into this category. They can be generally categorized into four different categories [15] - Gradient based models, perturbation based models, surrogate methods and decomposition methods. Our GNN-DTI Explainer falls into the gradient category, since it map the gradients or hidden feature values and calculate the importance values of input graphs based accordingly.

3 Proposed Methodology

3.1 Problem Statement

We propose a GNN-DTI model for predicting binding affinities that demonstrate strong predictive capabilities and can be considered an improvement over current state-of-the-art frameworks. However, the inherent complexity of the model poses significant challenges in terms of explainability. It is currently nearly impossible for a human to understand *why* a particular drug achieves better results than another drug. Although wet lab experiments require a longer duration of time and greater resources, scientists have complete control over every step of the process.

To overcome this challenge of a lack of transparency, we propose the model, GNN-DTI Explainer for our DTI prediction model, which aims to generate explanations for the model's predictions by identifying the most important nodes and edges in the input graphs. By visualizing the graphs and pinpointing the atoms and the bonds between atoms that had the greatest influence on the result. it can provide valuable context for the model outputs, enhancing understanding of the protein-drug interactions. By integrating an Explainer into our DTI prediction framework, we aim to create a more transparent model that can facilitate more informed decision-making in drug development and research.

3.2 GNN-DTI Model

Pre-processing: Preliminary stage involves translating the molecule and target inputs into data configurations that can be inputted into the GNN and CNN blocks of the GNN-DTI model respectively. The chemical compounds are processed through the graph convolutional layers, necessitating their conversion into molecular graphs. Consequently, these compounds will be the exclusive focus of the subsequent GNN-DTI Explainer model analysis.

To convert the drug compounds into their appropriate graphical format, they are first represented using the SMILES format. They are then converted into molecular graphs while retaining the key information of each drug. Each node represents an atom in the molecule, and the edges of the graph display the bonds between these atoms. The node features are displayed in Table 1. The protein target compounds are obtained as FASTA sequences from the UniProt database, and sent to the next stage as a string. FASTA format is used to represent the amino acids that constitute a protein sequence. Since they are not being processed as a graph, they will not be considered by the GNN-DTI Explainer model.

Table 1. Node features for each atom

Atom Features	Description	Size
Atom Type	C, N, O, S, F, Si, P, Cl, Br, Mg, Na, Ca, Fe, As, Al, I, B, V, K, Tl, Yb, Sb, Sn, Ag, Pd, Co, Se, Ti, Zn, H, Li, Ge, Cu, Au, Ni, Cd, In, Mn, Zr, Cr, Pt, Hg, Pb, or "Unknown"	44
Degree	Number of directly bonded neighbours (one-hot)	11
Num of H	Number of hydrogens bound to the atom	11
Valence	Number of implicit hydrogens bound to the atom	11
Aromaticity	Whether the atom is aromatic or not	1
Total		**78**

GNN and CNN Layers: The graphs pass through the GNN block, which consists of three consecutive GENConv layers followed by a global add pooling layer. This particular graph neural network block has been greatly inspired by the DeepNC framework [13]. GENConv (Generalized Graph Convolution) is derived from the DeeperGCN framework [6] that focuses on tackling problems that arise in deep GCNs with many layers (>15), such as vanishing gradients and over-fitting. The generalized message aggregation function developed as a graph representation learning solution is what we utilize in our model. A GCN message passing framework [5] is used for message passing between nodes and their neighbouring nodes, and the message aggregation function is used to aggregate the set of messages accumulated. We employ the use of the SoftMax Aggregation with the final message passing function while building our GENConv layers. The final equation can be represented as:

$$x_i' = MLP(x_i + Softmax\{(ReLU(x_j + e_{ij}) + \varepsilon | j \epsilon N(i))\}) \qquad (1)$$

The target compound strings are passed through a simple CNN block consisting of Conv1D layers to extract features, followed by a max pooling layer for dimensionality reduction.

Calculating Binding Affinities: The drug and protein features are concatenated and passed through two fully connected layers. The output of this process is a positive real number that we refer to as the binding affinity of the drug-target pair (Figs. 1 and 2).

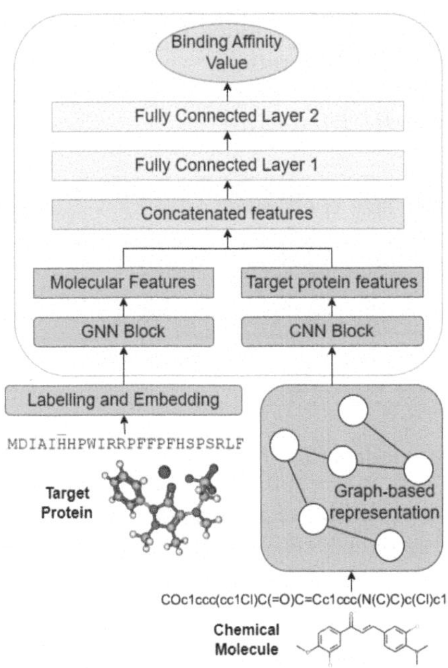

Fig. 1. GNN-DTI Model

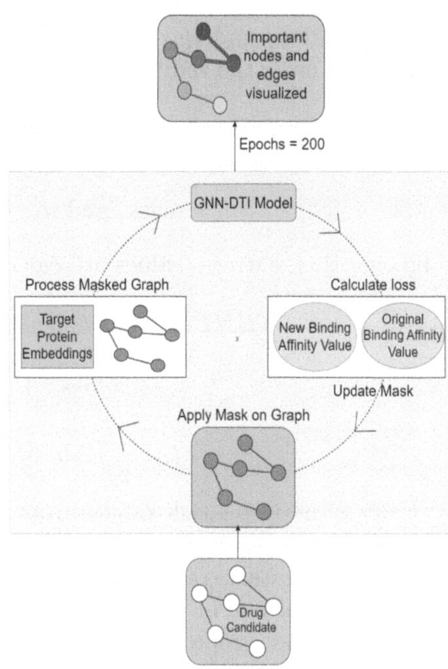

Fig. 2. GNN-DTI Explainer Model

3.3 GNN-DTI Explainer

The GNN block takes in pre-processed drug compound data, which includes node features, edge indices, and target values. This same data is also fed into the GNN-DTI Explainer block. This Explainer utilizes concepts derived from the GNNExplainer [14] framework, to explain DTI predictions. The goal of the GNN-DTI Explainer model is to identify nodes and edges that are most crucial to the GNN's prediction.

Using the node and edge attributes, the model initializes node and edge masks as a trainable parameter. This masked graph is passed through the GNN-DTI model. After obtaining a new binding affinity value, it is compared to the original binding affinity obtained from the GNN-DTI model and passed through a Mean-Square Error loss function. The gradients obtained from this loss function are then used to update the node and edge masks. This process is repeated for each epoch before outputting the final node and edge mask values.

$$MSE = \frac{1}{n}\sum_{i=1}^{n}(y_i - \hat{y_i})^2 \tag{2}$$

$$(n = \text{length of mask} = \text{number of node features})$$

The mask values undergo a reduction operation, that is added to the loss function. In our case, the reduction function is a mean operation.

$$loss_\alpha = MSE(y, \hat{y}) + n_k * REDUCE(m) \tag{3}$$

$$(m = \text{mask, and } n_k = \text{node size coefficient})$$

Following this, entropy values are calculated and added:

$$ENT = -m * log(m) - (1 - m) * log(1 - m) \tag{4}$$

$$loss = loss_\alpha + n_e * AVG(ENT) \tag{5}$$

$$(n_e = \text{node feature coefficient})$$

Each value in a mask ranges from 0 to 1. A higher value indicates a more influential node. After the final node and edge masks are returned, we use Matplotlib to generate a graph with the mask values as color gradients. The darkness of the color gradient of the node/edge is directly proportional to its mask value. The darkest nodes represent atoms that have the highest mask values and therefore, the highest relevance while calculating the binding affinity, and vice versa. Similarly, the darkest edges represent atom bonds that play the most crucial role in this process. This type of visualization makes it simple to observe the most important nodes and edges as identified by the GNN-DTI Explainer.

4 Results and Discussion

In this section we discuss each aspect of the results obtained, including the datasets and hyperparameters used during testing and the performance metrics used to assess the results. Following this, we present analyses of the GNN-DTI model and the explanations generated by the GNN-DTI Explainer.

4.1 Datasets

The Davis dataset, introduced in Davis et al. [2], analyzes the interactions between kinase proteins and kinase inhibitor compounds. It measures the binding affinity (dissociation constant) as K_d. The dataset contains 72 compounds that are tested against 448 proteins for a total of 30056 drug-target pairs. The Kiba dataset was established in Tang et al. [12], and named after the scoring system introduced in the aforementioned paper that is used to measure the binding affinities between drug-target pairs. The ligand compounds are all extracted from the CHEMBL database. It has 118254 drug-target pairs.

4.2 Performance Metrics

To evaluate the performance of the DTI model and make comparisons with other baseline models in the same domain, we have used the Concordance Index (CI) and MSE values 2 obtained from the model. CI is used to evaluate whether the performance ranking of the continuous drug-target affinity outputs are accurate and can be computed as:

$$CI = \frac{1}{Z} \sum_{\delta_x > \delta_y} h(b_x - b_y) \qquad h(x) = \begin{cases} 1 \text{ if x} > 0 \\ 0.5 \text{ if x } = 0 \\ 0 \text{ if x } < 0 \end{cases} \tag{6}$$

where δ_x and δ_y are the larger and smaller affinity values respectively, and b_x and b_y are the corresponding predicted values from the model. Z is the normalization constant. $h(x)$ is the step function as described.

4.3 Hyperparameter Settings

The hyperparameter values used for training the GNN-DTI model and the Explainer model have been mentioned below in Table 2. Most of the GNN-DTI hyperparameters have been derived from the hyperparameters used in the DeepNC [13] model. The GNN-DTI Explainer hyperparameters have been optimized and selected through extensive testing.

Table 2. Hyperparameters used in the proposed models

GNN-DTI model		GNN-DTI Explainer	
Hyperparameters	Settings	Hyperparameters	Settings
Epoch	200	Epoch	2000
Learning Rate	0.01	Batch Size	512
Edge Size Coefficient	0.005	Optimizer	Adam
Node Feature Size Coefficient	1.0	Learning Rate	0.0005
Edge Entropy	1.0	No. of GenConv layers	3
Node Feature Entropy	0.1	No. of Conv1D layers	1
Edge Reduction Aggregator	Sum		
Node Feature Reduction Aggregator	Mean		

4.4 Comparative Analysis of GNN-DTI Model

To compare the results of a propose GNN-DTI model to other existing DTI architectures, we have run multiple model on the two benchmark datasets, Davis and Kiba. The results are presented in Table 3. These tables highlight the different methods used to represent the drugs and protein targets. SimBoost uses the Smith-Waterman method to represent the protein compounds and the PubChem Sim descriptor for drugs [4]. DeepDTA uses CNNs to represent both proteins and drugs molecules, while DeepGS, GraphDTA and GEN model use graphs to describe drugs. Since the GNN-DTI model shares many similarities with the GraphDTA framework, similar hyperparameter settings have been replicated to provide a fair comparison.

Table 3. Comparison of different models on the Davis and Kiba datasets

Model	Drug and Protein	Davis		Kiba	
		CI	MSE	CI	MSE
SimBoost [3]	Pubchem Sim & S-W	0.869	0.282	0.836	0.222
DeepDTA [17]	CNN & CNN	0.876	0.262	0.861	0.199
GraphDTA [9]	Graph & CNN	0.858	0.308	0.871	0.176
DeepGS [7]	Graph & CNN	0.880	0.252	0.862	0.188
GNN-DTI	Graph & CNN	0.877	0.233	0.869	0.138

Table 3 highlight the results obtained by all models on the Davis and Kiba dataset. It can be observed that the GNN-DTI model provides the best results overall, with the best MSE values on both datasets (0.233 & 0.138). The GNN-DTI model displays matches the best CI values on both datasets (0.877 & 0.871), and thus evaluating both metrics together demonstrates that the GNN-DTI model provides the best results.

Our goal is to extract potential drug candidates, that refers to drug-target pairs that display a binding affinity value above a certain threshold, which is different for each dataset. After thorough examination of the datasets and cross-referencing with other DTI datasets, we have assigned a threshold value of $K_d >$ 6.5 for the Davis dataset, and $K_d >$ 11.5 for the Kiba dataset. 7.8% DTI pairs (395 of 5010) crossed the threshold in the Davis test subset, while 6.4% DTI pairs (1263 of 19709) crossed the threshold in the Kiba test subset. These selected DTI pairs are exclusively considered for highlighting node and edge explanations in the next section.

4.5 Node and Edge Explanations

After the identification of a target protein, the next step of drug discovery tends to be the discovery and selection of lead drug compounds [11]. This is done by referring to compound libraries, virtual screening methods and in-vitro pharmacology. GNN-DTI Explainer adds additional contextual information that can be used to cross-validate the validity of the results with wet lab experiments, and can also be used the next time experiments on the same protein target are conducted. Although explanation evaluation metrics such as fidelity and sparsity [1] exist for classification predictions, we have yet to come across any such metric for explainability evaluation for regression tasks, such as ours. Therefore, we have presented a visualization of our findings and our subjective interpretation of it.

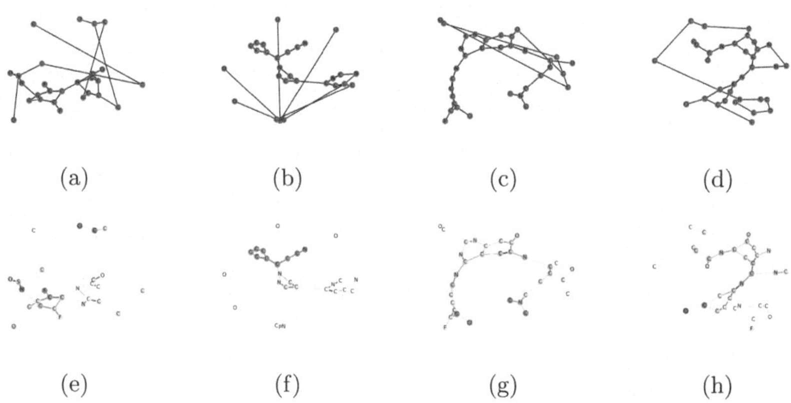

Fig. 3. (a), (b) and (c), (d) represent original drug graphs while (d), (e) and (f), (g) show their respective explanations of Davis and Kiba Datasets respectively

Figure 3 represents generated graphs from the Davis and Kiba datasets. After evaluating the results, we observe certain recurring patterns in the explanation of graphs with certain characteristics. The graphs that are to be presented in this section have been selected since they can effectively highlight these characteristics. However, we have been unable to obtain any Explainability datasets with existing ground-truth explanations. This makes it impossible to provide any numerical performance metrics for the GNN-DTI Explainer.

Graphs that contain one or more carbon ring structures tend to have high explainability values for atoms and atom bonds connecting the cyclic structure, as observed in Fig. 3f. In some cases, even if the nodes themselves do not contribute greatly to the overall explanation, the edges hold great importance (Fig. 3h). We hypothesize that this is due to the interconnected nature of carbon rings. If a single bond in the ring is removed, the cyclic ring is broken and the chemical composition of the drug molecule is changed to a large degree. Another key observation is that the model gives high importance to Nitrogen (N) atoms, especially their bonds with Carbon (C) atoms. The C-N bonds almost always have high edge mask values, and the subject N atom also tends to carry a high node mask value (Fig. 3g). Finally, Chlorine (Cl) atoms present in graphs typically have high node mask values (Figs. 3g, 3h). This implies that Cl plays an important role in the binding process and its removal negatively affects the model's predictions to a notable degree. This importance is often observed in other halogen atoms (F, Br) as well, but there are too many exceptions present for it to be presented as a reoccurring pattern here.

5 Conclusion

In this study, we tackled the challenge of DTI prediction by leveraging the power of GNNs to represent the intricate molecular structures of drug compounds and proteins. The complexity of GNN-based model initially posed interpretability challenges, so to overcome this, we used the GNN-DTI Explainer algorithm, which enabled us to generate explanations for the model's predictions by identifying the most important nodes and edges within the molecular compound graphs. These insights can significantly contribute to the progress of drug development and research. One shortcoming of our experiment is that we focus exclusively on converting chemical compounds into molecular graphs, leaving protein structures in their original string representations. To create a more comprehensive and nuanced DTI prediction framework, in future research we plan to explore new models that convert proteins also into graph structures.

Acknowledgements. This research is supported by the Science and Engineering Research Board (SERB) (Sanction order File No. SRG/2023/000941), Department of Science and Technology (DST), Government of India.

References

1. Agarwal, C., Queen, O., Lakkaraju, H., Zitnik, M.: Evaluating explainability for graph neural networks. Sci. Data **10**(1), 144 (2023)
2. Davis, M.I., et al.: Comprehensive analysis of kinase inhibitor selectivity. Nat. Biotechnol. **29**(11), 1046–1051 (2011)

3. He, T., Heidemeyer, M., Ban, F., Cherkasov, A., Ester, M.: SimBoost: a read-across approach for predicting drug-target binding affinities using gradient boosting machines. J. Cheminform. **9**, 1–14 (2017)
4. Kim, S., et al.: Pubchem 2019 update: improved access to chemical data. Nucleic Acids Res. **47**(D1), D1102–D1109 (2019)
5. Kipf, T.N., Welling, M.: Semi-supervised classification with graph convolutional networks. arXiv preprint arXiv:1609.02907 (2016)
6. Li, G., Xiong, C., Thabet, A., Ghanem, B.: DeeperGCN: all you need to train deeper GCNs. arXiv preprint arXiv:2006.07739 (2020)
7. Lin, X., Zhao, K., Xiao, T., Quan, Z., Wang, Z.J., Yu, P.S.: DeepGS: deep representation learning of graphs and sequences for drug-target binding affinity prediction. In: ECAI 2020, pp. 1301–1308. IOS Press (2020)
8. Longa, A., et al.: Explaining the explainers in graph neural networks: a comparative study. arXiv preprint arXiv:2210.15304 (2022)
9. Nguyen, T., Le, H., Quinn, T.P., Nguyen, T., Le, T.D., Venkatesh, S.: GraphDTA: predicting drug-target binding affinity with graph neural networks. Bioinformatics **37**(8), 1140–1147 (2021)
10. Retzlaff, C.O., et al.: Post-hoc vs ante-hoc explanations: XAI design guidelines for data scientists. Cogn. Syst. Res. **86**, 101243 (2024)
11. Salo-Ahen, O.M., et al.: Molecular dynamics simulations in drug discovery and pharmaceutical development. Processes **9**(1), 71 (2020)
12. Tang, J., et al.: Making sense of large-scale kinase inhibitor bioactivity data sets: a comparative and integrative analysis. J. Chem. Inf. Model. **54**(3), 735–743 (2014)
13. Tran, H.N.T., Thomas, J.J., Malim, N.H.A.H.: DeepNC: a framework for drug-target interaction prediction with graph neural networks. PeerJ **10** (2022)
14. Ying, Z., Bourgeois, D., You, J., Zitnik, M., Leskovec, J.: GNNExplainer: generating explanations for graph neural networks. In: Advances in Neural Information Processing Systems **32** (2019)
15. Yuan, H., Yu, H., Gui, S., Ji, S.: Explainability in graph neural networks: a taxonomic survey. IEEE Trans. Pattern Anal. Mach. Intell. **45**(5), 5782–5799 (2022)
16. Zhang, Z., et al.: Graph neural network approaches for drug-target interactions. Curr. Opin. Struct. Biol. **73** (2022)
17. Öztürk, H., Özgür, A., Ozkirimli, E.: DeepDTA: deep drug-target binding affinity prediction. Bioinformatics **34**(17), i821–i829 (2018)

Enhancing Cardiovascular Disease Prediction Among Middle-Aged Individuals Using Reinforcement Learning Dynamic Ensemble Selection with Customizable Actions and Exploration-Exploitation Balance

S. R. Divyasri(✉) and A. Rama Prasath

Hindustan Institute of Technology and Science, Kelambakkam 603103, Tamil Nadu, India
sr.divyasriragu@gmail.com, rprasath@hindustanuniv.ac.in

Abstract. After the COVID-19 pandemic, Cardiovascular Disease (CVD) has arisen as a notable health concern in middle-aged individuals. Discovering the CVD earlier and taking necessary precautions is much needed to reduce the risk of CVD complications [1]. This study explores various reinforcement learning (RL) techniques to improve CVD forecasting among people aged 40 to 60 years. We present a novel framework in its integration of customizable actions within reinforcement learning (RL), allowing dynamic ensemble selection that adapts to real-time feedback. This approach differs from traditional static methods by continuously optimizing classifier selection based on evolving data. We used preprocessing techniques such as label encoding, one-hot encoding, and numerical feature standardization by employing a heart disease dataset [2] comprising 696 rows and 14 attributes. Hyperparameter tuning was conducted using GridSearchCV across 10 classifiers. Classifier performance was evaluated through cross-validation with metrics including accuracy, precision, recall, F1 score, ROC AUC score, and average precision score. The RL framework dynamically selected ensemble strategies by majority voting, weighted voting, and best-performing classifier from different classifiers to balance exploration and exploitation using techniques like Upper Confidence Bound (UCB), epsilon-greedy, Thompson Sampling, and SoftMax. The comparisons and observations prove that Thompson Sampling performed the best with an optimized accuracy of 98% and 100% precision. The results highlight the effectiveness of RL-based dynamic ensemble selection in enhancing prediction for middle-aged people at risk of CVD, concluding a reliable forecasting system for improving early detecting techniques and preventing CVD prior to its occurrences [3].

Keywords: Heart Disease prediction · Middle-aged patients · Reinforcement Learning Dynamic Ensemble Selection · Action discretization · Balancing exploration and exploitation

A. Dasgupta et al. (Eds.): BDA 2024, LNCS 15526, pp. 326–335, 2025.
https://doi.org/10.1007/978-3-031-81821-9_20

1 Introduction

The significant public health issue that primarily affects middle-aged people in recent times is cardiovascular disease (CVD). Because of their recent lifestyles, early detection and intervention strategies are needed to lessen the effects of CVD in middle-aged people, as they are more likely to have CVD than older adults than other age groups [4]. Our study investigates how reinforcement learning (RL) can improve the forecasting of cardiovascular disease in this middle-aged people population.

Adults in their middle years frequently juggle a number of responsibilities in addition to dealing with health issues like high blood pressure (high BP), diabetes, and weight gain [4]. If we can catch the warning symptoms early, the individuals can adjust their lifestyle and seek medical assistance to prevent serious cardiac difficulties. In the field of medical sciences machine learning (ML) has come out as a much-needed tool for managing, diagnosing, and predicting illnesses. When predicting cardiovascular disease in middle-aged individuals, these ML techniques can integrate multiple data sources and highlight subtle risk factors that traditional statistical methods might overlook.

Algorithms that use reinforcement learning acquire its skills using the trial and error method [5]. Every time they receive feedback on their actions, which makes them get better over time. Reinforcement learning is used in the context of heart disease prediction to enable algorithms to adjust to new data and improve their forecasts in response to actual patient and physician feedback. The dynamic ensemble selection used in our work improves prediction by combining various forecasting models. This approach ensures that our model relies on more than one method, which can give unbiased results [6].

Traditional models often fail to capture the complexity of patient data due to their reliance on single classifiers and fixed parameters. We suggest an intensified approach including, the advantage of Reinforcement Learning (RL) for dynamic ensemble selection in order to overcome these drawbacks. This method used in this paper maximizes classifier performance by means of adaptable actions and a well-balanced exploration-exploitation approach. Finding the perfect balance between exploring and sticking with what works, exploitation is crucial in reinforcement learning dynamic ensemble selection [7]. Our model will continue to look for novel approaches to heart disease prediction which constantly improving without getting stuck in old patterns to harness the power of machine learning and lock out the right balance between exploration and exploitation to develop a predictive system that can help people live healthier lives.

2 Literature Review

Cardiovascular disease is a pressing global health issue which has the urgent need for precise forecasting models for early identification and intervention [8]. While extensive research has addressed cardiovascular disease prediction, there exists a noticeable gap in studies specifically targeting middle-aged individuals. The existing literature mainly focuses on broader populations but middle-aged adults constitute a pivotal demographic where proactive interventions could get significant benefits in averting cardiovascular risks and enhancing long-term outcomes. However, a review of existing literature, including the dataset under examination, exposes a scarcity of studies explicitly dedicated to cardiovascular disease prediction among middle-aged individuals. Despite this, several noteworthy papers as mentioned below have contributed to our understanding of predictive modelling in cardiovascular health with the dataset that concentrates on heterogenous ages. Other than that, the research conducted on various Heart disease dataset using reinforcement learning are highlighted in this section.

Mohd Syafiq Asyraf Suhaimi [9] discussed a heart disease prediction method using an ensemble of K-nearest neighbor, random forest, and logistic regression, achieving 82.42% accuracy, with an area under the ROC of 0.903 and an F1 score of 0.843, matching Random Forest accuracy. The study aims to use smartwatch datasets for future research. Stephen Akatore Atimbire et al. [10] used the Whale Optimization Algorithm (WOA) for feature selection, improving performance across ten classification models with five datasets, showing WOA's effectiveness. Ghulab Nabi Ahamad et al. [11] analyzed six machine learning algorithms for early-stage heart disease prediction, with extreme gradient boosting achieving 99.03% accuracy.

Prabavathi G. T. and Sindhu M. [12] introduced DNFH, a hybrid method using Weighted Transform K-Means Clustering and ensemble feature selection, significantly improving accuracy over SVM and LR models on the Cleveland dataset. Murat Lüy and Nuri Alper Metin [13] used Particle Swarm Optimization and Grey Wolf Optimizer for heart disease prediction with K-Nearest Neighbor. Xiao Yan Gao et al. [14] employed ensemble learning methods and evaluated bagging, with decision tree achieving the best performance with linear discriminant analysis (LDA) and principal component analysis (PCA) for feature selection.

Khaled Mohamad Almustafa's research [15] highlighted feature selection's role in improving classifier accuracy for heart disease prediction, especially with K-Nearest Neighbor, Decision Tree, and JRip classifiers. Kamepalli S.L. Prasanna et al. [5] proposed a reinforcement learning model using Q-learning to predict heart disease, outperforming traditional methods like KNN and DT. Thota Radha Rajesh et al. [16] introduced the Penguin Search Optimization Algorithm with Multi-agent Reinforcement Learning for disease prediction and recommendation (PSOAMRL-DPR), combining data from networks and cloud, showing improved performance. Wenqi Li et al. [6] proposed a Combined Reinforcement Multitask Progressive Time-Series Network (CRMPTN) for coronary heart disease forecasting, using deep reinforcement learning and progressive time-series networks, showing superior results.

Prominent contributions from multiple studies highlight how well ensemble approaches, optimization algorithms, and reinforcement learning frameworks predict CVD. There are still a few gaps in spite of these developments. The need for the identification of CVD with dynamic nature of cardiovascular risk factors is one significant gap. Many of the models in use today are based on static patient data and do not account for the ways in which risk variables interact and change over time. Forecasting accuracy can be improved by instigating dynamic modeling techniques like time-series analysis and frameworks for reinforcement learning that can conform according to new data.

The exploration of reinforcement learning techniques approaches that balance exploration and exploitation and dynamically select the most relevant models or features could significantly improve predictive performance and clinical utility. With these above observations, the significance of the proposed study becomes important. By introducing a novel approach which is Reinforcement Learning Dynamic Ensemble Selection with Customizable Actions and Exploration-Exploitation Balance, our study aims to bridge the existing gaps in the literature and compare the best model suitable for the selected dataset. With introduction of reinforcement learning techniques with customizable actions and a balanced exploration-exploitation strategy, our study tries to enhance prediction accuracy and intervention strategies, thereby facilitating early detection and intervention to mitigate cardiovascular risks.

3 Methodology

3.1 Data Description, Analysis and Preprocessing

The filtered heart disease dataset [2] under this study focuses mainly on middle-aged people specifically those aged between 40 and 60, containing 696 rows and 14 columns. The attributes include demographic details (age, sex), clinical measurements (blood pressure, cholesterol, heart rate), and diagnostic results (fasting blood sugar, ECG results, angina, ST depression, thalassemia). The target variable indicates the presence of cardiovascular disease (CVD).

Key statistics reveal an average patient with a higher proportion of males (73.99%) than females (26.01%). Chest pain type shows typical angina in 52.01% of cases. The mean blood pressure is 130.04 mm Hg, and serum cholesterol averages 243.47 mg/dl. Diagnostic results indicate 15.80% have elevated fasting blood sugar, 47.99% show ST-T wave abnormalities, and 34.34% experience exercise-induced angina. The slope of the peak exercise ST segment is flat in 62.79% of cases, and 60.06% have no major vessels colored. Thalassemia reveals 55.75% have a reversible defect, while 53.59% of patients are diagnosed with heart disease.

The dataset contains no missing values, with categorical variables encoded and numerical features standardized. Features were separated from the target variable, and the dataset was split into training and testing sets using an 80–20 ratio [17]. This preprocessing step is crucial for creating a robust model that generalizes well to unseen data.

3.2 Classifier Initialization and Hyperparameter Tuning

We initialized a diverse set of classifiers to ensure comprehensive coverage of various machine learning methods. The selected classifiers include Random Forest, K-Nearest Neighbors (KNN), Logistic Regression, Support Vector Classifier (SVC), Gaussian Naive Bayes, Decision Tree, Gradient Boosting, Multi-Layer Perceptron (MLP), Quadratic Discriminant Analysis (QDA), and XGBoost. Each classifier was chosen for its unique algorithmic approach and potential contribution to an effective ensemble model. For instance, SVC excels in high-dimensional spaces, KNN is a straightforward yet powerful instance-based learning technique, and XGBoost is known for its high performance across tasks. Our study aims to capture a wide spectrum of decision boundaries and identify their strengths within the ensemble.

Hyperparameter tuning is a critical step to enhance the performance of each classifier [11]. We employed GridSearchCV, a robust technique that systematically searches a predefined hyperparameter grid for each classifier, ensuring optimal hyperparameter identification to maximize performance. For Random Forest, hyperparameters include the number of trees (n_estimators) and maximum tree depth (max_depth). For KNN, the parameters consist of the number of neighbors (n_neighbors) and distance weighting (weights). Each classifier had a tailored grid of hyperparameters reflecting its specific needs.

3.3 Cross Validation and Evaluation Metrics

In our research, a five-fold cross-validation method was used to ensure the reliability and generalizability of the machine learning models. The dataset was split into five subsets, with each iteration training on four and validating on one, reducing the variance associated with a single train-test split. This process provides a comprehensive performance assessment for each classifier before integration into the ensemble model.

To evaluate the classifiers, we used metrics such as accuracy, precision, recall, F1 score, ROC AUC, and average precision [18]. These metrics offered a detailed understanding of classifier performance, especially in scenarios with class imbalances. By analyzing the mean and standard deviation across folds, the study identified the best-performing classifiers, informing their selection and weighting in the ensemble. The performance data from cross-validation also laid the foundation for the reinforcement learning framework, guiding dynamic ensemble selection based on reliable performance data. This approach ensures improved cardiovascular disease prediction among middle-aged individuals (Fig. 1).

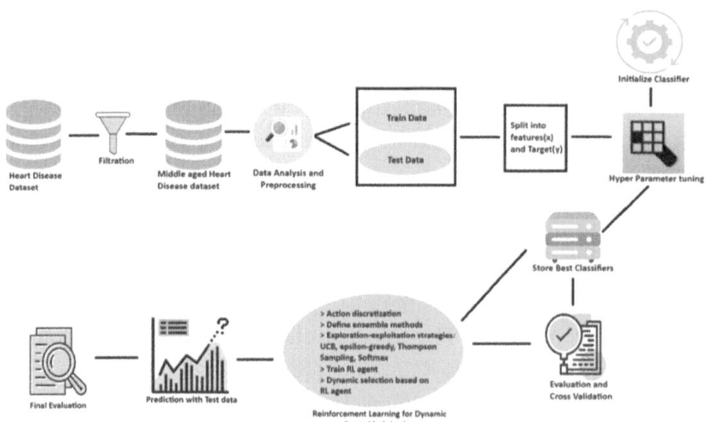

Fig. 1. Process Architecture

3.4 Reinforcement Learning for Dynamic Ensemble Selection

We used reinforcement learning (RL) to enhance how we choose the best ensemble methods for our model. Instead of manually selecting strategies, we approached this as a sequential decision-making problem where the RL agent learns from past performance to pick the most effective ensemble technique. The choices, or actions, available to the RL agent include methods like majority voting, weighted voting, and choosing the top-performing classifier. The exploration-exploitation strategy is crucial in reinforcement learning, as it allows the agent to test new methods while capitalizing on known successful techniques, ensuring an optimal blend of innovation and reliability in model performance.

A crucial part of our approach is simplifying the action space for ensemble selection. We do this by defining a set range of ensemble methods that the RL agent can select from, such as majority voting, weighted voting based on classifier performance, and picking the best single classifier. This discretization makes the decision-making process more manageable for the RL agent, allowing it to focus on a few well-defined strategies. It's essential for effectively exploring and leveraging various ensemble methods, ensuring that the most suitable strategy is used for each task. Dynamic ensemble selection refers to the process where, instead of relying on a single predefined method, the RL agent dynamically adjusts its strategy based on real-time performance feedback. This enables the model to handle changing patterns in the data and adapt to different circumstances.

To balance exploring new strategies with exploiting known ones, we use different techniques within the RL framework, including Upper Confidence Bound (UCB), epsilon-greedy, Thompson Sampling, and Softmax. The UCB method helps the RL agent explore less-tried strategies by attributing higher uncertainty to them. Epsilon-greedy allows for occasional random actions while primarily focusing on the best-known methods, preventing the agent from getting stuck in less optimal strategies.

Our dynamic selection process is designed to handle the unpredictability of medical data effectively. The RL agent continuously updates its knowledge with new data, fine-tuning its strategy based on the latest patterns. Feedback from ensemble predictions and their performance metrics helps the agent adjust and improve the classifier choices over time [7]. By combining various RL techniques to balance exploration and exploitation, our approach ensures that the ensemble model remains effective and responsive to new data, ultimately improving the prediction of cardiovascular disease among middle-aged individuals.

4 Results and Discussion

The results of the dynamic ensemble selection strategy using different approaches for balancing exploration and exploitation provide significant insights into the performance of various methods as mentioned in Table 1. The evaluation metrics include accuracy, precision, recall, F1 score, ROC AUC score, and average precision score, and they were obtained through rigorous cross-validation and hyperparameter tuning applied across ten different machine learning algorithms.

The Upper Confidence Bound (UCB) approach achieved an accuracy of 0.9799, with a precision of 1.0, recall of 0.9571, F1 score of 0.9781, ROC AUC score of 0.9786, and average precision score of 0.9773. The perfect precision indicates no false positives, while the recall shows the model correctly identified 95.71% of true positives, contributing to its strong F1 score. This performance highlights UCB's effective balance between exploration and exploitation in optimizing model learning.

Table 1. Results of experiment

Classifier	Accuracy	Precision	Recall	F1-Score	ROC AUC	PR-AUC
UCB	0.979	1	0.957	0.978	0.9786	0.977
Epsilon-Greedy	0.968	1	0.950	0.969	0.9693	0.970
Thompson Sampling	0.98	1	0.962	0.981	0.9813	0.983
Naïve Bayes	0.965	1	0.948	0.966	0.9667	0.967

The Epsilon-Greedy approach had slightly lower results, with an accuracy of 0.9680, precision of 1.0, recall of 0.9505, F1 score of 0.9689, ROC AUC score of 0.9693, and average precision score of 0.9705. Although it maintained perfect precision, the slightly lower recall resulted in a reduced F1 score, indicating a trade-off in performance.

The Thompson Sampling approach performed the best, achieving an accuracy of 0.9800, precision of 1.0, recall of 0.9625, F1 score of 0.9809, ROC AUC score of 0.9813, and average precision score of 0.9825. Its higher recall signifies better identification of true positives, leading to the highest F1 score. Thompson Sampling's effective balance between exploration and exploitation allowed it to leverage the dataset's potential fully (Table 2).

Table 2. Performance Metrics for Various Classifiers used in Thompson sampling

Classifier	Accuracy (Mean ± Std)	Precision (Mean ± Std)	Recall (Mean ± Std)	F1-Score (Mean ± Std)
Random Forest	0.9933 ± 0.0097	0.9970 ± 0.0060	0.9909 ± 0.0182	0.9938 ± 0.0090
KNN	0.9883 ± 0.0085	0.9881 ± 0.0112	0.9909 ± 0.0074	0.9895 ± 0.0077
Logistic Regression	0.8748 ± 0.0174	0.8662 ± 0.0228	0.9154 ± 0.0123	0.8900 ± 0.0145
SVC	0.8681 ± 0.0214	0.8521 ± 0.0364	0.9245 ± 0.0164	0.8861 ± 0.0153
Gaussian NB	0.8281 ± 0.0307	0.8209 ± 0.0289	0.8824 ± 0.0367	0.8501 ± 0.0268
Decision Tree	0.9867 ± 0.0113	0.9970 ± 0.0061	0.9788 ± 0.0206	0.9846 ± 0.0130
Gradient Boosting	0.9900 ± 0.0097	0.9912 ± 0.0118	0.9909 ± 0.0182	0.9909 ± 0.0090
MLP	0.8447 ± 0.0276	0.8487 ± 0.0437	0.8790 ± 0.0452	0.8710 ± 0.0165
QDA	0.8715 ± 0.0285	0.8538 ± 0.0329	0.9275 ± 0.0145	0.8889 ± 0.0231
XGB	0.9917 ± 0.0091	0.9941 ± 0.0073	0.9909 ± 0.0182	0.9924 ± 0.0085

The Softmax approach had the lowest performance metrics among the four strategies, with an accuracy of 0.9654, a precision of 1.0, a recall of 0.9479, an F1 score of 0.9663, an

ROC AUC score of 0.9667, and an average precision score of 0.9679. Despite achieving a precision of 1.0, the recall was the lowest at 0.9479, indicating it missed more true positives. This resulted in the lowest F1 score of 0.9663 among the methods evaluated.

Given the relatively small size of the dataset, achieving high precision and recall values across the methods is noteworthy as shown in Fig. 2. Smaller datasets often pose challenges such as increased performance variance and overfitting risks. However, Thompson Sampling and UCB demonstrated robust performance, suggesting their effectiveness even with limited data. The high precision across all methods indicates accurate identification of true positives, crucial for predicting conditions where false positives can be costly. High recall values, particularly in Thompson Sampling and UCB, show these methods capability to capture true positive instances effectively, indicating that the models were well-tuned to identify relevant patterns without missing critical instances. Overall, Thompson Sampling emerged as the best approach for this dataset, achieving the highest performance across multiple metrics, followed closely by UCB. The accomplishment of these approaches can be attributed to the careful hyperparameter tuning and robust cross-validation strategies employed, ensuring optimal model performance and generalization.

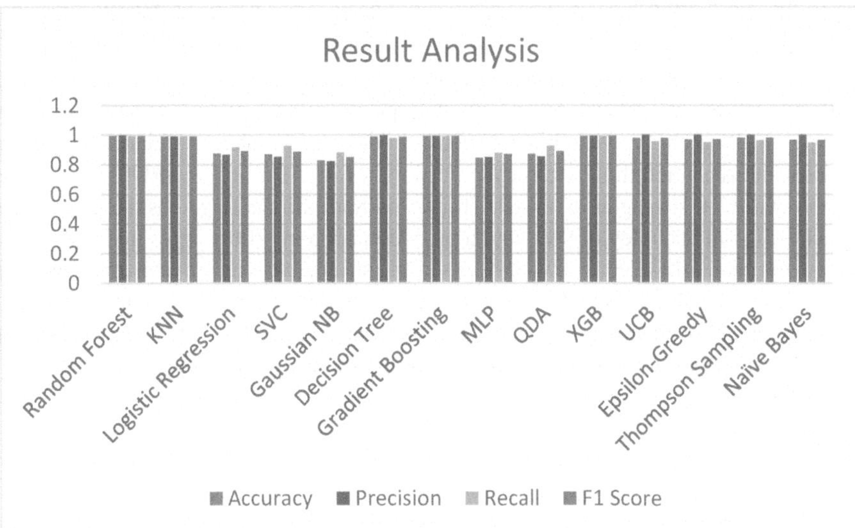

Fig. 2. Result Analysis

5 Future Enhancements

We have a small dataset, so applying data augmentation techniques for structured medical data, such as synthetic data generation using GANs or statistical methods, could artificially expand the dataset, providing more training examples. Transfer learning can overcome data limitations by pre-training models on larger datasets, improving initial performance and robustness of classifiers when used in an ensemble.

Additionally, future work can incorporate innovative ensemble methods and RL strategies tailored for small datasets. Bayesian optimization could fine-tune hyperparameters efficiently. Semi-supervised learning, combining labeled and unlabeled data, can improve training, especially where acquiring labeled medical data is challenging. Robust cross-validation techniques like nested cross-validation and stratified k-fold cross-validation should be considered to improve model generalizability and prevent overfitting. Another improvement is incorporating explainable AI (XAI) to interpret the RL-driven ensemble selection process, enhancing model trust and acceptance in clinical settings.

While our study broadly predicts CVDs, focusing on specific conditions like coronary artery disease (CAD) or hypertension could yield more impactful clinical results. Each CVD has distinct risk factors that affect prediction. Future research will develop specialized models for specific CVDs, improving prediction accuracy and clinical relevance.

6 Conclusion

This study highlights the reinforcement learning (RL) techniques in improving cardiovascular disease (CVD) prediction among middle-aged individuals. We developed a novel framework that employs dynamic ensemble selection with customizable actions and a balanced exploration-exploitation strategy, leading to significant enhancements in classifier performance. Through rigorous data preprocessing and hyperparameter tuning using GridSearchCV, we achieved robust model performance, validated by various evaluation metrics, including accuracy, precision, recall, F1 score, and ROC AUC score.

Among the RL strategies assessed, Thompson Sampling emerged as the most effective, achieving an accuracy of 98% and a precision score of 100%. This underscores its ability to dynamically select and weight ensemble strategies to improve prediction accuracy for those at risk of CVD. Overall, the integration of RL-based dynamic ensemble methods into CVD forecasting systems enhances early detection and risk assessment, allowing for timely interventions leading to improved health outcomes.

References

1. Aljanabi, M., Qutqut, M.H., Hijjawi, M., Al-Janabi, M.I.: Machine learning classification techniques for heart disease prediction: a review. Int. J. Eng. Technol. **7**, 5373–5379 (2018). https://doi.org/10.14419/ijet.v7i4.28646
2. Heart Disease Dataset Comprehensive. https://ieee-dataport.org/open-access/heart-disease-dataset-comprehensive
3. Swathy, M., Saruladha, K.: A comparative study of classification and prediction of Cardio-Vascular Diseases (CVD) using Machine Learning and Deep Learning techniques. ICT Express **8**, 109–116 (2022). https://doi.org/10.1016/j.icte.2021.08.021
4. Huang, R.C., Lillycrop, K.A., Beilin, L.J., et al.: Epigenetic age acceleration in adolescence associates with BMI, inflammation, and risk score for middle age cardiovascular disease. J. Clin. Endocrinol. Metab. **104**, 3012–3024 (2019). https://doi.org/10.1210/jc.2018-02076
5. Prasanna, K.S.L., Challa, N.P., Nagaraju, J.: Heart disease prediction using reinforcement learning technique. In: 2023 3rd International Conference on Advances in Electrical, Computing, Communication and Sustainable Technologies, ICAECT 2023 (2023)

6. Li, W., Zuo, M., Zhao, H., et al.: Prediction of coronary heart disease based on combined reinforcement multitask progressive time-series networks. Methods **198**, 96–106 (2022). https://doi.org/10.1016/j.ymeth.2021.12.009
7. Crepinsek, M., Liu, S.H., Mernik, M.: Exploration and exploitation in evolutionary algorithms: a survey. ACM Comput. Surv. **45** (2013)
8. Biswas, N., Ali, M.M., Rahaman, M.A., et al.: Machine learning-based model to predict heart disease in early stage employing different feature selection techniques. Biomed. Res. Int. **2023**, 6864343 (2023). https://doi.org/10.1155/2023/6864343
9. Suhaimi, M.S.A., Ramli, N.A., Muhammad, N.: Heart disease prediction using ensemble of k-nearest neighbour, random forest and logistic regression method. In: AIP Conference Proceedings (2024)
10. Atimbire, S.A., Appati, J.K., Owusu, E.: Empirical exploration of whale optimisation algorithm for heart disease prediction. Sci. Rep. **14**, 4530 (2024). https://doi.org/10.1038/s41598-024-54990-1
11. Ahamad, G.N., Shafiullah, F.H., et al.: Influence of optimal hyperparameters on the performance of machine learning algorithms for predicting heart disease. Processes **11**, 734 (2023). https://doi.org/10.3390/pr11030734
12. Prabavathi, G.T., Sindhu, M.: DNFH: data normalization and feature selection using hybrid methods for heart disease prediction. J. Namibian Stud. 35–36 (2023)
13. Lüy, M., Metin, N.A.: Classification of heart disease dataset with k-NN optimized by PSO and GWO algorithms. J. Comput. Electr. Electron. Eng. Sci. **1** (2023). https://doi.org/10.51271/jceees-0009
14. Gao, X.Y., Amin Ali, A., Shaban Hassan, H., Anwar, E.M.: Improving the accuracy for analyzing heart diseases prediction based on the ensemble method. Complexity **2021**, 6663455 (2021). https://doi.org/10.1155/2021/6663455
15. Almustafa, K.M.: Prediction of heart disease and classifiers' sensitivity analysis. BMC Bioinform. **21**, 1–18 (2020). https://doi.org/10.1186/s12859-020-03626-y
16. Rajesh, T.R., Rajendran, S., Alharbi, M.: Penguin search optimization algorithm with multi-agent reinforcement learning for disease prediction and recommendation model. J. Intell. Fuzzy Syst. **44**, 8521–8533 (2023). https://doi.org/10.3233/JIFS-223933
17. Rácz, A., Bajusz, D., Héberger, K.: Effect of dataset size and train/test split ratios in QSAR/QSPR multiclass classification. Molecules **26**, 1111 (2021). https://doi.org/10.3390/molecules26041111
18. Naidu, G., Zuva, T., Sibanda, E.M.: A review of evaluation metrics in machine learning algorithms. In: Lecture Notes in Networks and Systems (2023)

eLearning

Evaluating Deep Learning Embedding Techniques for Code Smell Detection

Praveen Singh Thakur[1(✉)], Mahipal Jadeja[1], Satyendra Singh Chouhan[1], and Santosh Singh Rathore[2]

[1] Department of CSE, MNIT Jaipur, Jaipur, India
{2021rcp9036,mahipaljadeja.cse,sschouhan.cse}@mnit.ac.in
[2] Department of Computer Science and Engineering, ABV-IIITM Gwalior, Gwalior, India
santoshs@iiitm.ac.in

Abstract. Detecting code smells in software systems is crucial for maintaining and evaluating their quality. Code smells are structural issues in the code that can make it harder to understand, modify, and ultimately lead to bugs. Many researchers have used machine learning (ML) techniques for automatic code smell detection. However, the effectiveness of these techniques depends majorly on the representation of the source code. This paper evaluates various code representation techniques, such as CodeBERT, CuBERT, and Code2vec, for the representation of the source code for effective code smell detection. These techniques convert source code into numerical vectors to understand the code's structure and semantics. We then employed various ML methods to build models that predict code smells. We assessed the presented techniques on four publicly available code smell datasets within the MLCQ repository. The results suggest that Code2vec embeddings were generally more accurate, achieving the highest F1 scores (ranging from 0.690 to 0.726) for Datasets 1, 2, and 3. However, Dataset 4 showed an exception. Here, CodeBERT performed better, achieving an F1-score of 0.774 compared to Code2vec's 0.669.

Keywords: Code Smell Detection · Software Maintainability · Empirical Study · Imbalance Learning · Source Code Representation

1 Introduction

Code smells are not errors or bugs within a software system but rather signify suboptimal software design and coding practices in software development. While software with code smells might still function correctly, these smells can make the code harder to understand and modify. If code smells are not handled, then they can lead to software failures. Generally, code smell occurs when programmers do not adopt the proper standard format for programming. Figure 1 illustrates an example of the God Class code smell. In Fig. 1(a), we can observe the presence

A. Dasgupta et al. (Eds.): BDA 2024, LNCS 15526, pp. 339–350, 2025.
https://doi.org/10.1007/978-3-031-81821-9_21

of the God code smell. Here, we can see that all of the work has been done by one class, "Library_Management". To resolve this issue, we can transfer some of the similar methods (represented with blue and green colors) into another class. Figure 1(b) illustrates the corrected version of the code, where the programmer has relocated methods (green and blue) into their respective class "Item" and "Catalog". Previous studies have shown that code smells can negatively impact developers' ability to maintain and understand software systems design [1].

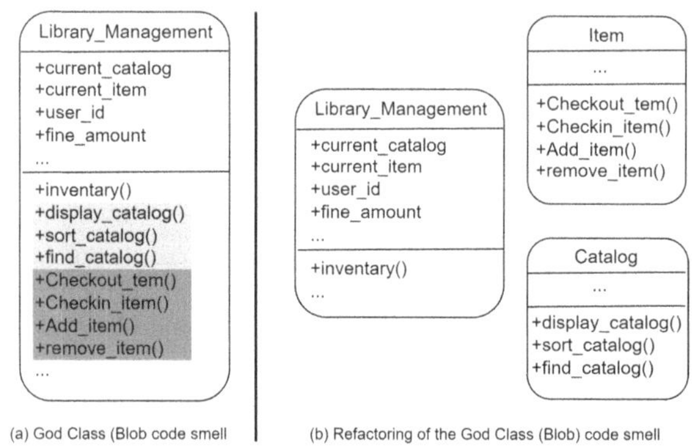

Fig. 1. Example to illustrate the God Class code smell (Color figure online)

Various research has been conducted to detect code smells in software systems. Some authors like Fernandes et al. [2], and Almashfi et al. [3] used various tools to detect code smells. They used inFusion, JDeodorant, PMD, and JSpIRIT tools to detect God Class, God Method, and Feature Envy code smells. These techniques extract metrics from the source code and, based on threshold values, segregate the smelly and non-smelly instances. However, the threshold value can differ for each type of code smell, affecting the performance of the code smell detectors. To address these issues, researchers have applied Machine Learning (ML) and Deep Learning (DL) based code smell detection techniques [4]. ML approaches, such as those by Fontana et al. [5], map relationships to detect code smells, while Pecorelli et al. [6] addressed data imbalance with oversampling techniques like SMOTE. Shen et al. [7] and Dewangan et al. [8] optimized classifiers by tuning hyper-parameters for various code smells. DL models (e.g., by Hui et al. [9] and Lin et al. [10]) automate feature extraction, creating complex mappings for accurate detection. Sharma et al. [11] incorporated CNN and RNN models for context-based learning, while Madeyski et al. [12] evaluated real-world datasets using the MCC metric. ML and DL techniques learn the data distribution based on the provided training instances. However, ML-based techniques rely on manually extracted metrics from the software. Some researchers have used DL techniques to automatically extract features from the source code to overcome this issue.

Our study aims to examine the effectiveness of different source code representations in code smell detection, particularly with the growing application of deep learning in software engineering. By comparing various embeddings and their performance with traditional machine learning models, we offer insights to guide future research in selecting suitable techniques for code analysis tasks. This paper explores DL-based source code representation techniques like CodeBERT, CuBERT, and Code2vec, which transform source code into vector representations used to train code smell detection models. We used nine machine learning techniques (detailed in Sect. 3.4) to build the detection models, evaluating them with the MLCQ repository across four code smell datasets [13]. Each dataset corresponds to a code smell type: `Blob` (excessive class responsibilities), `Data Class` (inadequate class features), `Feature Envy` (method dependence on another class's attributes), and `Long Method` (excessive method lines). Results indicate that Code2vec achieved the highest F1-scores (0.690 to 0.726) on Datasets 1, 2, and 3, while CodeBERT outperformed Code2vec on Dataset 4, achieving an F1-score of 0.774 compared to Code2vec's 0.669.

The remainder of the paper is structured as follows: Sect. 2 outlines the proposed framework in detail. Section 3 describes the experimental setup and implementation specifics. Section 4 presents an analysis of the experimental results. Finally, Sect. 5 concludes the paper and suggests future research directions.

2 Proposed Framework for Code Smell Detection

This section describes the proposed framework for code smell detection using ML and DL techniques. The proposed framework is divided into three steps: 1) Data preprocessing, 2) Numerical Representation of the Source Code (NRSC), and 3) Code smell detection model development. In Data Preprocessing, source codes are extracted from GitHub and stored locally. In the NRSC step, various code representation techniques (e.g., CuBERT, CodeBERT, code2vec) convert source code into vector form. These vectors are then used in ML techniques to build the code smell detector. The framework's graphical representation is shown in Fig. 2. The details of each step are given below.

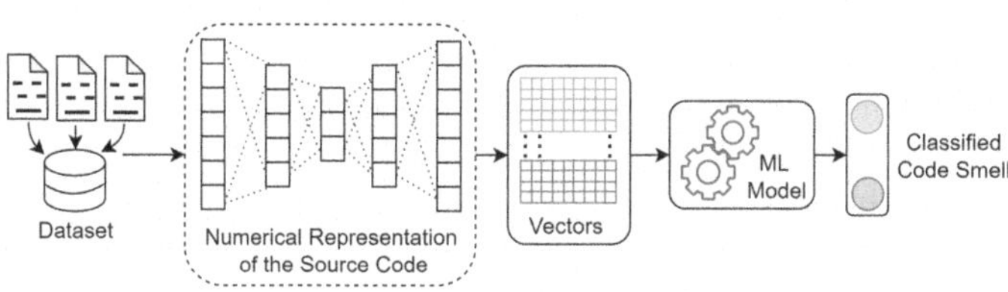

Fig. 2. Overview of the code smell detection framework

2.1 Data Preprocessing

We extract datasets from GitHub using the links provided by L. Madeyski [13], organizing them into four sub-directories for each code smell type: Data Class, Blob, Feature Envy, and Long Method. Each sub-directory contains True (with the corresponding code smell) and False (without the code smell) sub-directories. Non-functional links were skipped, leading to slightly different source code counts from the original dataset.

2.2 Numerical Representation of the Source Code (NRSC)

The numerical representation of source code is a crucial task, as the performance of code smell detectors largely depends on the quality of these representations. This step uses the CuBERT, CodeBERT, and Code2vec techniques to generate a vector (numerical representation) for each source code instance. A brief description of each technique used in the NRSC step is provided below.

1. **CodeBERT:** CodeBERT is a pre-trained model developed to understand both natural language (NL) and programming languages (PL) such as Python, Java, and JavaScript [14]. It connects NL and PL meanings, creating versatile representations. Built on a multilayer Transformer model, CodeBERT combines masked language modeling and replaced token detection during training to enhance learning. Trained on GitHub code from six languages with accompanying descriptions, CodeBERT can interpret these languages without requiring explicit language labels. In our framework, we use a pre-trained version of CodeBERT to produce embeddings. Each source code snippet from Step 1 is processed, generating a $768-dimensional$ vector.
2. **CuBERT:** CuBERT (Code Understanding BERT) is a BERT-based model trained on code snippets rather than natural language. Using the BERT Large model architecture with 16 attention heads over four layers, it generates $1024 - dimensional$ embeddings for tokens. Initially trained on Python by Kanade et al. [15], CuBERT was later adapted for Java code, which we employ in our experiments. After preprocessing Java code with a custom tokenizer, we pass it into CuBERT. Unlike CodeBERT, CuBERT blends NL and PL tokens during training without distinction. For each code sample, CuBERT generates a $1024 - dimensional$ embedding.
3. **Code2vec:** This technique uses a deep learning-based model to create a 384-dimensional vector for each code snippet [16]. Code2vec parses source code into an Abstract Syntax Tree (AST) and extracts paths between leaf nodes. Each path is a sequence of AST nodes connected by up and downlinks, capturing the syntactic structure. Code2vec processes these paths through an LSTM network to learn the structural relationships within code. After training on source code samples, it generates a $384-dimensional$ embedding for each code snippet. These vectors are subsequently used to train ML models for code smell detection.

2.3 Code Smell Detector

In this step, we used various ML techniques, described in Sect. 3.4, to build a code smell detector. The code smell detector used the vectors obtained from the NRSC step to train the ML models. Further, the trained model will used to detect code smells from the source code. The performance of the code smell detector depends on how real the NRSC provides the numerical (vector) representation of the source code.

3 Experimental Setup

This section provides information about the dataset, Machine Learning methods for developing code smell detectors, performance measures, and details about the experimental procedure. To guide the evaluation of the proposed framework, we formulated the following research questions and sought their answers.
RQ1. *Which machine learning techniques perform best with the embedding techniques CodeBERT, CuBERT, and Code2vec?*
RQ2. *Which embedding technique produces the most accurate results among the best-performing machine learning techniques?*

Table 1. Description of the datasets

Dataset	Types of Code smell	Actual #instances		After preprocessing	
		Smelly	Non-Smelly	Smelly	Non-Smelly
Dataset 1	Data Class	981	2700	465	1561
Dataset 2	God Class	876	2803	156	1864
Dataset 3	Feature Envy	392	2545	229	2367
Dataset 4	Long Method	668	2285	196	2414

3.1 Dataset Used

We conducted experiments using publicly available code smell datasets from L. Madeyski [13], which include links to source code categorized by severity and code smell type. The dataset consists of around 14,000 reviews from 26 software developers, covering 5,000 code snippets. These are classified into four code smells (Blob, Data Class, Feature Envy, Long Method) and four severity levels (none, minor, major, critical). For binary classification, we labeled minor, major, and critical severity instances as smelly (label 1) and none as non-smelly (label 0). The extracted source code was stored locally, with non-functional links excluded, resulting in adjusted source code counts from the original dataset. The original counts of smelly and non-smelly instances are detailed in Table 1.

Table 2. Description of the used performance measures

Performance Measures	Description
Precision	It shows how precise the code smell detection model is for smelly and non-smelly items. $Precision = \frac{T_{ss}}{T_{ss}+F_{ss}}$
Recall	Recall measures the validity of the code smell detection model by determining the counts of the correctly identified smelly items out of the total tested smelly items. $Recall = \frac{T_{ss}}{T_{ss}+F_{nss}}$
F1-Score	The F1-score is the harmonic mean of the precision and recall. It shows the overall performance of the code smell detection model with respect to both smelly and non-smelly items. $F1 - score = 2 \times \frac{precision \times recall}{precision + recall}$

3.2 Experimental Procedure

To evaluate the proposed framework, we conducted experiments on four code smell datasets, each with two classes: smelly and non-smelly (see Sect. 3.1). We generated source code vectors using embedding techniques such as CodeBERT, CuBERT, and Code2vec. For code smell detection, various machine learning techniques, detailed in Sect. 3.4, were applied using default parameters. We split the datasets into training (80%) and testing (20%) sets using random stratified sampling, ensuring class balance. To reduce bias, we repeated the experiment ten times for each dataset and reported the mean performance metrics as the final results.

3.3 Performance Measures

To evaluate our code smell detection model, we used metrics such as recall, precision, and F1-score, deliberately excluding accuracy. Accuracy is not included as it is unsuitable for performance evaluation with imbalanced datasets.

The performance measures are described in Table 2, based on the following confusion matrix parameters for the code smell prediction model: F_{nss} (False Non-smelly Samples), T_{nss} (True Non-smelly Samples), F_{ss} (False Smelly Samples), and T_{ss} (True Smelly Samples).

3.4 Techniques to Build Code Smell Detection Model

The Logistic Regression (LR) model establishes a relationship between dependent and independent variables to predict class probabilities [17]. Support Vector Machine (SVM) constructs a hyperplane to maximize the margin between classes for classification or regression [17]. Decision Tree (DT) forms a tree to classify instances [17], while Random Forest (RF) builds multiple trees from random samples [17]. Naive Bayes (NB) uses Bayes' theorem, assuming feature independence across datasets [18]. K-Nearest Neighbors (KNN) classifies instances based on the closest neighbors [17]. AdaBoost (AdaB) adjusts misclassified instance weights to build the model [17], while Gradient Boosting (GB) combines weak classifiers [19]. Extreme Gradient Boosting (XGB), an advanced GB, handles missing data and avoids overfitting [19].

4 Result Analysis and Discussion

In this section, we will analyze the results to find out the answers to the formulated research questions (RQs).

4.1 Result Analysis for RQ1

RQ1: Which machine learning techniques perform best with the embedding techniques CodeBERT, CuBERT, and Code2vec?

Tables 3, 4, and 5 shows the results of the built code smell detection models on all four datasets. Based on the results, the following inferences can be drawn. Here, the overall performance has been drawn based on the F1 score.

- Table 3 shows the results of the code smell prediction models using CodeBERT embedding techniques. The highest precision values are 0.581 and 0.570, produced by the LR and KNN classifiers for Datasets 1 and 3, respectively. For Datasets 2 and 4, the RF classifier produced the highest precision values of 0.494 and 0.798, respectively. The highest recall values of 0.566 and 0.586 were produced by the NB classifier for Datasets 1 and 3. The recall values of 0.500 and 0.754 were recorded for Datasets 2 and 4 using the RF classifier. Additionally, the highest F1-score for Dataset 1 is 0.543 using the NB classifier, while the highest F1-scores for Datasets 2, 3, and 4 are 0.497, 0.535, and 0.774, respectively, produced by the RF classifier.
- Table 4 shows that CuBERT produced the highest precision of 0.624, 0.556, and 0.718 using the NB, GB, and RF classifiers for Datasets 1, 3, and 4, respectively. However, for Dataset 2, CuBERT produced the highest precision of 0.493 using the SVM, DT, RF, AdaB, GB, and XGB classifiers. Additionally, the highest precision of 0.727, 0.485, 0.635, and 0.811 has been recorded using the LR, DT, NB, and LR classifiers for Datasets 1, 2, 3, and 4, respectively. Similarly, the highest F1-scores of 0.623, 0.489, 0.562, and 0.746 were recorded for Datasets 1, 2, 3, and 4 using the ML techniques LR, DT, GB, and RF, respectively.
- Table 5 presents the results recorded using the Code2vec embedding techniques. It shows that the RF classifier produced the highest precision values of 0.759, 0.733, and 0.695 for Datasets 1, 3, and 4, respectively. However, for Dataset 2, Code2vec achieved a precision of 0.726 using the LR and SVM classifiers. Similarly, the highest recall values for Datasets 1 and 4 are 0.679 and

Table 3. ML techniques performance using the CodeBERT embedding techniques

Metrics	LR	SVM	DT	RF	NB	KNN	AdaB	GB	XGB
	Dataset 1								
Precision	**0.581**	0.563	0.494	0.525	0.541	0.551	0.548	0.531	0.489
Recall	0.528	0.527	0.498	0.514	**0.566**	0.524	0.533	0.515	0.496
F1-score	0.524	0.526	0.483	0.512	**0.543**	0.522	0.535	0.511	0.481
	Dataset 2								
Precision	0.493	0.493	0.493	**0.494**	0.492	0.493	0.493	0.493	0.493
Recall	0.500	0.491	0.487	**0.500**	0.486	0.489	0.487	0.487	0.489
F1-score	0.496	0.492	0.490	**0.497**	0.426	0.491	0.490	0.490	0.491
	Dataset 3								
Precision	0.452	0.485	0.545	0.548	0.502	**0.570**	0.538	0.529	0.555
Recall	0.495	0.495	0.524	0.531	**0.586**	0.522	0.522	0.514	0.526
F1-score	0.473	0.484	0.527	**0.535**	0.406	0.523	0.524	0.513	0.525
	Dataset 4								
Precision	0.711	0.730	0.713	**0.798**	0.447	0.555	0.708	0.724	0.739
Recall	0.629	0.717	0.700	**0.754**	0.500	0.526	0.714	0.730	0.704
F1-score	0.658	0.723	0.706	**0.774**	0.462	0.530	0.710	0.727	0.719

Table 4. ML techniques performance using the CuBERT embedding techniques

Metrics	LR	SVM	DT	RF	NB	KNN	AdaB	GB	XGB
	Dataset 1								
Precision	0.627	0.623	0.529	0.563	**0.624**	0.514	0.593	0.598	0.558
Recall	**0.727**	0.720	0.532	0.580	0.721	0.523	0.642	0.653	0.577
F1-score	**0.623**	0.618	0.530	0.568	0.620	0.498	0.598	0.605	0.562
	Dataset 2								
Precision	0.492	**0.493**	**0.493**	**0.493**	0.491	0.491	**0.493**	**0.493**	**0.493**
Recall	0.473	0.478	**0.485**	0.482	0.395	0.396	0.478	0.482	0.484
F1-score	0.482	0.485	**0.489**	0.487	0.438	0.439	0.485	0.487	0.488
	Dataset 3								
Precision	0.538	0.525	0.484	0.544	0.548	0.494	0.523	**0.556**	0.532
Recall	0.584	0.553	0.471	0.562	**0.635**	0.484	0.542	0.593	0.545
F1-score	0.527	0.513	0.471	0.549	0.494	0.375	0.519	**0.562**	0.534
	Dataset 4								
Precision	0.666	0.671	0.673	**0.718**	0.583	0.621	0.685	0.674	0.694
Recall	**0.811**	0.786	0.693	0.786	0.738	0.808	0.791	0.774	0.780
F1-score	0.706	0.708	0.683	**0.746**	0.587	0.647	0.722	0.709	0.726

Table 5. ML techniques performance using the Code2vec embedding techniques

Metrics	LR	SVM	DT	RF	NB	KNN	AdaB	GB	XGB
	Dataset 1								
Precision	0.733	0.733	0.617	**0.759**	0.646	0.708	0.694	0.747	0.717
Recall	0.585	0.611	0.623	0.675	0.667	**0.679**	0.634	0.648	0.674
F1-score	0.581	0.619	0.619	**0.695**	0.650	0.689	0.646	0.665	0.687
	Dataset 2								
Precision	**0.726**	**0.726**	0.613	0.701	0.643	0.640	0.679	0.705	0.675
Recall	0.663	**0.673**	0.621	0.644	0.688	0.617	0.636	0.657	0.654
F1-score	0.680	**0.690**	0.617	0.659	0.630	0.624	0.648	0.672	0.662
	Dataset 3								
Precision	0.674	0.717	0.645	**0.733**	0.645	0.673	0.692	0.710	0.719
Recall	0.691	**0.730**	0.656	0.720	0.669	0.709	0.710	0.723	0.716
F1-score	0.680	0.723	0.650	**0.726**	0.648	0.674	0.698	0.715	0.717
	Dataset 4								
Precision	0.648	0.662	0.579	**0.695**	0.641	0.673	0.630	0.688	0.680
Recall	0.673	0.682	0.596	0.704	0.660	0.700	0.651	**0.709**	0.702
F1-score	0.652	0.667	0.570	**0.699**	0.645	0.678	0.633	0.694	0.686

0.709, recorded using the KNN and GB ML techniques, respectively. Whereas the highest recall values for Datasets 2 and 3 are 0.673 and 0.730, respectively, recorded using the SVM classifier. The highest F1-scores of 0.695, 0.726, and 0.699 are recorded using the RF classifier for Datasets 1, 3, and 4, respectively. However, for Dataset 2, the highest F1-score of 0.690 is recorded using the SVM classifier.

Outcomes of RQ1: *For the CodeBERT, the NB classifier for Dataset 1 and the RF classifier for Datasets 2, 3, and 4 have been selected as the best code smell detectors. For CuBERT, the ML techniques LR, DT, GB, and RF produced the best results for Datasets 1, 2, 3, and 4, respectively. For Code2vec, the RF classifier produced the best results for Datasets 1, 3, and 4. Whereas, for Dataset 2, SVM is selected as the best classifier.*

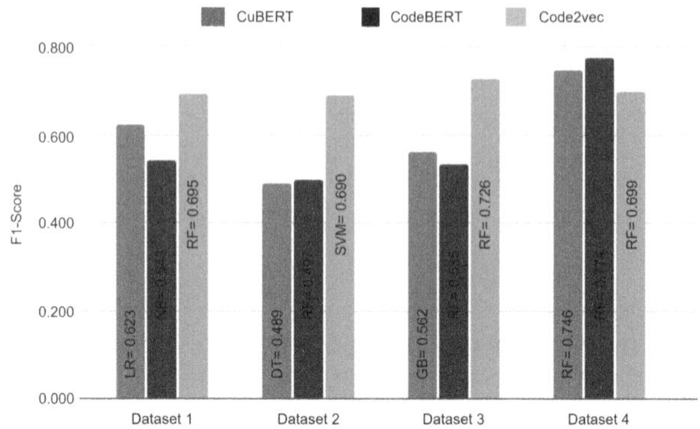

Fig. 3. Performance comparison of the CuBERT, CodeBERT, and Code2vec embedding techniques with their best-performing ML techniques for each dataset

4.2 Result Analysis for RQ2

RQ2: Which embedding technique produces the most accurate results among the best-performing machine learning techniques?

To find the supporting evidence to answer this research question, we selected the best ML techniques for each dataset with respect to each embedding technique (CodeBERT, CuBERT, and Code2vec). The overall performance of the ML techniques was measured using the F1-score. The final comparison between all three embedding techniques, based on their best-performing ML techniques, is shown in Fig. 3.

For Dataset 1, Code2vec produced the highest F1-score of 0.695 using the RF classifier. Similarly, for Dataset 3, Code2vec achieved an improved F1-score of 0.726 by employing the RF classifier. For Dataset 2, Code2vec again produced an improved F1-score of 0.690 using the SVM classifier, compared to the other embedding techniques. However, for Dataset 4, CodeBERT produced a more improved F1-score of 0.774 compared to the others.

Outcomes of RQ2: *Code2vec emerged as the best-performing embedding technique for Datasets 1, 2, and 3. However, for Dataset 4, CodeBERT produced the most improved results.*

4.3 Discussion

The experimental results show that Code2vec outperforms CuBERT and CodeBERT in generating more accurate embeddings. Code2vec uses AST representation to capture the structure of the source code and employs an LSTM-based model to understand node contexts. In contrast, CodeBERT and CuBERT treat the source code as plain text, which omits the flow and structural details, leading to less representative embeddings. However, for larger datasets with a lower imbalance ratio (Dataset 4), CodeBERT's pre-trained model proves more effective in capturing contextual information.

5 Conclusion

This paper presented an evaluation of embedding techniques combined with different ML techniques to detect code smells. We proposed a framework that leverages the advantages of various embedding techniques like CodeBERT, CuBERT, and Code2vec to generate vectors from source code. The framework employs various ML techniques to build a code smell detection model, which was tested on four publicly available MLCQ imbalanced datasets. The experimental results show that the Code2vec embedding technique produced more accurate embeddings, achieving the highest F1-scores ranging from 0.690 to 0.726 for Datasets 1, 2, and 3. However, for Dataset 4, Code2vec produced an F1-score of 0.669, slightly lower than CodeBERT, which achieved a higher F1-score of 0.774 for this dataset. In the future, we will explore different embedding techniques to provide better numerical source code representations.

References

1. Santos, J.A.M., Rocha-Junior, J.B., Prates, L.C.L., Do Nascimento, R.S., Freitas, M.F., De Mendonça, M.G.: A systematic review on the code smell effect. J. Syst. Softw. **144**, 450–477 (2018)
2. Fernandes, E., Oliveira, J., Vale, G., Paiva, T., Figueiredo, E.: A review-based comparative study of bad smell detection tools. In: Proceedings of the 20th International Conference on Evaluation and Assessment in Software Engineering, pp. 1–12 (2016)
3. Almashfi, N., Lu, L.: Code smell detection tool for Java script programs. In: 2020 5th International Conference on Computer and Communication Systems (ICCCS), pp. 172–176. IEEE (2020)
4. Lewowski, T., Madeyski, L.: Code smells detection using artificial intelligence techniques: a business-driven systematic review. Dev. Inf. Knowl. Manag. Bus. Appl. **3**, 285–319 (2022)
5. Fontana, F.A., Zanoni, M.: Code smell severity classification using machine learning techniques. Knowl.-Based Syst. **128**, 43–58 (2017)
6. Pecorelli, F., Di Nucci, D., De Roover, C., De Lucia, A.: On the role of data balancing for machine learning-based code smell detection. In: Proceedings of the 3rd ACM SIGSOFT International Workshop on Machine Learning Techniques for Software Quality Evaluation, pp. 19–24 (2019)
7. Shen, L., Liu, W., Chen, X., Gu, Q., Liu, X.: Improving machine learning-based code smell detection via hyper-parameter optimization. In: 2020 27th Asia-Pacific Software Engineering Conference (APSEC), pp. 276–285. IEEE (2020)
8. Dewangan, S., Rao, R.S., Mishra, A., Gupta, M.: A novel approach for code smell detection: an empirical study. IEEE Access **9**, 162869–162883 (2021)
9. Liu, H., Jin, J., Zhifeng, X., Zou, Y., Yifan, B., Zhang, L.: Deep learning based code smell detection. IEEE Trans. Softw. Eng. **47**(9), 1811–1837 (2019)
10. Lin, T., Fu, X., Chen, F., Li, L.: A novel approach for code smells detection based on deep leaning. In: Applied Cryptography in Computer and Communications: First EAI International Conference, AC3 2021, Virtual Event, 15–16 May 2021, Proceedings 1, pp. 171–174. Springer, Cham (2021)

11. Sharma, T., Efstathiou, V., Louridas, P., Spinellis, D.: Code smell detection by deep direct-learning and transfer-learning. J. Syst. Softw. **176** (2021)
12. Madeyski, L., Lewowski, T.: Detecting code smells using industry-relevant data. Inf. Softw. Technol. **155**, 107112 (2023)
13. Madeyski, L., Lewowski, T.: MLCQ: industry-relevant code smell data set. In: Proceedings of the 24th International Conference on Evaluation and Assessment in Software Engineering, pp. 342–347 (2020)
14. Feng, Z., et al.: CodeBERT: a pre-trained model for programming and natural languages. arXiv preprint: arXiv:2002.08155 (2020)
15. Kanade, A., Maniatis, P., Balakrishnan, G., Shi, K.: Learning and evaluating contextual embedding of source code. In: International Conference on Machine Learning, pp. 5110–5121. PMLR (2020)
16. Alon, U., Zilberstein, M., Levy, O., Yahav, E.: code2vec: learning distributed representations of code. Proc. ACM Program. Lang. **3**(POPL):1–29 (2019)
17. Jain, S., Saha, A.: Improving performance with hybrid feature selection and ensemble machine learning techniques for code smell detection. Sci. Comput. Program. **212**, 102713 (2021)
18. Berrar, D.: Bayes' theorem and Naive Bayes classifier. Encyclopedia Bioinform. Comput. Biol.: ABC Bioinform. **403**, 412 (2018)
19. Aydin, Z.E., Ozturk, Z.K.: Performance analysis of XGBoost classifier with missing data. Manchester J. Artif. Intell. Appl. Sci. (MJAIAS) **2**(02), 2021 (2021)

FedNorm: Optimised Federated Learning Strategy Using Hybrid Regularisation Norm

Binu John[1,2(✉)], Varun Menon[1,2], and Mainak Adhikari[3]

[1] Department of Computer Science and Engineering, SCMS School of Engineering and Technology, Ernakulam 683576, Kerala, India
{binujohn,varunmenon}@scmsgroup.org
[2] APJ Abdul Kalam Technological University, Thiruvananthapuram 695016, Kerala, India
[3] School of Data Science, Indian Institute of Science Education and Research, Vithura, Thiruvananthapuram 695551, Kerala, India
mainak@iisertvm.ac.in

Abstract. Federated Learning (FL) has emerged as a game-changing method for training machine learning (ML) models on decentralized data sources without centralizing the data itself. Conventional FL algorithms frequently exhibit performance instability and inefficiency, especially when working with non-Independent and Identically Distributed (non-IID) data distributions. To overcome these difficulties, we present FedNorm, a new FL method that uses hybrid regularisation approach to improve model stability and accuracy by balancing sparsity and smoothness, to increase generalization across varied data distributions. Our experimental setup uses the MNIST dataset to train models in both Independent and Identically Distributed (IID) and non-IID situations. The findings show that FedNorm routinely outperforms classic algorithms like FedAvg, FedSGD, FedNova and FedSAM, with minimal test loss and high accuracy. FedNorm has an exceptional accuracy rating of 0.98 on both IID and non-IID data. FedNorm addresses the constraints of conventional FL algorithms, resulting in a strong and dependable solution for FL in diverse situations.

Keywords: Federated Learning · Deep Learning · Regularization Norm · Collaborative Machine Learning

1 Introduction

FL has emerged as a potential model for collaborative ML that eliminates the necessity of raw data exchange. This method secures data privacy [1] by allowing multiple users to train models locally on their own data and then integrate the results. FL is especially beneficial in situations where data is spread across several

V. Menon and M. Adhikari—These authors contributed equally to this work.

A. Dasgupta et al. (Eds.): BDA 2024, LNCS 15526, pp. 351–360, 2025.
https://doi.org/10.1007/978-3-031-81821-9_22

sources and cannot be centralized for privacy, regulatory, or technological reasons [2]. It has been effectively used in a variety of fields, including healthcare, banking, and mobile apps. Federated Averaging (FedAvg) [3], a foundational approach in FL, involves clients training local models on their own datasets and then sending the resulting model updates to a central server for aggregation. Despite their simplicity and efficacy, FedAvg and comparable algorithms encounter various obstacles, such as data heterogeneity, communication cost, and scalability.

1.1 Related Work

FL has emerged as a promising approach to training models on decentralized data while maintaining privacy. McMahan et al. [3] introduced the FedAvg algorithm, which is one of FL's core algorithms. It works by training local models on each client's data and then averaging the model updates (weights) to create a global model. In [4] Kairouz et al. give a thorough review of FL, stressing FedAvg as a foundational method in the area. The article discusses several developments in FL, including algorithmic enhancements and practical implementation issues. Sattler et al. [5] investigate methods to enhance the robustness of model aggregation in FL, building upon the FedAvg algorithm by improving the stability and reliability of the FL process. The comparative analysis by Bektemyssova and Bakirova [6] evaluates algorithms like FedAdam, FedYogi, and FedSparse, noting that FedAvg is fundamental to many FL algorithms. Song et al. [7] introduce a new FL method named Federated Biased Learning (FedBiased) which emphasizes individual model knowledge by weighting the parameters of local models based on their performance (accuracy or error). Su et al. [8] provide an FL strategy that enhances model compression and convergence time using quantization and sparsification approaches.

FedNova [9], a federated optimization method, uses normalized average approach to solve issues in FL induced by variability in local datasets and processing capacity among clients. Li et al. [10,11] suggests FedProx, an extension of the FedAvg method that includes a proximal term to tackle heterogeneity. Smith et al. [12] investigate a Federated Multi-Task Learning (FMTL) system that enables each client to learn a personalized model while sharing information with other clients, hence improving overall performance. Giuseppe et al. [13] provides a decentralized FL technique designed specifically for hospital networks. Federated Stochastic Gradient Descent (FedSGD) is an adaptation of the classic Stochastic Gradient Descent (SGD) [14] for federated contexts. FedSGD computes gradients based on local data and updates their models accordingly. G. Huang and C. Lee [15] proposed P-PFedSGD (Personalised Privacy-Preserving Federated SGD) improves tool wear estimates by adding privacy-preserving approaches and personalized model adjustments, increasing prediction accuracy while decreasing communication costs.

1.2 Motivation and Contributions

According to the related study, the main issue with the FL environment is the objective inconsistency problem which refers to a mismatch between the model's global objective and its local objectives. Motivated by the key issues identified from the related work the main contributions of this research are:

- To solve the issue of objective inconsistency we propose the FedNorm algorithm for the FL environment. A novel hybrid regularization term which is a combination of L1 and L2 regularization is added to the loss function to penalize substantial departures from the global model, which helps to control model drift.
- Application of L2 regularization during local client training helps to mitigate overfitting and thereby enhance the global model's generalization capabilities.
- We conducted a thorough comparison of our proposed method against four classic FL algorithms like FedAvg, FedSGD, FedSAM, and FedNova by computing key metrics like test accuracy, loss, latency, communication overhead, and training time.

The FedNorm algorithm is intended to reduce the effects of data heterogeneity by including personalized model updates. Each client trains a local model and uses hybrid norm regularisation to reduce the difference between local and global models. This guarantees that the global model gains from the distinct properties of each local dataset while preserving overall performance.

The rest of the paper is organized as follows: Sect. 2 discusses the system model and problem formulation. Section 3 discusses the proposed FedNorm model. Section 4 is about the experimental environment, results, and discussions and Sect. 5 concludes our work with future directions.

2 System Model and Problem Formulation

The following section details the system model followed by the problem formulation of the proposed strategy.

2.1 System Model

We propose an FL framework that consists of N clients and a server. For every $t \in [N]$, the t-th client holds d training samples $\xi_t = (X_i, Y)$ drawn from local dataset D_t where X_i is the input feature vector and Y is the label corresponding to X_i. D_t may vary from client to client, as in the real-world scenario data heterogeneity is a challenge faced by FL. Let $F_t(w)$ be the training loss function of the client t, as given in Eq. 1.

$$F_t(w) := \mathbb{E}_{\xi_t \sim D_t}\left[L_t(w, \xi_t)\right] \tag{1}$$

where $\mathbb{E}_{\xi_t \sim D_t}[\cdot]$ computes the expected (average) value of the loss function over all possible data points ξ_t that follow the distribution D_t, $L_t(w, \xi_t)$ is the per data loss function. The most commonly used loss function is Mean Squared Error (MSE) [16].

Latency refers to the time delay between sending a request and receiving a response, and it is calculated using the expression provided in Eq. 2.

$$L_i = \sum_{i=1}^{N} T_{\text{local},i} + \sum_{i=1}^{N} (T_{\text{upload},i} + T_{\text{download},i}) + T_{\text{sync}} \tag{2}$$

where the local computation time for client i is denoted as $T_{\text{local},i}$, the time taken to upload this local model from client i to the server is indicated by $T_{\text{upload},i}$, similarly, $T_{\text{download},i}$ refers to the time required to download the global model from the server to the client i. Additionally, T_{sync} represents the synchronization delay, which is the time delay involved in synchronizing the operations across the clients and the server. Communication overhead [17] for client i is computed using the expression given in Eq. 3.

$$CO_i = \sum_{i=1}^{N} \left(|\theta_i| + |\theta^G|\right) \tag{3}$$

where $\|\theta_i\|$, represents local parameter and $\|\theta_G\|$ refers to the global model parameter.

2.2 Problem Formulation

In our research, we have tried to optimize the key issues identified from the related study as follows.

- **Problem 1**: Existing FL frameworks lack proper handling of the heterogeneous nature of data distribution (non-IID data) across all the clients which may lead to model drift where the local model drifts significantly from the global model resulting in instability and slow model convergence So our main aim is to enhance the model performance by improving the model accuracy by minimizing loss function.

$$\min_{\theta_t} L_t(\theta_t, R(\theta_i, \theta_t^G)) \tag{4}$$

  where θ_i is the local model weights for the i-th client, and θ_t^G is the global model weights after aggregating local models at round t.
  This in turn minimizes the latency and communication overhead as the hybrid norm regularization can implicitly promote sparsity in the model updates (through the L1 component) so only fewer parameters need to be communicated.
- **Problem 2**: Secondly we have tried to optimize the hyperparameters used for model tuning. Hyperparameter adjustment in FL is primarily intended to increase the overall performance of the global model. This entails adjusting parameters like the learning rate, batch size, number of local training epochs, and regularisation terms.

$$\begin{aligned} &\min_{\theta_t} L_t(\theta_t, h_t) \\ &\text{Subject to: } \mathrm{opt}(h_t) \\ &h_t \in H \end{aligned} \tag{5}$$

where H represents the optimal hyperparameters.

- **Problem 3**: The third challenge was to minimize latency and communication overhead by reducing the size of the model parameters exchanged between the client and server and also by reducing the training time.

$$\min\left(LT_k, C_k, TT_k\right) \tag{6}$$

where LT_k, C_k, TT_k are latency, communication overhead, and training time for the kth round.

Subject to:

$$\begin{aligned} LT_k &= \sum_{i=1}^{N} T_{\text{local},i} + \sum_{i=1}^{N} (T_{\text{upload},i} + T_{\text{download},i}) + T_{\text{sync}} \\ C_k &= \sum_{i=1}^{N} \left(|\theta_i| + |\theta^G|\right) \end{aligned} \tag{7}$$

3 FedNorm: Optimised Federated Learning Strategy

In this study, we present the FedNorm algorithm with hybrid norm regularisation to solve the shortcomings of current FL methods. We use a Multi-Layer Perceptron (MLP) architecture combined with a hybrid norm regularisation strategy to increase model performance and resilience. The hybrid norm combines L1 and L2 regularisation to balance the trade-off between sparsity and weight decay, allowing the model to generalize more effectively across different datasets. The L2 norm ensures that models do not deviate too far from the global model, maintaining consistency, while the L1 norm encourages sparsity, which can help in adapting to client-specific data. This helps in achieving better performance on non-IID data by allowing some flexibility for local adaptations while keeping the overall model coherent.

Algorithm 1 shows the training, updation, and aggregation process of the proposed framework. We compute the hybrid regularization using the following equation in our proposed algorithm.

$$R(\theta_i, \theta_t^G) = \lambda_2 \|\theta_i - \theta_t^G\|_2^2 + \lambda_1 \|\theta_i - \theta_t^G\|_1 \tag{8}$$

where $\lambda_2 \|\theta_i - \theta_t^G\|_2^2$ is the L2 norm regularization term, and $\lambda_1 \|\theta_i - \theta_t^G\|_1$ is the L1 norm regularization term. The values of λ_1 and λ_2 determine the strength of the regularization. When the values of λ_1 and λ_2 are too small, they will have little effect on the weights, allowing the model to fit the training data more

Algorithm 1. FedNorm with Hybrid Norm Regularization

Initialize:
Global model θ^G, Local models θ_i for each client i, Regularization parameters λ_2 for L2 norm and λ_1 for L1 norm, Learning rate α, Number of clients N
for each round t of federated learning **do**
  Server Side:
  Distribute the global model θ_t^G to each client i
  for each client i **do**
    Client Side:
    Initialize local model $\theta_i = \theta_t^G$
    for each local update k **do**
      Compute the hybrid regularization term:

$$R(\theta_i, \theta_t^G) = \lambda_2 \|\theta_i - \theta_t^G\|_2^2 + \lambda_1 \|\theta_i - \theta_t^G\|_1$$

      Total loss:

$$L_i^{\text{total}}(\theta_i) = L_i(\theta_i) + R(\theta_i, \theta_t^G)$$

      Update local model parameters:

$$\theta_i \leftarrow \theta_i - \alpha \nabla_{\theta_i} L_i^{\text{total}}(\theta_i)$$

    end for
    After K local updates, send the updated local model θ_i to the server
  end for
  Server Side:
  Aggregate local models to update the global model:

$$\theta_{t+1}^G = \frac{1}{N} \sum_{i=1}^{N} \theta_i$$

end for
Repeat until convergence

closely. Moderate values can help prevent overfitting while still allowing the model to capture important patterns in the data. When the model is overfitting (i.e., it performs well on training data but poorly on validation data), large regularization parameters will penalize large weights more heavily, reducing the model's capacity to overfit the training data. The regularization term is added to the loss function to calculate the total loss as expressed in Eq. 9.

$$L_i^{\text{total}}(\theta_i) = L_i(\theta_i) + R(\theta_i, \theta_t^G) \tag{9}$$

The model update formula for the local-global model is given as follows.

$$\theta_i \leftarrow \theta_i - \alpha \nabla_{\theta_i} L_i^{\text{total}}(\theta_i) \tag{10}$$

The server aggregates the model updates using the expression given in the following equation.

$$\theta_{t+1}^{G} = \frac{1}{N} \sum_{i=1}^{N} \theta_i \tag{11}$$

4 Experimental Analysis

The following section describes the experimental setup for the research work.

4.1 Experimental Baseline

For all instances during our experimentation, the Modified National Institute of Standards and Technology (MNIST) dataset [18] was used, which contains 60,000 training and 10,000 testing samples of handwritten digits ranging from 0 to 9. The network design for the IID configuration includes a Multi-Layer Perceptron (MLP) with two hidden layers of 128 neurons each, whereas the non-IID arrangement employs an MLP with three hidden layers with 256,256,128 neurons respectively. Both setups use the ReLU activation function for the hidden layers and the Softmax function for the output layer. The learning rate (η) was set at 0.001 in both scenarios. The batch size was 64 for both IID and non-IID data environments. Each configuration was trained in 20 rounds, with the IID setup running 5 epochs every round and the non-IID setup running 1 epoch per round. The regularisation parameters λ_1 and λ_2 were set at 0.001. The categorical cross-entropy loss function was applied to both IID and non-IID cases.

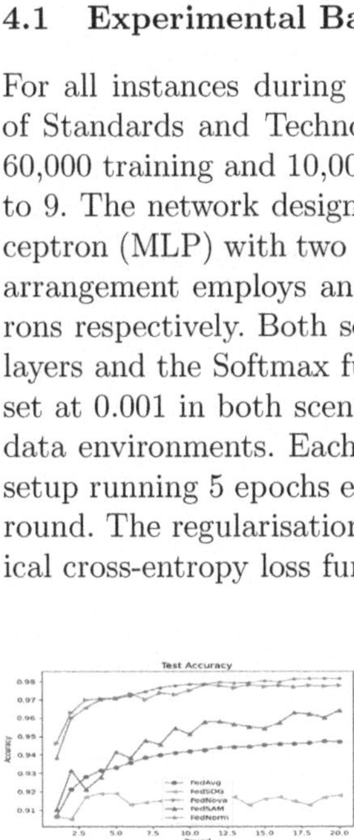

Fig. 1. Test Accuracy

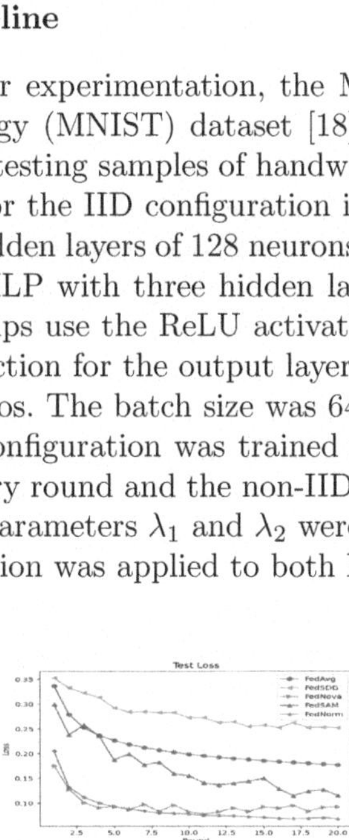

Fig. 2. Test Loss

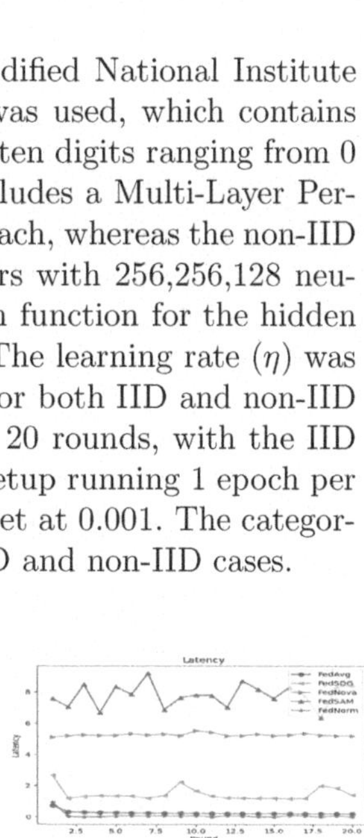

Fig. 3. Latency

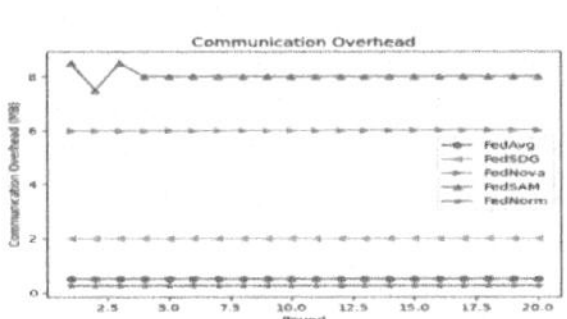

Fig. 4. Communication overhead

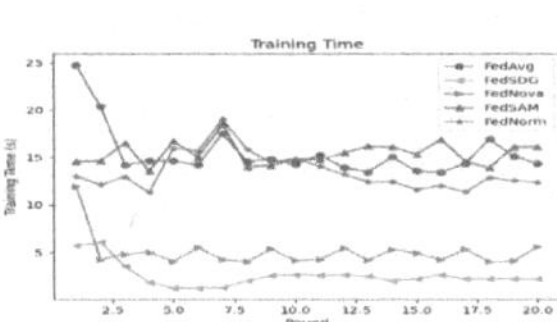

Fig. 5. Training time

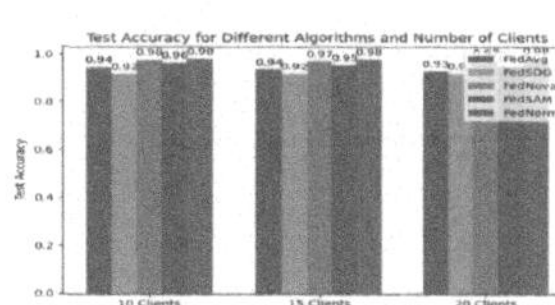

Fig. 6. Model Accuracy with number of clients = 10, 15, 20

4.2 Results and Discussions

This section discusses the performance of the proposed algorithm on the MNIST dataset in both IID and non-IID environments.

4.2.1 Results on IID Data

In this part, we provide a complete analysis of the simulation results gained by analyzing four well-known federated learning algorithms: FedAvg, FedSDG, FedNova, FedSAM, and the suggested FedNorm. The simulation results show that FedNorm outperforms test accuracy and loss, indicating its efficacy in improving model convergence and predictive power. However, these advantages come at a cost in terms of higher training time and communication overhead.

The test accuracy graph in Fig. 7 shows that FedNorm consistently outperforms the other algorithms, with an accuracy close to 0.98 by round 20. The test loss graph in Fig. 2 depicts that the FedNorm has the most efficient reduction in test loss, starting below 0.10 and converging to a minimum loss value by the conclusion of the rounds. The latency graph in Fig. 3 measures the time delay in communication between the central server and the client devices. FedAvg and FedNorm have the lowest latency, suggesting efficient round completion times. Figure 4 depicts the comparison of communication overhead. The graph shows that FedAvg and FedNorm have the lowest communication overhead, with constant and low values throughout the training rounds. The training time graph in Fig. 5 shows the computing efficiency of each approach. FedNorm has a training time higher than FedSGD and FedNova but less than the other two algorithms as the training progresses. The graph in Fig. 6 depicts the test accuracy of five federated learning algorithms at three distinct client configurations (10, 15, and 20 clients).

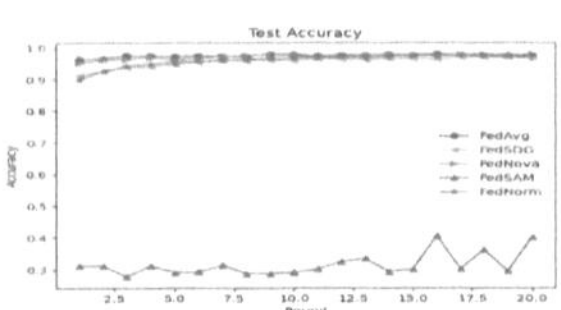

Fig. 7. Test Accuracy

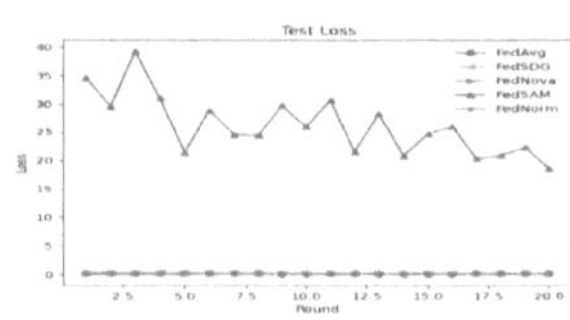

Fig. 8. Test Loss

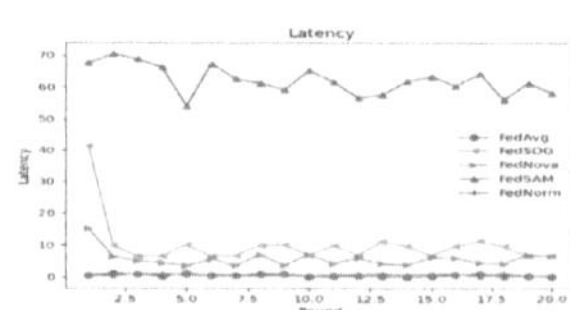

Fig. 9. Latency

Fig. 10. Communication overhead

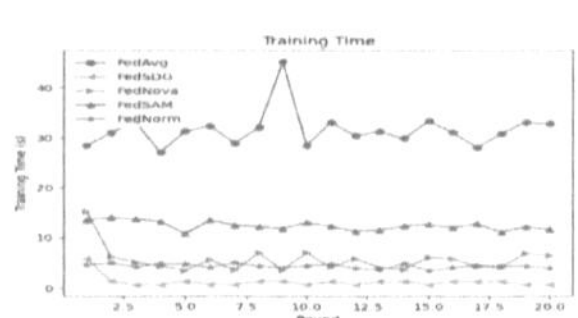

Fig. 11. Training time

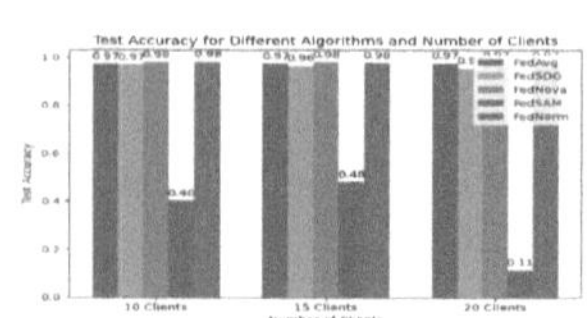

Fig. 12. Model Accuracy with number of clients = 10, 15, 20

4.2.2 Results on Non-IID Data

Figure 7 shows that FedNorm is very dependable in making correct forecasts, but FedSAM has substantially lower and variable accuracy, demonstrating FedNorm's resilience in handling non-IID data. The test loss results of the five federated algorithms on non-IID data are depicted in Fig. 8. FedNorm maintains a consistently low test loss across rounds, showing excellent error minimization during training. Figure 9 shows the latency plot of different algorithms. FedNorm has moderate latency, equivalent to FedSGD and FedNova, whereas FedAvg has the lowest and FedSAM the greatest. FedNorm has one of the lowest training times each round, along with FedSGD and FedNova, while FedAvg has the most as shown in Fig. 10. FedNorm has a consistent and minimal communication overhead throughout all rounds, as in Fig. 11. This demonstrates that FedNorm is efficient in terms of communication costs, making it appropriate for federated learning situations. FedNorm maintains good test accuracy over varying customer numbers (10, 15, 20), which is comparable to FedAvg, FedSDG, and FedNova. Despite a modest drop in accuracy with 20 clients, FedNorm's performance remains strong as projected in Fig. 12.

5 Conclusion and Future Scope

In our research, we tackled the fundamental issues provided by current FL algorithms, including their performance instability and inefficiency when dealing with non-IID data distributions. To address these difficulties, we presented the FedNorm algorithm with hybrid norm regularisation to solve the shortcomings of current FL methods. Our empirical results show that FedNorm consistently maintains minimal test loss and excellent test accuracy, with an exceptional accuracy rate of 0.97 on both IID and non-IID data. Furthermore, FedNorm has minimal latency and training timeframes, much surpassing FedSAM in terms of processing speed and resource utilization. Future research could focus on improving FedNorm's scalability and efficiency in more complicated FL contexts with varied client capabilities and network circumstances.

References

1. Zhang, C., Xie, Y., Bai, H., Yu, B., Li, W., Gao, Y.: A survey on federated learning. Knowl.-Based Syst. **216**, 106775 (2021)
2. Zahri, S., Bennouri, H., Chehri, A., Abdelmoniem, A.M.: Federated learning for IoT networks: enhancing efficiency and privacy. In: 2023 IEEE 9th World Forum on Internet of Things (WF-IoT), pp. 1–6 (2023). https://doi.org/10.1109/WF-IoT58464.2023.10539528
3. McMahan, B., Moore, E., Ramage, D., Hampson, S., Arcas, B.A.: Communication-efficient learning of deep networks from decentralized data. In: Artificial Intelligence and Statistics, pp. 1273–1282. PMLR (2017)
4. Kairouz, P., et al.: Advances and open problems in federated learning. Found. Trends® Mach. Learn. **14**(1–2), 1–210 (2021)

5. Sattler, F., Wiedemann, S., Müller, K.R., Samek, W.: Robust and communication-efficient federated learning from non-IID data. IEEE Trans. Neural Netw. Learn. Syst. **31**(9), 3400–3413 (2019)
6. Bektemyssova, G., Bakirova, G.: Comparative analysis of federated machine learning algorithms. Sci. J. Astana IT Univ. 57–67 (2024)
7. Song, M., Zhao, X., Pedrycz, W.: A robust federated biased learning algorithm for time series forecasting (2024)
8. Su, X., Zhou, Y., Cui, L., Sheng, Q.Z., Wang, Y., Guo, S.: Fast-convergent wireless federated learning: a voting based topk model compression approach. IEEE J. Sel. Areas Commun. 1 (2024). https://doi.org/10.1109/JSAC.2024.3431568
9. Wang, J., Liu, Q., Liang, H., Joshi, G., Poor, H.V.: A novel framework for the analysis and design of heterogeneous federated learning. IEEE Trans. Signal Process. **69**, 5234–5249 (2021). https://doi.org/10.1109/TSP.2021.3106104
10. Li, T., Sahu, A.K., Zaheer, M., Sanjabi, M., Talwalkar, A., Smith, V.: Federated optimization in heterogeneous networks. Proc. Mach. Learn. Syst. **2**, 429–450 (2020)
11. Hazra, A., Adhikari, M., Nandy, S., Doulani, K., Menon, V.G.: Federated-learning-aided next-generation edge networks for intelligent services. IEEE Netw. **36**(3), 56–64 (2022)
12. Smith, V., Chiang, C.K., Sanjabi, M., Talwalkar, A.S.: Federated multi-task learning. In: Advances in Neural Information Processing Systems, vol. 30 (2017)
13. Giuseppi, A., Manfredi, S., Menegatti, D., Poli, C., Pietrabissa, A.: Decentralised federated learning for hospital networks with application to Covid-19 detection. IEEE Access **10**, 92681–92691 (2022). https://doi.org/10.1109/ACCESS.2022.3202922
14. Stich, S.U.: Local SGD converges fast and communicates little. arXiv preprint arXiv:1805.09767 (2018)
15. Huang, G.Y., Lee, C.H.: Industrial federated learning algorithm (P-PFedSGD) for tool wear estimation. Futur. Gener. Comput. Syst. **158**, 150–157 (2024)
16. Zheng, J., Zhang, H., Kang, J., Gao, L., Ren, J., Niyato, D.: Covert federated learning via intelligent reflecting surfaces. IEEE Trans. Commun. **71**(8), 4591–4604 (2023). https://doi.org/10.1109/TCOMM.2023.3281880
17. Lee, S., Zhang, T., Avestimehr, A.S.: Layer-wise adaptive model aggregation for scalable federated learning. In: Proceedings of the AAAI Conference on Artificial Intelligence, vol. 37, pp. 8491–8499 (2023)
18. Deng, L.: The MNIST database of handwritten digit images for machine learning research [best of the web]. IEEE Signal Process. Mag. **29**(6), 141–142 (2012)

Efficient Self-Supervised Contrastive Learning with Representation Mixing

Yash Kumar Sharma[1(✉)], Akshay Badola[1], Vineet Padmanabhan[1], Wilson Naik[1], and Abdul Sattar[2]

[1] School of Computer and Information Sciences, University of Hyderabad, Hyderabad 500046, Telangana, India
{21mcpc11,badola,vineetnair,rathore}@uohyd.ac.in

[2] Institute for Integrated and Intelligent Systems, Griffith University, Nathan Campus, Brisbane, QLD 4111, Australia
a.sattar@griffith.edu.au

Abstract. Self-Supervised Learning (SSL) has emerged as a powerful method for pre-training image and language models for downstream tasks. SSL does not require labeled data for learning large models, whose features can then be adapted with supervised learning for required tasks, which makes it particularly amenable for cases with a paucity of labeled data. SSL in the context of images, typically relies on defining a pretext task, such as image inpainting, image shuffling, image contrastive learning, etc. A big challenge for SSL, however, is the computational requirements. Training with SSL takes much longer than with Supervised Learning and hence requires more computation power than Supervised Learning. This is an area of concern with the energy consumption for *deep learning* rising exponentially over the last few years. In this work, we develop a novel Representation Mixing method for Self-Supervised Contrastive Representation Learning that trains much faster than standard Contrastive Self-Supervised Learning methods. We utilize a feature transformation strategy by combining augmented and normalized views at the end of the network to form feature representations that are further apart in the learned embedding space. We conduct experiments on various benchmark datasets to validate our claims and show equivalent downstream accuracy while reducing pre-training. We also release our code for other researchers (https://github.com/Yash-Kumar-Sharma/Efficient-Self-Supervised-Contrastive-Learning-with-Representation-Mixing).

Keywords: Self-Supervised Learning · Contrastive Learning · Feature Representation · Data Augmentation

1 Introduction

Self-Supervised Learning (SSL) [9,11,17]. SSL is a paradigm for estimating statistical models with explicitly labeled data. Self-Supervised Representation Learning (SSRL) is a *Representation Learning* method where the aim is to learn

A. Dasgupta et al. (Eds.): BDA 2024, LNCS 15526, pp. 361–371, 2025.
https://doi.org/10.1007/978-3-031-81821-9_23

useful (usually deep) representations of data without labels which can be utilized in some downstream task.

In the context of Image Representation Learning a pretext task like in-painting or un-shuffling an image is often used. In the absence of any labels, vision SSRL models are usually trained for a large number of epochs (e.g., SIM-CLR [3] report results on 1000 epochs for their experiments while supervised training in ResNet [8] is for about 100 epochs). The slow rate of convergence can be attributed to the weak training signal due to a lack of labels. We focus on this particular aspect in our work. We propose a novel *representation mixing* method which enhances the distance between *positive pairs* and *negative pairs* for Self-Supervised Image Classification. We refer to this property as a *score gap* which creates a stronger training signal and helps in model convergence improving the sample efficiency of training.

2 Related Work

Seminal work in this area is by Chen et al. [3] which generates augmented positive/negative pairs from a set of images, by taking multiple views of the same image as positive versions and a random selection of images from the same training batch as negatives. Other work includes Caron et al. [2], who forego the positive/negative pairs and instead cluster the representations and contrast them with prototypes which they compute online. Momentum Contrast [4,7] combine contrastive learning with a dictionary learning approach. Tian et al. [14] try to learn representations that maximize mutual information between differing views of a given image. Van den Oord et al. [16] attempt to identify a unifying contrastive learning strategy with autoregressive modeling.

Our approach differs from the above as:

1. Ours is a simple method which is as efficient as Simclr [3] but generates better positive and negative samples implicitly via mixing.
2. Unlike [10,15], we generate **both good positive and negative pairs** via representation mixing.
3. Unlike Tian et al. [15] **we only employ standard augmentations** at the input stage.
4. Unlike Kalantidis et al. [10] we **do not attempt to rank the hard negatives** which adds complexity to the training procedure.

2.1 Our Approach

We propose to generate contrastive pairs for Self-Supervised Image Representation Learning by mixing representations before the output layer, such that, 1) The *positive* pairs are closer and 2) the *negative* embedding pairs are further.

For an anchor image and a counter-image $(\mathbf{x}_i, \mathbf{x}_j) \overset{i \neq j}{\sim} X$, we create two separate views with standard Imagent augmentation [8] and normalization. These feature representations are then mixed at the pre-final layer of a network f_{net}

to create a **hard positive** and a **hard negative** representation as following, which accelereates the training procedure:

$$\mathbf{h}_p = f_{net}(A_g(\mathbf{x}_i)) + f_{net}(A_n(\mathbf{x}_j))$$
$$\mathbf{h}_n = f_{net}(A_g(\mathbf{x}_j)) + f_{net}(A_n(\mathbf{x}_i))$$

More details of our approach are given in Sect. 4.

2.2 Our Contributions

We summarize our contributions below:

1. We propose a novel representation mixing strategy which speeds up Self-Supervised image representation learning.
2. We demonstrate our approach with various standard image datasets like CIFAR-10, CIFAR-100 and Imagenet.
3. We analyze the generated representations to show that the positive and negative pairs are farther on the unit sphere, which makes them harder to classify and leads to more efficient training.
4. We release the code for our experiments.

3 Efficient Self-Supervised Contrastive Representation Learning

Standard SSCRL aims to learn the weights θ of some Neural Network, such that for multiple views $\mathbf{x}_{i_k}$ of $\mathbf{x}_i$ generated by a data augmentation pipeline (say $A_g, \mathbf{x} = A_g(x)$), the inter-class distance of the ith sample and *rest of the data* is maximized and intra-class distance i^{th} sample is minimized. As mentioned in Sect. 1, the lack of labels requires many more training iterations than for supervised learning. The performance of Self-Supervised Learning is also critically dependent on the type of data augmentation and the number of positive views and positive-negative pairs generated.

The intuition behind our approach relies on a recently proposed data augmentation strategy called *mixup* proposed by Zhang et al. in [18]. *Mixup* interpolates a *labeled* dataset $(x, y) \in \{(X, Y)\}$ as a convex combination of the two inputs and labels as $x_{new} = \lambda x_i + (1-\lambda)x_j$, $y_{new} = \lambda y_i, (1-\lambda)y_j$ where $\lambda \in (0, 1)$. Our method uses *Mixup* at the representation level which aids in SSCRL. We ensure that while mixing our method *increases* the cosine similarity distance between the mixed pairs, to guide the network into learning better representations.

For an ith image sampled from the data, the positive score is the sum over all its K views, $\sum_{i_k} e^{\text{sim}(\mathbf{p}_{i_k}/\tau)} = \mathbf{P}_s$, and similarly, the negative score is $\sum_i \sum_j e^{\text{sim}(\mathbf{n}_{ij}/\tau)} = \mathbf{N}_s$. Intuitively we want to increase the similarity between sum of all positive pairs $\mathbf{P}_s = \sum_{i_k} e^{\text{sim}(\mathbf{p}_{i_k}/\tau)}$ and decrease that of all negative pairs $\mathbf{N}_s = \sum_i \sum_j e^{\text{sim}(\mathbf{n}_{ij}/\tau)}$, which can be achieved by maximizing the ratio $\mathbf{P}_s/\mathbf{N}_s$. As minimizing $\mathcal{L}_{\mathcal{N}_{mix}}$ is equivalent to maximizing $\log\left(\frac{\mathbf{P}_s}{\mathbf{N}_s}\right)$, due to monotonicity of log:

$$\text{minimize}\,\mathcal{L}_{\mathcal{N}_{mix}} = -\log\left(\frac{\sum_{i_k} e^{\text{sim}(\mathbf{p}_{i_k}/\tau)}}{\sum_i \sum_j e^{\text{sim}(\mathbf{n}_{ij}/\tau)}}\right) = -\log\left(\frac{\mathbf{P}_s}{\mathbf{N}_s}\right)$$

Minimizing $\mathcal{L}_{\mathcal{N}_{mix}}$ ensures that $\mathbf{P}_s > \mathbf{N}_s$.

A positive augmented and negative normalized representation at the pre-final layer and vice versa increased the *score gap* consistently. That is:

$$|f_{net}(A_g(x_i)), f_{net}(A_n(x_2))| > |f_{net}(A_n(x_1)), f_{net}(A_g(x_2))|$$

where A_g, A_n are the augmentation and normalization transformations on the raw image x_i and f_{net} is a neural network. Details of the proposed method are in the following section.

4 Proposed Method: $\mathcal{N}$mix

We term our approach as $\mathcal{N}$mix. The intuition for our approach relies on the following observations:

1. The rate of the contrastive learning objective with cosine similarity is directly proportional to the distance between the positive and negative representations.
2. The *score gap* between augmented and non-augmented image pairs is greater than the distance between two augmented images alone i.e.,

$$|f_{net}(A_g(x_i)), f_{net}(A_n(x_2))| > |f_{net}(A_n(x_1)), f_{net}(A_g(x_2))|$$

Keeping above in mind, we propose $\mathcal{N}$-mix, which mixes the samples at the pre-final layer, instead of before the input. Specifically, for each image $x_i \sim X$, we sample a *counter* image $x_j \sim X$ and mix their representations such that $\mathbf{p}_i = A_g(x_i)+A_n(x_j)$, $\mathbf{n}_i = A_n(x_i)+A_g(x_j)$, where $\mathbf{p}_i, \mathbf{n}_i$ are the ith positive/negative pair. We then use the loss function given in Eq. (1) on the positive negative pairs to learn the feature network f_{net}. Our target is to minimize the contrastive loss formed after mixing. We achieve that with the following Eq. (1)

$$\mathcal{L}_{\mathcal{N}mix} = -\log\left(\frac{\sum_{i_k} e^{\text{sim}(\mathbf{p}_{i_k}/\tau)}}{\sum_i \sum_j e^{\text{sim}(\mathbf{n}_{ij}/\tau)}}\right) \tag{1}$$

Where $\mathbf{p}_{i_k}$ are the positive pairs and $\mathbf{n}_{ik}$ the negative pairs. They are generated by our method as following:

$$\begin{aligned}\mathbf{p}_{i_k} &= \big(A_g(x_{i_k}) + A_n(x_{s\neq i}), A_n(x_{i_k}) + A_g(x_{s\neq i})\big),\\ \mathbf{n}_{ij} &= \big(A_g(x_{i_k}) + A_n(x_{s\neq i}), A_g(x_{j\neq i}) + A_n(x_{s'\neq i})\big)\end{aligned} \tag{2}$$

This representation mixing proved extremely effective in increasing the score gap, which translated well to training convergence and downstream accuracy. An illustration of the *score gap* can be seen in Fig. 1. The algorithm for our approach is given in Algorithm 1 and a detailed illustration of the entire process is given in Fig. 2.

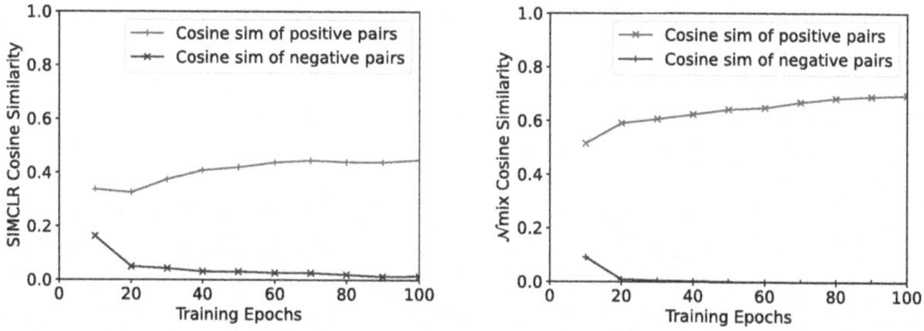

Fig. 1. Cosine Similarity positive and negative embeddings of the network f_{net}. The data is CIFAR-10 and model is Resnet-18 for 100 epochs. The figure on the left is for SIMCLR and the figure on the right is ours ($\mathcal{N}mix$). The gap between intra-class similarity and inter-class *dissimilarity* is larger than SIMCLR and keeps on increasing with number of epochs.

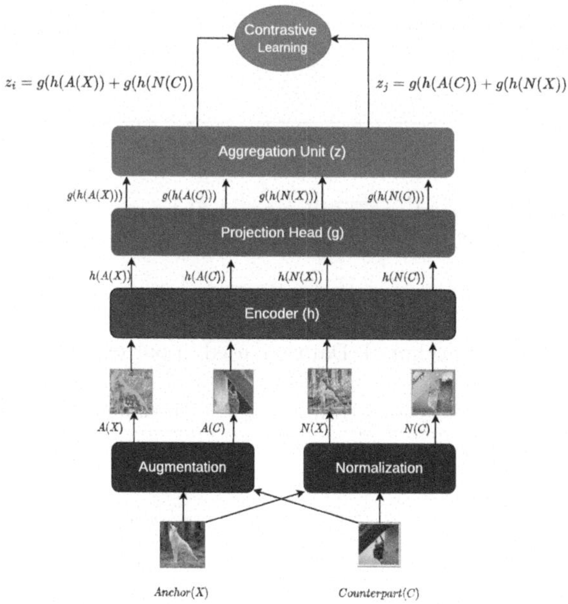

Fig. 2. Illustration of the representation mixing of our proposed approach

5 Experiments

In this section, we describe the various experiments conducted by us to evaluate the efficacy of our approach across different datasets along with different architectures. We have evaluated our method for downstream tasks of image classification, and domain transfer. All the metrics reported here are on the validation set after training for 200 epochs on each dataset with the specified architecture unless otherwise stated. All metrics are the average of 3 runs. The

Algorithm 1: $\mathcal{N}$mix algorithm

Input : batch size N, Temperature τ, backbone f_{net}, projection head g, Transformation A_g, Normalization A_n.
for *each mini-batch* $\{x\}_{k=1}^{N} \subset X$ **do**
  for $x_i \in \{x_1, x_2, x_3......x_N\}$ **do**
    $z_i^a \leftarrow g(f_{net}(A_g(x_i)))$ ▷ Augment Image and Project
    $z_i^n \leftarrow g(f_{net}(A_n(x_i)))$ ▷ Normalize Image and Project
    $x_i^c \sim \{x_1, x_2, x_3......x_N\} : x_i^c \neq x$ ▷ Sample Counter Image
    $z_i^{ca} \leftarrow g(f_{net}(A_g(x_i^c)))$ ▷ Augment Counter Image and Project
    $z_i^{cn} \leftarrow g(f_{net}(A_n(x_i^c)))$ ▷ Normalize Counter Image and Project
  end
  $Z_{i,0} \leftarrow z_a + z_{c_n},\ Z_{i,1} \leftarrow z_{c_a} + z_n$
  $\text{sim}_{i,i} = \frac{Z_{i,0}^T \cdot Z_{i,1}}{\|Z_{i,0}\| \|Z_{i,1}\|}$
  $\text{sim}_{i,j} = \sum_{k_1,k_2} \frac{Z_{i,k_1}^T \cdot Z_{j,k_2}}{\|Z_{i,k_1}\| \|Z_{j,k_2}\|}$
  Loss $\mathcal{L}_{i,j} = -\log \frac{\exp(\text{sim}_{i,i}/\tau)}{\sum_j \exp(\text{sim}_{i,j}/\tau)}$
  $\mathcal{L} \leftarrow \mathbb{E}[\mathcal{L}_{ij}]$
end
return backbone network $f_{net}(.)$

data augmentations we used are similar to [3,8] to get the augmented view of an image. The details of the datasets we used in our experiments are given in Table 1.

Table 1. Benchmark Datasets used in our experiments

Dataset	Categories	Train images	Test images	Image Size
Cifar-10	10	50,000	10,000	32×32
Cifar-100	100	50,000	10,000	32×32
STL-10	10	5,000 + 1,00,000	8,000	96×96
Tiny Imagenet	200	1,00,000	10,000	64×64
Imagenet	1000	1,281,167	50,000	224×224

5.1 Self-Supervised Image Classification

For Self-Supervised Image Classification, we have followed the methodology of [3]. In the first phase, which is the Self-Supervised learning phase, the backbone CNN network f_{net} is appended with a two-layer MLP as the projection head, and is trained on unlabeled data. In the second phase, the MLP head is discarded, and a linear layer is placed on the top of CNN. The CNN weights are frozen, and only the linear layer is trained for classification. The results of the training are given in Tables 2 and 3. Accuracy on CIFAR-10 with models Resnet18 and Resnet20 after every 10 epochs is given in Fig. 3 to demonstrate how the methods behave progressively.

Table 2. Comparison of Self-Supervised training without labels (phase-1) and then training final layer (phase-2) on Resnet variants. Both phases were trained for 200 epochs. Items in **bold** indicate the highest accuracy except for *supervised*. – indicates the results were not available. The values in braces (.) below model names denotes the number of parameters of the model in millions

Datasets	Model	Resnet20 (.2M)	resnet32 (.5M)	Resnet18 (11M)	Resnet50 (24M)
CIFAR-10	Supervised	88.21	87.81	94.63	91.95
	Simclr	73.43	75.36	84.27	80.63
	Moco	65.56	65.99	77.72	79.71
	Ours	**74.27**	**76.00**	**84.71**	**85.36**
CIFAR-100	Supervised	51.69	54.42	76.38	68.76
	Simclr	39.40	41.53	54.15	**57.06**
	moco	32.91	36.95	49.68	47.34
	Ours	**40.99**	**42.84**	**56.67**	53.55
STL-10	Supervised	–	–	–	–
	Simclr	65.76	69.08	77.57	**80.28**
	Moco	63.75	65.85	75.29	78.88
	Ours	**66.80**	**68.97**	**77.73**	76.52
Tiny Imagenet	Supervised	–	–	–	–
	Simclr	22.45	23.94	27.45	**33.17**
	Moco	**25.03**	**24.16**	28.18	32.44
	Ours	22.93	23.41	**30.00**	32.08

Table 3. Linear Evaluation (phase-2) accuracy on Imagenet using Resnet-50. The values in **bold** indicate the highest values.

Resnet50	
Supervised	76.2
SIMCLR [3]	61.7
Moco [6]	60.6
Colorization [19]	39.6
jigsaw [13]	45.7
BigBigGAN [5]	56.6
SeLa [1]	61.5
PIRL [12]	61.7
Ours	**63.40**

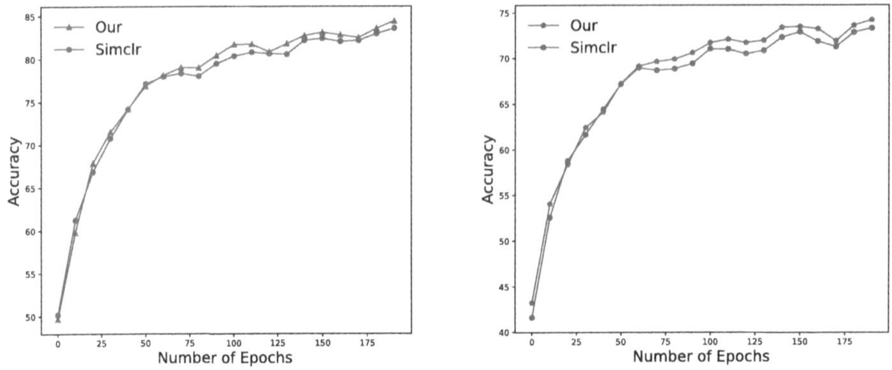

Fig. 3. Accuracy after every 10 epochs for Resnet18 (left) and Resnet20 (right) with Our method and Simclr

5.2 Domain Transfer

Domain Transfer involves pre-training the backbone network with SSCRL on one dataset and then fine-tuning on another dataset. We followed [3] for pre-training, then trained on the final linear layer on the target dataset, and validated on the test set. The results are given in Table 4.

For domain transfer we get a mixed performance with our method as compared to Simclr. For Resnet18, our method performs better than Simclr while for others it is not conclusive. At this point, we cannot conclusively state why this discrepancy arises, and we leave this to future work.

Table 4. Results of Domain Transfer from dataset $A \rightarrow B$. Pre-training on source domain and target domain was for 200 epochs. Values in **bold** indicate the highest accuracy

Datasets	Model	Resnet20 (.2M params)	Resnet32 (.5M params)	Resnet18 (11M params)
CIFAR-10 → 100	Simclr	**36.00**	36.00	47.45
	Ours	35.44	**36.24**	**49.32**
CIFAR-100 → 10	Simclr	**64.99**	**63.88**	74.26
	Ours	63.69	63.30	**76.06**

5.3 Convergence Efficiency

The key part of our study is the convergence efficiency. The representation mixing creates a larger gap between the representations and allows the network to learn the relevant features faster. We report results on Image Classification with similar experimental methodology as in Sect. 5.1. The results are given in Table 5. We can clearly see that our method yields better results with the same number of epochs.

Table 5. Model Accuracy with number of epochs on Imagenet and variants of Resnet. The values in **bold** indicate the highest values.

Resnet18 (11M params)			
Epochs	SIMCLR [3]	MoCo [6]	Ours
100	44.59	–	**48.50**
200	48.8	–	**49.86**
Resnet50 (24M params)			
Epochs	SIMCLR [3]	MoCo [6]	Ours
200	61.7	60.6	**63.40**

5.4 Effect of Number of Augmented Views (K)

For Self-Supervised Contrastive Representation Learning methods, the convergence and accuracy improve with the number of augmented views K. All the reported results were conducted with $K = 1$. We observed a similar increase in convergence speed upon increasing K, at the cost of increased GPU memory requirement. However, because of the efficiency of our method, $K = 1$ was sufficient for us. We describe the effect of increasing K with Resnet18 and CIFAR-10 in Table 6.

Table 6. Top-1 Accuracy for Self-Supervised Image Classification on Resnet18 and CIFAR-10 with varying K. Both Self-Supervised training (phase-1) and downstream training (phase-2) were performed for 100 epochs.

Model	$K = 1$	$K = 2$	K = 4	K = 8
Ours	75.34	81.76	85.73	88.24

6 Conclusion and Future Work

In this paper, we have presented an efficient and simple method for Self-Supervised Contrastive Representation Learning, which performs competitively with the state-of-the-art methods and does so with lesser computational requirements, with fewer augmented views necessary and in fewer epochs. While our method does not explicitly search for or optimize for *hard negatives*, it does generate pairs for contrastive learning which are empirically harder to classify as demonstrated by cosine distance between them. We leave the explicit optimization for harder negatives and better positive examples to future work. While we have studied a particular variation of mixing method, there may be more variants which may yield better results. There is also the question of what kind of features do the mixing methods learn that they can work on downstream tasks more efficiently. We leave all this to future work.

References

1. Asano, Y.M., Rupprecht, C., Vedaldi, A.: Self-labelling via simultaneous clustering and representation learning. arXiv preprint arXiv:1911.05371 (2019)
2. Caron, M., Misra, I., Mairal, J., Goyal, P., Bojanowski, P., Joulin, A.: Unsupervised learning of visual features by contrasting cluster assignments. Adv. Neural. Inf. Process. Syst. **33**, 9912–9924 (2020)
3. Chen, T., Kornblith, S., Norouzi, M., Hinton, G.: A simple framework for contrastive learning of visual representations. In: International Conference on Machine Learning, pp. 1597–1607. PMLR (2020)
4. Chen, X., Fan, H., Girshick, R.B., He, K.: Improved baselines with momentum contrastive learning. CoRR, abs/2003.04297 (2020)
5. Donahue, J., Simonyan, K.: Large scale adversarial representation learning. Advances in Neural Information Processing Systems, vol. 32 (2019)
6. He, K., Fan, H., Wu, Y., Xie, S., Girshick, R.: Momentum contrast for unsupervised visual representation learning. In: Proceedings of the IEEE/CVF Conference on Computer Vision and Pattern Recognition, pp. 9729–9738 (2020)
7. He, K., Fan, H., Wu, Y., Xie, S., Girshick, R.B.: Momentum contrast for unsupervised visual representation learning. In: 2020 IEEE/CVF Conference on Computer Vision and Pattern Recognition, CVPR 2020, Seattle, WA, USA, 13–19 June 2020, pp. 9726–9735. Computer Vision Foundation/IEEE (2020)
8. He, K., Zhang, X., Ren, S., Sun, J.: Deep residual learning for image recognition. In: Proceedings of the IEEE Conference on Computer Vision and Pattern Recognition (CVPR), June 2016
9. Jing, L., Tian, Y.: Self-supervised visual feature learning with deep neural networks: a survey. IEEE Trans. Pattern Anal. Mach. Intell. (2020)
10. Kalantidis, Y., Sariyildiz, M.B., Pion, N., Weinzaepfel, P., Larlus, D.: Hard negative mixing for contrastive learning. Adv. Neural Inf. Process. Syst. **33**, 21798–21809 (2020)
11. Mishra, I.: Self supervised learning pretext task. https://atcold.github.io/pytorch-Deep-Learning/en/week10/10-1/. Accessed 30 Sept 2020
12. Misra, I., van der Maaten, L.: Self-supervised learning of pretext-invariant representations. In: Proceedings of the IEEE/CVF Conference on Computer Vision and Pattern Recognition, pp. 6707–6717 (2020)
13. Noroozi, M., Favaro, P.: Unsupervised learning of visual representations by solving jigsaw puzzles. In: European Conference on Computer Vision, pp. 69–84. Springer, Cham (2016)
14. Tian, Y., Krishnan, D., Isola, P.: Contrastive multiview coding. In: Proceedings of ECCV 2020. Lecture Notes in Computer Science, vol. 12356, pp. 776–794. Springer, Cham (2020)
15. Tian, Y., Sun, C., Poole, B., Krishnan, D., Schmid, C., Isola, P.: What makes for good views for contrastive learning? In: Proceedings of Advances in Neural Information Processing Systems, vol. 33 (2020)
16. van den Oord, A., Li, Y., Vinyals, O.: Representation learning with contrastive predictive coding. CoRR, abs/1807.03748 (2018)

17. Mishra, I., Lecun, Y.: Self supervised learning. http://web.archive.org/web/20080207010024/http://www.808multimedia.com/winnt/kernel.htm. Accessed 30 Sept 2020
18. Zhang, H., Cisse, M., Dauphin, Y.N., Lopez-Paz, D.: mixup: beyond empirical risk minimization. arXiv preprint arXiv:1710.09412 (2017)
19. Zhang, R., Isola, P., Efros, A.A.: Colorful image colorization. In: Proceedings of ECCV 2016, Part III 14, pp. 649–666. Springer, Cham (2016)

Smart E-learning Development Tool Using Generative AI and Retrieval-Augmented Approach

Tilottama Goswami[1(✉)], Gummalla Bhavana[2], Chebolu V. S. Anirudh[3], Leena Reddy[1], and K. Ram Mohan Rao[1]

[1] Department of Information Technology, Vasavi College of Engineering (A), Hyderabad, India
tgoswami@staff.vce.ac.in
[2] S&P Global, Hyderabad, India
[3] Capital Quant Solutions, Hyderabad, India

Abstract. Generative AI is a technology that uses data to produce something new in response to a prompt entered by a human. Large Language Models (LLM) is a type of generative AI, the latest language model that can understand and interpret natural language. LLMs are very large deep-learning models which are pre-trained on huge amounts of existing data. LLMs can be used for text generation, code generation, translation, summarization, conversational AI and chatbots. With the advent of the latest technologies, one of the challenges faced in LLM is that the model pre-trained may become obsolete in future. The Retrieval Augmented Generation (RAG) technique is helpful in such cases as it augments the existing data as an external source and can be integrated with the pre-trained LLM. The paper uses Generative AI using LLM and RAG for the development of smart e-learning course content in the field of learning science to assist subject matter experts (SME) or instruction designers. The tool simplifies presenting complex information without compromising accuracy. The SME finds it easy to balance content creation with their other professional responsibilities. The scope of the paper is to generate content consisting of course objectives, course outcomes, and questionnaires for assessment. The course content thus generated in a certain required format is assessed by the subject matter expert to fine-tune.

Keywords: Generative AI · Large Language Model · Retrieval Augmented Generation · Mistral 8x22B · BM25 Retriever

1 Introduction

The paper aims to leverage Generative AI, specifically Large Language Models (LLMs), along with Retrieval Augmented Generation (RAG) techniques for creating e-learning course content in the domain of learning science. LLMs, advanced AI models trained on vast datasets, offer capabilities like text generation, code creation, and conversational AI. However, a challenge arises as pre-trained models may become outdated over time. RAG addresses this by integrating external data sources to enhance the model's knowledge.

A. Dasgupta et al. (Eds.): BDA 2024, LNCS 15526, pp. 372–381, 2025.
https://doi.org/10.1007/978-3-031-81821-9_24

The problem statement revolves around developing a system that can generate course content, including objectives, outcomes, and assessment questionnaires of various levels of Bloom's Taxonomy, through Generative AI techniques. This content will be evaluated by subject matter experts, and their feedback will be used to refine and improve the generated material. The ultimate goal is to create a robust system capable of producing high-quality educational content in the domain of learning science. The report will delve into the challenges, methodologies, and outcomes of implementing Generative AI with LLMs and RAG for e-learning course development.

2 Literature Survey

The role of cognitive tutors in integrating technology with learning sciences has been explored, highlighting their potential in designing AI-driven e-learning systems tailored to individual student needs [1]. Sequence-to-sequence learning, a foundational concept in neural network architectures, is crucial for understanding sequence generation, which is key for implementing AI techniques in e-learning course design [2]. Evidence-based principles for designing effective multimedia learning experiences have been outlined, providing a foundation for incorporating AI techniques to enhance course design in learning science [3]. The Transformer architecture, which is instrumental in natural language processing tasks, serves as the foundation for many modern language models used in e-learning content generation [4, 5]. A technique integrating retrieval-based methods with generative AI has been proposed to enhance the relevance of generated content, offering insights into its application in automatically generating content [6]. The application of a pre-trained language model in conversational AI has been explored [7]. The few-shot learning capabilities of language models have been investigated, highlighting how they can adapt to new tasks with limited training data [8]. Generative Pre-trained Transformer (GPT) is a class of Generative AI models developed by OpenAI, characterized by its Transformer architecture. These models are pre-trained on vast amounts of text data, enabling them to generate coherent and contextually relevant text. GPT models, such as GPT-3, excel in various natural language processing tasks, including text generation, language translation, summarization, and question-answering. In the context of e-learning course content creation, GPT can be leveraged to generate objectives, learning outcomes, assessment questions, and even entire lesson materials based on prompts or input provided by users. Bing, Microsoft's search engine, serves as one of the potential external data sources that can be integrated with RAG for knowledge retrieval. Bing's vast index of web pages, academic papers, research articles, and other online resources provides a rich repository of information that can augment the knowledge base of Generative AI models. Additionally, other external data sources such as academic databases, digital libraries, educational websites, and open-access repositories can be leveraged to further enrich the knowledge retrieval process. The paper focuses on the development of e-learning content in an efficient manner to assist the Subject Matter Expert in overcoming the challenges faced. The up-to-date information extraction in a prescribed format can be achieved using a chain of thought approach of prompt engineering These external data sources complement the capabilities of Generative AI models like GPT by providing access to diverse, up-to-date, and domain-specific information, thereby enhancing the quality and relevance of the generated e-learning course content. By incorporating methodologies such as GPT-based text generation and RAG-enabled knowledge augmentation, along with leveraging external data sources like Bing, the proposed system aims to overcome the challenges of outdated models and produce high-quality educational content in the domain of learning science.

3 Proposed Method

The proposed software tool uses a method that utilizes a large language model (LLM) which is Mixtral 8x22B to generate e-learning course content. The detailed architecture diagram is given in Fig. 1 and explained step by step below.

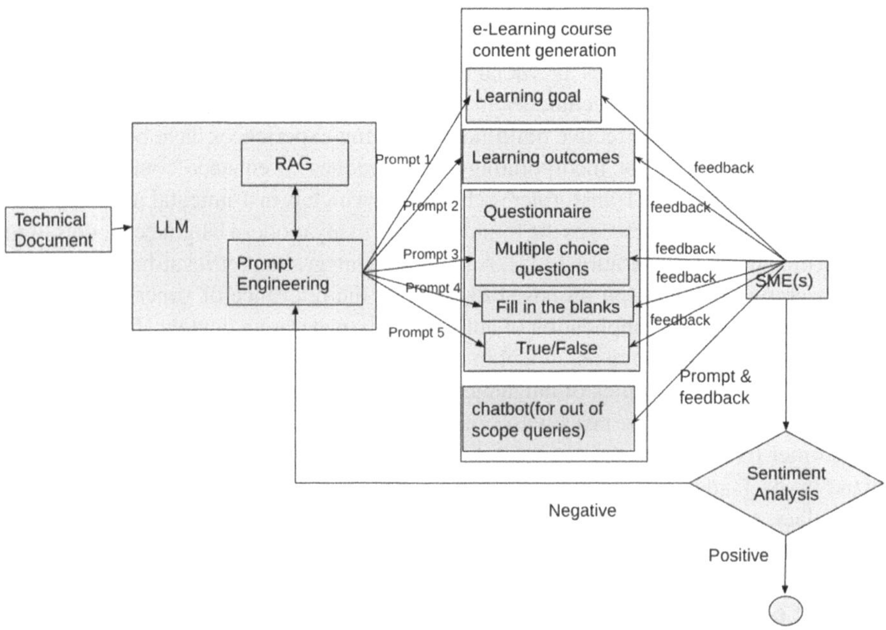

Fig. 1. Architecture of Smart E-Learning Development

Input: Customer Technical Document. The process begins by providing a customer technical document for the Mistral 8x22B model. This document serves as the foundational knowledge base from which the model generates course content. The technical document can include detailed information on the subject matter, intended learning outcomes, and specific areas of focus.

Content Generation. Mistral 8x22B is employed to generate various outputs for the course which include:

1. Learning Goals: High-level objectives that the course aims to achieve
2. Learning Outcomes: Specific skills or knowledge that learners are expected to acquire
3. Assessment Questions: Multiple-choice questions, fill-in-the-blank questions, and true/false statements are designed to assess learner comprehension. While framing the assessment of various levels of Bloom's Taxonomy, appropriate verbs are used for designing the prompts. The six levels are namely Remembering, Understanding, Applying, Analyzing, Evaluating, and Creating in an order of increasing order of difficulty. The instruction designer can create more effective and meaningful evaluations that not only measure rote memorization but also assess deeper understanding, critical thinking, and the ability to apply knowledge in diverse contexts.

The generated content is produced in a structured manner, ensuring coherence and alignment with the provided technical document. Prompt engineering in the Mistral 8x22B LLM Model plays a crucial role in giving the model enough context to understand the topic and set the tone. The model has used 5 prompts as shown in Fig. 1. Each prompt has its objective to produce the appropriate content. The challenge is to avoid ambiguous prompts and difficult-to-understand prompts. Testing the prompt and renting it is an essential component of the proposed architecture of the software tool, discussed in the next subsection.

Review and Feedback Loop. Subject Matter Experts (SMEs) or instruction designers play a critical role in the review process. SME evaluates the generated content for accuracy, relevance, and pedagogical soundness. Feedback from SMEs can be either positive or negative, providing insights into areas that need improvement. For each of the 5 prompts, the content generated is evaluated qualitatively by the SME and is stored.

Handling Feedback

1. Storage: All feedback, along with the generated content, is stored systematically for future reference.
2. Admin Module: The feedback is forwarded to the admin module, where administrators can view and take appropriate actions. This might involve adjusting prompts, refining content, or providing additional instructions to the model.
3. Feedback Processing: To enhance the quality of the generated content, we employ prompt engineering techniques, such as the Chain of Thought pattern. This method encourages the model to articulate its reasoning process, leading to more accurate and comprehensive outputs. The feedback loop is integral to the iterative improvement of the system.

This section is responsible for generating the prompt completion task depending on the scope of the customer technical document provided as input. To augment the knowledge base other than the input document, the model is integrated with Retrieval Augmented Generation (RAG) [9, 10] discussed in the subsequent subsection.

Integration of RAG Model with BM25 Retriever To supplement the knowledge base provided by the custom technical document, our system integrates the RAG model with the BM25 Retriever. This component accesses external resources and retrieves relevant information in a ranked order based on the query or technical document provided. In literature, there are two types of retrievers - dense and sparse. BM25 uses sparse representation, in contrast to BERT which is a type of dense retriever using continuous vector representations. The authors choose BM25, as it is computationally efficient for large-scale text retrieval tasks. It also leverages probabilistic modelling based on term frequency and distribution.

Few Challenges with Solutions

1. Authenticity of Sources: Ensuring the reliability of retrieved information is critical. This is addressed by implementing a vetting process where retrieved sources are cross-verified with reputable databases and publications.
2. Bias in RAG Outputs: The RAG model might favor information aligned with recent trends or popular content, potentially leading to biased outputs. To mitigate this, the process incorporates a diversity check that ensures a balanced representation of perspectives and prioritizes sources with established credibility.
3. Tokenize in LLM: The token limit restricts the amount of context the model can consider when it is taking input from the customer data. It leads to information loss and is unable to generate comprehensive responses. The authors solved this by replacing an 8x7B configuration with a Mistral 8x22B which significantly increased the token capacity.

Example: For producing content or assessment of higher levels of Bloom's Taxonomy comprising of something extra other than the input customer technical document, one may need an extensive knowledge base. This may be out of scope as seen in Fig. 2. The solution is provided using the RAG model as shown in Fig. 3. The Page Replacement Algorithms have been taken from an external source and displayed as a response.

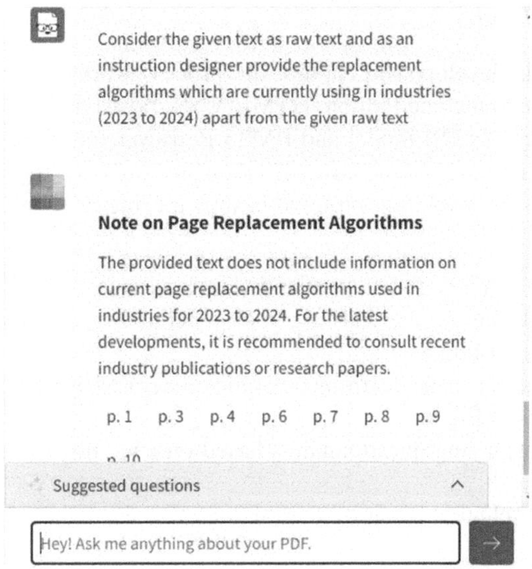

Fig. 2. The response generated for out-of-scope queries, the solution is not in the input document (without using the RAG model)

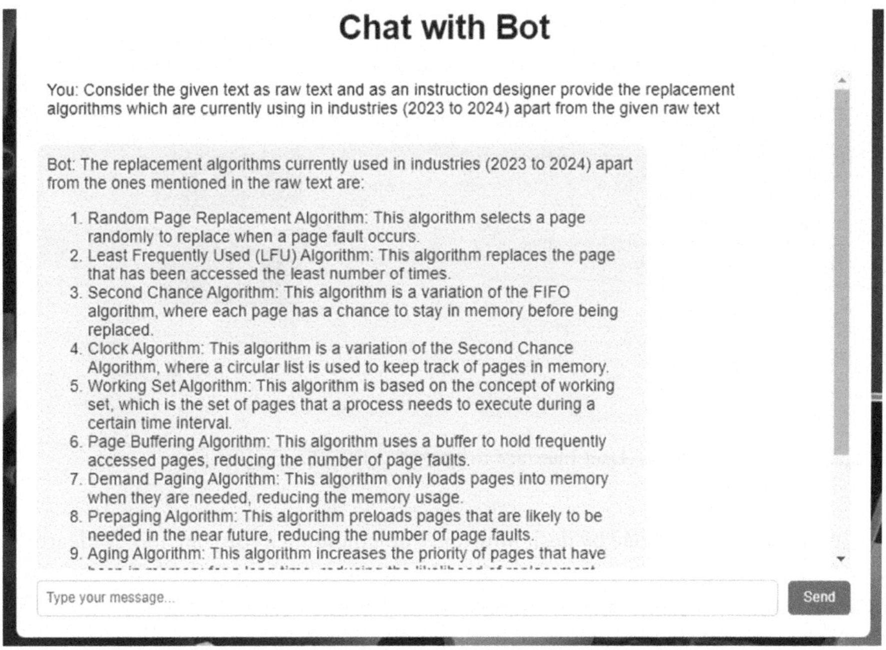

Fig. 3. The response generated with the RAG model for out-of-scope queries when the solution is not in the input document

3.1 Experiment Results

The software tool is developed to capture the process workflow as shown in 1. The software specifications include Python, HTML, CSS, Javascript, and Flask for the user interface. The Mistral 8x22B model and BM25 Retriever for the RAG framework are used for content generation. To increase the token size capacity, the 7B model has been replaced with 22B. The workflow captured is shown using the graphical user interface screenshots -

1. User interface to generate the content and view it as seen in Fig. 4
2. User interface displaying the learning goal generated by LLM from the customer input data as seen in Fig. 5.
3. User interface displaying the learning outcomes generated by LLM from the customer input data as seen in Fig. 6.
4. User interface displaying questionnaires based on Bloom's level taxonomy as seen in Fig. 7
5. User interface for chatbot using RAG as seen in Fig. 3

3.2 Conclusion and Future Work

This paper explores how Generative AI, such as Large Language Models (LLM) and Retrieval-Augmented Generation (RAG), can enhance e-learning courses.

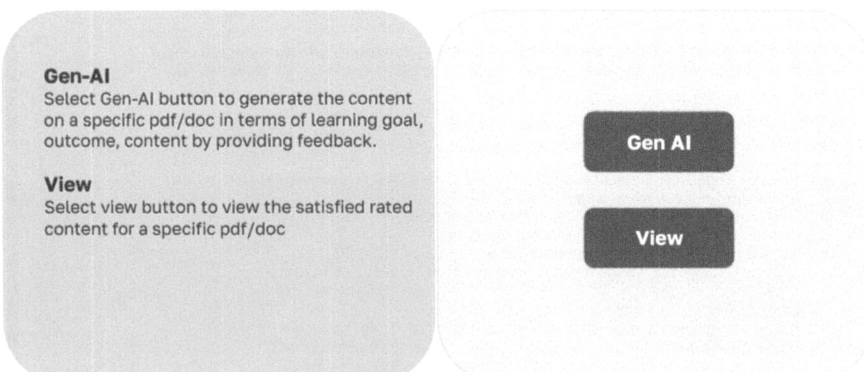

Fig. 4. User Interface to generate content and view it.

Content creation is done by designing and implementing a software tool for learning science and effectively assisting the subject matter expert or instructor designer. The software tool uses Python libraries and models such as Large Language Model (LLM) Mistral 8x22B, Retrieval Augmented Generation (RAG) Model BM25 Retriever. The goal is to generate course objectives, course outcomes, and assessment questionnaires based on Bloom's taxonomy using these advanced techniques. The outputs generated by prompts are evaluated by SMEs using a feedback mechanism. Furthermore, the tool

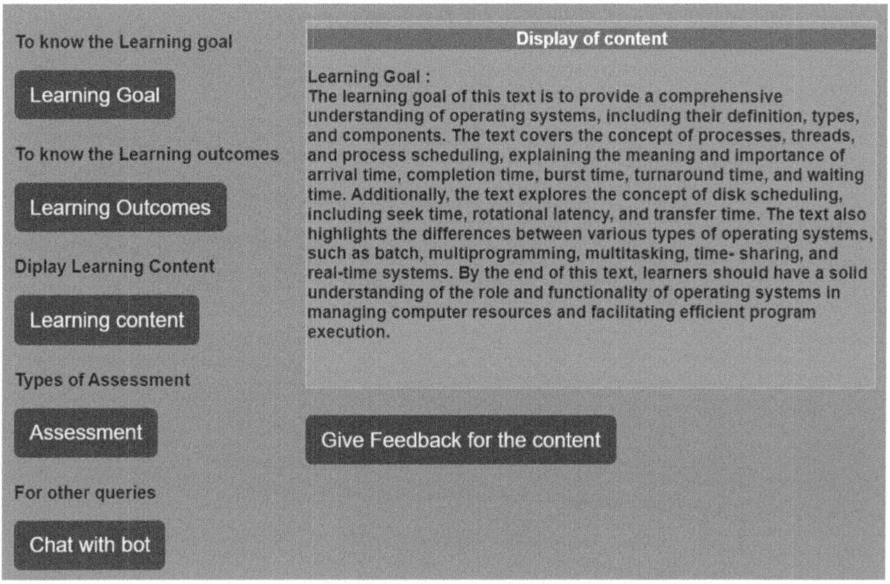

Fig. 5. User Interface to generate content and view it.

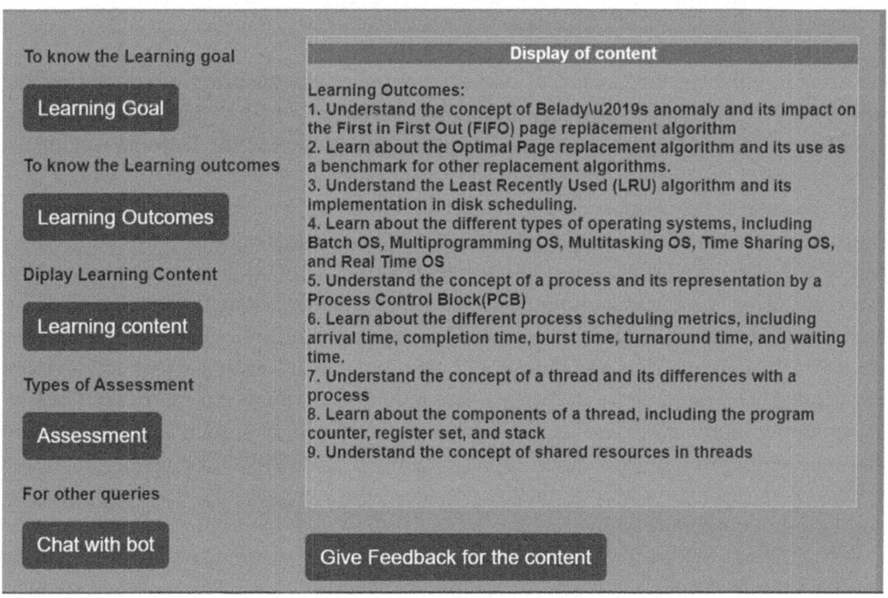

Fig. 6. User Interface to generate content and view it.

includes implementing a personal chatbot to address out-of-scope questions using RAG. This chatbot refers to external sources of knowledge to provide responses, continuously learning and improving its answers to offer comprehensive user support. Ultimately, the

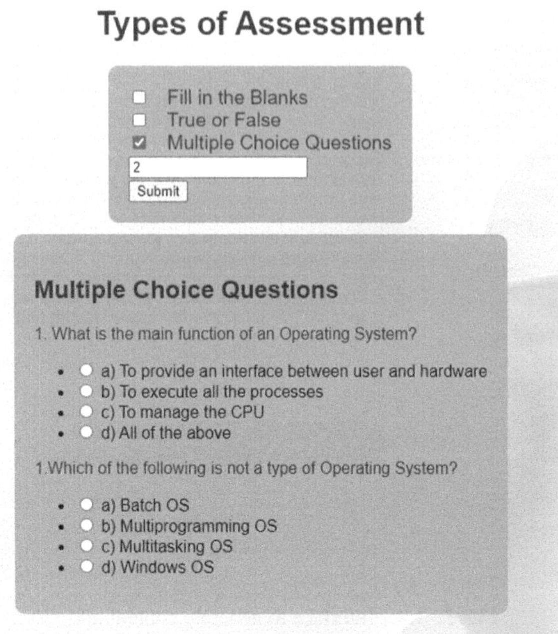

Fig. 7. User Interface to generate questionnaire using Bloom's taxonomy

tool aims to enhance the quality of educational materials, ensuring effective learning experiences for students worldwide. Through collaboration with experts and the utilization of innovative AI techniques, the instructor strives to redefine learning, making it more engaging, accessible, and impactful.

In the future, the authors will enhance the collaborative learning environment by incorporating tools for student interaction and teamwork, fostering a sense of community and peer-to-peer learning. The feedback-driven content generation process will be refined using advanced techniques to iteratively improve content based on SME or instructional designer feedback. Additionally, integrating multimodal learning resources like interactive simulations and videos will cater to diverse learning styles and increase engagement. The future work will also aim to reduce over-reliance on SME feedback by incorporating more automated methods for content refinements, such as coherence and relevance scoring, alongside expert input, creating a more efficient content generation process. Personalization will be further enhanced through adaptive learning technologies, tailoring content to individual needs. Integration with external Learning Management Systems (LMS) will broaden access and facilitate seamless use within existing educational infrastructures. Lastly, expanding the system's application to new subjects and languages will promote inclusive education and global reach.

Personal Chatbot Assistance. To further support users, the proposed software tool includes a personal chatbot capable of addressing queries that extend beyond the scope of the project. The chatbot utilizes both the Mistral 8x22B model and the RAG framework, providing users with precise and contextually relevant answers.

Acknowledgments. The authors would like to acknowledge Vasavi College of Engineering, Hyderabad for supporting the work.

References

1. Alon, U., Xu, F., He, J., Sengupta, S., Roth, D. Neu-big, G.: Neuro symbolic language modeling with automaton-augmented retrieval. In: The Proceedings of the 39th International Conference on Machine Learning, PMLR, vol. 162, pp. 468–485, Maryland USA (2022)
2. Feng, Z., Feng, X., Zhao, D., Yang, M., Qin, B.: Retrieval-generation synergy augmented large language models, arXiv preprint, arXiv:2310.05149 (2023)
3. Packer, C., Fang, V., Patil, G., Lin, K., Wooders, S., Gonzalez, J.: MemGPT: Towards LLMs as Operating Systems, arXiv preprint, arXiv:2310.08560 (2023)
4. Clark, R., Mayer, R.: e-Learning and the Science of Instruction: Proven Guidelines for Consumers and Designers of Multimedia Learning, 1st edn. Wiley (2020)
5. Li, N., Zhou, M.: Artificial Intelligence in Education: Promises, Challenges and Impacts. Springer (2020)
6. Radford, A., Wu, J., Child, R., Luan, D., Amodei, D., Sutskever, I.: Language Models are Unsupervised Multitask Learners. arXiv preprint, arXiv:1901.02860 (2019)
7. Vaswani, A., et al.: Attention is all you need. In: Proceedings of Advances in Neural Information Processing Systems, Long Beach, California, USA (2017)
8. Bajpai, A., Nangia, N.: A Comprehensive Guide to Building and Deploying Conversational AI with BERT. Apress (2020)
9. Lewis, P., Neumann, M., Dinan, E., Fan, A., Willems, L., Riedel, S.: Retrieval-Augmented Generation for Knowledge-Intensive NLP Tasks. arXiv preprint, arXiv:2005.11401 (2021)
10. Wang, A., et al.: Language Models are Few-Shot Learners, arXiv preprint, arXiv:2105.12492 (2021)

Prediction and Forecasting

An Efficient Predictive Big Data Analytics for Supply Chain Demand Forecasting Through Ensemble Learning

S. Navadersh and S. Vengadeswaran(✉)

Indian Institute of Information Technology Kottayam, Kottayam, Kerala, India
vengadesh@iiitkottayam.ac.in

Abstract. The increasing attention to Big Data Analytics (BDA) in Supply Chain Management (SCM) stems from its versatile applications, encompassing customer behaviour analysis, trend analysis, and demand prediction. This paper aims to explore the potential of predictive BDA applications in supply chain demand forecasting and to pinpoint gaps in existing knowledge. In this proposed work, a time series predictive big data analytics using SARIMA, Prophet and ensemble model is proposed. Further to achieve an improved prediction accuracy, optimal parameters were identified using manual and grid search methods. In addition, geographical feature is extracted using KMeans clustering to understand any underlying patterns. Finally, an ensemble model is proposed that integrates heterogeneous models (SARIMA, Prophet), to achieve improved performance. The experiments were evaluated using five node Spark clusters deployed in the cloud. The results exhibit that the proposed ensemble approach using Linear regression achieves the lowest RMSE of 30.76.

Keywords: Big Data Analytics · K-Means · SARIMA · Prophet · Ensemble

1 Introduction

Supply Chain Management (SCM) is the backbone for the smooth flow of products, services, and information. However, it faces challenges in terms of capacity, supply, and demand. Also, this adds up in production, storage, and delivery costs. The significance of dealing with these uncertainties can't be understated. This emphasises the crucial impact of demand forecasting on increasing supply chain performance [1–4]. Traditionally, two approaches have been employed: a forward-looking approach, anticipating potential demand over the next several years, and a backward-looking approach, relying on past or ongoing capabilities to respond to demand. However, traditional solutions, often relying on spreadsheet models and statistical methods like moving averages, face limitations in scalability for large-scale data and struggle to address the complexities and uncertainties inherent in supply chain management (SCM) [5]. Historically,

A. Dasgupta et al. (Eds.): BDA 2024, LNCS 15526, pp. 385–394, 2025.
https://doi.org/10.1007/978-3-031-81821-9_25

statistical analysis techniques like time-series and regression analysis have been employed for demand forecasting in SCM [6]. The limitations of conventional methods become apparent in the context of supply chain demand forecasting, where numerous parameters influence demand, often not captured by traditional approaches. Conventional methods tend to provide a partial understanding of demand variations and are less adept at handling the non-linear behaviours prevalent in supply chain dynamics.

However, the evolving landscape of information technologies and computational efficiencies have given rise to the prominence of big data analytics (BDA) [7]. BDA not only enables more precise predictions aligned with customer needs but also contributes to the assessment of SCM performance in efficiency improvement, reduced reaction time, and enhanced risk assessment. The advent of big data and high computing analytics presents a transformative opportunity, enabling scalability, efficiency and ease of data processing, for data-driven demand forecasting and planning [8,9]. Leveraging different Time series algorithms, BDA enables the extraction of valuable insights, leading to highly accurate demand forecasting models scalable for application in SCM. In this work, a study on the performance of KMeans clustering (geospatial clustering) in the dataset to understand any underlying patterns or features which can increase the efficiency of demand forecasting is done. For demand forecasting, we are considering mainly two time series algorithms, SARIMA and PROPHET.

2 Literature Review

A review of literature spanning from 2005 to 2023 reveals a growing trend in publications related to supply chain demand forecasting, with a focus on BDA applications. Notable techniques identified in this review include neural networks, regression, time-series forecasting (ARIMA), support vector machines, and decision trees. These techniques showcase the increasing utilization of BDA in SCM demand forecasting, reflecting a departure from conventional statistical forecasting approaches. The methodology, merits and demerits of the literature are discussed in Table 1.

3 Research Objective

An ensemble-based demand forecasting model is proposed. The objective is to

- **To study** the adaptability of big data demand forecasting models (SARIMA, Prophet) to nonlinear demand variations in the input dataset.
- **To extract** Geospatial Feature using K-Means clustering
- **To identify** the optimal parameters for training model using Grid search.
- **To design** ensemble architectures that integrate heterogeneous models (SARIMA, Prophet), to achieve superior predictive accuracy and generalization capabilities across diverse features considered.
- **To harness** Big Data processing platforms (SPARK), and cloud to experiment the big data demand forecasting.

Table 1. Summary of Literature Review

S.No	Title	Proposed Methodology	Merits	Demerits
1	Improved supply chain management based on hybrid demand forecasts [10]	Neural Networks, Random Forests, Time-Series Forecasting (ARIMA)	Improved accuracy in demand forecasting.	Limited scalability for large-scale data
2	Comparative Analysis of Machine Learning Algorithms for Predictive Modeling in Various Domains [11]	Relevance vector machine (KNN), Support Vector Machine, Decision Tree, Genetic Algorithms (GA), LSTM	Contributes to advancing the understanding of machine learning algorithm performance across diverse application domains, facilitating more accurate and robust predictive models	Time-Series Forecasting (ARIMA, SARIMA) are not considered for seasonal trend data
3	Big data analytics in supply chain management: A state-of-the-art literature review [12]	Data-Mining Algorithms, machine learning, predictive modelling	Understanding of the potential benefits and challenges associated with implementing big data analytics solutions in SCM contexts.	Scope of Ensemble Modelling for enhancing the forecasting accuracy
4	Demand Forecasting for Textile Products Using Statistical Analysis and Machine Learning Algorithms [13]	Regression models, Decision Trees, and Neural Networks	Improved prediction accuracy	Widely used but not scalable for large-scale data, and limited capabilities in handling uncertainties
5	Daily retail demand forecasting using machine learning with emphasis on calendric special days [14]	Machine Learning Techniques, Optimization	Improved forecast accuracy and resource allocation by capturing the nuanced effects of calendric special days.	Computational intensity due to the use complex algorithms
6	A Comparative Study of Demand Forecasting Models for a Multi-Channel Retail Company: A Novel Hybrid Machine Learning Approach [15]	A hybrid framework (ARIMA and LSTM ML Models) for demand forecasting in multi-channel retail	Improved accuracy through model combination, adaptable to diverse retail scenarios.	Requires hyperparameter tuning, potential complexity in model integration
7	An Ensemble-Learning-Based Method for Short-Term Water Demand Forecasting [16]	Ensemble method	Enhanced accuracy and robustness to fluctuating demand patterns.	Complexity in model fusion and potential overfitting with excessive algorithm combinations

4 Methodology

In this proposed work, a time series predictive big data analytics using deep learning models like SARIMA, Prophet and ensemble model is proposed Fig. 1. Further to achieve improved prediction accuracy, optimal parameters were identified using manual and grid search methods.

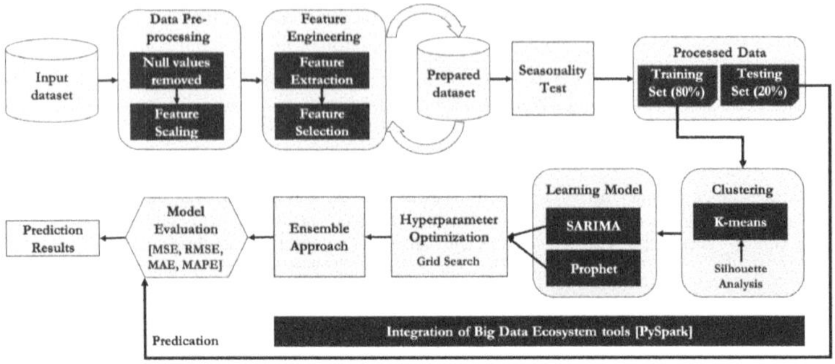

Fig. 1. Methodology Overview

4.1 Dataset Characteristics

The dataset utilized in this study is IOWA_LIQUOR_SALES [17], which includes 11 years of information on spirits purchases made by Iowa Class "E" liquor licensees, categorized by product and purchase date. The total size of the dataset is 13.2 GB, containing 29 million rows of records. The dataset contains 24 features including information about orders, products, sales, order quantity, spatial coordinates (longitude and latitude) and shipping details.

4.2 Data Pre-processing

Initially, the IOWA_LIQUOR_SALES dataset is preprocessed by removing null values using the Pandas library. The following preprocessing steps were done to improve the accuracy of the models in our proposed work. Initially, the Date feature is extracted from the input dataset. The code extracts various features from the DATE column, such as year, month, day, and weekday. These features capture different aspects of time, like seasonal variations and weekly trends. Incorporating date features enriches the dataset and aids in time-based analysis. Then the potential outliers are identified and removed using Z-score. Further to maintain distribution, reduce the impact of any outliers and simplify interpretation where each data point is scaled linearly to fit within a specific range using Min-Max Scaling. Finally, the data is aggregated at different time intervals, such as monthly and weekly. Aggregating data allows for summarizing information and reducing the dataset's dimensionality while retaining essential insights. It provides a broader perspective on trends and patterns over time.

4.3 Feature Engineering

To identify potential features affecting prediction, a subset of the dataset containing only selected features was created, enabling focus on specific variables of interest. For each selected feature, the Analysis of Variance test is used to generate F-values and p-values. Features are deemed more significant for analysis based on higher F-values and lower p-values. Subsequently, dataset was refined with only necessary features (Order_Item_Quantity,

Category_Name, Category_ID, Product_Name, Days_For_Shipping, Order_Date, Year_Month, Year_Week) for the multivariate experiments.

In addition, geographical similarities in demand patterns are identified by extracting new feature Clusters. It helps in understanding spatial patterns in the distribution of orders and can aid in optimizing logistics and resource allocation. Before clustering, the latitude and longitude coordinates in the dataset are standardized. This process ensures that all coordinates have the same scale, which is important for accurate clustering. Then standardised geographic data (latitude and longitude) of orders are clustered using the K-means algorithm. The Euclidean distance is used to measure the distance between each pair of data blocks (d_i, d_j) from the dataset D $((d_i, d_j) \subseteq D)$.

$$Dist_{eq}(d_i, d_j) = \sqrt{(x_{d_i} - x_{d_j})^2 + (y_{d_i} - y_{d_j})^2} \tag{1}$$

The optimal number of clusters is determined using the Silhouette score [18]. The K-means algorithms [19] segregate the geographical coordinates into clusters, which are then included as a new feature for training the model.

4.4 Seasonal Decomposition

The seasonal decomposition is performed to gain insights into the distribution and trends of various variables within the dataset. It explores the distribution of order item quantities, investigates temporal trends in total orders over time, and analyzes the distribution of orders by different categories such as region, category, and order status.

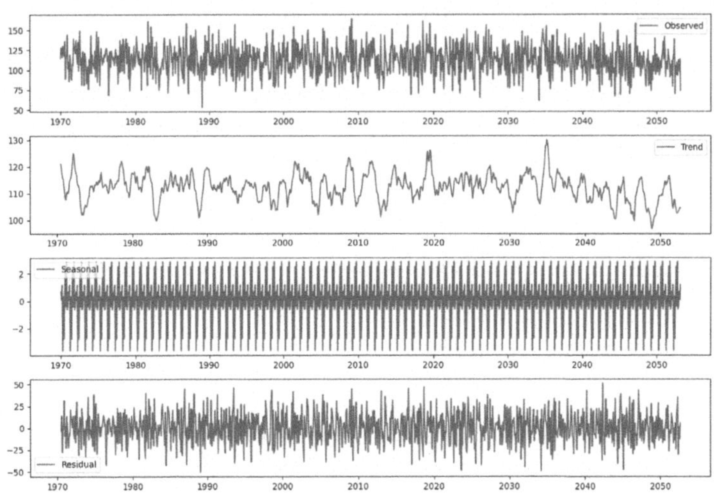

Fig. 2. Decomposition plots for seasonal analysis

From Fig. 2 both the seasonal component and residuals appear as not static in the decomposition plot. It suggests that the data exhibit strong seasonality and residual variability, indicating seasonal fluctuations.

4.5 Learning Models

The daily order quantities are aggregated and performed time series analysis to understand patterns and trends. For training the model, input data is split into training and testing sets in an 80-20 ratio. **Model Selection:** Based on preliminary studies [10–14], SARIMA and Prophet models are used to generate forecasts for future order based on historical data.

SARIMA. For demand forecasting, the SARIMA (Seasonal Auto Regressive Integrated Moving Average) model is applied to training data having cluster information. SARIMA is particularly effective for time-series data, considering seasonality and trends over time. This model takes into account the historical demand patterns within each cluster to provide accurate demand forecasts.

$$SARIMA(p, d, q)(P, D, Q, s) \tag{2}$$

- AR(p): Autoregressive component of order p
- MA(q): Moving average component of order q
- I(d): Integrated component of order d
- Seasonal AR(P): Seasonal autoregressive component of order P
- MA(Q): Seasonal moving average component of order Q
- Seasonal I(D): Seasonal integrated component of order D
- s: Seasonal period

Prophet. Prophet, a robust forecasting tool developed by Facebook is used in this work. It handles various sources of uncertainty in time-series data, including holidays and special events. By applying Prophet after clustering, the model tailors its forecasting approach to the specific demand patterns identified within each cluster, leading to more accurate predictions.

$$y(t) = g(t) + s(t) + h(t) + \varepsilon(t) \tag{3}$$

- g(t) describes a piecewise-linear trend (or "growth term"),
- s(t) describes the various seasonal patterns,
- h(t) captures the holiday effects,
- $\varepsilon(t)$ is a white noise error term.

Ensemble Model. Ensembling in machine learning is a potent strategy that combines multiple models to improve predictive performance. The results from individual models SARIMA and Prophet models are integrated using Weighted Average, Simple Average, and Linear Regression-based ensemble modelling. The Models (SARIMA, Prophet) are ensembled by combining predictions and actual order quantities of the two models, resulting in improved model efficiency.

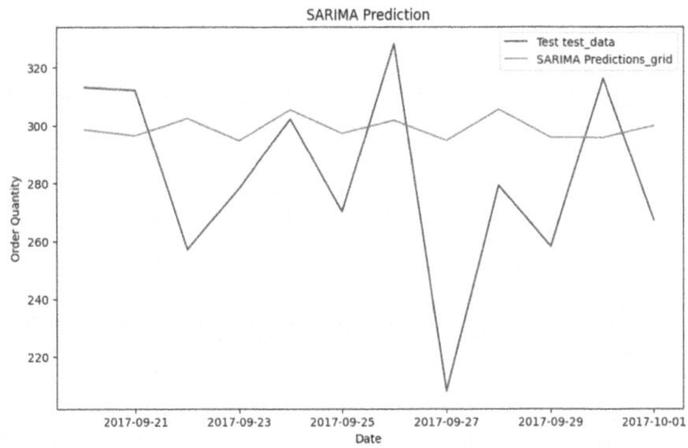

Fig. 3. Actual vs. Predicted Plot for SARIMA Model with Grid Search

5 Experimental Results and Analysis

For experimentation, five node SPARK clusters were deployed in the cloud environment. The cluster of nodes is installed with SPARK 3.3.4. Each node has a total memory capacity of 3 TB. One of the nodes, serving as the name node, records the location of all files. The data nodes in the cluster are chosen with DS4v2 instances which include 8 vCPU, and 28 GB RAM. The analysis starts by importing necessary libraries such as FileIO from io, SparkSession from pyspark.sql, os, and tensorflow from tf.SparkSession is initialized to work with Apache Spark. It' s configured with increased executor memory to handle large-scale data processing efficiently. Then the input dataset is loaded into a Spark Dataframe and preprocessed by standardizing column names, handling missing values, and converting data types as needed.

Table 2. Performance comparison of the SARIMA with and without Hyperparameter

Models/Metrics	SARIMA	Hyperparameter optimization	
		SARIMA [Manual] (0,1,1)(0,1,1,12)	SARIMA [Grid Search] (0,0,0),(0,1,1,12)
MSE	1520.44	1480.83	1375.39
RMSE	43.44	39.48	37.08
MAE	36.68	32.62	30.37
MAPE	12.99	11.99	11.44

In this proposed work, the historical data is trained using SARIMA and Prophet models to generate forecasts for future orders. The performance of the SARIMA and Prophet models is evaluated using metrics viz MSE, RMSE, MAE

Table 3. Performance comparison of PROPHET with and without Hyperparameter

Models/Metrics	PROPHET (Default)	Hyperparameter optimization
		PROPHET [Grid Search]
MSE	11553.28	1333.76
RMSE	411.42	36.52
MAE	34.18	29.59
MAPE	12.15	11.16

and MAPE, as depicted in Tables 2 and 3. Further, Fine-tuning the parameters of SARIMA and Prophet models could potentially lead to improved accuracy and better generalization of unseen data. Techniques such as grid search and manual search are explored in this work to efficiently navigate the high-dimensional parameter space and identify optimal configurations. Both manual and grid search methods were employed to determine the optimal parameters. The SARIMA (0, 1, 1)(0, 1, 1, 12) configuration achieved the lowest MSE among the SARIMA models, suggesting its effectiveness in capturing the underlying patterns in the data. The results depicted in Table 2 exhibit that the differences in performance metrics between grid search and manual methods were marginal, indicating that the grid search might be more practical due to its automation (Fig. 3).

The default configuration of the Prophet model yielded relatively higher MSE compared to the SARIMA models. After hyperparameter tuning through grid search, the configuration 'multiplicative', 0.01, 0.01, 0.01, 0.95, 30, TRUE, TRUE, TRUE achieves improved performance, resulting in lower MSE, RMSE, MAE, and MAPE values (Table 3). The comparison of actual and predicted values for the Prophet Model with Grid Search is represented in Fig. 4.

Finally in this ensemble model using weighted Average, Simple Average and Linear regression are experimented. Ensemble methods enhance prediction accuracy by combining the strengths of multiple models rather than depending on a single model. In this work, the combined predictions from SARIMA and Prophet models are used for ensemble modelling. The trained the ensemble model on the combined predictions and actual order quantities to leverage the strengths

Table 4. Performance comparison of the various Ensemble approaches

Models/Metrics	Ensemble Model		
	Weighted Average	Simple Average	Linear Regression
MSE	1341.26	1341.60	946.64
RMSE	35.50	35.20	30.76
MAE	28.86	28.76	24.43
MAPE	10.26	10.24	8.78

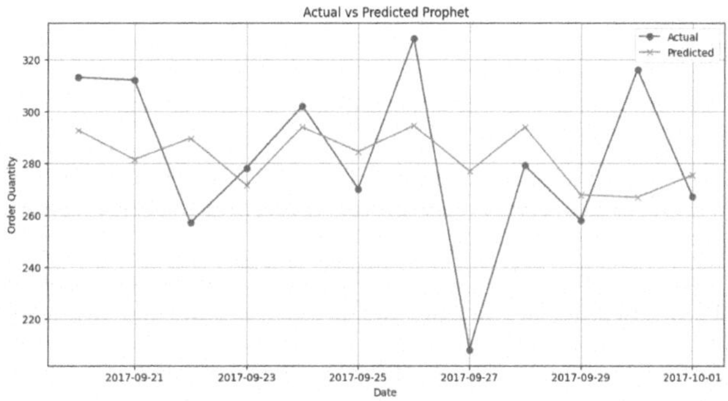

Fig. 4. Actual vs. Predicted Plot for Prophet Model with Grid Search

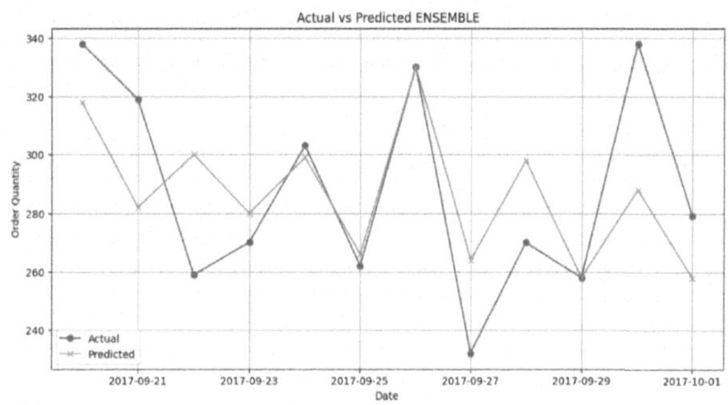

Fig. 5. Actual vs. Predicted Plot for ENSEMBLE Model using Linear Regression

of both models. The results tabulated in Table 4 exhibit the linear regression ensemble model shows less RMSE, it outperforms especially when the relationships between individual model predictions and the target variable are linear (Fig. 5).

6 Conclusion and Future Work

In this proposed work, a time series predictive big data analytics is proposed. The time series and deep learning models like SARIMA, Prophet and ensemble model are used to predict the sales demand from the IOWA_LIQUOR_SALES dataset. In addition, geographical feature is extracted using KMeans clustering and used in the predictive modelling. The experiments were evaluated using five node Spark clusters deployed in the cloud environment. The finding underscores the efficacy of ensemble modelling in augmenting forecasting accuracy, particularly when integrating clustering-based segmentation with SARIMA and Prophet models. In future studies, ensemble methods such as stacking, boosting, and bagging will be investigated to construct more robust ensemble frameworks, aiming to improve predictive performance.

References

1. Huang, H., Zhang, Z., Song, F.: An ensemble-learning-based method for short-term water demand forecasting. Water Resour. Manage **35**(6), 1757–1773 (2021)
2. Jahani, H., Jain, R., Ivanov, D.: Data science and big data analytics: a systematic review of methodologies used in the supply chain and logistics research. Ann. Oper. Res. (2023)
3. Huang, L., Xie, G., Zhao, W., Gu, Y., Huang, Y.: Regional logistics demand forecasting: a BP neural network approach. Complex Intell. Syst. **9**(3), 2297–2312 (2023)
4. Hosseinnia Shavaki, F., Ebrahimi Ghahnavieh, A.: Applications of deep learning into supply chain management: a systematic literature review and a framework for future research. Artif. Intell. Rev. **56**(5), 4447–4489 (2023)
5. Bohanec, M., Borštnar, M.K., Robnik-Šikonja, M.: Explaining machine learning models in sales predictions. Expert Syst. Appl. **71**, 416–428 (2017)
6. Constante, F., Silva, F., Pereira, A.: DataCo smart supply chain for big data analysis (2019)
7. Gholizadeh, H., Tajdin, A., Javadian, N.: A closed-loop supply chain robust optimization for disposable appliances. Neural Comput. Appl. (2018)
8. Chaturvedi, S., Mishra, N., Govindan, K., Cheng, T.C.E.: Big data analytics and application for logistics and supply chain management. Transp. Res. Part E Logistics Transp. Rev. **114**, 343–359 (2018)
9. Rejeb, A., Keogh, J.G., Rejeb, K.: Big data in the food supply chain: a literature review. J. Data Inf. Manag. **4**(1), 33–47 (2022)
10. Aburto, L., Weber, R.: Improved supply chain management based on hybrid demand forecasts. Appl. Soft Comput. **7**(1), 136–144 (2007)
11. Mahmoud, N.: Comparative analysis of machine learning techniques for predictive modeling in social and infrastructural systems. Emerg. Trends Mach. Intell. Big Data **15**(11), 36–42 (2023)
12. Nguyen, T., Li, Z.H.O.U., Spiegler, V., Ieromonachou, P., Lin, Y.: Big data analytics in supply chain management: a state-of-the-art literature review. Comput. Oper. Res. **98**, 254–264 (2018)
13. Lorente-Leyva, L.L., Alemany, M.M.E., Peluffo-Ordóñez, D.H., Araujo, R.A.: Demand forecasting for textile products using statistical analysis and machine learning algorithms. In: Asian Conference on Intelligent Information and Database Systems, pp. 181–194. Springer, Cham (2021)
14. Huber, J., Stuckenschmidt, H.: Daily retail demand forecasting using machine learning with emphasis on calendric special days. Int. J. Forecast. **36**(4), 1420–1438 (2020)
15. Mitra, A., Jain, A., Kishore, A., Kumar, P.: A comparative study of demand forecasting models for a multi-channel retail company: a novel hybrid machine learning approach. Oper. Res. Forum **3**(4), 58 (2022)
16. Huang, H., Zhang, Z., Song, F.: An ensemble-learning-based method for short-term water demand forecasting. Water Resour. Manag. **35**, 1757–1773 (2021)
17. https://data.iowa.gov/Sales-Distribution/Iowa-Liquor-Sales/m3tr-qhgy/data
18. Kodinariya, T.M., Makwana, P.R.: Review on determining number of cluster in K-means clustering. Int. J. **1**(6), 90–95 (2013)
19. Sinaga, K.P., Yang, M.S.: Unsupervised K-means clustering algorithm. IEEE Access **8**, 80716–80727 (2020)

Enhanced Stock Price Prediction with DES-ED-Bi-GRU Using Smooth Maximum Unit Activation Function

Talabathula Jayanth and A. Manimaran(✉)

Department of Mathematics, School of Advanced Sciences, VIT-AP University, Amaravati 522237, Andhra Pradesh, India
{talapaphula22phd7106,manimaran.a}@vitap.ac.in

Abstract. Investors have always been keen on predicting stock market movements, often calculating their potential return on investment beforehand. This study evaluates stock price predictability using statistical and Deep Learning (DL) models on General Electric (GE) daily, weekly, and monthly closing stock prices. A hybrid model is proposed to address both short-term and long-term dependencies, combining an Encoder-Decoder based Bi-directional Gated Recurrent Unit (Encoder-Decoder Bi-GRU) network with the Double Exponential Smoothing (DES) technique. The Encoder-Decoder Bi-GRU forms a nonlinear prediction model using original data, nonlinear residuals, and linear prediction results, while DES develops a linear model for forecasting linear components. The Smooth Maximum Unit (SMU) Activation Function is mainly incorporated to boost model performance. Applied to GE's dataset, the DES-ED-Bi-GRU hybrid model shows consistently high R^2 values (0.998) with high accuracy (98.7%), and low error values MAE (1.415), MSE (3.579), RMSE (1.891), MAPE (1.449), and Theil's U-Statistic (TUS) (0.007). Results show that the DES-ED-Bi-GRU model, enhanced with the SMU activation function, surpasses other well-known techniques. Its bi-directional learning captures historical and future data points, providing reliable predictions for investors and financial experts. Future work will involves applying the DES-ED-Bi-GRU hybrid model to a broader range of stocks and sectors to validate its generalization.

Keywords: Double Exponential Smoothing (DES) · Encoder-Decoder (ED) · Bi-directional GRU (Bi-GRU) · Smooth Maximum Unit (SMU)

1 Introduction and Literature Review

The buying and selling of bonds and stocks by publicly traded companies occurs on the stock market via exchanges and open trading. It entails the issuance and exchange of publicly traded company shares. The purpose of forecasting stock prices is to anticipate the future stock value of a company, which makes it a challenging undertaking. Stock market operations are a significant area of interest in finance and academia because of their inherent risks and benefits. Because there is a lot of information and uncertainty,

A. Dasgupta et al. (Eds.): BDA 2024, LNCS 15526, pp. 395–405, 2025.
https://doi.org/10.1007/978-3-031-81821-9_26

stock market prediction is difficult. Many variables impact the market, including individual behavior, political decisions, corporate performance, and the environment. The fluctuation of the global economy is influenced by stock markets, underscoring their significance to investors. Economic growth in a nation is frequently correlated with stock market performance.

In recent years, advancements in Deep Learning (DL) and Machine Learning (ML) have offered new avenues for improving the accuracy of stock price predictions [13–20]. One such advancement is the introduction of the Smooth Maximum Unit (SMU) activation function [21–25], which has shown promise in enhancing the performance of neural networks. The SMU activation function combines the strengths of traditional activation functions, offering a smooth transition and better gradient propagation, which leads to more efficient learning and improved model accuracy. By incorporating proposed SMU into hybrid models such as the Double Exponential Smoothing Encoder-Decoder based Bi-directional GRU (DES-ED-Bi-GRU), aim to capture both short-term and long-term dependencies in stock price data, thereby providing more reliable forecasts. This study explores the integration of SMU with DES-ED-Bi-GRU hybrid model to enhance stock price prediction accuracy. Demonstrating the potential of SMU to address the complexities and uncertainties inherent in stock market operations.

Several studies have explored various models for stock price forecasting. [1] Shak Shahid et al. (2020) investigated models like Simple Exponential Smoothing (SES), Holt's Exponential Smoothing (HES), and Holt-Winter Exponential Smoothing (HWES), achieving RMSE, MAPE, and MAE accuracy scores of 72.84, 6.26%, and 57.64, respectively. [2] Quang Thanh Tran et al. (2019) used models such as Seasonal ARIMA, ETS, SES, and Double Exponential Smoothing (DES), recommending Holt-Winters Multiplicative seasonal (HWMS) for better performance. [3] Xin He et al. (2024) introduced HW-GRU, evaluated with PICP, PINRW, and Time metrics, achieving a PICP value of 95.53%.

[4, 6] Dheeraj Kodati et al. (2024) proposed the Context-ABT-BiLSTM-CNN model, outperforming baseline models with 0.973 accuracy. [5] Gitanjali Kumari, [7] Rina Kumari introduced MSEM + MFB for text-image tasks. Other works like CoDyn-BMHSA-CNN and BERT + ResNet also showed high performance.

[8] Renzhuo Wan (2019) and [9] Shun-Yao Shih (2019) focused on neural network models like M-TCN and TPA-LSTM, respectively, and assessed performance using RMSE, MAE, and MAPE. Several other studies ([10] Benhaddi, [11] Loannis E, [12] Kang Wanga) proposed models like CNN-LSTM, MLSTM-FCN, and inception Time. Recent works by [21] Koushik Biswa (2021), [22] Marian Adriana Mercioni (2023), [23] Kang Kang Song, [24] Youngmin Seo, and [25] Ruohuan Liu et al. (2024) highlighted the superior performance of the Smooth Maximum Unit (SMU) activation function in reducing errors like MAE across models, demonstrating its potential for enhancing model accuracy and robustness.

To summarize, the literature review presents various research works that have examined distinct forecasting models and suggested innovative methods to enhance prediction precision. These studies used a variety of metrics, including accuracy, correlation, RMSE, MAE, MSE and more, to assess their models. Research also investigates hybrid models and blends of conventional techniques with cutting-edge neural network methodologies to obtain better forecasting performance.

2 Proposed Methodology

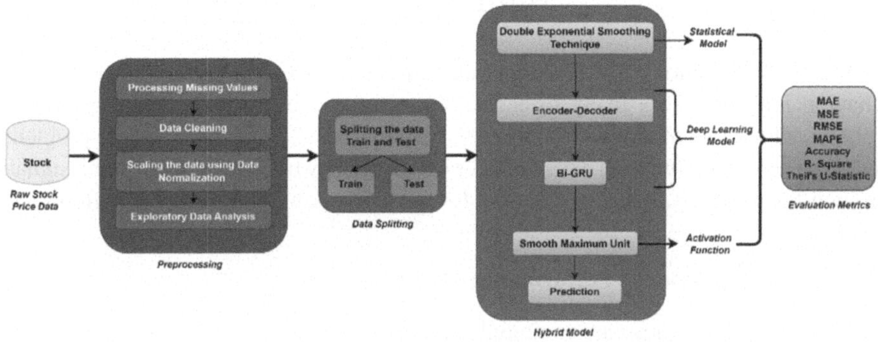

Fig. 1. Flow chart of the Proposed Methodology (DES-ED-Bi-GRU-SMU)

From Fig. 1, flowchart represents a stock price forecasting model using a combination of data preprocessing techniques and advanced Deep Learning models. Here's a step-by-step explanation:

Data Preprocessing: The process starts with raw stock data i.e., General Electric (GE). Initial steps include handling missing values, cleaning the data, and preforming normalization to scale the data, ensuring consistency. Exploratory Data Analysis (EDA) is performed to understand patterns and insights in the data.

Data Splitting: The data is dividing into two parts training and testing set. The training set helps the model learn patterns from historical data, while the test set evaluates the model's accuracy.

Double Exponential Smoothing (DES): DES is applied to the stock price data. This method is used to capture trends in time series data and smoothness out fluctuations, making predictions more reliable.

Model Architecture: The core prediction model involves an Encoder-Decoder architecture combined with a Bi-GRU neural network. The Smooth Maximum Unit (SMU) is used as the activation function to enhance model performance.

Prediction and Evaluation: The model generates stock price predictions. Key evaluation metrics like MAE, MSE, RMSE, MAPE, Accuracy, R-Squared, and Theil's U-Statistics (TUS) are used to measure prediction accuracy and model performance.

2.1 Smooth Maximum Unit (SMU)

The Smooth Maximum Unit (SMU) activation function is a type of activation function used in neural networks. It is designed to combine the benefits of the Rectified Linear Unit (ReLU) and the soft maximum functions, providing a smoother alternative that can help with training stability and performance. The SMU can be represented as:

$$SMU(x) = \frac{x}{1 + e^{-x}} \tag{1}$$

The SMU function [21–25] smoothly transitions between linear and non-linear behavior, making it less likely to encounter issues related to the vanishing gradient problem, which is common with sigmoid or tanh functions, or the dying ReLU problem. The SMU activation function works in neural networks by transforming the input signals into non-liner outputs, which is essential for the network to learn complex patterns.

2.2 Forecasting Model's Performance Measures

Forecasting error measures are commonly employed to assess the efficacy of forecasting techniques. Theil's U-Statistic (TUS), Mean Absolute Error (MAE), Mean Squared Error (MSE), R^2, Root Mean Squared Error (RMSE), Mean Absolute Percentage Error (MAPE), and Accuracy are the error metrics used in this study.

MAE: MAE measures the average absolute difference between the actual values and predicted values. It provides an intuitive measure of how close the predictions are to the actual values.

$$MAE(y_n^*, y_n) = \frac{1}{N}\sum\nolimits_{n=1}^{N} |y_n - y_n^*| \tag{2}$$

MSE: MSE is a commonly used metric that measures the average squares difference between the actual and predicted values. It penalizes larger errors more than smaller errors.

$$MSE(y_n^*, y_n) = \frac{1}{N}\sum\nolimits_{n=1}^{N} (y_n - y_n^*)^2 \tag{3}$$

RMSE: RMSE is the square root of MSE. It is one of the most popular evaluation metrics as it is in the same units as the actual values and is more interpretable than MSE.

$$RMSE(y_n^*, y_n) = \sqrt{\frac{1}{N}\sum\nolimits_{n=1}^{N} (y_n - y_n^*)^2} \tag{4}$$

MAPE: MAPE measures the percentage error between the predicted and actual values, providing a normalized view of how far off the predictions are in percentage terms.

$$MAPE = \frac{1}{N}\sum\nolimits_{n=1}^{N} \left|\frac{y_n - y_n^*}{y_n}\right| x100 \tag{5}$$

Accuracy: Accuracy is the percentage of correct predictions or how close the predictions are to the actual values.

$$Accuracy = \frac{1}{N}\sum\nolimits_{n=1}^{N} 1(\left|\frac{y_n - y_n^*}{y_n}\right| \leq \in x100 \tag{6}$$

$\mathbf{R^2}$**:** R-square measures the proportion of the variance in the actual values that is predictable from the predicted values. It ranges from 0 to 1, with higher values indicating a better model fit.

$$R^2 = 1 - \frac{\sum (y_n - y_n^*)^2}{\sum (y_n - \overline{y_n})^2} \tag{7}$$

TUS: Theil's U-Statistic is a relative accuracy measure used to evaluate the forecasting performance of a model, typically the previous period's actual value. It helps assess whether the forecasting model performs better than random guessing or a naïve approach.

$$TUS = \sqrt{\frac{\sum_{n=1}^{N} (y_n - y_n^*)^2}{\sum_{n=1}^{N} y_n^2}} \tag{8}$$

2.3 Our Contribution

Innovative Combination: The true innovation of the model comes from combining the strengths of the following different methods DES for trend smoothing, Encoder-Decoder for sequence prediction, Bi-GRU for bidirectional learning, and SMU for smoother optimization. These techniques have rarely been combined in this specific configuration for stock price prediction.

Enhanced Accuracy: By using DES to preprocess the data and fee it into a powerful Bi-GRU network with SMU activation, the model can make more robust and accurate predictions, particularly in highly volatile stock markets.

Flexibility and Efficiency: The model's ability to handle both short-term fluctuations (via DES) and long-term dependencies (via Bi-GEU) makes it flexible for various stock price patterns, while the SMU activation enhances computational efficiency.

The novelty lies in how these components are integrated, and the innovation stems from the unique synergy they create, resulting in a model that is more accurate, efficient, and effective for stock price forecasting.

3 Data Analysis and Results

A correlation heatmap visually represents the correlation between variables in a dataset using colors.

From Fig. 2, the high positive correlation coefficients close to 0.60, that the variables move in the same direction, indicating a strong positive relationship. High negative

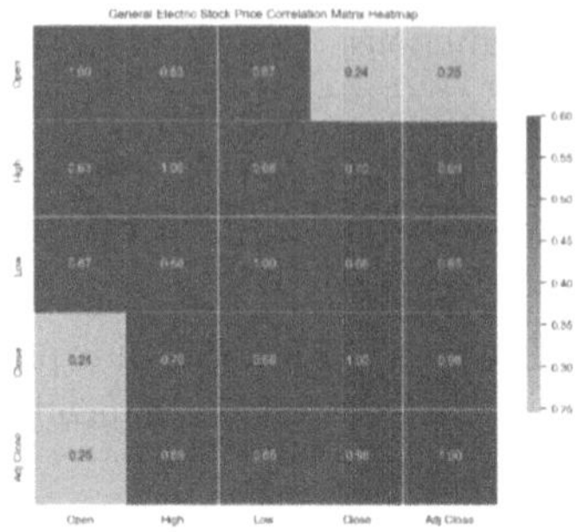

Fig. 2. Correlation Matrix Heatmap to the General Electric Stock Data.

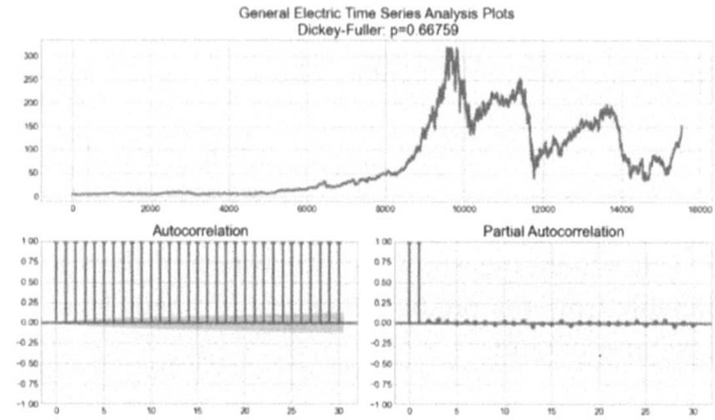

Fig. 3. Autocorrelation and Partial Autocorrelation plots using General Electric Stock Price Data

correlation coefficients close to −0.25, that the variables move in opposite directions, indicating a strong negative relationship. Correlation coefficients close to 0, indicate no linear relationship between the variables. The above correlation heatmap provide a visual summary of the relationship between different variables in the stock price data, helping analysts and investors easily understand the interdependencies and potential patterns in the data.

Autocorrelation plot (ACF), and Partial Autocorrelation (PACF) plots are used in time series analysis to examine the relationship between a series and its lagged values. They help identify patterns like seasonality or trends, guiding model selection for forecasting. ACF shows how a series is correlated with its past values at different lags. For stock prices, the ACF plot reveals if and how much the current price is influenced by past prices. Significant correlations suggest the stock price is affected by its previous values.

PACF measures the direct relationship between a series and its lagged values, removing the effect of intervening lags. In stock price data, significant partial autocorrelations at certain lags show a direct link between current and past prices, ignoring intermediate values. Figure 3, shows that the GE stock with ACF and PACF plots suggest non-stationarity. A high p-value (e.g., 0.66759) indicates weak evidence of autocorrelation, helping to select appropriate forecasting models.

Training, testing, and validation are essential for developing and evaluating the hybrid DES-ED-Bi-GRU model for GE stock price prediction. As shown in Fig. 4(B), the model is trained using historical stock prices to learn patterns like trends and seasonality through optimization algorithms (e.g., Adam). The model adjusts its parameters to minimize prediction errors. In Fig. 4(C), the testing set contains unseen data that evaluates the model's performance and generalization, showing how well it predicts future prices. Figure 4(D) highlights the validation set, used to fine-tune hyperparameters and prevent overfitting. Monitoring performance on this set helps optimize the model and avoid overfitting during training.

Table 1, provides performance metrics for the GE Daily stock price data. The dataset has three subsets; training, testing, and validation of the performance metrics are included. The best model for the dataset can be determined based on the performance metrics across training, testing, and validation sets. Generally, lower MAE, MSE, and

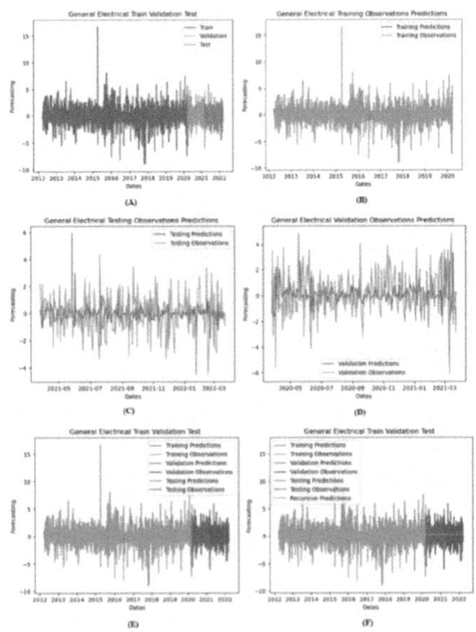

Fig. 4. Training, Validation, and Testing to the General Electric Stock Price Data.

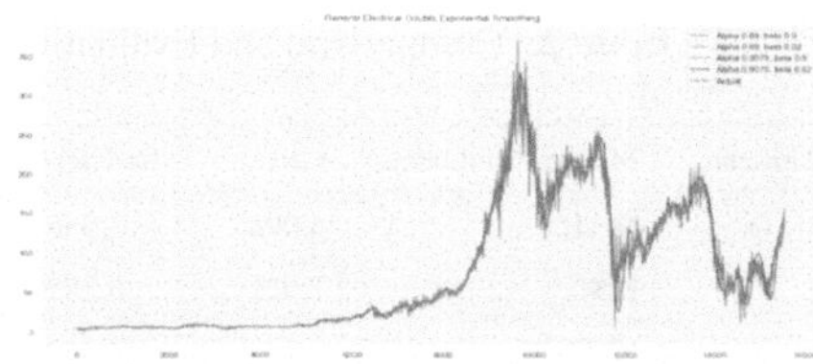

Fig. 5. General Electric Daily, Weekly, and Monthly Stock Price Prediction Using Encoder-Decoder Bi-directional GRU with Double Exponential Smoothing Technique Using Smooth Maximum Unit activation function

Table 1. Evaluation Metric for GE stock price dataset using train, test, and validation.

Dataset	GE-D		
Metrics	MAE	MSE	RMSE
Training	1.166	2.582	1.607
Testing	1.204	2.481	1.57
Validation	1.396	3.45	1.857

RMSE values indicate better model performance. Analysis of the GE dataset based on Table 1.

GE-D: The model performs relatively consistently across training and testing, but the metrics worsen slightly in the validation set.

Figure 5, shows the stock price GE-D, plotted using DES with alphas and betas specified (alphas = [0.09, 0.0079] and betas = [0.9, 0.02]), which can provide insights into stock price trends.

The evaluation metrics provided assess the performance of different models in predicting General Electric (GE) stock prices over daily (GE-D), weekly (GE-W), and monthly (GE-M) intervals as shown in Table 2.

GE-D (Daily): Figure 6(A), shows that the proposed model DES-ED-Bi-GRU performs best across all metrics with the lowest MAE (1.415), MSE (3.579), RMSE (1.891), MAPE (1.449), and TUS (0.007). It also achieves the highest R^2 (0.998), and Accuracy (98.7%) indicating its effectiveness in predicting daily stock prices.

Table 2. Comparison and Evaluation Metrics of General Electric Stock Price

Datasets	Metrics/Methods	GRU	Bi-GRU	ED-Bi-GRU	DES	DES-ED-Bi-GRU-SMU
GE-D	MAE	4.098	3.956	3.198	2.844	1.415
	MSE	7.921	6.289	4.297	6.385	3.579
	RMSE	3.149	2.919	2.389	4.378	1.891
	MAPE	4.119	3.331	3.498	2.947	1.449
	R^2	0.989	0.995	0.997	0.995	0.998
	Acc%	89%	92.8%	94.97%	92.1%	98.7%
	TUS	0.093	0.089	0.086	0.015	0.007
GE-W	MAE	7.433	4.920	4.737	4.289	3.189
	MSE	108.8	74.360	52.955	30.052	18.031
	RMSE	10.43	8.623	7.277	5.482	4.246
	MAPE	6.177	5.709	4.982	4.324	3.382
	R^2	0.894	0.912	0.957	0.985	0.991
	Acc%	90.7%	92.01%	92.8%	96.2%	97.1%
	TUS	0.085	0.040	0.025	0.022	0.017
GE-M	MAE	13.50	11.73	9.261	8.059	7.899
	MSE	151.9	137.07	114.15	102.19	99.974
	RMSE	15.87	12.549	10.684	10.109	9.974
	MAPE	13.60	11.632	10.856	8.891	8.701
	R^2	0.846	0.894	0.923	0.951	0.952
	Acc%	90.3%	91.56%	92.12%	92.8%	93.01%
	TUS	0.067	0.061	0.051	0.047	0.041

GE-W (Weekly): From Fig. 6(B), DES-ED-Bi-GRU outperforms other models with the lowest MAE (3.189), MSE (18.031), RMSE (4.246), MAPE (3.382) and TUS (0.017). It has a high R^2 (0.998), and Accuracy (97.1%) showcasing its strong performance in forecasting weekly stock prices.

GE-M (Monthly): From Fig. 6(C), DES-ED-Bi-GRU remains the top performer with the lowest MAE (7.899), MSE (99.974), RMSE (9.974), MAPE (8.701), and TUS (0.041). It also achieves the highest R^2 (0.952), and Accuracy (93.01%) demonstrating its efficiency in predicting monthly stock prices.

To improve the performance of the proposed hybrid model DES-ED-Bi-GRU for General Electric (GE) stock price forecasting, incorporating the SMU activation can be highly beneficial.

$$SMU(x) = \frac{1}{\beta} log(1 + e^{\beta x})$$

where β is a hyperparameter that controls the smoothness of the function. SMU is a smooth alternative to the ReLU function, with smoother transitions at zero. This helps improve the training process by ensuring more stable gradients and better convergence. Unlike ReLU, which has sharp changes, SMU's smoothness allows for more efficient

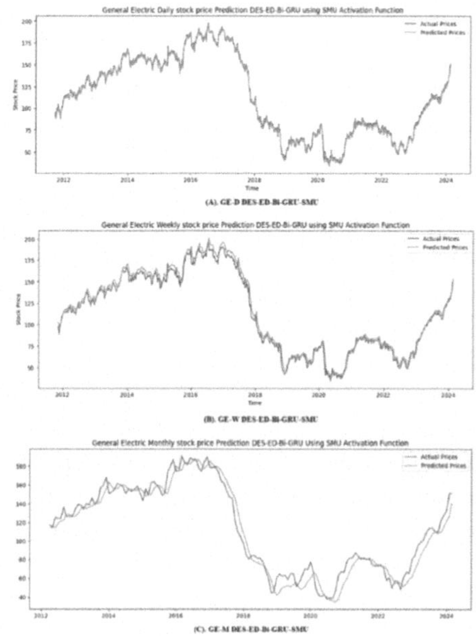

Fig. 6. GE-D Double Exponential Smoothing (DES) Plot

optimization, avoiding issues like vanishing or exploding gradients common in deep networks.

By using SMU instead of ReLU or sigmoid in the DES-ED-Bi-GRU model, the result is improved gradient flow, faster convergence, and more accurate stock price predictions, particularly for complex and noisy data like GE stock prices.

Table 3. Evaluation Metric Theil's U-Statistic

Theil's U-Statistic	GRU	Bi-GRU	ED-Bi-GRU	DES	DES-ED-Bi-GRU-SMU
GE-Daily	0.093	0.089	0.0886	0.015	0.007
GE-Weekly	0.085	0.040	0.025	0.022	0.017
GE-Monthly	0.067	0.061	0.051	0.047	0.041

TUS is shown in the Table 3 for each of the three dataset's models. TUS for the GE-D dataset is lowest for DES-ED-Bi-GRU (0.007), suggesting that this model best matches the maximum prediction accuracy. Once more, DES-ED-Bi-GRU has the lowest TUS (0.017) for the GE-W dataset, indicating that it is the best-suited model. TUS for the GE-M dataset is lowest for DES-ED-Bi-GRU (0.041), meaning it outperforms all other models. To sum up, the results indicate that DES-ED-Bi-GRU consistently displays the lowest TUS compared to other models. This indicates that GE-D stock is the model with the highest forecast accuracy.

4 Conclusion

In conclusion, the research validates the DES-ED-Bi-GRU hybrid model's excellent performance in forecasting General Electric (GE) stock values over a range of time periods. The model effectively combines linear and nonlinear patterns in stock price data by integrating the Encoder-Decoder Bidirectional Gated Recurrent Unit (ED-Bi-GRU) network with the Double Exponential Smoothing (DES) approach and utilizing the Smooth Maximum Unit (SMU) activation function. Comparing this hybrid strategy to other well-established forecasting techniques yields lower error rates (MAE, MSE, RMSE, MAPE, and TUS) and higher R^2 values, and Accuracy percentage. The bidirectional learning mechanism of the model, enhanced by the SMU activation function, produces reliable and strong forecasts by capturing both past and future data points more effectively. The smooth and differentiable nature of SMU ensures better gradient flow, leading to improved training convergence and overall model performance. The model's capacity to maintain accuracy over time and adapt to abrupt changes in the market is improved by this dual perspective and the advanced properties of the SMU activation function. Because of its flexibility and adaptability, the DES-ED-Bi-GRU hybrid model is a valuable tool for financial professionals and investors looking for dependable information on stock price patterns in erratic financial markets. Its thorough method of identifying intricate patterns in stock price movements, coupled with the benefits of the SMU activation function, makes it an indispensable tool for well-informed investing choices.

References

1. Shahid, S., Rahaman, S.A.: Exponential smoothing methods for detection of the movement of stock prices. Int. J. Recent Technol. Eng. (IJRTE) **8**(5), 1420–1422 (2020)
2. Tran, Q.T., Hao, L., Trinh, Q.K.: Comprehensive research on exponential smoothing methods in modeling and forecasting cellular traffic. Concurr. Comput. Pract. Exp. **32**(23), e5602 (2020)
3. He, X., et al.: A hybrid prediction interval model for short-term electric load forecast using Holt-Winters and Gate Recurrent Unit. Sustain. Energy Grids Netw. **38**, 101343 (2024)
4. Kodati, D., Dasari, C.M.: Negative emotion detection on social media during the peak time of COVID-19 through deep learning with an auto-regressive transformer. Eng. Appl. Artif. Intell. **127**, 107361 (2024)
5. Kumari, G., Bandyopadhyay, D., Ekbal, A.: EmoffMeme: identifying offensive memes by leveraging underlying emotions. Multimedia Tools Appl. **82**(29), 45061–45096 (2023)
6. Kodati, D., Tene, R.: Emotion mining for early suicidal threat detection on both social media and suicide notes using context dynamic masking-based transformer with deep learning. Multimedia Tools Appl. 1–24 (2024)
7. Kumari, R., et al.: Identifying multimodal misinformation leveraging novelty detection and emotion recognition. J. Intell. Inf. Syst. **61**(3), 673–694 (2023)
8. Wan, R., et al.: Multivariate temporal convolutional network: a deep neural networks approach for multivariate time series forecasting. Electronics **8**(8), 876 (2019)
9. Shih, S.-Y., Sun, F.-K., Lee, H.: Temporal pattern attention for multivariate time series forecasting. Mach. Learn. **108**, 1421–1441 (2019)

10. Benhaddi, M., Ouarzazi, J.: Multivariate time series forecasting with dilated residual convolutional neural networks for urban air quality prediction. Arab. J. Sci. Eng. **46**, 3423–3442 (2021)
11. Livieris, I.E., Pintelas, E., Pintelas, P.: A CNN–LSTM model for gold price time-series forecasting. Neural Comput. Appl. **32**, 17351–17360 (2020)
12. Wang, K., et al.: Multiple convolutional neural networks for multivariate time series prediction. Neurocomputing **360**, 107–119 (2019)
13. Laubscher, R.: Time-series forecasting of coal-fired power plant reheater metal temperatures using encoder-decoder recurrent neural networks. Energy **189**, 116187 (2019)
14. Hu, J., Zheng, W.: Multistage attention network for multivariate time series prediction. Neurocomputing **383**, 122–137 (2020)
15. Karim, F., et al.: Multivariate LSTM-FCNs for time series classification. Neural Netw. **116**, 237–245 (2019)
16. Jiang, W.: Time series classification: Nearest neighbor versus deep learning models. SN Appl. Sci. **2**(4), 721 (2020)
17. Chen, R., et al.: DA-Net: dual-attention network for multivariate time series classification. Inf. Sci. **610**, 472–487 (2022)
18. Yang, C., et al.: Attentional gated Res2Net for multivariate time series classification. Neural Process. Lett. **55**(2), 1371–1395 (2023)
19. Tripathi, A.M., Baruah, R.D.: Multivariate time series classification with an attention-based multivariate convolutional neural network. In: 2020 International Joint Conference on Neural Networks (IJCNN). IEEE (2020)
20. Cheng, X., et al.: A novel channel and temporal-wise attention in convolutional networks for multivariate time series classification. IEEE Access **8**, 212247–212257 (2020)
21. Biswas, K., et al.: SMU: smooth activation function for deep networks using smoothing maximum technique. arXiv preprint arXiv:2111.04682 (2021)
22. Mercioni, M.A., Holban, S.: A brief review of the most recent activation functions for neural networks. In: 2023 17th International Conference on Engineering of Modern Electric Systems (EMES). IEEE (2023)
23. Song, K.K., et al.: An improved bearing defect detection algorithm based on yolo. In: 2022 International Symposium on Control Engineering and Robotics (ISCER). IEEE (2022)
24. Seo, Y., Kim, J., Park, U.: Swish-T: Enhancing Swish Activation with Tanh Bias for Improved Neural Network Performance. arXiv preprint arXiv:2407.01012 (2024)
25. Liu, R., et al.: Prediction of cement slurry formulation based on RMAPCN. In: 2024 IEEE 2nd International Conference on Control, Electronics and Computer Technology (ICCECT). IEEE (2024)

Real-Time Pothole Detection System: A Deep Learning Approach with SSD

B. Leelavathy[1(✉)], Ram Mohan Rao Kovvur[1], Khateeb Mohammed Zahid Parvez[1], Bathini Ajay Goud[1], and Rathan Kumar Chenoori[2]

[1] Vasavi College of Engineering, Ibrahim Bagh, Hyderabad, India
leelapallava@staff.vce.ac.in
[2] KMIT, Hyderabad, India

Abstract. Roadway maintenance and safety are paramount concerns for commuters and transportation authorities, with potholes posing considerable risks to vehicles and causing accidents, damage, and increased maintenance expenses. Conventional pothole detection methods have been largely inefficient and lacking in real-time analytics, impeding proactive problem-solving. This study introduces an innovative, centralized platform for pothole detection and management, featuring an intuitive interface enriched with interactive tools. Utilizing a Single Shot Detector (SSD) ensures efficient and accurate pothole detection. Complementing the SSD, the Neo6M GPS Module and ESP32 Camera Module provide precise spatial mapping and data collection. Beyond detection, the platform integrates real-time mapping and route optimization, suggesting alternative pathways when damages are identified in-route. Additionally, the platform triggers automated email alerts to the concerned authorities every 24 h until the pothole is repaired, ensuring timely intervention. Furthermore, robust analytical tools allow users to sift through pothole data by criteria such as location and size. The results demonstrate effective pothole identification with varying detection accuracy across different configurations, highlighting the importance of optimal parameter tuning. The system enhances road safety by providing timely information and route suggestions, aiding in proactive road maintenance and improved commuter safety.

Keywords: Pothole detection · CNN · SSD · YOLO · ResNet34 · SSD · Augmentations

1 Introduction

Potholes are a major hazard on roads worldwide, causing significant financial burdens and serious safety risks to motorists. These road crashes cause billions in damages and around 425,000 deaths worldwide each year. Despite the serious consequences, there is still a worrying lack of timely notification and preventive measures to mitigate these hazards [7]. The lack of real-time pothole detection systems on vehicles increases the risks associated with these traffic hazards. Traditional reporting methods often rely on manual audits or public complaints, resulting in delays in problem resolution. Such inefficiency not only threatens road safety but also increases vehicle damage and maintenance

A. Dasgupta et al. (Eds.): BDA 2024, LNCS 15526, pp. 406–415, 2025.
https://doi.org/10.1007/978-3-031-81821-9_27

costs [8]. The main aim of this paper is to propose a robust and efficient real-time hole detection system using a convolutional neural network (CNN) based SSD (Single Shot Detector) detector. The purpose of this system is to warn drivers of rapidly developing potholes, allowing them to take a different approach and keep their vehicles safe [18–20]. This research involves the design, development, and evaluation of an SSD-based detection model, focusing on its efficiency, accuracy, and applicability under different roads and weather conditions. This article makes an important contribution by introducing an advanced real-time pothole detection system using SSD model. Our proposed system shows superior performance in terms of accuracy and real-time detection capabilities compared to existing solutions. Using advanced deep learning techniques, our approach not only increases road safety but also helps reduce vehicle damage and related maintenance costs.

2 Related Works

Object detection is a fundamental task in computer vision, aiming to identify and locate objects within images or videos [3, 4]. Over the years, various techniques have been developed to address this challenge, ranging from traditional methods like Haar cascades and Histogram of Oriented Gradients (HOG) to more advanced deep learning-based approaches [6]. Conventional methods often rely on handcrafted features and classifiers to detect objects but may struggle with complex scenes and variations in object appearance [1]. Deep learning-based approaches [21], particularly Convolutional Neural Networks (CNNs), have revolutionized object detection by automatically learning hierarchical features from data. These methods can be broadly categorized into two-stage and one-stage detectors. Two-stage detectors, such as Faster R-CNN and R-FCN, first propose regions of interest (ROIs) and then classify and refine these regions [9] Single Shot Detector (SSD) is a leading one-stage object detection framework that has gained significant attention due to its efficiency and performance [5]. SSD predicts detections at multiple scales within a single network using default boxes or anchor boxes. This allows SSD to capture objects of various sizes and aspect ratios, making it robust to scale variations. One of the key advantages of SSD is its balance between speed and accuracy. With a frame rate of up to 60 FPS, SSD offers real-time object detection capabilities without compromising much on accuracy [10]. Additionally, SSD's architecture is relatively simple compared to two-stage detectors, making it easier to train and deploy [22]. Its single-shot prediction approach eliminates the need for complex region proposal networks, resulting in faster inference times and reduced computational costs. The SSD framework has been extensively studied and improved upon since its introduction. Liu et al. first introduced SSD [12], demonstrating its effectiveness in real-time object detection across various datasets. Subsequent works have focused on optimizing the SSD architecture, enhancing its performance, and adapting it to specific applications [19]. Several studies have explored different aspects of SSD, such as feature extraction techniques, anchor box design, and loss function optimization. For instance, Zhang et al. proposed an improved version of SSD by incorporating context information into the feature maps to enhance object localization accuracy [25]. Other works have investigated the integration of SSD with other techniques, such as attention mechanisms and data augmentation strategies, to further boost its performance in challenging scenarios [23].

3 Methodology

3.1 Data Collection and Preprocessing

The dataset utilized for this project comprises over 847 images scaled up to 2333 images augmentation sourced from Kaggle as shown in Fig. 1, are renowned platform for machine learning datasets and competitions [13–17]. These images are annotated using LabelMe, a widely used tool for annotating images with bounding boxes. The annotations, detailing the location of objects within the images, are stored in XML format. The Properties of the dataset are shown in Table 1 below.

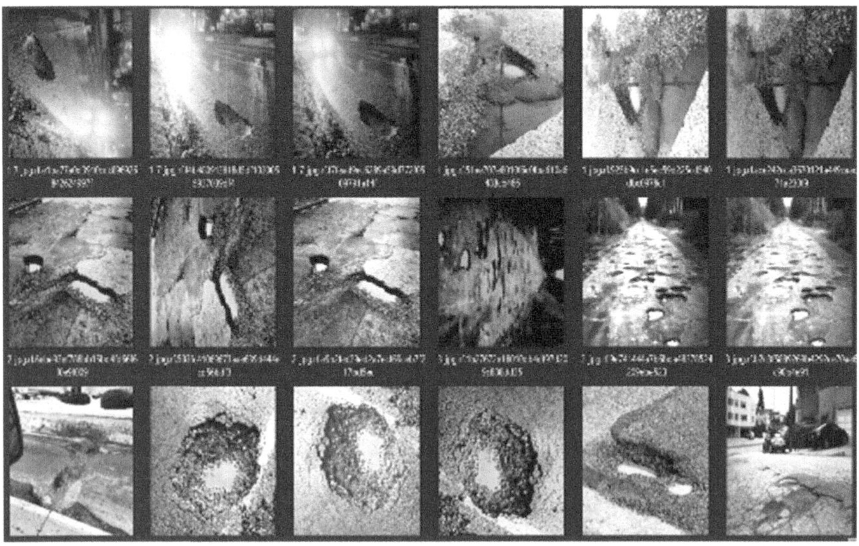

Fig. 1. Augmented Images in Dataset

Model Architecture

Single Shot Detector (SSD) as shown in Fig. 2 is a state-of-the-art object detection framework that enables the detection of multiple objects within an image in a single forward pass of the network. Unlike traditional two-stage detectors, SSD integrates object localization and classification into a unified single-stage architecture, making it more efficient and faster. SSD achieves this by predicting bounding boxes of various aspect ratios and scales at multiple feature map layers. This multi-scale prediction allows SSD to detect objects of different sizes and aspect ratios, making it highly versatile for various object detection tasks. The ResNet34 architecture serves as the backbone of our SSD model. ResNet (Residual Network) is a deep convolutional neural network architecture that introduces residual connections or skip connections. These skip connections enable the model to learn residual mappings, allowing for the training of deeper networks without encountering the vanishing gradient problem.

Table 1. Properties of the dataset before and after augmentation.

Property	Before Augmentation	After Augmentation
Number of Images	847	2333
Class Distribution	Balanced(Pothole, No pothole)	Balanced(Pothole, No pothole)
Data Augmentation	N/A	Rotation, Scaling, Flipping, Brightness Adjustment
Average Image Brightness	Standardized	Varies due to brightness augmentation
Image Variety	Limited variations in angle lighting	Increased diversity in image conditions
Annotation	Bounding boxes for potholes	Bounding boxes for potholes

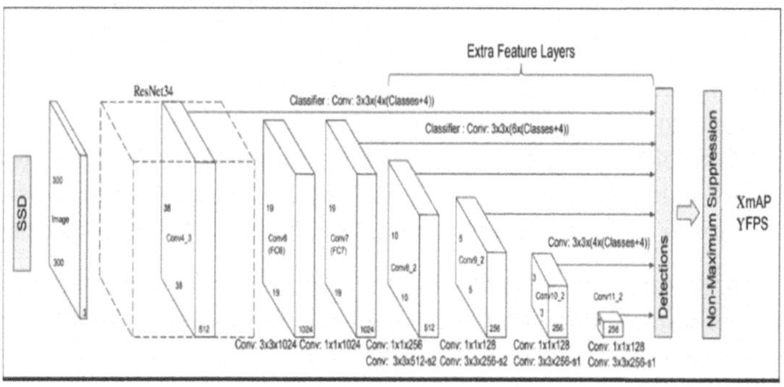

Fig. 2. SSD Architecture.

The SSD architecture employs a Default Box Generator and an SSDHead to predict bounding boxes and class scores across various feature map layers.

Default Box Generator

This component generates default bounding boxes of various aspect ratios and scales across different feature map layers. These default boxes serve as reference boxes for predicting the final bounding boxes of objects within the image.

SSDHead

The SSDHead consists of a series of convolutional layers responsible for predicting the offsets and confidences for each default box. For each default box, the SSDHead predicts the offsets to adjust the box to better fit the object's true location and the confidence scores representing the likelihood of an object belonging to a particular class.

3.2 Training and Evaluation

The training process involved fine-tuning the SSD model with 70% of our dataset. We experimented with different hyperparameters, including Momentum, Learning Rate, and Loss Function, to optimize the model's performance. The training was conducted using the PyTorch framework, which offers flexibility, ease of use, and strong community support.

Momentum: We experimented with different momentum values to accelerate convergence and enhance training stability. Momentum values like 0.1 and 0.9 were tested to observe their impact on the model's performance.

Learning Rate: Various learning rates, such as 0.0005, 0.01, and 0.001, were tested to find the optimal rate that facilitates faster convergence without compromising the model's accuracy.

Loss Function: Both Stochastic Gradient Descent (SGD) Loss and Accelerated Gradient Descent (AGD) Loss functions were employed during training to evaluate their impact on the model's performance.

Optimal Parameters: The configuration characterized by a momentum of 0.1, a learning rate of 0.0005, and an SGD optimizer emerged as the most productive, achieving an outstanding mAP@50 of 88.676 and a corresponding training loss of 0.6764, as illustrated in Table 2. After training the SSD model, we evaluated its performance using several evaluation metrics to assess its accuracy, precision, and recall in detecting potholes.

mAP@50 (mean Average Precision at IoU 0.5): This metric measures the average precision of the model across different Intersection over Union (IoU) thresholds. A mAP@50 score provides insights into the model's ability to precisely locate and classify objects within the images.

By employing these evaluation metrics and criteria, we were able to quantify the performance of our SSD-based pothole detection system accurate as shown in Fig. 3.

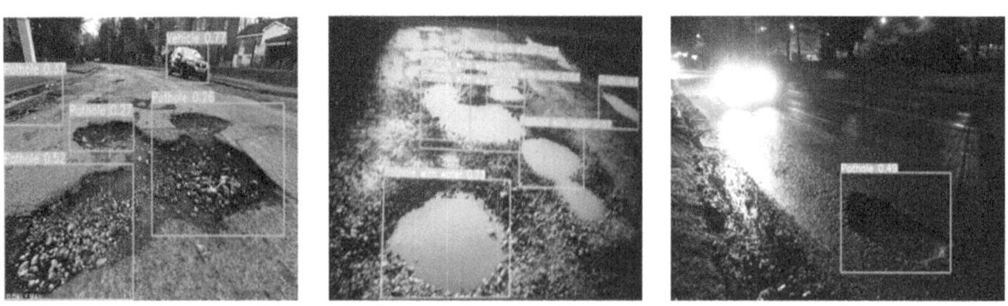

Fig. 3. Detection of different potholes & Detection of False Objects like Trees, Shadows, Vehicles, Healthy Roads.

3.3 Optimal Route Suggestions

The website is a sophisticated platform designed for users to access and analyse pothole detection results comprehensively. With an emphasis on user-centric design, it features an intuitive interface complemented by interactive functionalities, facilitating effortless exploration and interpretation of the aggregated data [11, 26]. Incorporating the Neo6M GPS Module and ESP32 Camera Module for mapping, the platform provides users with a spatially contextualized view of detected potholes. Leveraging the real-time data from the Neo6M GPS Module and ESP32 Camera Module, it dynamically evaluates the road conditions and pothole severity to recommend optimal detours. This proactive approach not only minimizes the risk of vehicular damage but also enhances user safety and convenience. In terms of data interpretation, the platform organizes pothole detection results in a structured manner, allowing users to filter and sort based on parameters like location, size, and detection confidence. This analytical flexibility empowers users to discern trends, identify patterns, and prioritize maintenance areas effectively.

Furthermore, the real-time integration with the Neo6M GPS Module and ESP32 Camera Module ensures that the map interface always reflects the most recent pothole detection results as shown in Fig. 4. This continuous update mechanism provides users with up-to-date information, facilitating informed decision-making and enabling prompt response to identified issues.

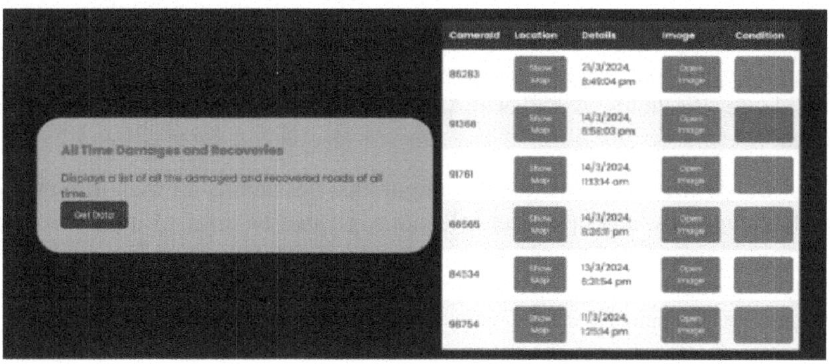

Fig. 4. Comprehensive website integrated with Google Maps API for analyzing results for concerned officers to take needed actions.

4 Experimental Results

4.1 Performance Evaluation

Impact of Configurations: The diverse configurations, characterized by varying momentum values, learning rates, and optimizer selections, yielded a spectrum of training loss values, ranging from a minimal 0.545 to a peak of 39.944. This variance underscores the profound impact of these parameters on the model's learning dynamics and convergence rate.

Optimal Configuration for training loss: The configuration characterized by a momentum of 0.1, a learning rate of 0.0005, and an ADAM optimizer emerged as the most efficacious, attaining a training loss of 0.545, as depicted in Table 2.

Table 2. Results obtained while training the SSD model under various hyperparameters.

Momentum	Learning Rate	Optimizer	Achieved mAP@50	Achieved Train Loss
0.1	0.0005	SGD	88.676	0.6764
0.5	0.0000002	SGD	13.15	39.944
0.9	0.01	SGD	85.85	0.976
0.9	0.02	SGD	88.24	0.994
0.9	0.001	ADAM	49.9	1.892
0.1	0.001	ADAM	25.769	0.545
0.1	0.01	ADAM	14.2	5.026
0.1	0.02	ADAM	3.5	7.5

4.2 mAP Scores Evaluation

Significance: The mAP (mean Average Precision) scores encapsulate the model's detection accuracy by considering the precision and recall of the detected objects across varying confidence thresholds. A higher mAP score indicates superior detection accuracy and efficacy.

Impact of Configurations: The tested configurations, spanning a gamut of momentum values, learning rates, and optimizer selections, yielded an array of mAP@50 scores, ranging from a meager 3.5 to an exceptional 88.676. This divergence accentuates the criticality of parameter selection and optimization in enhancing the model's detection accuracy and performance.

4.3 Comparative Analysis

In terms of detection accuracy, measured by the mean Average Precision at a 50% overlap (mAP@50), the SSD (Single Shot MultiBox Detector) model displays a wide variation in its performance. The achieved mAP@50 values for SSD span from a modest 3.5 to an impressive 88.676. This broad range suggests that the effectiveness of the SSD model can significantly differ depending on the specific configuration and conditions under which it operates. The mAP values across the YOLOv8 and YOLOv5 models, from "YOLOv8n" to "YOLOv8x", remain consistent, ranging between 82.3 and 85.2. This suggests that enhancements made in these versions maintain a stable level of accuracy. On the other hand, the YOLO V5 models, with "YOLOv5n6" leading at 82.5 and "YOLOv5m6" trailing at 80.8. When compared to these YOLO models, the SSD (Single

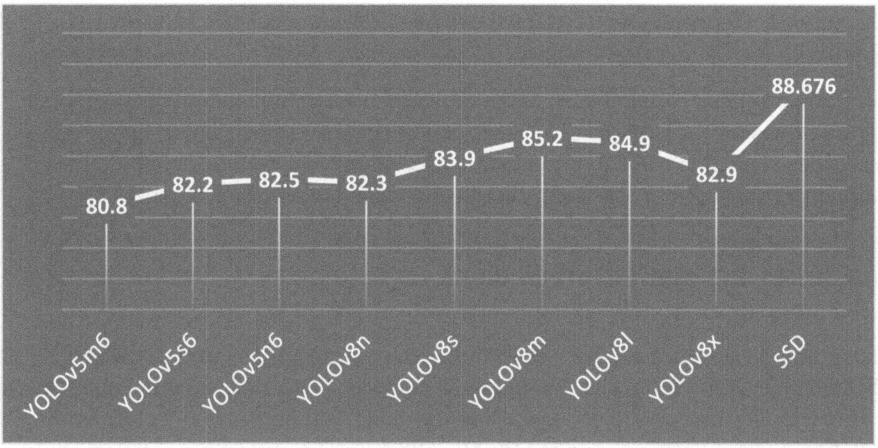

Fig. 5. Comparison between different object detection models with SSD.

Shot MultiBox Detector) model's performance range, spanning from 3.5 to 88.676 in mAP@50, surpasses the YOLO V8's accuracy levels as depicted in Table 2 and Fig. 5.

In summary, while YOLO models [2, 24, 27] offer consistent but relatively lower detection accuracies, SSD demonstrates a broader and potentially higher range of detection performance, indicating its superior capability for pothole detection tasks.

Momentum Variation

Insights: The momentum values of 0.5, 0.1, 0.9 were rigorously evaluated to ascertain their impact on detection performance. The superior mAP@50 scores attained with a momentum of 0.1 substantiate the hypothesis that higher momentum values facilitate enhanced detection accuracy by enabling smoother and more consistent weight updates.

Recommendations: Opting for default momentum value, i.e. 0.1, can potentially augment the model's detection accuracy and robustness, albeit necessitating meticulous parameter tuning and optimization to avert potential overfitting or oscillations.

Learning Rate Variation

Insights: The exploration of learning rates ranging from 0.0000002 to 0.02 revealed that an optimal balance is pivotal for achieving peak detection performance. The efficacy of a learning rate of 0.01 / 0.02 / 0.0005 underscores the detrimental impact of excessively low or high learning rates on the model's learning dynamics and convergence rate.

Recommendations: Endeavoring to strike an optimal balance in learning rate selection, preferably around ~0.01 or 0.0005, can potentially optimize the model's learning trajectory, enhance convergence, and augment detection accuracy, albeit necessitating iterative experimentation and validation to ascertain the most efficacious learning rate.

Optimizer Variation

Insights: The comparative evaluation between SGD and ADAM optimizers unveiled the former's superiority in optimizing the model for heightened detection accuracy. The discernible performance gap between the two optimizers accentuates the pivotal role of optimizer selection in determining the model's efficacy, performance, and robustness.

Recommendations: Opting for an SGD optimizer can potentially optimize the model's learning dynamics, convergence, and detection accuracy, albeit necessitating meticulous experimentation, validation, and parameter tuning to ascertain the most efficacious configuration tailored to the specific application and dataset characteristics.

5 Conclusion and Future Scope

The project on a real-time pothole detection system using an enhanced CNN model with SSD architecture revealed that fine-tuning hyperparameters, such as setting momentum at 0.1, learning rate at 0.0005, and using the SGD optimizer, significantly improved detection accuracy, achieving an mAP@50 of 88.676. This system has practical applications in infrastructure monitoring, providing timely alerts and facilitating proactive road maintenance to improve safety. Future research could broaden the dataset to cover diverse conditions, explore advanced architectures, and integrate IoT and cloud-based processing to support scalable, real-time monitoring. Collaborative efforts with academia, industry, and government agencies may further drive innovation in automated infrastructure management.

References

1. Alamgeer, M., et al.: Optimal Fuzzy Wavelet Neural Network Based Road Damage Detection (2023)
2. Her, A.Y., Yew, W.K., Yew, P.J., Ying, M.C.J.: Realtime pothole detection system on vehicle using improved YOLOv5 in Malaysia (2022)
3. Liu, W., Wang, Z., Liu, X., Zeng, N., Liu, Y., Alsaadi, F.E.: A survey of deep neural network architectures and their applications. Neurocomputing (2020)
4. Alom, M.Z., et al.: A stateoftheart survey on deep learning theory and architectures. Electronics (2021)
5. Darma, Y., Karim, M.R., Abdullah, S.: An analysis of Malaysia road traffic death distribution by road environment. Sadhana Acad. Proc. Eng. Sci. **42**(9), 1605–1615 (2017). https://doi.org/10.1007/s1204601706949
6. Malaysia's Minister Khairy Jamaluddin injures from fall after bicycle hits pothole. The Straits Times (2020). https://www.straitstimes.com/asia/seasia/malaysiasministerkhairyjamaluddin injuredfromfallafterbicyclehitspothole. Accessed 31 May 2022
7. Perimbanayagam, K.: 75-year-old man killed after crashing into the pothole. New Straits Times (2021). https://www.nst.com.my/news/nation/2021/01/654169/75yearoldmankilledaft ercrashingpothole. Accessed 31 May 2022
8. Thiruppathiraj, S., Kumar, U., Buchke, S.: Automatic pothole classification and segmentation using android smartphone sensors and camera images with machine learning techniques. In: 2020 IEEE Region 10 Conference (TENCON), pp. 1386–1391 (2020). https://doi.org/10.1109/TENCON50793.2020.9293883
9. Fan, R., Ozgunalp, U., Hosking, B., Liu, M., Pitas, I.: Pothole detection based on disparity transformation and road surface modeling. IEEE Trans. Image Process. **29**, 897–908 (2020). https://doi.org/10.1109/TIP.2019.2933750

10. Baek, J.W., Chung, K.: Pothole classification model using edge detection in road image. Appl. Sci. **10**(19) (2020). https://doi.org/10.3390/APP10196662
11. Jo, Y., Ryu, S.: Pothole detection system using a blackbox camera. Sensors **15**(11), 29316–29331 (2015). https://doi.org/10.3390/s151129316
12. Liu, S., Qi, L., Qin, H., Shi, J., Jia, J.: Path Aggregation Network for Instance Segmentation (2018)
13. GitHub: GitHub ultralytics/yolov5. https://github.com/ultralytics/yolov5. Accessed 31 May 2022
14. https://docs.ultralytics.com/models/#featuredmodels
15. https://www.kaggle.com/datasets/atulyakumar98/potholedetectiondataset
16. https://www.kaggle.com/datasets/andrewmvd/potholedetection
17. https://www.kaggle.com/datasets/rajdalsaniya/potholedetectiondataset
18. Andika, F., Bandung, Y.: Road damage classification using SSD mobilenet with image enhancement. In: 2023 International Conference on Computer Science, Information Technology and Engineering (ICCoSITE). IEEE (2023)
19. Pawar, S., Nahar, S., Shaikh, M.D., Meher, V., Narwane, S.: Cloud based single shot detector model for speed breaker detection. In: 2023 International Conference on Emerging Smart Computing and Informatics (ESCI). IEEE (2023)
20. Bratulescu, R.A., Vatasoiu, R.I., Sucic, G., Mitroi, S.A., Vochin, M.C., Sachian, M.A.: Object detection in autonomous vehicles. In: 2022 25th International Symposium on Wireless Personal Multimedia Communications (WPMC). IEEE (2022)
21. Hasan, M.M., Sakib, S., Deb, K.: Road damage detection and classification using deep neural network. In: 2022 4th International Conference on Electrical, Computer & Telecommunication Engineering (ICECTE). IEEE (2022)
22. Khan, M.A.A., et al.: Road damages detection and classification using deep learning and UAVs. In: 2022 2nd Asian Conference on Innovation in Technology (ASIANCON). IEEE (2022)
23. Wei, C., Zhang, Q., Zhang, X., Yang, Y., Zhang, J., Zhai, D.: YOLOv5s-BSS: a novel deep neural network for crack detection of road damage. In: 2023 IEEE International Conference on Big Data (BigData). IEEE (2023)
24. Pham, V., Nguyen, D., Donan, C.: Road damage detection and classification with YOLOv7. In: 2022 IEEE International Conference on Big Data (Big Data). IEEE (2022)
25. Zhang, L., Wan, F., Lei, G., Xu, L.: A road damage detection method based on improved YOLOV8s. In: 2023 5th International Conference on Frontiers Technology of Information and Computer (ICFTIC). IEEE (2023)
26. Tarun, R., Esther, B.P.: Real-time regional road sign detection and identification using Raspberry Pi. In: 2023 International Conference on Networking and Communications (ICNWC). IEEE (2023)
27. Dharneeshkar, J., Soban Dhakshana, V., Aniruthan, S.A., Karthika, R., Parameswaran, L.: Deep learning based detection of potholes in Indian roads using YOLO. In: 2020 International Conference on Inventive Computation Technologies (ICICT). IEEE (2020)

Data Analysis with Model Explainability for Air Quality and Pollution Prediction of Urban Areas in Edge Networks

Akansha Singh[1(✉)], Soumendu Chakraborty[1], and Mainak Adhikari[2]

[1] Department of Computer Science, Indian Institute of Information Technology, Lucknow, Lucknow, India
{RCS24101,soumendu}@iiitl.ac.in

[2] School of Data Science, Indian Institute of Science Education and Research, Thiruvananthapuram, Thiruvananthapuram, India
mainak@iisertvm.ac.in

Abstract. Air quality significantly impacts human health and economic conditions, making precise and timely assessment crucial in urban areas. Existing studies often fail to predict pollution accurately in smaller areas due to varying meteorological conditions. Most existing prototypes rely on standard ML models, trained with generic, publicly available datasets, which often lack local dynamics, adaptability, and synchronization. To address these challenges, in this paper, we propose a new Hybrid Edge-Cloud Air Quality Monitoring System. For initial model training a robust testbed is created by Data Fusion of air quality data from diverse sources like Application Programming Interfaces and public datasets. The system trains and refines ML models in the cloud, identifies key features to reduce feature dimensions, and leverages local decision-making using site-specific data collected by sensors on edge devices. Sensors monitor local environmental parameters for real-time analysis and decision-making, ensuring accurate predictions for specific urban locations while optimizing performance and minimizing resource usage on edge devices. Explainable AI techniques improve transparency and confidence in decisions. The model is validated by simulating various ML models on the testbed, with and without feature extraction.

Keywords: Edge Networks · Data Analysis · Data Fusion · Explainable Artificial Intelligence · Machine Learning · Pollution Prediction

1 Introduction

Air Pollution has emerged as a growing concern in recent years, primarily driven by rapid industrialization and urbanization [1,11]. Increasing urbanization worsens air pollution, harming the environment and human health [12]. For instance, exposure to Particulate Matter contributes to cardiovascular morbidity and mortality. According to a report by the WHO (2002), Geneva, Switzerland, estimated

A. Dasgupta et al. (Eds.): BDA 2024, LNCS 15526, pp. 416–425, 2025.
https://doi.org/10.1007/978-3-031-81821-9_28

that PM2.5 contributes to approximately 800000 premature deaths annually, making it the 13th leading cause of mortality worldwide [2]. Human activities significantly degrade air quality and such activities emit a variety of dangerous pollutants, such as carbon monoxide (CO), carbon dioxide (CO_2), PM, nitrogen dioxide (NO_2), sulfur dioxide (SO_2), ozone (O_3), ammonia (NH_3), and lead (Pb), into the environment [5,7]. Pollutant concentrations vary geographically, requiring localized monitoring to identify harmful pollutants and enable targeted air quality management. Air quality prediction involves managing extensive real-time data, but models often show inconsistent accuracy due to the complexity of air pollutants [9].

Recent research based on ML/DL models to analyze pollution levels in localized areas using large-scale public datasets. Public datasets may not reflect local conditions, and training on resource-constrained edge devices is challenging due to limited computational capacity. Ram et al. have developed an XGboost model to forecast the air quality [10]. Similarly, Kekulanadara et al. employed DT, SVM, and RF for classifying the different pollutant concentrations [4]. Similarly, in the papers [3,14], various ML models have been used to predict air quality using limited pollutants. Data and metrics from the local area must be used to achieve precise analysis and recommendations, considering fluctuating pollutant levels.

Another major challenge is training models with all features without identifying key parameters, which makes them unsuitable for cloud and edge deployment due to computational and memory constraints. This can be removed by feature extraction. In [8], the paper aims to improve feature extraction accuracy using the k-means algorithm to select the best feature subset, enhancing accuracy. It highlights the importance of optimal features for classification or regression tasks. Paper [6] discussed the problems related to high dimensional data in classification models due to high computational cost and memory usage. Another paper by Xu et al. [15] presents a framework for feature extraction in learning tasks involving multiple variables. However, it does not explore specific performance metrics and lacks a detailed discussion on the framework's scalability to large datasets and complex learning tasks.

Local data collection with edge devices reduces latency but transferring data to the cloud can drain batteries and strain bandwidth. The paper by Song et al. [13] emphasizes the essential role of edge devices in tasks like ML-based image recognition, data verification, and model retraining. However, it focuses only on indoor mapping through door sign recognition.

Addressing the challenges outlined above, this paper presents an intelligent, edge-centric prototype for air quality monitoring and analysis utilizing sensors for localized data collection and decision-making. The key contributions of this work are as follows:

- Data fusion is applied to data collected from APIs and publicly available datasets with similar features to create a consistent, reliable, error-free testbed. Datasets undergo preprocessing to remove noise before being used for cloud-based ML model training.

- The centralized approach utilizes cloud processing for large-scale data and ML training, employing feature extraction to enhance model performance and identify optimal models.
- Trained ML models and key parameters are deployed on edge devices for real-time decision-making while training leverages cloud power. SHAP enhances transparency and trust through interpretable explanations.

The remaining parts of the paper are organized as follows. Section 2 presents the system model and problem statement, Sect. 3 details the proposed methodology, Sect. 4 discusses experimental analysis, and Sect. 5 concludes the work.

2 System Model and Problem Formulation

This section highlights the system model of the proposed Hybrid Edge-Cloud prediction model followed by the problem statements associated with the research.

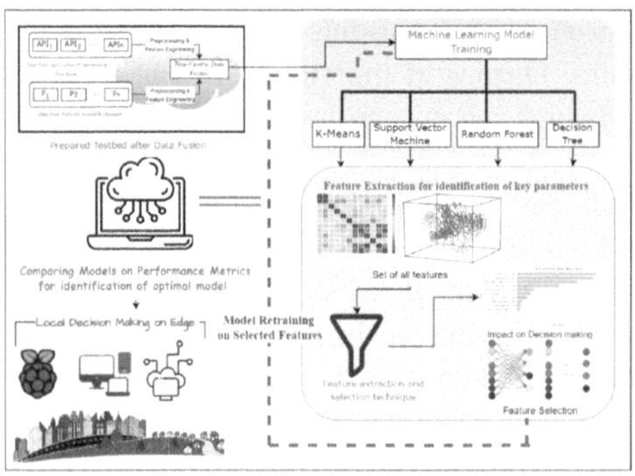

Fig. 1. Proposed edge-centric air quality and pollution prediction framework

2.1 System Model

The proposed Hybrid Edge-Cloud air quality prediction model, illustrated in Fig. 1, utilizes local edge devices like Raspberry Pis for data analysis and a private cloud workstation for testbed preparation and initial training. The model is trained on fused data from APIs (e.g., RapidAPI, Open-Meteo) and public datasets (e.g., Kaggle) after preprocessing to remove errors and noise. The testbed includes various meteorological features such as SO_2, NO_2, CO_2, CO, PM10, and PM2.5. Integrating diverse data sources into a unified testbed is challenging due to varying time intervals. Cloud-based data fusion aggregates preprocessed data to ensure quality. Standard ML models (KNN, SVM, RF, DT) are trained on the private cloud, and feature extraction identifies key parameters affecting performance. The optimal model is deployed on resource-constrained edge devices that gather local data with multiple sensors, further fine-tuning the model to improve prediction accuracy.

2.2 Problem Formulation

Based on the review of existing research the following problem statements are identified.

Problem I: Insufficient datasets in ML often lead to inaccurate predictions that do not reflect local variations. Data augmentation can expand datasets but usually with repetitive values, resulting in minimal improvement. Relying on a single API or public dataset introduces bias. The prediction error E can be expressed as follows.

$$E = \frac{1}{N}\sum_{i=1}^{N}(y_i - \hat{y}_i)^2 \quad (1)$$

Problem II: The testbed created by fusing data from APIs and public datasets executes on a high-dimensional feature space, imposing significant storage and computational demand. Deploying these models and all the features to resource-constrained edge devices poses further challenges. The original model M_{full} is trained on the full feature set X and the model with reduced testbed $M_{reduced}$ is trained on the extracted feature set $\hat{X}$. The prediction errors for both models can be represented as follows.

$$E_{full} = \frac{1}{n}\sum_{i=1}^{n}(y_i - \hat{y}_{i,full})^2 \quad (2)$$

Problem III: Each microenvironment has unique contaminants and pollutants, and making cloud-based decisions using public data is insufficient for accurate regional pollution. Decision-making based solely on cloud servers using publicly available data may not provide accurate decisions for regional areas. To demonstrate the improvement in the model with and without feature extraction is defined as follows.

$$\Delta E = E_{\text{public}} - E_{\text{edge}} \quad (3)$$

3 Hybrid Edge-Cloud Air Quality Monitoring

The proposed model addresses challenges such as data insufficiency, regional data limitations, and ensures accurate and timely predictions. This section details the working methodology of the proposed model.

3.1 Data Collection and Data Fusion Strategy

ML models require high-quality, large datasets for accurate predictions. Data insufficiency can lead to suboptimal results, as explained in Eq. (1). Incorporating diverse datasets broadens the range of values and environmental conditions. We integrated data from APIs and public datasets to ensure comprehensive environmental coverage. The datasets typically include critical pollutants like sulfur dioxide (SO_2), nitrogen dioxide NO_2, carbon dioxide (CO_2), carbon monoxide (CO), PM ($PM10$), and fine PM ($PM2.5$). The data undergoes thorough preprocessing, including handling missing values, to minimize errors and ensure optimal condition before data fusion, enhancing the reliability and accuracy of the final dataset. Data Fusion from different sources is elaborated through Eq. (4).

$$D_{\text{total}} = D_{\text{API}} \cup D_{\text{public}} \tag{4}$$

Given that data is collected from numerous sources, there is a possibility that these sources may produce data values at different time intervals. To address varying time intervals for producing data from different sources, we defined a standardized time interval of 4 h for data fusion that inference that data fusion occurs 6 times a day, ensuring that the collected data is synchronized and accurate given in Eq. (5). Let T be the set of 4-hour time intervals within a day, i.e., $T = \{t_1, t_2, \ldots, t_6\}$. Let F be the set of features from different datasets. For each 4-hour time interval $t \in T$, aggregate features $f_i \in F$ to create fused features f_i^*. The fused dataset D_{fused} is represented as follows.

$$D_{\text{fused}} = \{(t, f_1^*, f_2^*, \ldots, f_m^*) \mid t \in T \tag{5}$$

For each interval, features $f_i \in F$ from different datasets are aggregated to form a fused dataset D_{fused}.

3.2 Cloud-Centric Model Training and Feature Extraction

Cloud-based model training offers scalability, computational power, and storage, enabling efficient processing of large datasets. The standard ML models KNN, SVM, RF, and DT are trained on the cloud on the fused testbed having features X and a target variable y. Training models on the full set of features would require significantly more resources making it less feasible for real-time decision-making on resource constrained edge device. Therefore, after training the models and evaluating their performance metrics for future comparisons, feature extraction techniques are applied to identify the most influential features, using mutual information as feature extraction. Feature extraction is a critical step in reducing the dimension of the data while retaining the most informative features. Mutual information measures the amount of information obtained about one variable through another variable, helping to select the most relevant features that can be utilized on resource com trained devices. The mutual information between X and y features is calculated using Eq. (6).

$$I(X; y) = \sum_{x \in X} \sum_{y \in Y} p(x, y) \log \left(\frac{p(x, y)}{p(x)p(y)} \right) \tag{6}$$

In this approach, the top k features with the highest mutual information scores are selected. Let $\hat{X}$ be The reduced feature set after feature extraction, which is represented as $\hat{X} = \{x_i \mid I(x_i; y)$ is among the top key features$\}$. After identification of the prime features from the testbed, the models are retained over the filtered feature set, represented as $M_{\text{reduced}} = \text{train\_model}(\hat{X}, y)$. The prediction error for the reduced model is calculated using Eq. (7).

$$E_{\text{reduced}} = \frac{1}{n_{\text{test}}} \sum_{i=1}^{n_{\text{test}}} (y_{\text{test},i} - \hat{y}_{\text{reduced},i})^2 \tag{7}$$

where n_{test} is the number of samples in the test set, $y_{\text{test},i}$ is the true value for the ith sample, and $\hat{y}_{\text{reduced},i}$ is the predicted value for the ith sample. To evaluate the improvement resulting from feature extraction, we compare the initial and reduced models. This comparison is represented by Eq. (8), which is formulated based on prediction error and performance metrics.

$$\Delta E = E_{\text{initial}} - E_{\text{reduced}} \tag{8}$$

After comparing the models based on performance metrics, and CPU and memory utilization, the best model is identified. The selected features and the best-performing model are then deployed to edge devices for localized decision-making.

3.3 Edge Deployment and Explainable Artificial Intelligence

Air quality varies across microenvironments, necessitating localized data collection for accurate predictions. Cloud-based training offers scalability but suffers from latency and prediction errors Eq. (3). Sensor-based edge stations collect real-time data for precise and localized results. Let X_{local} be the feature matrix and y_{local} be the target variable collected by the edge devices. The optimal model M_{optimal} is deployed to the edge devices. The edge devices use this model to make predictions on the local data X_{local} as $\hat{y}_{\text{edge}} = M_{\text{optimal}}(X_{\text{local}})$. To ensure interpretability and transparency, we use SHAP (SHapley Additive exPlanations) for explainable AI (XAI). SHAP quantifies each feature's influence, providing insights into model behavior. Analyzing SHAP values helps validate the model and identify impactful features for informed decision-making. Leveraging SHAP enhances the reliability and trustworthiness of predictions. The SHAP values ϕ_i for each feature i are calculated using Eq. (9).

$$\phi_i = \sum_{S \subseteq F \setminus \{i\}} \frac{|F|!}{|S|!(|F| - |S| - 1)!} [f(S \cup \{i\}) - f(S)] \tag{9}$$

For local interpretation, the SHAP values for individual predictions are calculated as the contributions of each feature to a single prediction $\hat{y}_j$ is given by Eq. (10).

$$\hat{y}_k = \phi_0 + \sum_{i=1}^{m} \phi_{i,k} \tag{10}$$

4 Experimental Analysis

This section evaluates the effectiveness of the proposed Hybrid Edge-Cloud air quality model through various performance metrics.

4.1 Experimental Setup and Dataset

For the experimental setup and simulation, a system equipped with an Intel i7 12th Gen CPU, 4GB of graphics memory, and 8GB of RAM is utilized. The simulation, implemented in Python, used Numpy and Pandas for data preprocessing and normalization. ML models, including KNN, SVM, RM, and DT, were employed with performance evaluation, conducted using the Sklearn module to measure metrics such as F1 score, accuracy, precision, and recall. SHAP was applied to attribute feature importance for individual predictions to ensure model transparency and validation.

The dataset, comprising 3000 records, is generated by fusing publicly available data from Kaggle and API data from RapidAPI and Open-Meteo. This dataset was divided into training and testing subsets, including features such as CO_2, CO, NO_2, SO_2, $PM10$ (particulates less than 10 microns), and meteorological variables like humidity, precipitation, temperature, soil moisture, and wind speed. Models are initially trained on a complete dataset and post-feature extraction. For local decision-making, data is collected using sensors on edge devices, such as Raspberry Pi, for localized prediction, with SHAP ensuring model transparency and validation throughout the process.

4.2 Results Discussion

To validate feature extraction in the proposed model, performance metrics are computed to identify the most effective model for deployment on resource-constrained edge devices. This ensures optimal performance and resource efficiency in limited computational environments.

The performance metrics for each classifier, evaluated both before and after feature extraction, are depicted through a series of visual graphs. Figure 2a illustrates the accuracy of the models, showcasing improvements after retraining on a refined testbed with high-impact features. Similarly, Fig. 2b focuses on precision, revealing the effect of feature extraction on this metric. Figure 2c presents the F1-score, reflecting the balance between precision and recall following feature extraction. Lastly, Fig. 2d displays the recall, highlighting the model's ability to identify relevant instance variations.

The results indicate a notable improvement in model performance following feature extraction as it effectively improved the performance of classifiers, confirming its value in optimizing model accuracy and reliability. Analyzing performance metrics assesses the model's generalization, accuracy in identifying positives, and balance between precision and recall. Table 1 illustrates the positive impact of feature extraction, reflected in the performance metrics CPU and memory utilization were also evaluated to validate efficiency for deployment on

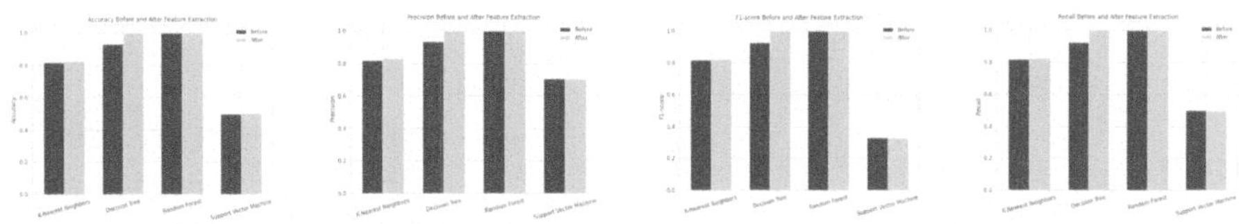

(a) Accuracy(%) (b) Precision(%) (c) F1 Score (d) Recall

Fig. 2. Comparative Analysis before and after Feature Extraction

Table 1. Performance Analysis Before and After Feature Selection

Classifier	Before Feature Selection				After Feature Selection			
	Accuracy	Precision	Recall	F1-score	Accuracy	Precision	Recall	F1-score
KNN	0.8151	0.8141	0.8151	0.8136	0.8235	0.8271	0.8235	0.8201
DT	0.971	0.972	0.9701	0.9721	1.0000	1.0000	1.0000	1.0000
RF	0.9861	0.9812	0.9889	0.9721	1.0000	1.0000	1.0000	1.0000
SVM	0.5084	0.7501	0.5084	0.3427	0.5084	0.7501	0.5084	0.3427

resource-constrained edge devices. Figure 3 shows significant improvements in CPU utilization, training time, and memory usage. The Random Forest model is the best candidate for edge device deployment due to its superior performance metrics and stability. It offers a balance between accuracy and resource efficiency, making it suitable for constrained environments. Its robustness to noisy data and overfitting ensures reliable predictions in varying conditions.

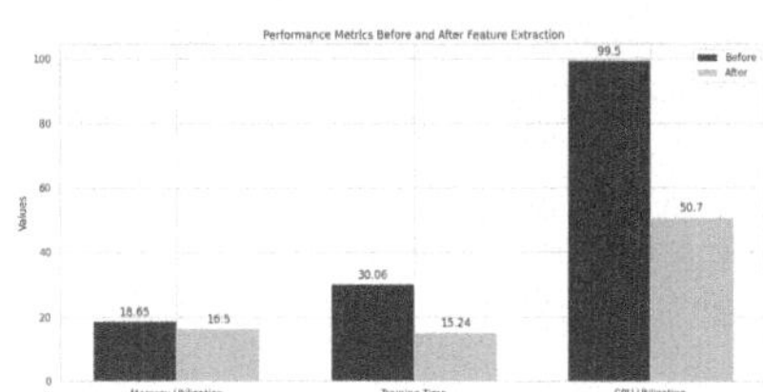

Fig. 3. CPU and Memory Usage and Training Time for Validating Resource Efficiency

To further validate and ensure transparency in the model and justify the feature extraction of high-impact features, SHAP has been integrated into the proposed model. Figure 4a represents the SHAP interaction values to elucidate the contribution and significance of each feature in the model's predictions. Besides that, Fig. 4b shows feature extraction using the mutual information method for the RF classifier. This visualization highlights the impact of individual features, providing a clear understanding of their role and validating the selection of high-impact features through feature extraction.

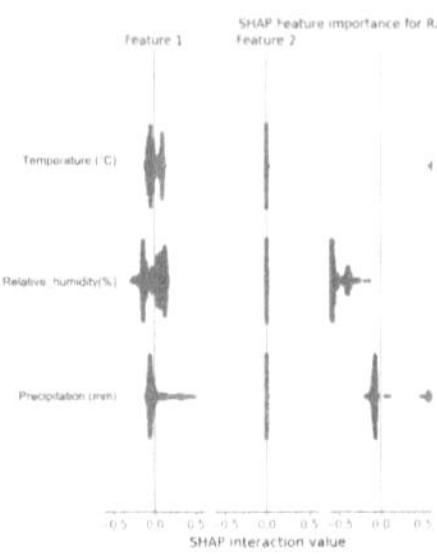

(a) SHAP interaction Plot for Random Forest Classifier

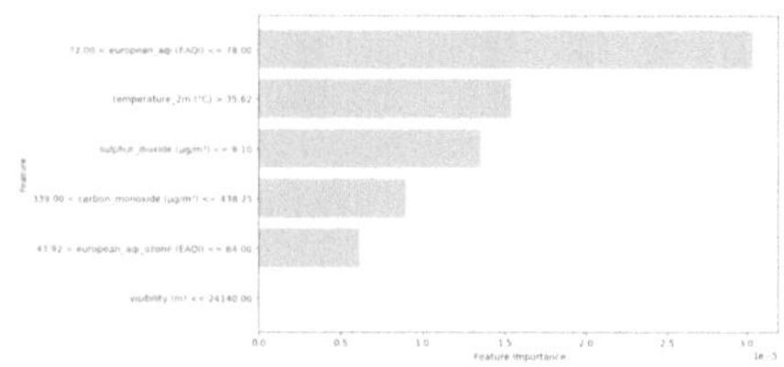

(b) Feature importance for Random Forest classifier using mutual information technique

Fig. 4. Feature contributions comparison during Model Training using mutual information and SHAP-XAI Technique

5 Conclusion and Future Scope

This study outlines deploying a robust model for air quality prediction on resource-constrained edge devices in an urban location. Models are initially trained on a cloud-based architecture using a fused dataset from various APIs and public sources, with preprocessing to ensure data quality. Feature extraction identifies the key features, and models are retrained on the cloud. The optimal features and best-performing model are then deployed to edge devices for efficient local decision-making. This approach ensures accurate, unbiased predictions with minimal delays. Future advancements in air pollution control can leverage emerging technologies for new insights and solutions. Furthermore, we will develop a federated learning technique within edge networks, facilitating collaboration among local edge devices.

Acknowledgement. This work is supported by CSTUP, Government of Uttar Pradesh, under the Grant CSTUP/2023/2643.

References

1. Bekkar, A., Hssina, B., Douzi, S., Douzi, K.: Air-pollution prediction in smart city, deep learning approach. J. Big Data **8**(1), 1–21 (2021). https://doi.org/10.1186/s40537-021-00548-1
2. Brook, R.D., et al.: Particulate matter air pollution and cardiovascular disease: an update to the scientific statement from the American heart association. Circulation **121**(21), 2331–2378 (2010)
3. Gupta, S., Moledina, A., Athavale, S., Gajare, S., Kate, M.: Air quality prediction using machine learning: a comparative study. In: 2023 6th International Conference on Advances in Science and Technology (ICAST), pp. 485–489 (2023)

4. Kekulanadara, K.M.O.V.K., Kumara, B.S., Kuhaneswaran, B.: Comparative analysis of machine learning algorithms for predicting air quality index. In: 2021 From Innovation to Impact (FITI), vol. 1, pp. 1–5 (2021)
5. Kekulanadara, K.M.O.V.K., Kumara, B.T.G.S., Kuhaneswaran, B.: Machine learning approach for predicting air quality index. In: 2021 International Conference on Decision Aid Sciences and Application (DASA), pp. 622–626 (2021)
6. Khalid, S., Khalil, T., Nasreen, S.: A survey of feature selection and feature extraction techniques in machine learning. In: 2014 Science and Information Conference, pp. 372–378. IEEE (2014)
7. Kumar, K., Pande, B.P.: Air pollution prediction with machine learning: a case study of Indian cities. Int. J. Environ. Sci. Technol. **20**(5), 5333–5348 (2023)
8. Luan, B.: An optimization of k-mean algorithm for feature extraction based on computer programming. In: 2023 International Conference on Advances in Electrical Engineering and Computer Applications (AEECA), pp. 599–603 (2023)
9. Masood, A., Ahmad, K.: A review on emerging artificial intelligence (AI) techniques for air pollution forecasting: fundamentals, application and performance. J. Clean. Prod. **322**, 129072 (2021)
10. Sitha Ram, M., Reshmasri, C., Shahila, S., Saketh, J.V.P.: Air quality prediction using machine learning algorithm. In: 2023 International Conference on Sustainable Computing and Data Communication Systems (ICSCDS), pp. 316–321 (2023)
11. Seng, D., Zhang, Q., Zhang, X., Chen, G., Chen, X.: Spatiotemporal prediction of air quality based on LSTM neural network. Alexandria Eng. J. **60**(2) (2021)
12. Singh, K., Aggarwal, A., Singh, G.: Accurate air quality prediction through the use of statistical models and machine learning approach. In: 2023 6th International Conference on Contemporary Computing and Informatics (IC3I), vol. 6, pp. 665–670 (2023)
13. Song, Z., Cheng, J., Chauhan, A., Tilevich, E.: Pushing participatory sensing further to the edge. In: 2019 IEEE International Conference on Edge Computing (EDGE), pp. 24–26 (2019)
14. Vaishnavi, K., Sreya, G., Reddy, K.K., Anisha, P.R.: Machine learning for air quality prediction: random forest classifier. In: 2024 Fourth International Conference on Advances in Electrical, Computing, Communication and Sustainable Technologies (ICAECT), pp. 1–5 (2024)
15. Xu, X., Zheng, L.: Multivariate feature extraction. In: 2022 58th Annual Allerton Conference on Communication, Control, and Computing (Allerton), pp. 1–8 (2022)

Author Index

A. Dasgupta et al. (Eds.): BDA 2024, LNCS 15526, pp. 427–428, 2025.
https://doi.org/10.1007/978-3-031-81821-9